STATISTICS

AN INTRODUCTION

FOURTH EDITION

Mason Lind Marchal

STATISTICS

AN INTRODUCTION

FOURTH EDITION

Robert D. Mason
Douglas A. Lind
William G. Marchal

The University of Toledo

Saunders College Publishing
Harcourt Brace College Publishers

Fort Worth Philadelphia San Diego New York Orlando Austin San Antonio
Toronto Montreal London Sydney Tokyo

Publisher	Elizabeth Widdicombe
Director of Editing, Design, & Production	Diane Southworth
Acquisitions Editor	Scott Isenberg
Developmental Editor	Millicent Treloar
Assistant Developmental Editor	Van Strength
Project Editor	Sheila M. Spahn
Art Director	Pat Bracken
Production Manager	Mandy Manzano
Copy Editor	Christopher B. Nelson
Indexer	Kristina Sanfilipo-Devico
Compositor	Typo-Graphics
Text Type	10/12 Stone Serif
Cover Illustration	Ambrose Rivera

Address for Editorial Correspondence
The Dryden Press, 301 Commerce Street, Suite 3700, Fort Worth, TX 76102

Address for Orders
The Dryden Press, 6277 Sea Harbor Drive, Orlando, FL 32887
1-800-782-4479, or 1-800-433-0001 (in Florida)

ISBN: 0-03-096917-4

Library of Congress Catalog Card Number: 93-80130

Printed in the United States of America

4 5 6 7 8 9 0 1 2   0 3 2   9 8 7 6 5 4 3 2

The Dryden Press
Harcourt Brace College Publishers

To Anita, Jane, and Andrea

The Dryden Press Series in Management Science and Quantitative Methods

Costin
Readings in Total Quality Management

Etienne-Hamilton
Operations Strategies for Competitive Advantage: Text and Cases

Forgionne
Quantitative Management

Freed
Basic Business Statistics

Gaither
Production and Operations Management
Sixth Edition

Glaskowsky, Hudson, and Ivie
Business Logistics
Third Edition

Hamburg and Young
Statistical Analysis for Decision Making
Sixth Edition

Ingram and Monks
Statistics for Business and Economics
Second Edition

Lapin
Statistics for Modern Business Decisions
Sixth Edition

Lapin
Quantitative Methods for Business Decisions with Cases
Sixth Edition

Lee
Introduction to Management Science
Second Edition

Mason, Lind, and Marchal
Statistics: An Introduction
Fourth Edition

Miller and Wichern
Intermediate Business Statistics

Weiers
Introduction to Business Statistics
Second Edition

Zikmund
Business Research Methods
Fourth Edition

The Harcourt Brace College Outline Series

Lapin
Business Statistics

Pentico
Management Science

Rothenberg
Probability and Statistics

Tanis
Statistics I: Descriptive Statistics and Probability

Tanis
Statistics II: Estimation and Tests of Hypothesis

PREFACE

The Fourth Edition of *Statistics: An Introduction* is intended for use in an applied introductory statistics course. As in the previous editions, the principles of both descriptive and inferential statistics are discussed and illustrated in situations that are close to most students' own experience. Students with a background limited to basic algebra will be able to complete the necessary mathematical calculations.

This book is appropriate for use in any survey course in behavioral statistics. The material is organized so that it provides the greatest possible flexibility of use. Illustrations and exercises are drawn from disciplines as varied as sociology, education, business, sports, demography, meteorology, politics, and mathematics. Considerable latitude is built into the text, ensuring use in a one-semester, one-quarter, or two-quarter course. A one-semester course, for example, might include Chapters 1–10, 13, and 14. Time permitting, Chapters 11 and 16 are logical additions.

Special Features

Statistics: An Introduction, Fourth Edition, has a number of special features that will motivate and assist students as they progress through the material:

- Each chapter begins with a set of *learning objectives*—explaining what the student should be able to do at the completion of the chapter. These objectives act as advance organizers and motivators.
- Following the learning objectives is a chapter integrative problem that is solved in the discussion as the applicable concepts are covered. This opening problem is identified with a distinctive logo and the logo reappears in portions of the text where the solution is discussed.
- At the beginning of each chapter, the important concepts presented in the preceding chapter are reviewed in a brief *introduction* that explains how the earlier concepts are linked to the present chapter. Then, a chapter overview introduces the chapter's new topics.
- The text discussion of each concept is followed by a realistic *problem and solution.*
- Throughout the chapter are interspersed a number of *Self-Reviews,* each one closely patterned after the chapter problems that precede it. The Self-Reviews help students monitor their progress, and they provide students with constant reinforcement. Answers and methods of solution for Self-Reviews are

given at the end of the chapter.

- Many interesting, real-world *Exercises* are incorporated within and at the end of each chapter. Answers and methods of solution for all even-numbered exercises are given at the back of the book.
- Every important *new term and formula* is defined and has been placed in a box for easy reference.
- Each chapter includes a *Summary*.
- End-of-chapter *Achievement Tests* help students evaluate their comprehension of the material covered. Answers and methods of solution are given at the back of the book.
- A *Unit Review* is included after each of six major groups of chapters. Each includes a brief review, key concepts, key terms, key symbols, review problems, and a section called "Using Statistics," in which a situation is presented for which statistics can provide a solution, followed by a discussion of the situation and its resolution.
- Each chapter contains an interesting, challenging *Case*.
- A *Glossary* is included at the back of the book.
- The normal distribution and Student's *t* distribution are printed on the front and back endpapers for easy reference.

The Fourth Edition also includes:

- a large number of new chapter exercises.
- extensive use of statistical software in problem solving; this allows the student to focus on interpretation of the output.
- expanded use and interpretation of *p-values.*
- *stat bytes* scattered throughout the chapters. These provide interesting, and sometimes humorous, applications of statistical methods.
- *drill problems* scattered throughout the chapters. These are simple problems, with minimal calculation, that will offer the student practice and feedback on the topic just discussed.
- an expanded discussion of the underlying idea behind the coefficient of correlation.
- virtually a whole new chapter on the design of experiments and the construction of questionnaires.
- formulas in the chapters that are numbered. Reference is made in the chapter to these formula numbers, giving the student quick access to the appropriate formula.
- *data exercises* are the last few exercises in most chapters. Copies of the two data exercises are found in the back of the book. The first exercise contains the selling price, size, distance from the center of the city, and other data on 50 homes sold during 1992 in northwest Ohio. The second exercise refers to 75 middle managers in Sarasota, Florida. Included in the data exercise is information on the yearly salary, age, number of employees supervised, and other data.

Ancillary Materials

A complete ancillary package accompanies the Fourth Edition.

The comprehensive *Study Guide* is organized much like the textbook. Each chapter includes objectives, problems and solutions, exercises, a summary, and assignments. Ample space is provided for computations.

An *Instructor's/Solutions Manual* contains the complete solutions to all chapter exercises, Unit Review exercises, and Cases in the textbook and to the assignments in the Study Guide. Lecture materials and in-class lecture problems (plus solutions to these problems) developed by Dr. Denise McGinnis, are prepared as transparency masters.

An *EasyStat Guide* is available for the Fourth Edition. This guide, prepared by Dr. Toni M. Somers of Wayne State University, is a user-friendly program for any IBM-PC system. *EasyStat* is a computer program that works within the framework of a spreadsheet and includes a data set for 1,000 urban families, from which students can select random samples for carrying out exercises. A separate data disk contains the real estate and salary survey data listed at the back of the book.

The *Testbook* is organized by chapter with each chapter containing multiple choice, problem, and essay questions. The questions in a chapter may be used as a test over the entire chapter, selected questions may be chosen and given in class as a "pop" quiz, the entire chapter test may be assigned to be completed outside of class, or questions from several chapters may be combined to form a unit test, a midterm, or a final examination. The Testbook is also available computerized, in IBM, and Macintosh formats.

ACKNOWLEDGMENTS

We wish to express our gratitude to the reviewers of the First Edition: John M. Rogers (California Polytechnic State University, San Luis Obispo), Bayard Baylis (The King's College, New York), William W. Lau (California State University, Fullerton), Kenneth R. Eberhard (Chabot College), and David Macky (San Diego State University). Their contributions continue to enhance the textbook.

For the Second Edition, we wish to thank Francesca Alexander (California State University, Los Angeles), John R. Anderson (DePauw University), Daniel Cherwein (Cumberland County College), C. Philip Cox (Iowa State University), Geoffrey C. Crosslin (Kalamazoo Valley Community College), Larry Gausen (Bemidji State University), David Macky (San Diego State University), Bruce F. Sloan, Donald Hansen, and Donald Frazee (Bellevue College), Nick Watson (Kankakee Community College), and William W. Wood (Carroll College of Montana).

For the Third Edition, thanks to Dharam Chopra (The Wichita State University), Rudolph J. Freund (Texas A&M University), H.G. Mushenheim (University of Dayton), James C. Navarra (University of Maryland), and Samuel B. Thompson (University of Maryland) for their reviews of the revised manuscript and their many comments and suggestions.

We thank the following persons for their input to the Fourth Edition: W.P. Abeysinghe (Lock Haven University), Roger Champagne (Hudson Valley Community College), Dan Cherwien (Cumberland Community College), Mahmood Ghamsary (Long Beach City College), Robert Hale (St. Louis University), Mohamad Nayebpour (University of Houston-Clear Lake).

At The Dryden Press we thank Scott Isenberg, Millicent Treloar, Van Strength, Sheila M. Spahn, Pat Bracken, and Mandy Manzano.

Finally, we wish to acknowledge the valuable contributions made to this book by friends, students, colleagues, and collaborators throughout the development of the four editions.

We are grateful to the Literary Executor of the late Sir Ronald A. Fisher, F.R.S., to Dr. Frank Yates, F.R.S., and to Longman Group Ltd., London, for permission to reprint Table IV from their *Statistical Tables for Biological, Agricultural and Medical Research* (6th ed., 1974).

Robert D. Mason
Douglas A. Lind
William G. Marchal

CONTENTS

CHAPTER 4

CHAPTER 5

CHAPTER 6

Probability Distributions 203

CHAPTER 7

The Normal Probability Distribution 235

Unit Review Chapters 5, 6, and 7 269

CHAPTER 8

Sampling Methods and Sampling Distributions 277

CHAPTER 9

CHAPTER 10

CHAPTER 11

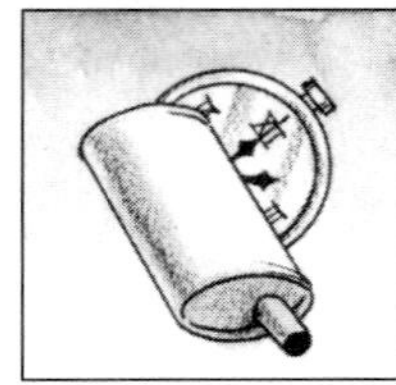

CHAPTER 12

CHAPTER 13

CHAPTER 14

CHAPTER 15

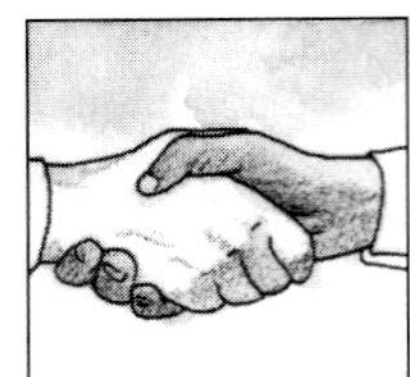

CHAPTER 16

CHAPTER 17

Nonparametric Methods: Analysis of Ranked Data 613

STATISTICS

AN INTRODUCTION

FOURTH EDITION

Mason Lind Marchal

CHAPTER 1

An Introduction to Statistics

OBJECTIVES

When you have completed this chapter, you will be able to

- explain what is meant by statistics;
- define the terms *descriptive statistics* and *inferential statistics;*
- cite some examples of the application of statistics from the social sciences, business, education, and other fields;
- explain the difference between nominal, ordinal, interval, and ratio levels of measurement.

CHAPTER PROBLEM **A Roller Coaster Salary**

During the 1980s, the chairman and president of Chrysler Corporation, Lee Iacocca, accomplished one of the most spectacular turnarounds in U.S. business history. As reflected in his salary, however, the transition was not always smooth. In what year was his income the largest? What was his income in 1990?

INTRODUCTION

When you watch a football game on television, listen to the stock market report on the radio, or read an article in *USA Today* or *Time,* you are sometimes overwhelmed by assorted facts and figures usually called "statistics." For example:

- A recent Harris Poll revealed that 64% of respondents think small-business owners have good moral and ethical standards. The results for a sample of other professions: journalists, 39%; business executives, 31%; Bush Administration officials, 29%; lawyers, 25%; and members of Congress, 19%.
- The National Center for Education Statistics reported that of the 64,465,000 students who use computers 28,662,000 are in high school and 10,661,000 are college undergraduates.
- According to the Defense Department's *Selected Manpower Statistics,* there are 112,000 first-time enlisted military personnel in the Army and 111,000 reenlistments.
- The Bureau of the Census projects the population of the United States to be 298,252,000 in the year 2025.
- According to the Department of Health and Human Services, the life expectancy of a newborn white female was 40.5 years in 1850; today it is 78.7 years.
- The Department of Labor's *Monthly Labor Review* reported that 249,770 new businesses were started and 57,253 failed in 1992. The following average hourly earnings were reported for April: mining, $14.53; manufacturing, $11.43; and retail trade, $7.16.

Each of the above figures is a **statistic.** A collection of more than one figure is referred to as **statistics** (plural).

Statistics can appear in graphic as well as in sentence form. A graph is often used to capture reader attention and portray a large body of data over a long period of time. For example, over 300 figures were used to construct Figure 1–1. Only a quick glance is needed to discover that the life expectancy of all groups has risen dramatically since 1900—especially that of blacks, which has increased by over 100%.

What happened to Lee Iacocca's income from 1982 to 1991?

Year	1982	1983	1984	1985	1986	1987	1988	1989	1990	1991
Salary (in millions)	.40	.40	1.1	11.4	20.6	17.9	3.7	4.0	4.6	3.3

We could cite his annual income for each year in sentence form, but by portraying the data in graphic form (see Figure 1–2 on page 4) we can quickly analyze the trend. Note that his million-dollar income peaked in 1986 and decreased throughout the rest of the decade.

We will examine graphic presentation of statistical data in Chapter 2.

Figure 1-1

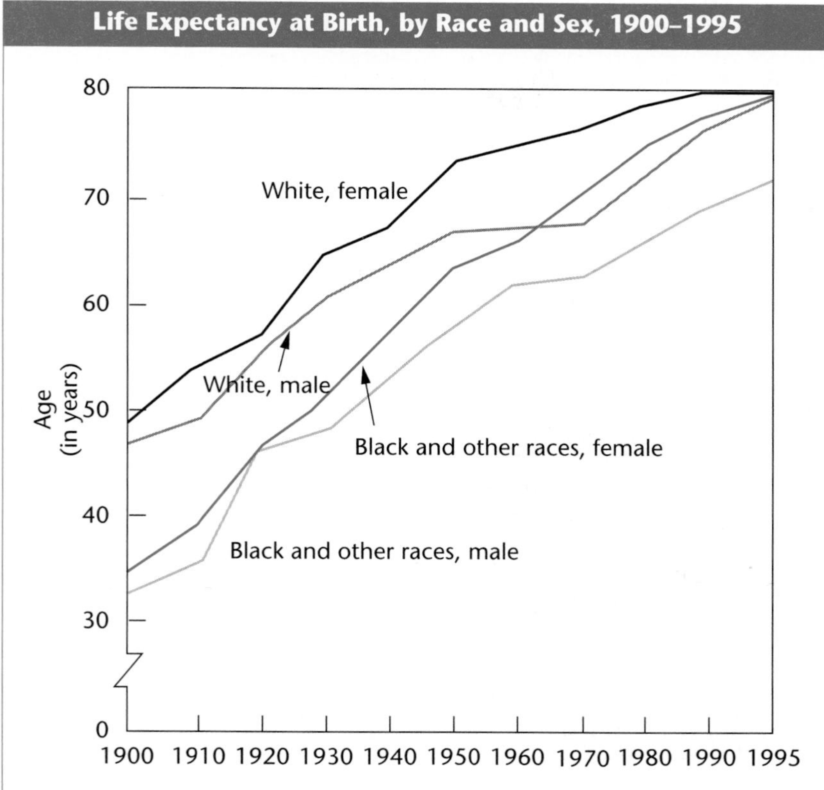

A DEFINITION OF STATISTICS

The life expectancy of females, the average hourly earnings in retail trade, and the percent of people who think members of Congress have good moral and ethical standards are all statistics. However, statistics also has a broader meaning. The following definition of statistics is the one we will use throughout this book:

> **Statistics** Techniques used to collect, organize, present, analyze, and interpret data to make better decisions.

As the definition indicates, the first step in investigating a problem is to collect pertinent data. The data must then be organized in some way and perhaps presented in a graph, such as Figure 1–1 or 1–2. Only then will we be ready to analyze and interpret the data.

Figure 1-2

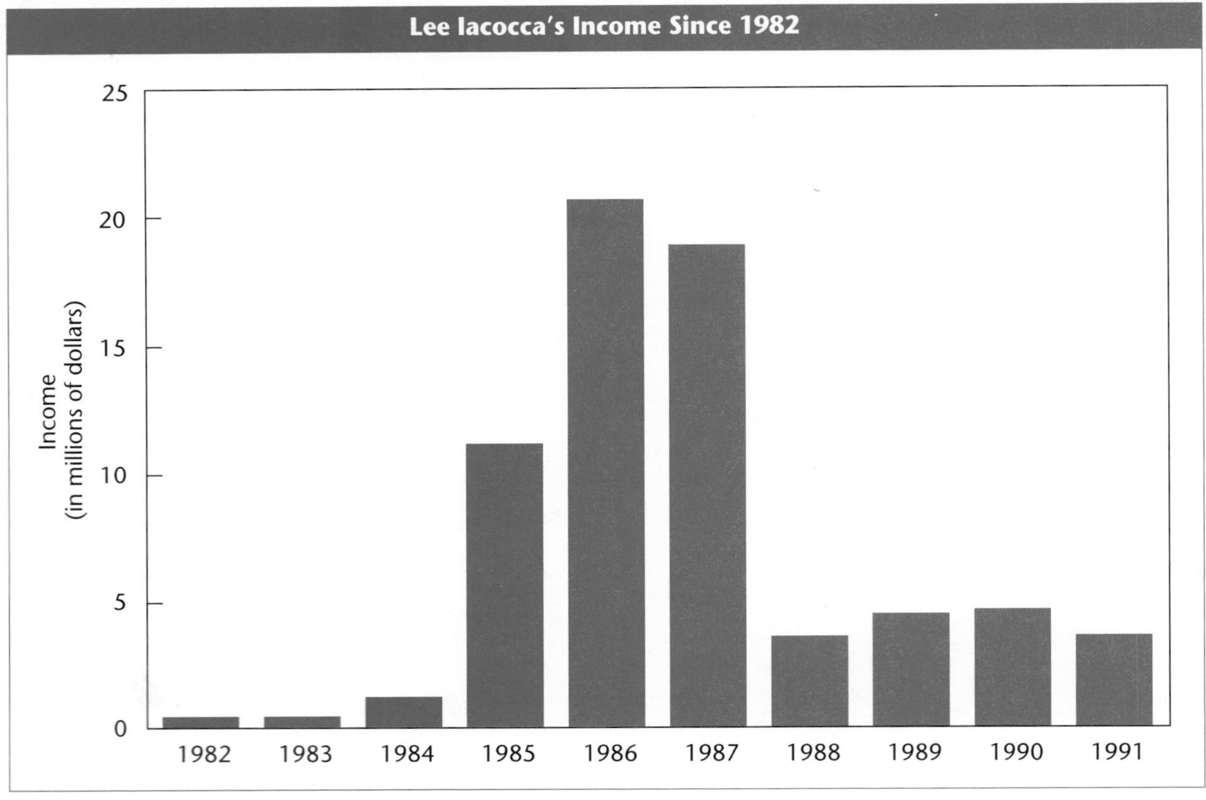

For example, suppose you are a life insurance actuary who must set premium rates for a new and unique term policy for newborn blacks, whites, males, and females. *Vital Statistics,* published by the Department of Health and Human Services' National Center for Health Statistics, would give you data on the life expectancy of these four groups. For each group, you would also need to know the expected commission for the underwriter, and so on.

WHO USES STATISTICS?

Just as life insurance companies collect, organize, analyze, and interpret data before setting a premium for term, whole life, and other types of policies, so too must automobile insurance companies. For example, what kind of information would be needed to set a premium for a Ford Taurus? How old the automobile is, the repair record, and the driver's age are some of the data needed. Some examples of how statistics are used in other areas:

- Law enforcement agencies are concerned with repeat offenders. Why do some convicted criminals commit a second or third crime, while others do not? If you were doing criminal research, what factors would you consider?
- In medical research, a common way to evaluate the effectiveness of a newly developed medicine is to administer it to one group of people or animals, and not to another group, called the "control group." Statistics are collected from both groups, organized, analyzed, and interpreted before making a decision regarding the effectiveness of the new medicine.
- Management in mass production firms must be assured that current production is satisfactory. The quality assurance department has the job of selecting, collecting, and evaluating a sample of observations. To make a determination, a quality control inspector might, for example, take a sample of five piston rings every hour, determine the average outside diameter of the rings, and plot it on what is called a "mean chart." Figure 1–3 is a mean chart that portrays the average outside diameters for the first seven hours of production.
- Note that there are an upper control limit (UCL) and a lower control limit (LCL). If the average outside diameter for the five piston rings is either above or below those limits, the process is "out of control." Such is the case at 1 p.m., when the average outside diameter was too large. The quality control engineer reported this to the production manager and corrective action was taken. By 2 p.m. the process was back "in control."

To solve many problems, it is often necessary to collect statistical data using a *questionnaire*. For example, suppose the Postal Service is concerned about its overall performance in Wilson,

STAT BYTE

Defining a problem, gathering the essential data, conducting pertinent statistical tests, and interpreting the results is a common procedure in medicine, law enforcement, business, and other areas. For example, a team of Finnish epidemiologists recently explored the relationship between the amount of iron in the body and the incidence of heart attacks. They studied 1,900 men from eastern Finland, ranging in age from 42 to 60, for serum levels of ferritin (an iron-storing protein) over a period of five years. At the end of the five-year period, 51 had suffered a heart attack. An analysis of the figures revealed that men with more than 200 micrograms of ferritin per liter of blood were twice as likely to have heart attacks than men with levels below 200 micrograms.

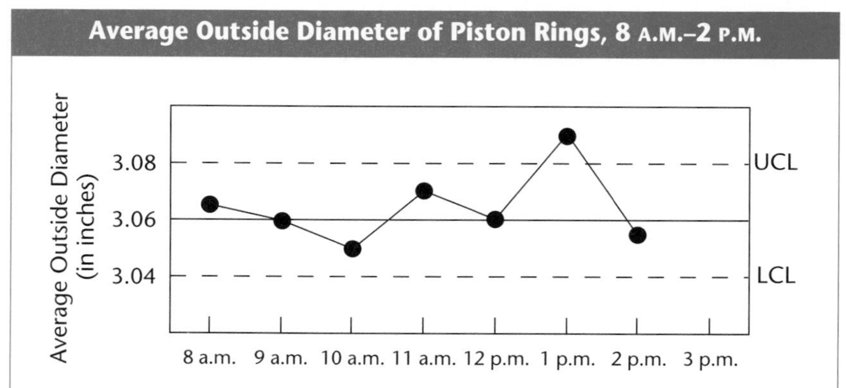

Figure 1-3

Wyoming. To evaluate it, they send each resident a questionnaire, part of which follows.

1. Overall Performance (in the past three months)

We would like your opinion of the U.S. Postal Service's performance **during the past three months** on some general topics. Use a seven-point scale, where 1 means "Poor," 4 means "Good," and 7 means "Excellent." Please remember that you can mark any box between 1 and 7 or the "Don't Know" box.

Please rate the U.S. Postal Service on . . .	(Poor—Fair) 1	2	3	Good 4	(Very Good–Excellent) 5	6	7	Don't Know
a. Its overall performance	☐	☐	☐	☐	☐	☐	☐	☐
b. Delivery of the mail in good condition (undamaged)	☐	☐	☐	☐	☐	☐	☐	☐
c. The length of time it usually takes a letter mailed **in your local area** to be delivered in your local area	☐	☐	☐	☐	☐	☐	☐	☐
d. The length of time it usually takes a letter mailed **in other parts of the country** to be delivered in your local area	☐	☐	☐	☐	☐	☐	☐	☐
e. Consistency of delivering **local** mail in the same number of days each time	☐	☐	☐	☐	☐	☐	☐	☐
f. Consistency of delivering mail **from outside your local area** in the same number of days each time	☐	☐	☐	☐	☐	☐	☐	☐
g. Having conveniently located mail deposit boxes where you can mail letters	☐	☐	☐	☐	☐	☐	☐	☐
h. Willingness to help customers	☐	☐	☐	☐	☐	☐	☐	☐
i. Providing services which are a good value for the price	☐	☐	☐	☐	☐	☐	☐	☐
j. Its ability to keep up with the times	☐	☐	☐	☐	☐	☐	☐	☐
k. Overall communication with customers	☐	☐	☐	☐	☐	☐	☐	☐

Some personal questions for the respondent are usually included to better evaluate the survey results. The Postal Service wants to determine if perceptions of its performance are the same for all ages, incomes, and levels of education (see sample on page 7).

Consider another illustration of how questionnaires are used to collect statistics. The Pantone Color Institute recently queried 3,370 consumers regarding their color preferences for various types of consumer products. The results of such a survey are especially important to automobile manufacturers, interior designers, furniture

24. Your age:

☐ Under 25 years ☐ 35–44 years ☐ 55–64 years
☐ 25–34 years ☐ 45–54 years ☐ 65 or older

25. Highest level of school you completed:

☐ Did not finish high school ☐ Some college/technical school/trade school
☐ High school graduate ☐ College graduate or beyond

26. Which of the following categories includes your total household income before taxes?

☐ Under $10,000 ☐ $20,000–$29,999 ☐ $40,000–$49,999 ☐ $75,000 or more
☐ $10,000–$19,999 ☐ $30,000–$39,999 ☐ $50,000–$74,999

Your answers to these questions will be kept confidential and will only be used to identify groups of similar people for statistical purposes. The United States Postal Service greatly appreciates your help in completing this questionnaire.

manufacturers, and other consumer-oriented industries. As reported in the *Chicago Tribune,* red is the favorite color for a sports car, followed by a new color, teal. The least liked color, by far, is a sickly yellow-green. Neon orange is also a no-no. There is a growing demand for colors in the red-violet range, such as fuchsia and magenta, which are expected to be the trend in swimwear and neckties. Light and dusty blues now dominate in the bathroom, but the future belongs to watery shades of misty green and aquamarine. Yellows will remain hot in the kitchen.

TYPES OF STATISTICS

Descriptive Statistics

As stated earlier, statistics involves the collection, organization, and presentation of numerical data. Masses of unorganized data stored in a computer, or collected by Gallup-type polls with respect to the preferences of voters, are usually of little value. Techniques are available, however, to organize such data into some meaningful form. These aids in organizing, analyzing, and describing (or summarizing) a large collection of numbers are collectively referred to as **descriptive statistics.**

Descriptive Statistics Methods used to describe the data that have collected.

STAT BYTE

Ratios are often used to express the relationship between two totals. For example, a recent Labor Department study reported that in the 1960s, 1 out of every 10 college graduates was working in a job that did not require a college degree; today, the ratio is 1 in 5 and—even more discouraging—is heading toward 1 in 3 by the end of the century.

We will discuss one summary technique, called a frequency distribution, in Chapter 2. In that chapter, we also examine how data are presented in graphic form. We have already graphically portrayed the life expectancy of males, females, whites, and blacks (Figure 1–1) and the income of Lee Iacocca (Figure 1–2). Figure 1–4 shows the trend in the number of new AIDS cases reported between 1983 and 1992 for both males and females.

An average, also called a measure of central tendency, is another descriptive statistic. There are a number of different averages, but each describes the tendency of a set of data to cluster about its central value. For example:

- According to *Golf Digest,* Fred Couples earned an average of \$85,117 per golf tournament in 1992. In some tournaments he earned more, in others less, but his earnings, clustered around \$85,117 per tournament.
- The *Monthly Labor Review* reported that in 1992 average hourly earnings in construction were \$14.04. Some construction workers earned more, and some less than \$14.04, but hourly earnings clustered around \$14.04.
- According to the Department of Agriculture, the typical size of a farm in the United States is 281 acres.
- The typical cost of operating an automobile, reports *Motor Vehicle Facts and Figures,* is 38.2 cents per mile. Those cars with a high finance charge, high insurance premiums, and so on, cost more

Figure 1-4

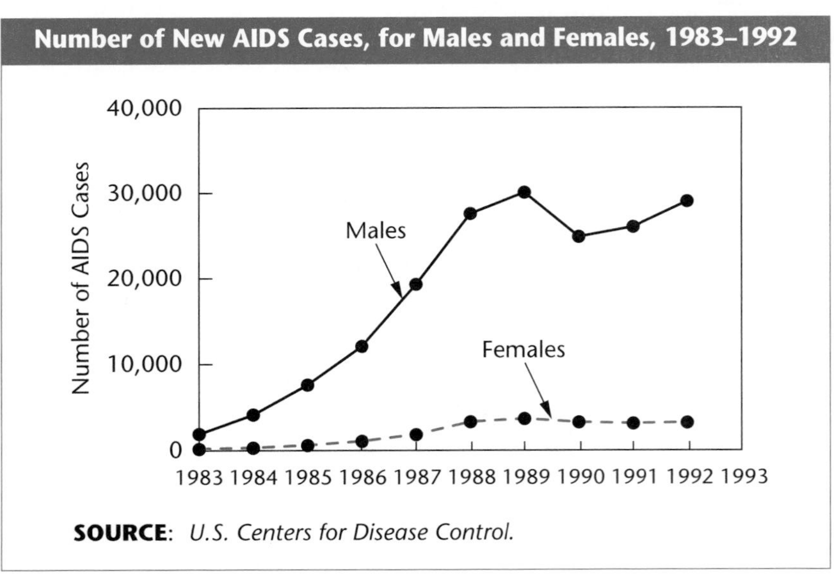

SOURCE: *U.S. Centers for Disease Control.*

than 38.2 cents per mile to operate. Others with low maintenance costs, no finance charges, and so on, cost less than 38.2 cents per mile to operate, but the costs cluster around the 38.2 cents-per-mile average.

In Chapter 4, we will describe the spread, or variation, in the data. For example, in Figure 1–5 the reading scores of 16 slow readers range from 50 to 90. One statistical measure, called the range, describes the spread in the reading scores. It is 40, found by subtracting 50 from 90. Other measures of variation will be discussed in that chapter.

Inferential Statistics

Descriptive statistics is only one facet of the science of statistics. Another is **inferential statistics** or **inductive statistics.** Inferential statistical methods are concerned with determining something about a **population.** Gallup, Harris, and other pollsters do just that when they are hired before an election to estimate how voters (the population) plan to vote on election day.

In the case of political polls, usually about 2,000 voters are sampled out of the population of registered voters in the United States. Based on the sample results, an inference is then made about the reaction of *all* voters on election day. Thus, we may define inferential statistics as appears on the following page.

Figure 1-5

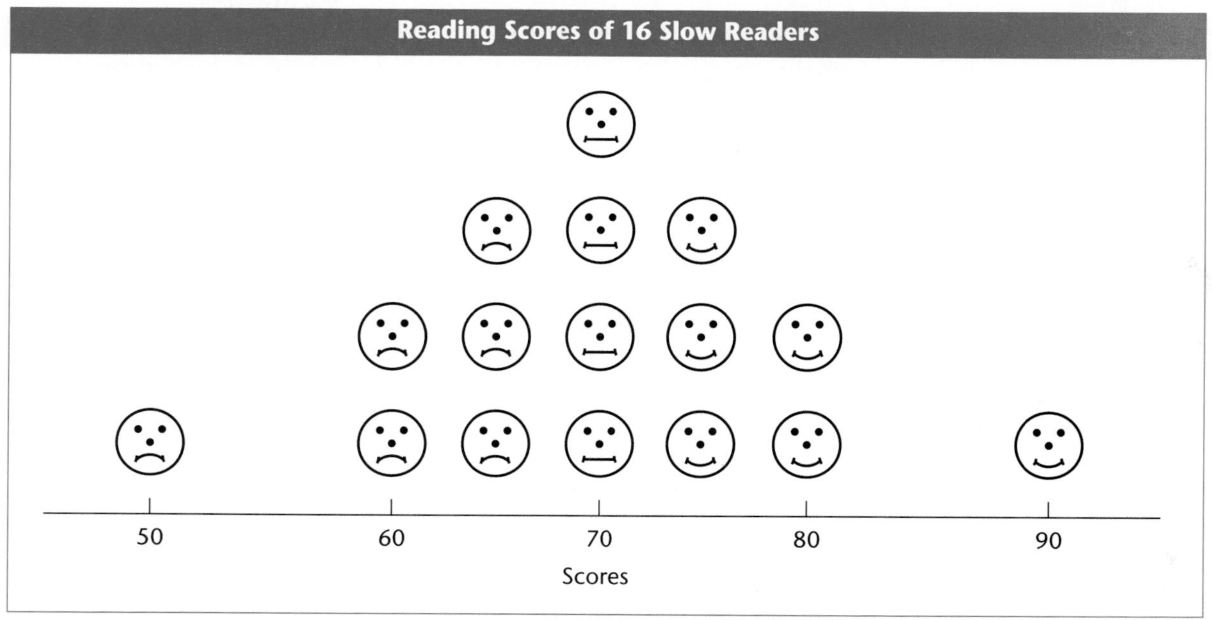

Inferential Statistics Techniques used to make a decision, estimate, prediction, or generalization about a population based on sample evidence.

What exactly do we mean by the words "population" and "sample," as used in our definition of inferential statistics? We normally think of a population as a large group of people: all the residents of Dade County, Florida, for example, or all the employees of the Ford Motor Company assembly plant in Atlanta, Georgia. However, a population can also consist of a group of objects, such as all the Mercury Sables produced in an eight-hour shift at the Ford plant in Atlanta. Some other populations: all the zebras in the Masa Mara game preserve in Kenya, all the *New York Times* newspapers sold last Sunday, and all the inmates at Attica Prison.

Population A collection, or set, of all individuals, objects, or measurements whose characteristics are analyzed.

It is almost impossible to contact every potential voter in Dade County to find out how he or she plans to vote on election day. Nor is it feasible to capture every zebra in the Masa Mara preserve to determine the status of its health or to interview every inmate at Attica regarding his plans after being released, and so on. We therefore select a portion, or part, of the population, called a **sample,** and concentrate our efforts on this group.

Sample A portion or part of the population of interest.

The diagram on page 11 shows the relationship between a population and a sample taken from that population.

The following are examples of inferential statistics:

- A random sample of 1,260 graduates of four-year colleges last year revealed that the mean starting salary was $22,674. We therefore infer that the mean salary of all college graduates during the last calendar year was $22,674.
- The Gorski Testing Laboratory tests all the culture dishes it purchases. Yesterday a shipment of 20,000 was received. A random sample of 300 from the shipment revealed 18 to be defective. We therefore estimate that 6% of the total shipment is defective.

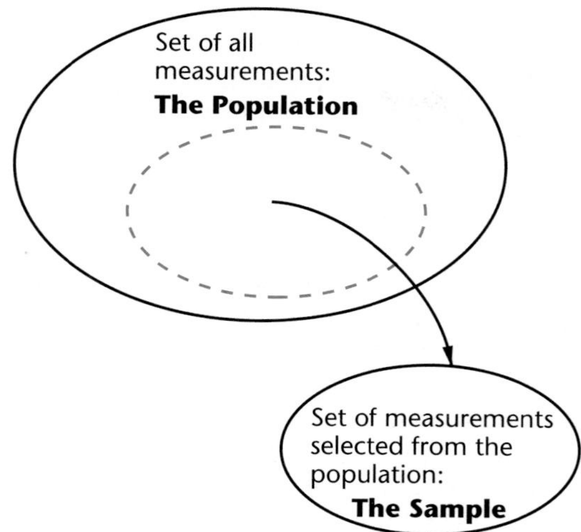

- A random sample of 846 prisoners released from the state prison in Smithville revealed that 27% were sent back to prison for committing another crime. We therefore conclude that 27% of all prisoners released from Smithville will commit another crime.

Self-Review 1-1

This is the first of many self-reviews you will encounter throughout the text. They are designed to help you determine immediately whether you comprehend the material. We suggest you complete these self-reviews and then check your answers against those provided at the end of the chapter.

Over a period of a week, the market research department of the Dairy Producers Association randomly selected 300 grocery shoppers and checked their shopping carts for dairy products. Of the 300 carts, 210 contained at least one dairy product.

a. What would the research department report to management regarding the buying habits of all grocery shoppers?
b. Is the 300 a sample or a population?
c. What is the population in this problem? all grocery shoppers
d. Is this an example of descriptive statistics or inferential statistics?

THE NATURE OF STATISTICAL DATA

Statistical data occur whenever measurements are made or observations are classified. The data may be the heights of female adults, the weights of newborn babies, the number of robberies committed in various districts of the city, the preference for a certain design of a new car, the measurement of a personality trait, or the "yes" or "no" response to the question "Have you ever been married?" Recall that we defined statistics as the study of *data. Data are more than just numbers*. Yet, in order to make measurement simple, or "quantifiable," we must convert data into numerical form. How then can we transform a "yes" or "no" response into numerical data? We simply record, for example, a 1 for the "yes" response and a 0 for the "no" response. In this artificial way, we can make this response into numerical data. Similarly, for the question "What is your marital status?" we could record a 1, 2, 3, or 4 to indicate the responses "single," "married," "widowed," or "divorced." In these artificial ways, data may be converted to a numerical form. If you are interested in a typical marital status, it would not make sense to average these values. For example, combining a married (2) and a divorced person (4) is not the same as having two widowed persons (3). The typical marital status is that category with the greatest number of observations—that is, "widowed"—because there are two observations in that category.

LEVELS OF MEASUREMENT

The Nominal Level of Measurement

Data are classified into four general types, or levels, of measurement. The four levels are nominal, ordinal, interval, and ratio. The information presented in Table 1–1 is an example of the nominal measurement scale. This level is the most "primitive," "lowest," or most limited type of measurement.

We use the words **nominal level of measurement,** or **nominal scaled,** when dealing with this type of data, which can only be classified into categories. Strictly speaking, the information in Table 1–1 is simply a count, or tabulation, of the number of Protestants (78,952,000), Roman Catholics (30,669,000), and so on.

Table 1-1 Religious Affiliation of the U.S. Population 14 Years Old and Over (self-reported)

Religion		Number
Protestant		78,952,000
Roman Catholic		30,669,000
Jewish		3,868,000
Other religion		1,545,000
No religion or religion not reported		3,195,000
	Total	118,229,000

Note that the arrangement of the religions can be changed. Roman Catholic can be listed first; Jewish, second; and so on. This indicates that for the nominal level of measurement *there is no particular order for the groupings.* Further, the categories are **mutually exclusive.**

Mutually Exclusive Categories Each individual or item, by virtue of being in one category, is excluded from others.

In this example, a person is not classified both as a Protestant and as having no religion.

To process on a computer the data on religious preference, sex, type of crime, and other nominal scaled data, the categories of interest are often coded 1, 2, 3, . . . to facilitate counting, with, say, 1 representing Protestant, 2 representing Roman Catholic, 3 representing Jewish, and so on. Of course, it does not make sense to perform arithmetic operations on such numbers!

Also, note that in the data on religious preference the categories are **exhaustive.**

Exhaustive Categories Each individual or item appears in at least one category.

If, for example, a person reports she is of the Greek Orthodox religion, she is counted in the "other religion" category. A person who refuses to name his religion is counted in the "no religion or religion not reported" category.

Tables 1–2 and 1–3 also illustrate a nominal level of measurement. Table 1–2 shows the results of a survey, conducted by The Higher Education Research Institute at UCLA, of first-time college freshmen as to their probable field of study.

Table 1–3 gives some of the characteristics of present smokers from 1965 to 1991, as reported by the Office of Smoking and Health.

To interpret Table 1–3: in 1965, 50.2% of males over 20 smoked; by 1991, this declined to 31.5%.

As we noted in Table 1–1, the order of the categories can be changed. In Table 1–2, social sciences can be listed first; education, second; and so on. In Table 1–3, race can be first, followed by sex.

Table 1-2

Probable Field of Study of College Freshmen

Probable field of study	Percent
Arts and humanities	9
Biological sciences	4
Business	25
Education	9
Engineering	10
Physical sciences	2
Social sciences	10
Professional	13
Technical	3
Other	15
Total	100

Table 1-3

Characteristics of Present Smokers, by Sex and Race, of Persons 20 Years Old and Over (in percent)

Characteristic	1965	1970	1974	1976	1977	1978	1979	1980	1983	1985	1991
Sex											
Male	50.2	44.3	43.4	42.1	40.9	39.0	38.4	38.5	35.5	33.2	31.5
Female	31.9	30.8	31.4	31.3	31.4	29.6	29.2	29.0	28.7	28.0	26.2
Race											
White	40.0	36.5	36.1	35.6	34.9	33.6	33.2	32.9	31.4	29.9	28.3
Black	43.0	41.4	44.0	41.2	41.8	38.2	36.8	37.2	36.6	36.0	33.5

SOURCE: *U.S. Centers for Disease Control.*

Exercises

Answers to the even-numbered exercises are at the back of the book.

1. Refer to Table 1–2.

a. Is this an example of a nominal level of measurement? If so, why?

b. Are the categories mutually exclusive? Why?

c. Are the categories exhaustive? Why?

2. Refer to Table 1–3.
   - **a.** Is this an example of a nominal level of measurement? If so, why?
   - **b.** Are the categories mutually exclusive? Why?
   - **c.** Are the categories exhaustive? Why?
   - **d.** Interpret the trend in smoking, by sex and race.

The Ordinal Level of Measurement

The information in Table 1–4 is an illustration of **ordinal scaling.** Observe that here one category is ranked higher, or more outstanding, than the next one. That is, "superior" is a higher rating than "good," a "good" rating is higher than a "fair" rating, and so on. When a category is higher than the preceding one, the information is of ordinal scale.

If we adopt a code in which superior is given a 4, good is given a 3, and fair is given a 2, a 4 ranking is obviously higher than a 3 ranking, and a 3 ranking is higher than a 2 ranking. However, one could not say a field worker rated 4 (superior) is twice as competent as a field worker rated 2 (fair). It can only be said that a rating of "superior" is better, or greater than, a rating of "good," and a "good" rating is better than a "fair" rating.

Tables 1–5, 1–6, and 1–7 also illustrate an ordinal level of measurement. Table 1–5 gives the number of enlisted military personnel on active duty in 1992, ranked from recruit (E-1) to sergeant major (E-9), as reported in the Department of Defense's *Selected Manpower Statistics*.

Table 1–6 gives the number of post-secondary degrees conferred in 1992 and those expected to be awarded in the year 2000. The data come from the National Center for Education Statistics' *Projections of Educational Statistics*. Note that the degrees are ranked from least advanced (associate's degree) to most advanced (law and medical degrees).

Every two years, *Business Week* ranks business schools on the basis of a poll of corporations, recruiters, graduates, and other factors. Table 1–7 gives the magazine's ranking for 1992.

Table 1-4

Ratings of 50 Field Social Workers by Supervisors

Rating	Number of Ratings
Superior	8
Good	20
Fair	11
Poor	9
No rating	2*

**Has been a field worker less than one month.*

Table 1-5 **Enlisted Military Rank of Active Duty Personnel, 1992**

Rank/Grade	Number (in thousands)
Total	1814.0
Recruit—E-1	128.9
Private—E-2	151.9
Pvt. 1st class—E-3	281.6
Corporal—E-4	456.8
Sergeant—E-5	362.1
Staff Sgt.—E-6	245.7
Sgt. 1st class—E-7	134.0
Master Sgt.—E-8	37.9
Sgt. Major—E-9	15.1

Table 1-6 **Post-Secondary Degrees Conferred**

	Number (in thousands)	
	1992	**2000***
Degree Conferred	1,959.7	1,960.3
Associate's	461.2	489.2
Bachelor's	1,062.5	1,036.4
Master's	323.2	327.1
Doctorate	36.1	36.3
First professional	76.8	71.3

**Estimated*

To review: The major difference between a nominal and an ordinal level of measurement is the "greater than" relationship between ordinal-level categories. Otherwise, the ordinal scale has the same characteristics as the nominal scale—that is, the categories are mutually exclusive and exhaustive.

Exercises

3. Refer to Table 1–5.

a. Why are the data considered an ordinal level of measurement?

Table 1-7

Ranking of Business Schools, 1992

Rank	School
1	Northwestern
2	Chicago
3	Harvard
4	Pennsylvania
5	Michigan
6	Dartmouth
7	Stanford
8	Indiana
9	Columbia
10	North Carolina

SOURCE: Business Week, *October 26, 1992, 62.*

b. Are the categories mutually exclusive? Explain.
c. Are the categories exhaustive? Explain.

4. Refer to Table 1–6.
a. Why are the data considered an ordinal level of measurement?
b. Are the categories mutually exclusive? Explain.
c. Are the categories exhaustive? Explain.
d. Interpret the distribution of degrees conferred.

The Interval Level of Measurement

For the **interval level** of measurement, the difference between numbers is a constant size. Temperature measured on either the Fahrenheit or the Celsius scale is an example. Suppose the Fahrenheit temperatures on three consecutive days were 38°, 45°, and 56°. These temperatures can be easily ranked, but in addition we can study the difference between readings, because 1° Fahrenheit represents a constant unit of measurement for all three days. It is important to note that the zero point is arbitrary and just another point on the scale. Zero degrees Fahrenheit does not represent the absence of heat. In addition, if the high temperature was 50°F today and 25°F yesterday, this does not mean that it is twice as hot today. If you doubt this, convert the temperature to the Celsius scale, and note that the relationship changes. Test scores, such as those obtained on the Scholastic Aptitude Test (SAT), or on a statistics or history examination, are also examples of the interval scale of measurement.

In addition to the equidistant—or constant size—characteristic, interval-scaled measurement has all the features of nominal and ordinal measurements. The temperatures are mutually exclusive; that is, the high temperature could not be both 50°F and 62°F on the same day. And the "greater than" feature of ordinal data permits the ranking of daily high temperatures. The interval scale of measurement also assumes the categories are exhaustive; that is, all the cases are included.

The Ratio Level of Measurement

The **ratio level** is the highest level of measurement. As in the interval scale, the observations are ordered and the distance between successive observations is measured. In addition, the ratio scale uses the number zero to indicate the absence of the characteristic being measured. Money is an example of the ratio scale of measurement. The zero point is meaningful—that is, at zero you have none!

Also, if you have $5 and your friend has $10, your friend has twice as much money as you. Hence, the ratio of two observations is meaningful. Weight is another example of the ratio scale of measurement. A 100-pound woman weighs half as much as a 200-pound man. The ratio scale has all the characteristics of the other scales of measurement.

As an example, the prices of a selection of compact disc players offered by Sound Advice are $239.95, $349.95, $369.95, and $429.95. Note that the distance between successive observations are ordered, the distance between successive observations can be measured, and the ratio of two CD changer prices is meaningful. That is, the price of a $400 changer is twice that of a $200 changer. Also, the zero point is meaningful. If the price of a CD changer is zero dollars, it indicates you pay nothing for it!

Self-Review 1-2

What is the level of measurement for each of the following? Give your reasoning.

a. Gallup Polls asked 1,000 adults, "Would you say you are financially better off now than you were a year ago?" The responses:

Better off	47%
Worse off	24%
Same	28%
No opinion	1%

b. Every four hours the quality control inspector at Cannon Mills selects 100 bedsheets at random and records the intensity of the blue color. There is some latitude but if the color is too deep, or too light, the sheet is "irregular." The irregular sheets are sold at a discount. The percent defective for the 10 checks Monday through Friday are 1%, 2%, 0%, 1%, 1%, 0%, 0%, 1%, 3%, and 0%.

c. The coaching staff rated each Oklahoma football player who participated in more than five plays in the game against Nebraska.

Rating	Number of Players
Outstanding	4
Excellent	10
Good	8
Fair	5
Poor	7

d. The Center for Education Statistics cited the following college enrollment figures for 1992 and 2000 (projected):

Sex	1992	2000
Male	6,276,000	6,468,000
Female	7,337,000	7,858,000

A WORD OF ENCOURAGEMENT

If you are a freshman or sophomore studying liberal arts or one of the social sciences, this may be your first college course with a quantitative orientation. Theory, and symbols such as Σ, σ, and μ are used extensively here. Also, formulas such as

$$\overline{X} = \frac{\Sigma x}{n}$$

are used throughout. Do not be intimidated by them. They are merely a shorthand that helps to condense the subject matter significantly. As a result, the material is slow reading, and only rarely will you feel you have a complete understanding of it, unless you have gone over it more than once.

One of the ways in which you can verify your understanding of the material, as you progress through the book, is to work through each of several **self-review problems** that are included in every chapter. Checking your answers against those provided at the end of the chapter allows you to determine immediately whether you understand previously covered subject matter. Further, doing the **exercises** is a very valuable learning tool. Answers to even-numbered exercises are at the back of the book. You will also want to work the **Chapter Achievement Test** found at the end of each chapter. This test will point you to topics you need to review further.

STAT BYTE

More and more pressure is being exerted to hire and promote women and minorities to top-level management positions. In December 1992, the federal government appointed a 21-member Glass Ceiling Commission to study ways to remove gender barriers to job advancement. The Commission found that, although there are 57 million working women, representing 45% of the total labor force, relatively few are in middle- or top-level positions. Another study by the Florida Consumers Federation studied the 750 Publix Supermarket stores and found that fewer than 3% of all managers and assistant managers are black and fewer than 2% are women.

COMPUTER APPLICATIONS

Computer use has accelerated greatly over the last few years, particularly in the field of statistics. Prior to 1940, most of the computations involving statistics were done by hand or on an adding machine. Extensive calculations, like those required in Chapter 15, "Multiple Regression and Correlation Analysis," were very time consuming, and the accuracy of the hundreds of necessary calculations was questionable.

The development of rotary calculators, by such companies as Marchant or Monroe, was the next step in problem solving. These have now been replaced by electronic hand calculators and computers. Many of you have computers at home, and most colleges and universities have PCs available to students. There are many statistical software packages available that will be useful for calculations involving larger data sets. Check with your instructor to see what packages are available to you. Some of the more popular software packages are SAS, SPSS, SPSS/PC + (Statistical Package for the Social Sciences), and MINITAB. We chose MINITAB for most of the computer applications in this textbook. We acknowledge the complimentary software provided by MINITAB, INC., for use in preparing this book. It is user friendly, meaning it is easy to operate, and does not require you to learn a programming language. To help you, we give the MINITAB commands at the top of each computer output and highlight them in a second color. Our goal is to show how the computer is used as a tool in statistical analysis. You should view the computer as an aid in performing the calculations. It does not, however, provide any interpretation. That is left to you.

SUMMARY

Statistics is often thought of as a collection of figures or data items. They might be the inflation rate in the United States (5.9%), the price of a Nikon N 90 ($842), the number of IBM employees (387,112), and their net earnings ($5.8 billion).

However, in this book we are concerned with statistics as the techniques applied in collecting, organizing, analyzing, and interpreting data for the purpose of making better decisions.

If we want to describe data, the techniques we use—such as frequency distribution in Chapter 2, averages in Chapter 3, or measures of dispersion in Chapter 4—are descriptive statistics. If we infer something about a population based on a sample taken from that population, the approach used is called inferential statistics.

There are four levels of measurement, namely: *nominal, ordinal, interval,* and *ratio*. The level of measurement must be identified so the appropriate statistical techniques can be applied. For example, to compute the standard deviation (Chapter 4) the data must be, at least, at interval level.

Exercises

5. A random sample of 300 executives out of 2,500 employed by a large firm revealed 270 would move to another location if it meant a substantial promotion. Based on these findings, write a brief note to management regarding all executives in the firm.
6. A random sample of 500 consumers was asked to test a new toothpaste. Out of the 500, 400 said it was excellent, 32 thought it was fair, and the remaining consumers had no opinion. Based on these sample findings, make an inference about the reaction of all consumers to the new toothpaste.
7. Using hypothetical examples, give illustrations of nominal-, ordinal-, and interval-level data.
8. Using actual figures from such publications as the *Statistical Abstract of the United States,* the *World Almanac, Forbes,* or your local newspaper, give examples of nominal-, ordinal-, and interval-level data.

For questions 9, 10, and 11, indicate whether the statement is true or false. If false, give the correct answer.

9. The National Cancer Institute reported the following new cases of cancer in 1992:

Type	Number (in thousands)
Lung	157
Breast	151
Colon	110
Prostate	106
Bladder	49

The above listing is an ordinal level of measurement.

10. The *Monthly Labor Review* reported the following average hourly earnings by industry: wholesale trade, $11.35; services, $10.52; tobacco products, $17.52; and textile mill products, $8.58. The collection of these data is referred to as statistics.
11. The questionnaire on page 22 from Kodalux Processing Services is designed to gather information from photographers about their developing services.

Dear Customer: Thank you for your order. In our continuing desire to provide the best possible service to our customers, we would greatly appreciate your answering the following questions, and mailing this postage-free card back to us. Many thanks for your help.

1. What date did you mail your film to us?____________
2. What date did you receive it?____________
3. Was it in good condition when received? Yes___ No___
4. What film type did you use? Slide Color Negative____________
5. What size film did you use? 35mm 110 Disc Other____________
6. Do you occasionally send film in multiple batches? Yes___ No___
7. Are your pictures for professional, industrial or commercial use? Yes___ No___
8. Are you satisfied with the service you received? Yes___ No___

Additional Comments:____________

____

____

PLEASE INCLUDE: City:____________ State:____________

Zip Code:____________

Kodalux
Processing Services

ROC

CHAPTER ACHIEVEMENT TEST

This is the first in a series of chapter achievement tests. These allow you to evaluate your understanding of the material. Answer all the questions. The answers are at the back of the book.

Indicate whether the statement is true or false. If false, give the correct answer.

1. A sample of 1,200 senior citizens rated a new cereal, OH Oats, as excellent, good, fair, or poor. The level of measurement is nominal.
2. Inferential and inductive statistics are the same.
3. Ratio-level data is the "lowest level" of measurement, and the data must be mutually exclusive.
4. During the 1991 NBA season, Michael Jordan averaged 38.8 minutes of playing time, 6.4 rebounds, 6.1 assists, and 30.1 points per game. He was voted the Most Valuable Player. This collection of data about Jordan is referred to as statistics.

5. A sample of 40 former prisoners was studied. Based on their responses, it was said that had all former prisoners been studied, 20% would be fully adjusted. This is an example of descriptive statistics.
6. There are 14,373 members in the Stone, Clay, and Glass Union. Randomly, 326 were selected for an opinion survey. The 326 can be considered as the population.
7. As reported by the Sarasota *Herald-Tribune,* gold prices soared past $400 an ounce on Tuesday. The $400 is a statistic.
8. In a broad sense, statistics is the collection, organization, presentation, analysis, and interpretation of data for the purpose of making better decisions.

ANSWERS TO SELF-REVIEW PROBLEMS

1-1 **a.** Of all grocery shoppers, 70% purchase at least one dairy product each time they shop for groceries.
b. A sample
c. All grocery shoppers
d. Inferential statistics, because sample results are being used to infer something about all grocery shoppers

1-2 **a.** Nominal level, because the categories can be rearranged, as in "worse off," "same," "better off," and "no opinion"
b. Ratio. The zero point is meaningful.
c. Ordinal
d. Nominal

CHAPTER 2

Summarizing Data: Frequency Distributions and Graphic Presentation

OBJECTIVES

When you have completed this chapter, you will be able to

- construct a frequency distribution;
- construct a stem-and-leaf chart;
- draw a histogram, a frequency polygon, and a cumulative frequency polygon;
- draw line graphs, bar graphs, and pie graphs.

CHAPTER PROBLEM **Constructing a Plan for Success**

The Reynolds Company wants to further develop the Rafter J Ranch Subdivision in Jackson Hole, Wyoming. Charles Reynolds is putting together a prospective that gives current information on the 80 homes in the subdivision. He checked each of the homes to determine its estimated selling price in today's market. Now he wants to know what a typical home will sell for, what the high and low prices are, where the prices tend to cluster, and so on.

INTRODUCTION

With this chapter, we begin our study of **descriptive statistics.** Recall from Chapter 1 that the purpose of descriptive statistics is to effectively summarize data. Two important elements of descriptive statistics are the arrangement and display of numerical information. Raw numbers alone provide little insight into the underlying pattern of the data from which conclusions can be drawn. For example, we may believe physicians generally have high incomes, or young people have lower incomes than old people. However, even if we were given access to Internal Revenue Service files, the income data would be of little value unless arranged, sorted, and condensed. This chapter introduces techniques used in arranging, sorting, and depicting relevant data—specifically, the **frequency distribution** and various statistical charts.

THE FREQUENCY DISTRIBUTION

A frequency distribution is defined as follows:

> **Frequency Distribution** A grouping of data into categories showing the number of observations in each of the nonoverlapping classes.

Tables 2–1 and 2–2 are examples of frequency distributions. In its *Motor Vehicle Facts and Figures,* the Motor Vehicle Manufacturers Association annually provides the ages of automobiles currently in use, grouped in classes "Under 3 years," "3–5 years," and so on. It also gives the number of automobiles in each class—27.1 million,

Table 2-1 **Ages of Automobiles Currently in Use, 1992**

Age	Number (in millions)
Under 3 years	27.1
3–5 years	30.5
6–8 years	20.7
9–11 years	21.1
12 years or more	14.0

SOURCE: *Motor Vehicle Manufacturers Association of the United States,* Motor Vehicle Facts and Figures, *1992.*

Table 2-2

Acquired Immune Deficiency Syndrome (AIDS) Deaths, 1992		
Age	**Number**	**Percent**
Under 5 years old	235	1.1
5–12 years old	52	0.2
13–29 years old	4,039	18.6
30–39 years old	9,865	45.5
40–49 years old	4,967	22.9
50–59 years old	1,764	8.1
60 years old or more	753	3.5

SOURCE: *U.S. Centers for Disease Control, Atlanta, Ga. unpublished data, 1992.*

30.5 million, and so on. Table 2–1 gives the figures for 1992. The frequency distribution in Table 2–2 presents the number and percent distribution of deaths due to Acquired Immune Deficiency Syndrome (AIDS), by age group, in 1992.

CONSTRUCTING A FREQUENCY DISTRIBUTION

The construction of a frequency distribution is illustrated by the following problem:

Problem Recall from the Chapter Problem that the Reynolds Development Company is putting together a prospectus for the further development of the Rafter J Ranch subdivision in Jackson Hole, Wyoming. Charles Reynolds, the developer, had each of the current homes in the Rafter J area appraised. The appraised value (the estimated selling price) of each of the 80 homes, rounded to the nearest thousand dollars, is shown in Table 2–3.

Table 2-3

Selling Prices of 80 Homes in the Rafter J Ranch Subdivision (in thousands)

$ 89	$112	$106	$ 89	$ 96	$ 97	$ 99	$ 85	$113	$ 84
95	102	99	91	102	93	97	98	103	95
105	96	88	90	101	99	101	100	98	100
113	98	95	98	93	86	97	91	108	93
86	85	91	82	94	93	86	98	83	87
105	98	92	98	96	103	93	91	106	94
106	101	98	97	104	83	99	114	91	97
81	107	104	99	104	93	103	99	97	86

lowest (81) highest (114)

The numerical information in Table 2–3 is usually called the **raw data.** As such, it has little meaning—that is, the raw data does not reveal much about the value of homes in the Rafter J Ranch area. Construct a frequency distribution to describe the selling prices.

Solution To construct a frequency distribution, the first step is to decide on the class limits. Note that in Table 2–3 the lowest selling price is $81 thousand. We let the lower limit of the first class be $80 thousand. The upper limit of that class is $84 thousand. The next class would be $85 through $89. These groupings are called **classes.**

$80–$84 ← a class

$85–$89 ← a class

Then, tally the raw data (selling prices in this problem) into the appropriate classes. For example, the price in the upper left-hand corner of Table 2–3 is $89 thousand. It is tallied in the $85–$89 class. A tally mark (/) is placed opposite the $85–$89 class. The same procedure is followed for each of the remaining amounts.

The number of tallies occurring in each class is the **class frequency** and usually is denoted by the letter *f*. For example, 5 is the class frequency for the lowest class of $80–$84. The frequency distribution of the selling prices of Rafter J Ranch homes is shown in Table 2–4.

STATED LIMITS. The lower and upper class limits in Table 2–4 are generally referred to as the **stated limits,** but sometimes they are called the "apparent class limits." The stated lower limit of the first class is $80 thousand and the stated upper limit is $84 thousand. The stated limits for the next class are $85 and $89 thousand.

TRUE LIMITS. The selling prices of the homes in the Rafter J Ranch subdivision have been rounded to the nearest thousand dollars. By conventional rounding rules, a selling price of $84,500 would be rounded *up* to $85,000. Any amount over $84,000 but under $84,500 would be rounded *down* to $84,000. Therefore, the **true limits** of the $80,000–$84,000 class are $79,500 and $84,500. The true limits of the next class are $84,500 and $89,500, and so on. Note that the width of each class is exactly $5,000. If we are dealing with data that are not rounded, the stated class limits are the same as the real class limits.

MIDPOINTS. The midpoint of a class, also called a "class mark," divides the class into two equal parts and is determined in two ways. First, it is halfway between the *true* class limits. Halfway between the true limits of $84,500 and $89,500 is $87,000. Second, the midpoint

Table 2-4

Frequency Distribution: Selling Prices of Homes in the Rafter J Ranch Subdivision

Classes		Class Frequencies
Selling Price (in thousands)	Tallies	Number of Homes (f)
$ 80–84	~~||||~~	5
85–89	~~||||~~ ~~||||~~	10
90–94	~~||||~~ ~~||||~~ ~~||||~~	15
95–99	~~||||~~ ~~||||~~ ~~||||~~ ~~||||~~ ~~||||~~ /	26
100–104	~~||||~~ ~~||||~~ ///	13
105–109	~~||||~~ //	7
110–114	////	4

of a class is determined from the *stated* class limits. For the second class, that would be halfway between $85,000 and $89,000, which is also $87,000. Later in this chapter we use midpoints to draw frequency polygons. They are also used in computing measures of central tendency (in Chapter 3) and measures of dispersion (in Chapter 4).

Describing the Data

Now that the raw data are organized into a frequency distribution, the pattern of the selling prices of homes in the Rafter J Ranch subdivision can be described. Table 2–4 shows that the lowest selling price is about $80 thousand, the highest about $114 thousand. Moreover, most of the homes cost between $90 thousand and $105 thousand, with the largest cluster in the $95–$99 thousand class. The single value of $97 thousand (the midpoint of the $95–$99 class) can be considered a "typical" selling price.

Suggestions for Constructing a Frequency Distribution

Here are some suggestions for constructing frequency distributions:

1. **Overlapping Classes.** Overlapping classes—such as $80–$85, $85–$90, and $90–$95—must be avoided. Otherwise, it is not clear where to tally a selling price of $85 thousand or $90 thousand. Assuming the groupings do not overlap, it is said that the categories are *mutually exclusive,* as they are in Table 2–4. In Chapter 1, we defined mutually exclusive as follows:

Mutually Exclusive Categories Each individual or item, by virtue of being included in one category, is excluded from all others.

> **STAT BYTE**
>
> Which U.S. cities are growing and which cities are declining? Among cities with a population over 100,000, Moreno Valley (a suburb of Riverside, California) has been the fastest growing city in the United States. Its population increased by a whopping 320% from 1980 to 1990. Mesa (a suburb of Phoenix, Arizona) was the second-fastest growing city during the 1980s with an increase of 89%. What cities had the largest percent decline during the same period? Gary, Indiana, lost 23% of its population and Newark, New Jersey, had a decline of 16%.

2. **Equal-Sized Classes.** If possible, class intervals should be equal. The use of equal intervals allows computation of averages and measures of dispersion discussed in the next two chapters. Classes of unequal size are sometimes necessary, however, in order to accommodate all the data, as Table 2–5 illustrates. Had the interval been kept at a constant size of $1,000, the frequency distribution would include at least 1,000 classes—obviously too many to make any meaningful analysis. (Incidentally, the Internal Revenue Service used the frequency distribution in Table 2–5 to show the adjusted gross income, before taxes, for over 67 million persons who filed income tax forms during the year.)
3. **Open-Ended Classes.** If possible, open-ended classes should be avoided. Table 2–5 has two such classes, namely, "Under $2,000" and "$1,000,000 and over." Strictly speaking, the class "Under $2,000" is not open-ended, because it has an implied lower limit of zero. However, we usually avoid classes like this because the class midpoint ($1,000) is not representative of all the values in the class. If a frequency distribution has an open end, a commonly used measure of central tendency called the arithmetic mean cannot be used. (More about the arithmetic mean in Chapter 3.)
4. **Number of Classes.** No fewer than 5 and no more than 20 classes should be used in the construction of a frequency distribution. Too few, or too many, classes give little insight into the distribution of the data. For example, the following age distribution contains only two classes. Very little can be said about the distribution of the ages of the inmates in Northeastern Prison.

Age of Prisoners	Number of Prisoners
20–39	432
40–59	431

5. **Class Size.** It is common practice to use intervals that are multiples of 5 or 10, such as 5, 10, 20, 100, or 1,000. If it is decided to have classes of equal size, we approximate the width of the interval by subtracting the lowest from the highest value and dividing by the number of classes. That is,

$$\text{Width of class interval} = \frac{\text{Highest value} - \text{Lowest value}}{\text{Number of classes}} \qquad \textbf{2-1}$$

Table 2-5

Adjusted Gross Income for Individuals Filing Income Tax Returns, 1990

Adjusted Gross Income Class	Number of Returns (in thousands)
Under $2,000	135
$2,000–2,999	3,399
3,000–4,999	8,175
5,000–9,999	19,740
10,000–14,999	15,539
15,000–24,999	14,944
25,000–49,999	4,451
50,000–99,999	699
100,000–499,999	162
500,000–999,999	3
$1,000,000 and over	1

Suppose it was decided that the price data in Table 2–3 should be placed into 7 classes. What would the width of the interval be?

$$\text{Width of interval} = \frac{\$114 - \$81}{7} = \$4.71$$

An interval of $4.71 is cumbersome to use. As shown in Table 2–4, a figure of $5 (thousand) is a more convenient interval.

6. **Class Limits.** Frequently, it is easy to make the lower limit of the first class a multiple of the class interval. In this instance, $80 thousand is a multiple of $5 thousand.

Self-Review 2-1

Answers to all Self-Review problems are at the end of the chapter. Police files reveal the following ages of persons arrested for purse snatching: 16, 41, 25, 21, 30, 17, 29, 50, 30, and 39.

a. Using 15 years as the lower limit for the first class, and an interval of 10 years, organize the age data into a frequency distribution.

b. What are the numbers 16, 41, 25, . . . , called?

c. Based on the data contained only in the frequency distribution, describe the age distribution of the purse snatchers.

Exercises

Answers to the even-numbered exercises are at the back of the book.

1. The following data represent the highest amount spent on a textbook for the current quarter based on a sample of 35 students:

$57	$34	$27	$41	$25	$18	$39
33	37	39	38	47	31	42
60	58	31	47	37	16	64
34	41	43	50	46	63	51
41	30	42	37	48	28	34

Starting with $15 as the beginning number for the first class, and using an interval of $10, organize the data into a frequency distribution. Briefly describe the distribution.

2. A survey of 50 amateur photographers included the question, "How many rolls of film did you expose during the past month?" The numbers of rolls reported were:

5	3	3	1	4	3	4	3	6	8
4	5	3	4	2	4	7	6	5	9
6	6	6	7	0	11	3	12	4	7
14	0	2	4	4	3	5	15	0	10
4	5	2	3	5	1	8	1	2	12

Starting with 0 as the beginning number for the first class, and using an interval of 3, organize the data into a frequency distribution. Briefly describe the distribution.

3. Schneider Nursery employs 30 people. The length of service, in years, for each employee is as follows:

4	1	4	12	10	5	8	14	4	1
8	8	1	4	1	8	8	1	7	3
3	2	7	12	13	11	10	6	2	3

Construct a frequency distribution. Use 0 as the lower limit of the first class and a class interval of 3 years.

4. The Leona Library (a small rural library) reported the following number of patrons using the library in the evening during the past 30 days:

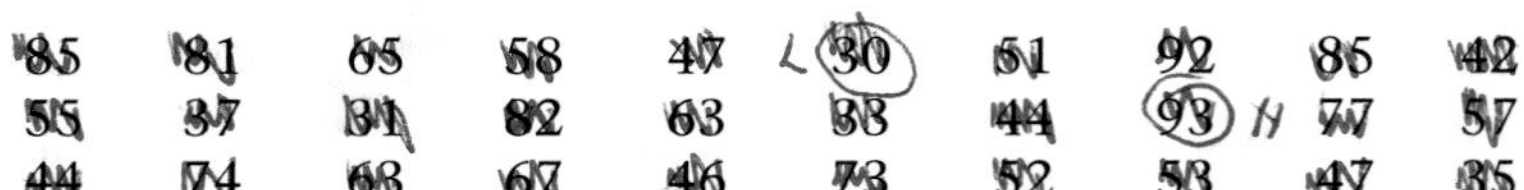

85	81	65	58	47	30	51	92	85	42
55	37	31	82	63	33	44	93	77	57
44	74	68	67	46	73	52	58	47	35

Construct a frequency distribution. Use a class interval of 10 and a lower limit of the first class of 25.

STEM-AND-LEAF DISPLAYS

Recently, a device known as a **stem-and-leaf display** has received considerable attention. It is a combination of sorting and graphing, but appears much like a frequency distribution.

The first step in constructing a stem-and-leaf display is to sort the data from smallest to largest. The stem is the *leading digit* or digits of the number, and the leaf is the *trailing digit*. To illustrate its construction, we will use a class interval of 10 or a multiple of 10. A digit is recorded instead of a tally. For example, the number 52 has a stem value of 5 and a leaf value of 2. The number 867 has a stem value of 86 and a leaf value of 7.

Problem A supermarket is studying the length of time, in minutes, that customers spend in the store. A sample of 20 observations revealed the following times:

34	28	32	24	38	16	8	24	46	26
12	18	22	42	36	26	2	26	32	28

Construct a stem-and-leaf chart.

Solution A quick glance at the data indicates that the times range from 2 minutes to 46 minutes. The first digit of each time is used as the stem and the second digit as the leaf. The first customer remained in the store 34 minutes. Hence, the stem value is 3 and the leaf value is 4. Here is the chart after only the first customer is recorded:

Leading Digit	Trailing Digit
0	
1	
2	
3	4
4	

The usual practice is to arrange the trailing digits (leaves) in each row from low to high. The final stem-and-leaf display would appear as follows:

Leading Digit Stem	Trailing Digit or Leaf
0	28
1	268
2	24466688
3	22468
4	26

As noted in Chapter 1, a number of computer software packages are available to make setting up and solving statistics problems easier. We use MINITAB to show how the burden of arranging raw data in a stem-and-leaf display is shifted to the computer. The selling prices of the homes in the Rafter J Ranch subdivision in Table 2-3 are used. We want the data to be categorized in increments of $5 thousand. The computer ranks the values in each row from low to high. For the first row the selling prices are, in thousands of dollars, 81, 82, 83, 83, and 84.

```
MTB >      stem c1;
SUBC >   increment 5.

Stem-and-Leaf of C1          N = 80
Leaf Unit = 1.0

                      STEM  LEAF
           (80–84)     8    12334
           (85–89)     8    5566667899
           (90–94)     9    011111233333344
           (95–99)     9    5556667777778888888899999
           (100–104)   10   0011122333444
           (105–109)   10   5566678
           (110–114)   11   2334
```

The stem-and-leaf display is very flexible. As an example, suppose that, during a two-week period, selected Burger King restaurants sold 6732, 6791, 6823, and 6752 Whoppers. The stem-and-leaf display would be:

Stem	Leaf
67	359
68	2

Notice that the stem includes the thousands and hundreds digits, and the leaf contains the tens digit. The units digit is omitted.

Self-Review 2-2

The police records referred to in Self-Review 2–1 showed the ages of those arrested for purse snatching to be 16, 41, 25, 21, 30, 17, 29, 50, 30, and 39. Develop a stem-and-leaf display.

Exercises

5. The annual payments to the 20 employees of a small firm (in thousands of dollars) are given below. Develop a stem-and-leaf display.

$48	$23	$27	$46	$24
28	17	39	43	35
53	45	14	24	21
41	31	23	18	19

6. The number of parking tickets issued for the last 25 days in downtown Oxford is shown below. Develop a stem-and-leaf display.

53	42	34	43	35
39	27	17	40	41
32	21	21	37	12
36	19	29	26	14
38	57	45	22	33

7. The Quick Change Oil Company advertises that it can change a car's motor oil in 10 minutes. However, business has been very slow lately. A study of the last 25 days revealed the following number of cars serviced each day:

65	98	55	62	79
59	51	90	72	56
70	62	66	80	94
79	63	73	71	85
93	68	86	53	90

Develop a stem-and-leaf display.

8. Robinson TV and Appliance employs 20 salespeople. The number of TVs each salesperson sold last month are given below. Develop a stem-and-leaf display.

21	34	38	12	6
22	18	36	13	28
19	40	8	14	7
20	18	23	32	14

PORTRAYING THE FREQUENCY DISTRIBUTION GRAPHICALLY

The popularity of magazines such as *Better Homes and Gardens, Time,* and *Ebony* and newspapers such as *USA Today* and *The Wall Street*

Journal is a tribute to the Chinese proverb that "a picture is worth a thousand words."

Pictures of a special variety are also used extensively to help hospital administrators, business executives, and consumers get a quick grasp of statistical reports. Such pictures are called *graphs* or *charts*. Three graphic forms commonly employed to portray a frequency distribution are the **histogram,** the **frequency polygon,** and the **cumulative frequency polygon,** often referred to as an **ogive.**

Histogram

A **histogram** is one of the most easily interpreted charts. We will illustrate its construction by using the distribution of the selling price of homes in the Rafter J Ranch subdivision.

The class frequencies (number of homes sold) are plotted on the vertical axis (Y-axis). The variable (selling price) is scaled on the horizontal axis (X-axis). Either the true limits of the classes or the stated limits of the classes may be used. We will use the true limits, as given in Table 2–6.

Five homes sold for between $79,500 and $84,500. The first step in constructing a histogram is to draw vertical lines from both $79,500 and $84,500 on the X-axis to points opposite 5 on the Y-axis, and then to connect the tops of these two lines to form a bar. The area of that bar represents the number of homes (5) sold in that class.

Figure 2–1 shows how the histogram would appear for the first two classes. The completed histogram is shown in Figure 2–2 on page 38.

MINITAB has a procedure to generate a histogram. It gives the midpoints and the class counts (class frequencies) and shows the bars in the form of asterisks. The data are the same Rafter J Ranch selling prices.

STAT BYTE

The percent of workers in the United States that hold down two jobs is higher today than at any time in the past three decades. In 1960, about 4.6% of the population maintained two jobs; by 1991, this figure had increased to 6.2%, an increase of almost 35%.

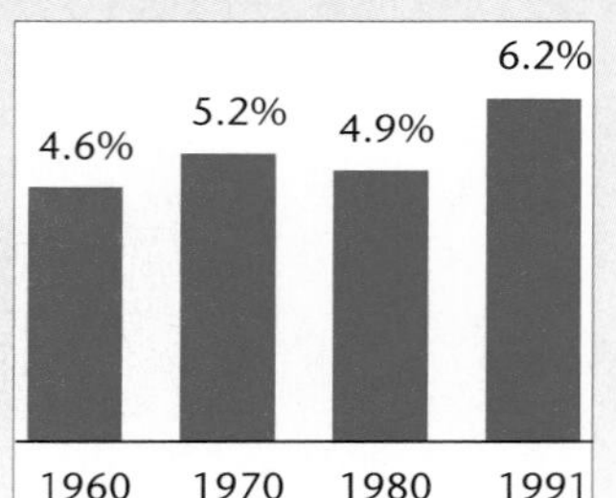


```
MTB      hist c1;
SUBC     start 82
SUBC     increment 5.

Histogram of C1    N = 80

Midpoint   Count
  82.00      5   *****
  87.00     10   **********
  92.00     15   ***************
  97.00     26   **************************
 102.00     13   *************
 107.00      7   *******
 112.00      4   ****
```

Table 2-6

Selling Prices of Homes in the Rafter J Ranch Subdivision

True Limits Selling Price	**Class Frequencies** Number of Homes (f)
$79,500–84,500	5
84,500–89,500	10
89,500–94,500	15
94,500–99,500	26
99,500–104,500	13
104,500–109,500	7
109,500–114,500	4

Figure 2-1

Histogram for the First Two Classes in Table 2-6

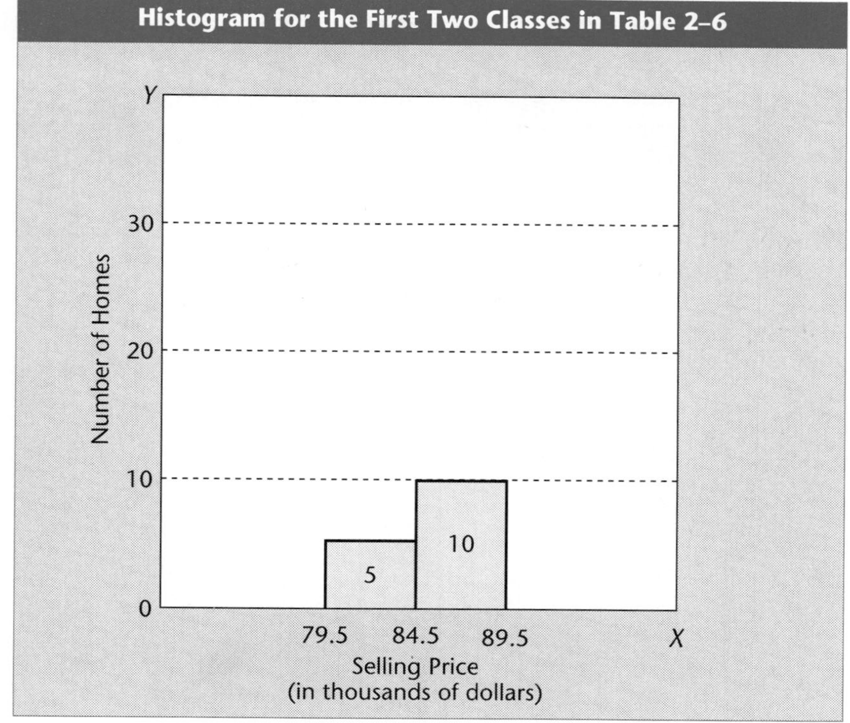

Self-Review 2-3

The age distribution of the prison guards hired by the Nebraska Corrections Bureau in the past two years is as follows:

Age	Number
20–29	2
30–39	13
40–49	20
50–59	12
60–69	3

Construct a histogram for this data.

The Frequency Polygon

To illustrate the construction of a **frequency polygon,** the frequency distribution that lists the selling prices of homes in the Rafter J Ranch subdivision is reintroduced (see Table 2–7).

The class frequencies (number of homes in this problem) are plotted on the vertical axis (Y-axis) and the class midpoints are on the X-axis.

Figure 2-2

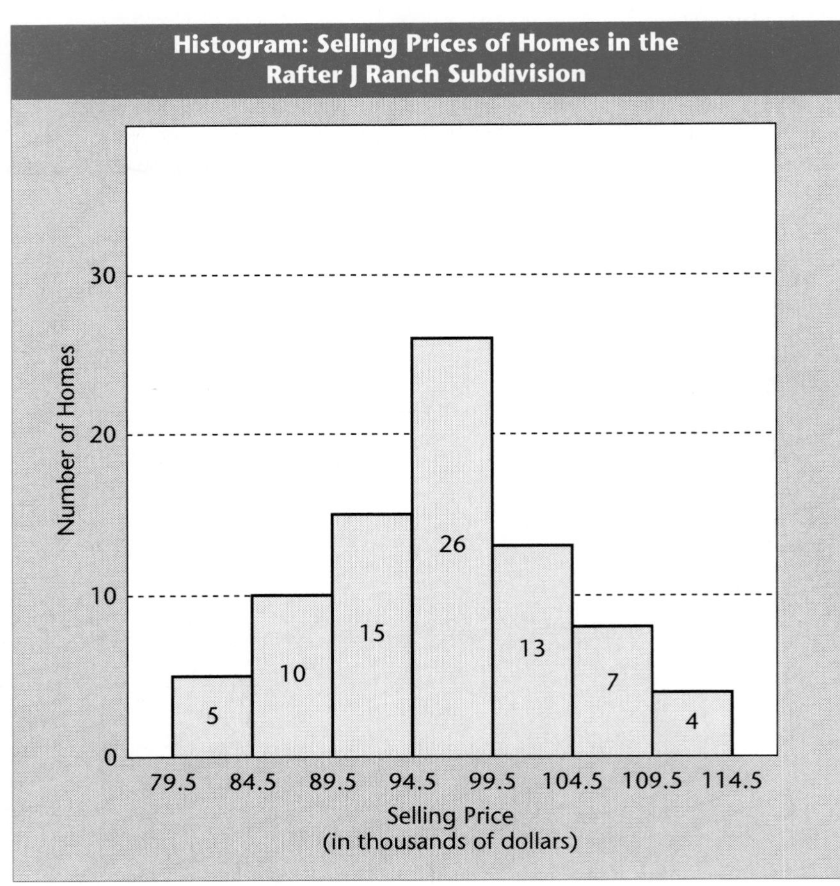

Table 2-7

Frequency Distribution: Selling Prices of Homes in the Rafter J Ranch Subdivision

True Class Limits Selling Price (in thousands)	**Class Midpoints**	**Number of Homes**
$ 79.5–84.5	$ 82	5
84.5–89.5	87	10
89.5–94.5	92	15
94.5–99.5	97	26
99.5–104.5	102	13
104.5–109.5	107	7
109.5–114.5	112	4

In Table 2–7, there were five homes sold in the $79,500 – $84,500 price range. The midpoint representing that class is $82,000. Therefore, the coordinates of the first plot are $X = \$82$, $Y = 5$. The coordinates of the next plot are $X = \$87$, $Y = 10$. The dots are connected in order by straight line segments, as shown in Figure 2-3.

Figure 2-3

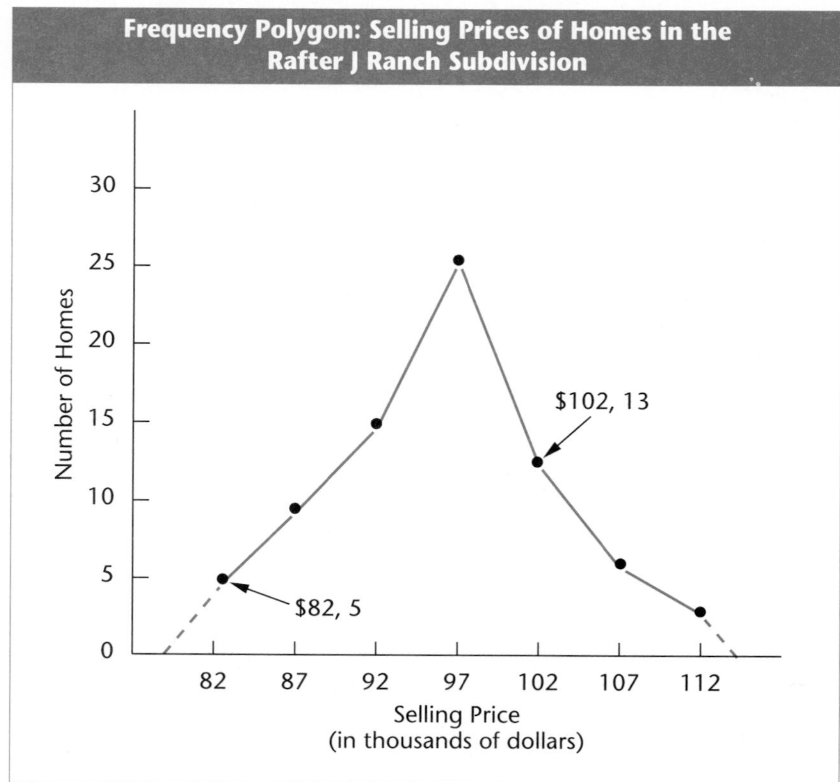

A technical note: The usual practice is to extend the two extremes of the frequency polygon to the X-axis. Conventionally, we show this by using dashed, rather than solid, lines for the extension. In Figure 2–3, the left end of the polygon extends to the midpoint of a class below the lowest nonzero class. The upper end of the polygon is treated in a similar way.

Self-Review 2-4

Refer to Self-Review 2–3. Portray the ages of the guards in the form of a frequency polygon.

Exercises

9. In Exercise 1, you constructed the following frequency distribution:

Amounts Spent on Textbooks	Number
$15–24	2
25–34	10
35–44	12
45–54	6
55–64	5

a. Portray the distribution in the form of a histogram.

b. Portray the distribution in the form of a frequency polygon.

10. Refer to Exercise 2. Draw a histogram and a frequency polygon using the distribution of the number of rolls of film.

11. Refer to Exercise 3. Draw a histogram and a frequency polygon for the length of service for the employees.

12. Refer to Exercise 4. Draw a histogram for the number of library patrons.

A frequency polygon is ideal for comparing two or more groups of values. As an illustration, recall that the selling prices of the homes in the Rafter J Ranch subdivision ranged from about $80 thousand to $114 thousand. Suppose that the broker at Reynolds Development Company wanted to compare the Rafter J Ranch selling prices with those in the Indian Paintbrush area. The frequency polygons are in Figure 2–4. It is readily apparent that the selling prices for the Indian Paintbrush homes are significantly higher than those for Rafter J Ranch. The typical price for a Rafter J Ranch home is about $97 thousand, for the Indian Paintbrush area $140 thousand.

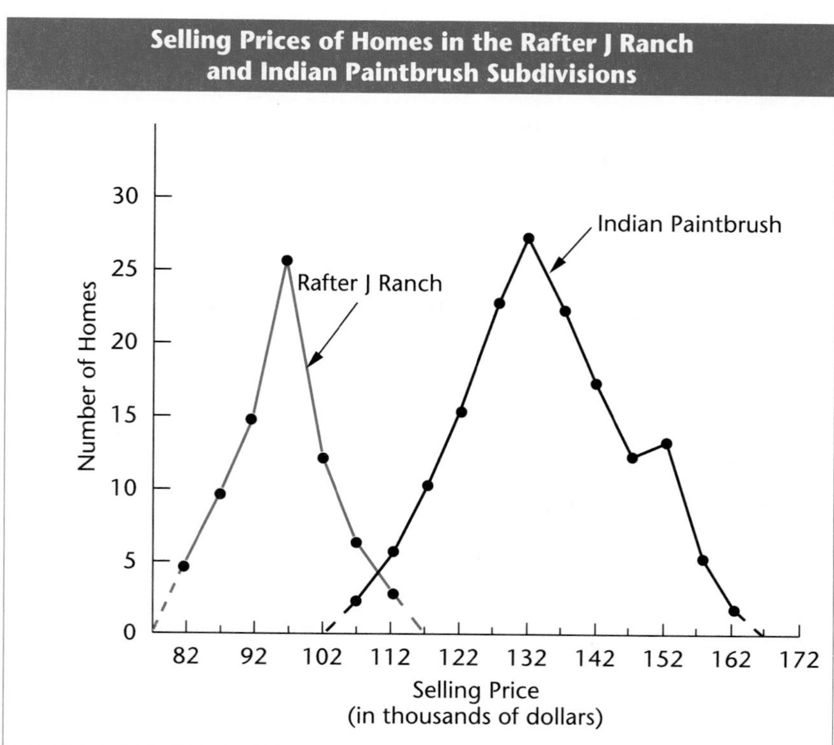

Figure 2-4

The Percent Frequency Distribution and The Percent Frequency Polygon

The number of homes sold in Rafter J Ranch and in the Indian Paintbrush area were about equal. This made the comparison of the two distributions relatively easy. It is difficult to make comparisons, however, if the size of one group is much larger than the size of the other. For example, 50 persons were surveyed on controversial subjects such as abortion and strip mining. Four years later, a much larger group was asked the same questions. The research report gave the age distributions for both groups (see Table 2–8).

Table 2-8

Age Distributions for the 1989 and 1993 Surveys

Age (in years)	**Number of Persons** 1989	1993
20–29	2	74
30–39	13	400
40–49	20	980
50–59	12	460
60–69	3	86
Totals	50	2,000

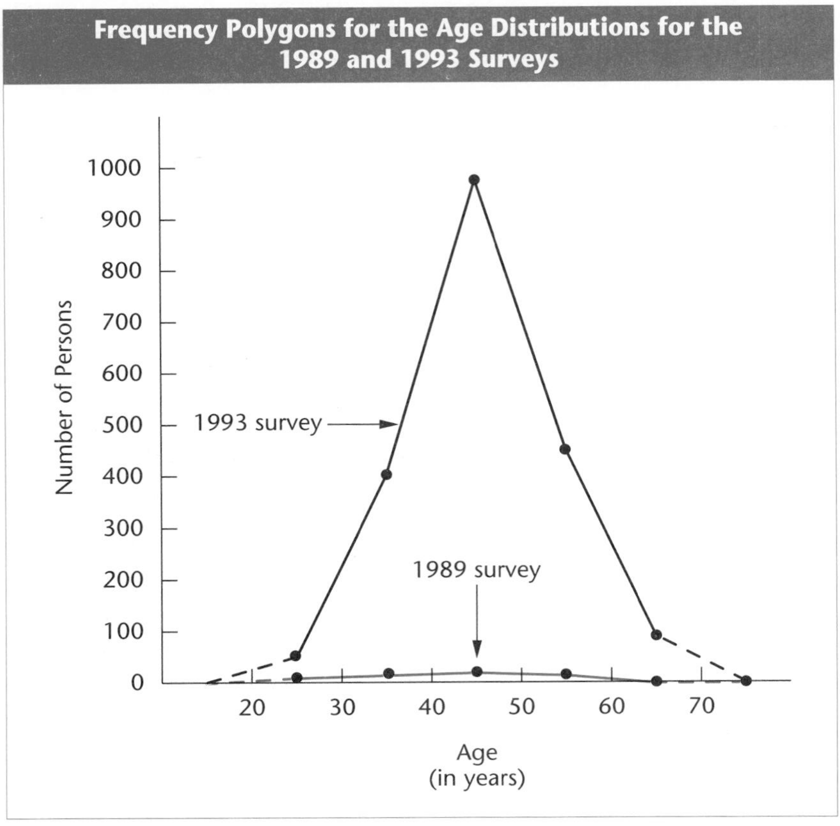

Figure 2-5

Because the number in the 1989 survey is small relative to the number in the 1993 survey, the accompanying frequency polygon is difficult to interpret. The 1989 results are compressed at the bottom of Figure 2–5.

A better approach is to convert the class frequencies to percentages of the total. For example, we can convert the class frequency in the 20–29 class for 1989 to a percent by dividing 2 by the total number of persons (2/50 = 0.04 = 4%). In terms of a formula, the relative frequency of a class is found by:

$$\text{Relative frequency of a class} = \frac{\text{Frequency of a class}}{\text{Total number of frequencies}} \qquad \textbf{2-2}$$

We can change the 1993 class frequency in the 50–59 class to a percent by:

$$\text{Relative frequency of a class} = \frac{460}{2000} = 0.23 = 23\%$$

The percents of the total are shown in columns 4 and 5 in Table 2-9. An analysis of the table and the accompanying chart (Figure 2–6 on page 44) reveals that when we use this method of conversion, the shapes of both age distributions are nearly identical. In Figure 2–6, the percentages of the total are plotted in the form of frequency polygons.

The Cumulative Frequency Distribution and the Cumulative Frequency Polygon (Ogive)

We may be interested in knowing the percent of observations that are less than a particular value. Or, the question might be: What percent of the observations are greater than a particular value? For example, what percent of the Rafter J Ranch selling prices are greater than $100,000? A **cumulative frequency distribution** or a **cumulative frequency polygon,** often referred to as an **ogive,** can be used to answer these questions.

As the name implies, a cumulative frequency distribution requires cumulative class frequencies. There are two cumulative frequency distributions—a "less-than" cumulative frequency distribution and a "more-than" cumulative frequency distribution. To show the construction of both types, the selling prices of the Rafter J Ranch homes are repeated in Table 2–10 on page 44.

Note that for the lowest class, 5 homes sold for less than the upper true class limit of $84,500. Those 5, plus the 10 in the next—lowest class, or 15, sold for less than $89,500. Then 5 + 10 + 15, or 30 homes sold for less than the upper true class limit of $94,500. This adding process is continued for all the frequencies.

Table 2-9

Age Distributions for the 1989 and 1993 Surveys

	Number of Persons		Percent of the Total	
Age	1989	1993	1989	1993
(1)	(2)	(3)	(4)	(5)
20–29	2	74	4.0	3.7
30–39	13	400	26.0	20.0
40–49	20	980	40.0	49.0
50–59	12	460	24.0	23.0
60–69	3	86	6.0	4.3
Total	50	2,000	100.0	100.0

Note: *The ages were rounded to the nearest year; that is, the age of a person 29 years 6 months was rounded up to 30 years. The age 29 years 5 months was rounded down to 29.*

Figure 2-6

Frequency Polygons for Two Age Distributions Calculated in Percents (1986 and 1993 Surveys)

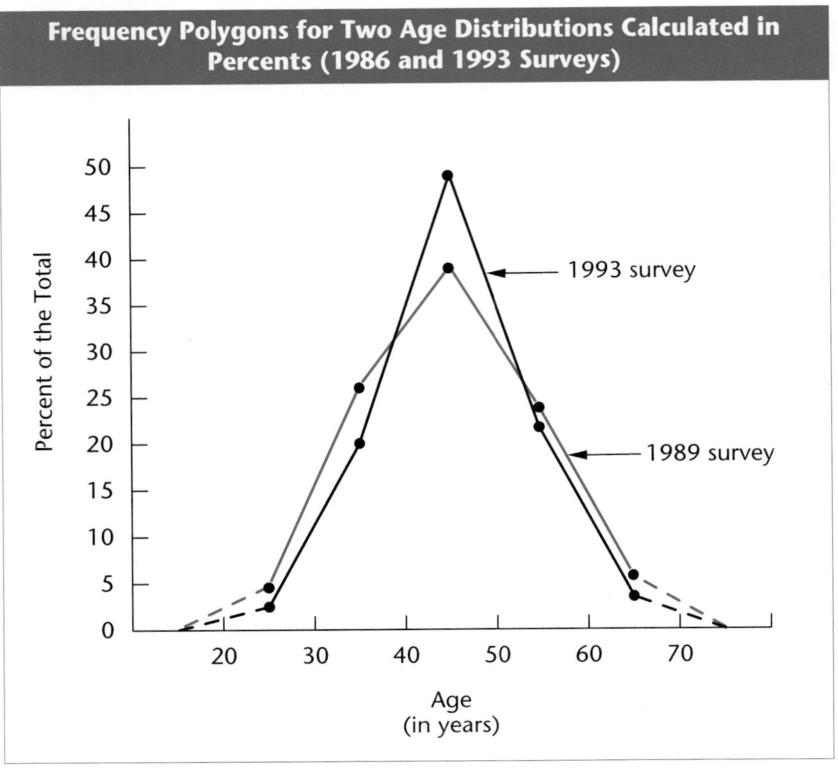

Table 2-10

Selling Prices of Homes in the Rafter J Ranch Subdivision

True Limits Selling Price	**Class Frequencies** Number of Homes (f)
$ 79,500–84,500	5
84,500–89,500	10
89,500–94,500	15
94,500–99,500	26
99,500–104,500	13
104,500–109,500	7
109,500–114,500	4

The complete set of cumulative frequencies are in Table 2-11.

The *upper true class limits* and the *cumulative frequencies* (Table 2–11) are plotted to portray the selling prices in the form of a *less-than cumulative frequency polygon*. The first three plots are $84,500

Table 2-11

Less-Than Cumulative Frequency Distribution for Selling Prices of Rafter J Ranch Homes

True Limits Selling Price	Frequency		Cumulative Frequency
$ 79,500–84,500	5	add down ↓	5
84,500–89,500	10		15
89,500–94,500	15		30
94,500–99,500	26		56
99,500–104,500	13		69
104,500–109,500	7		76
109,500–114,500	4		80

and 5, $89,500 and 15, and $94,500 and 30. These are plotted in Figure 2–7. As shown, one practice is to scale the cumulative frequencies on the left vertical axis and the cumulative percents on the right.

Based on the less-than cumulative frequency distribution and the accompanying ogive, we can make statements such as "about half

Figure 2-7

Less-Than Cumulative Frequency Polygon for the Selling Prices of Rafter J Ranch Homes

(50%) of the selling prices were less than $96,000." We find the $96 thousand as shown, by drawing a horizontal line from the 50% mark on the right-hand margin of the vertical axis to the polygon slope and then moving straight down to the X-axis to read the figure $96 thousand (approximately).

Because no values occur before the smallest class, the corresponding cumulative frequency for that class is zero. No (0) homes sold for less than $79,500 (see Figure 2–7).

We construct the *more-than cumulative frequency distribution* by adding the frequencies from the *highest* class to the *lowest class* (see Table 2-12).

The *lower true class limits* and the corresponding cumulative frequencies are plotted to construct the more-than cumulative frequency polygon (see Figure 2–8).

Self-Review 2-5

Presented below is the age distribution of guards from Self-Review 2–3.

Age	Number
20–29	2
30–39	13
40–49	20
50–59	12
60–69	3

a. Construct a less-than cumulative frequency distribution.
b. Draw a less-than cumulative frequency distribution.
c. Based on the graph, about half the guards hired by the Corrections Department were less than what age?

Table 2-12 **More-Than Cumulative Frequency Distribution for Selling Prices of Rafter J Ranch Homes**

True Limits Selling Price	Frequency		Cumulative Frequency
$ 79,500–84,500	5	↑	80
84,500–89,500	10		75
89,500–94,500	15		65
94,500–99,500	26	add	50
99,500–104,500	13	up	24
104,500–109,500	7		11
109,500–114,500	4		4

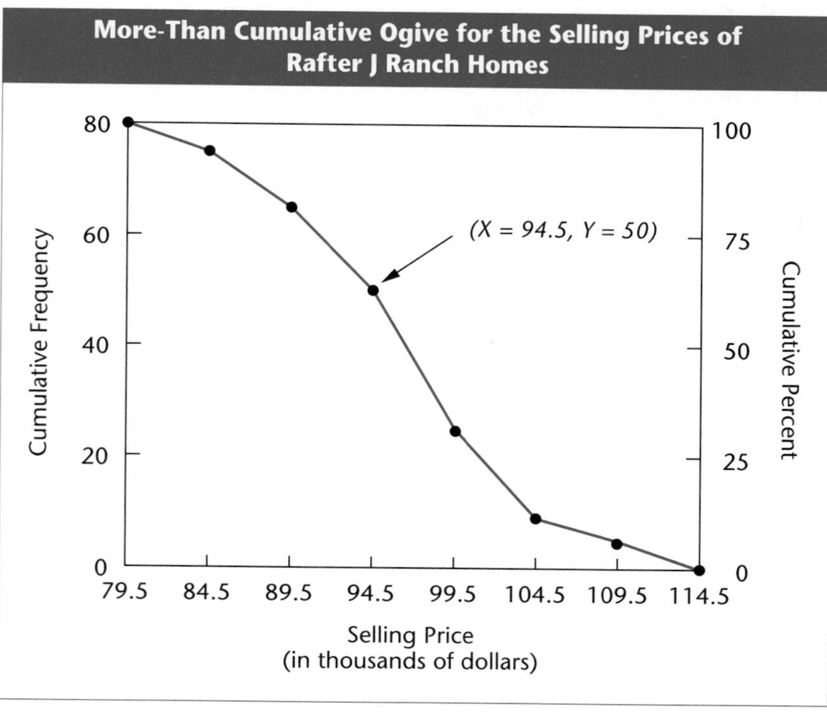

Figure 2-8

Exercises

13. A soft drink bottling machine is tested periodically to determine if it is functioning within certain limits. The frequency distribution below gives the results for 100 observations measured to the nearest hundredth of an ounce.

Class	Frequency
5.00–5.49	1
5.50–5.99	1
6.00–6.49	2
6.50–6.99	5
7.00–7.49	13
7.50–7.99	36
8.00–8.49	31
8.50–8.99	10
9.00–9.49	1

a. Construct a less-than cumulative frequency distribution for these data.

b. Draw a less-than cumulative frequency distribution.
c. Based on the graph, about one-fourth of the time the machine will yield less than what amount?

14. A psychologist is studying the length of time it takes mice to go through a maze and reach a reward at the end. Time is measured to the nearest tenth of a second, with the results given in the table below.

Time	Frequency
2.0–2.9	3
3.0–3.9	7
4.0–4.9	15
5.0–5.9	29
6.0–6.9	81
7.0–7.9	50
8.0–8.9	10
9.0–9.9	5

a. Construct a more-than cumulative frequency distribution for this data.
b. Draw a more-than cumulative frequency distribution.
c. Based on the graph, about 90% of the mice take more than what amount of time to complete the maze?

Other Graphic Techniques

The frequency polygon, the ogive, and the histogram have strong visual appeal. There are other graphs, however, that are commonly used in government reports, research reports, newspapers, and journals. Several are presented in this section. Others, such as the scatter diagram, are introduced in later chapters.

LINE CHARTS. A **line chart** is particularly useful in portraying data over a period of time. We will use the number of married women in the work force from 1960 to 1990 an example. Time (years in this case) is *always* plotted on the horizontal axis. The first plot is 1960, 12.3 million. The second plot is X=1970, Y=18.4 million; and so on (see Figure 2–9).

A line chart is relatively easy to interpret. In this case, the number of married women in the labor force has increased rapidly from about 12 million in 1960 to over 30 million in 1990.

Self-Review 2-6

The farm population of the United States from 1940 to 1990 is given at the top of page 49.

Year	Farm Population (in millions)
1940	30.5
1950	23.0
1960	15.6
1970	9.7
1980	7.9
1990	7.3

SOURCE: *U.S. Department of Agriculture, Agriculture Statistics Service, and* Historical Statistics, Colonial Times to 1970, *series K.*

a. Plot the farm population in the form of a line chart.
b. Interpret your chart.

Figure 2-9

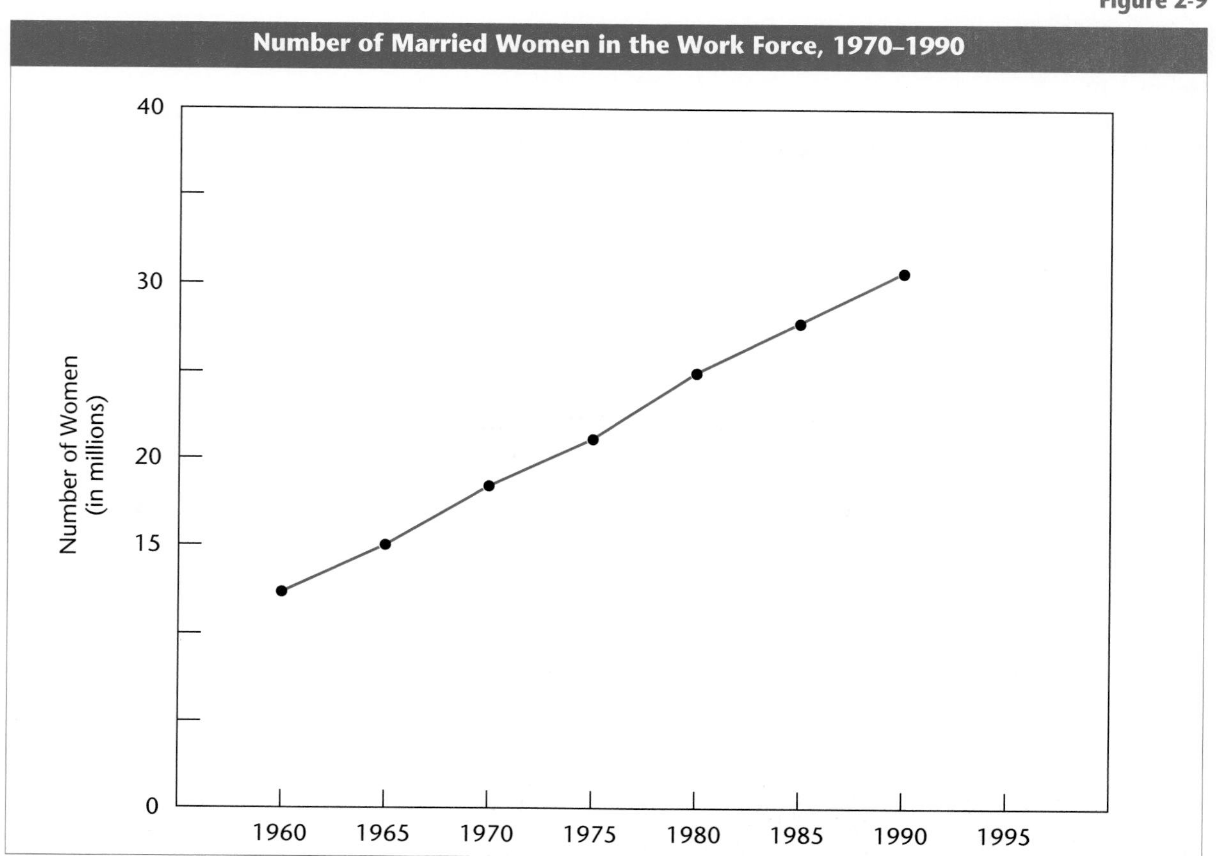

Several time series can be shown on the same chart. For example, Figure 2–10 traces the percent of persons below the poverty level, by race, from 1959 to 1992.

Self-Review 2-7 Interpret the trends in Figure 2–10.

BAR CHARTS. A **bar chart** is used to portray any one of the four levels of measurement—nominal, ordinal, interval, or ratio. The religion reported by the population of the United States 14 years old and over is used to illustrate the construction of a bar chart. The data in Figures 2–11 and 2–12 are nominal level. The chart may be organized so that the bars are either horizontal, as in Figure 2–11, or vertical, as in Figure 2–12.

Technical notes: As shown in these two bar charts, a small space usually separates each bar. Color is often used—especially in the annual reports of businesses and in magazines such as *Time.* Actual numbers may be placed on the graph, such as 79 million for Protestant, 31 million for Roman Catholic, and so on.

Figure 2-10

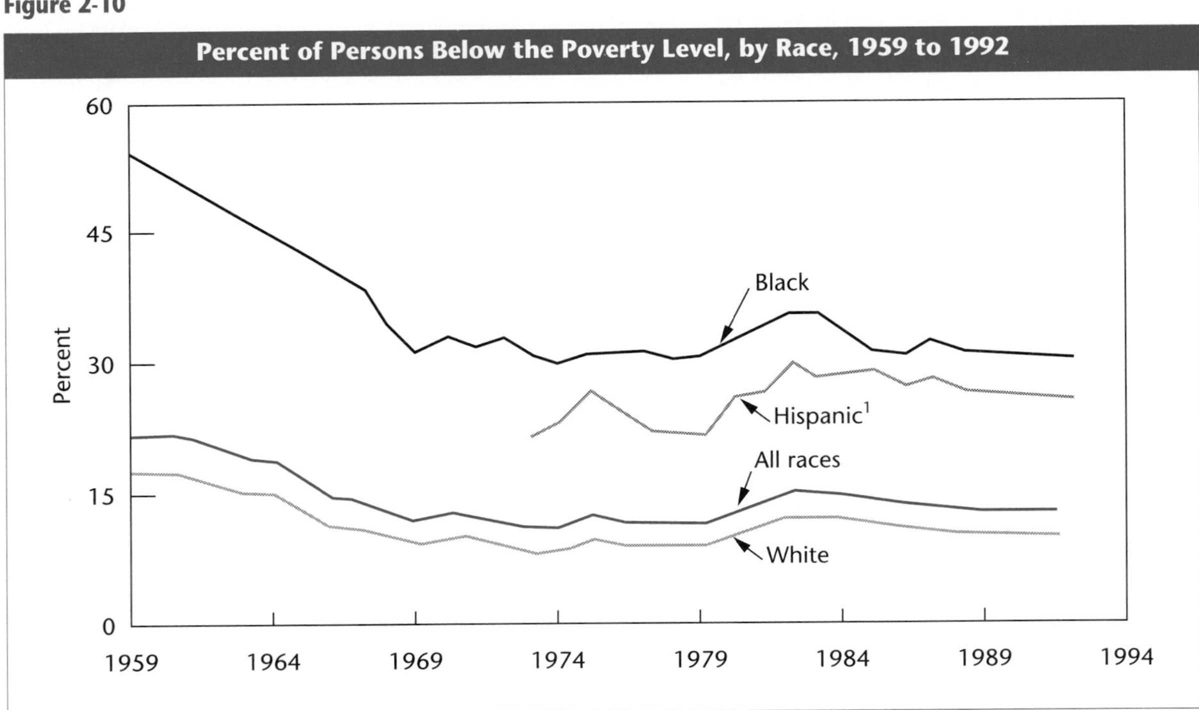

[1]Hispanic persons may be of any race.
SOURCE: *U.S. Bureau of the Census.*

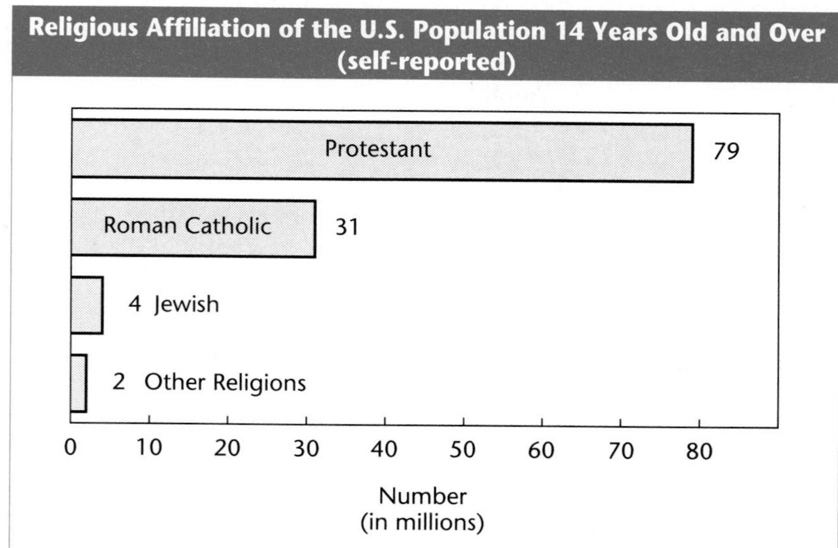

Figure 2-11

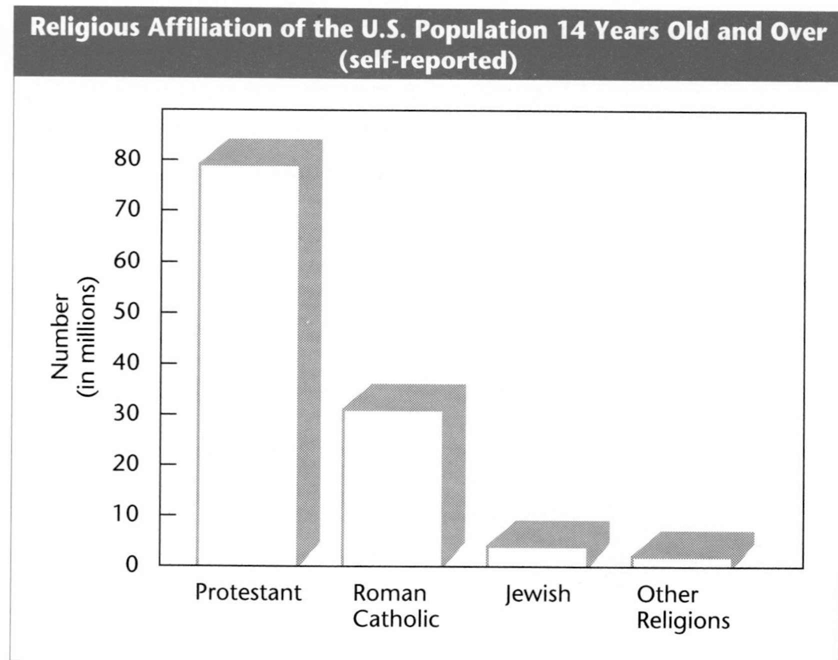

Figure 2-12

Self-Review 2-8

In Self-Review 2–6, we gave the farm population of the United States in the years 1940–1990 as:

Year	Farm Population (in millions)
1940	30.5
1950	23.0
1960	15.6
1970	9.7
1980	7.9
1990	7.3

a. Draw a bar chart placing the bars horizontally.
b. Draw a bar chart placing the bars vertically.

PIE CHARTS. A very popular graph used to portray parts of the total is called a **pie chart.** As an example, suppose students living on the campus of State University reported their annual expenditures on books, tuition, and so on. The figures were grouped, and typical amounts for each category were determined (see Table 2–13).

To represent tuition and other expenditures, each category must first be converted to percents of the total. Tuition, for example, accounts for 40% of the total, and we find it by computing \$3,200/\$8,000=0.40=40% (see Table 2–13).

To construct a pie chart, we first divide a circle (the pie) into equal "slices," as shown by the notches along the perimeter of the circle in Figure 2–13. Here the slices represent 5% each (the total area of the pie is 100%).

In constructing this pie chart, we arbitrarily decide to plot the 5% spent on books first. We draw a line from 0 to the center of the pie and another line from the center to 5 on the edge of the circle. The enclosed piece of the pie represents expenditures on books.

Next we plot expenditures for recreation (10%) by adding the 10% to the 5% for books, for a total of 15%. We draw a line from

Table 2-13 **Typical Amounts Spent on Tuition and Other Expenditures at State University**

Expenditure	Amount	Percent of Total
Tuition	\$3,200	40
Room and board	2,400	30
Transportation	1,200	15
Books	400	5
Recreation	800	10
Total	\$8,000	100

Figure 2-13

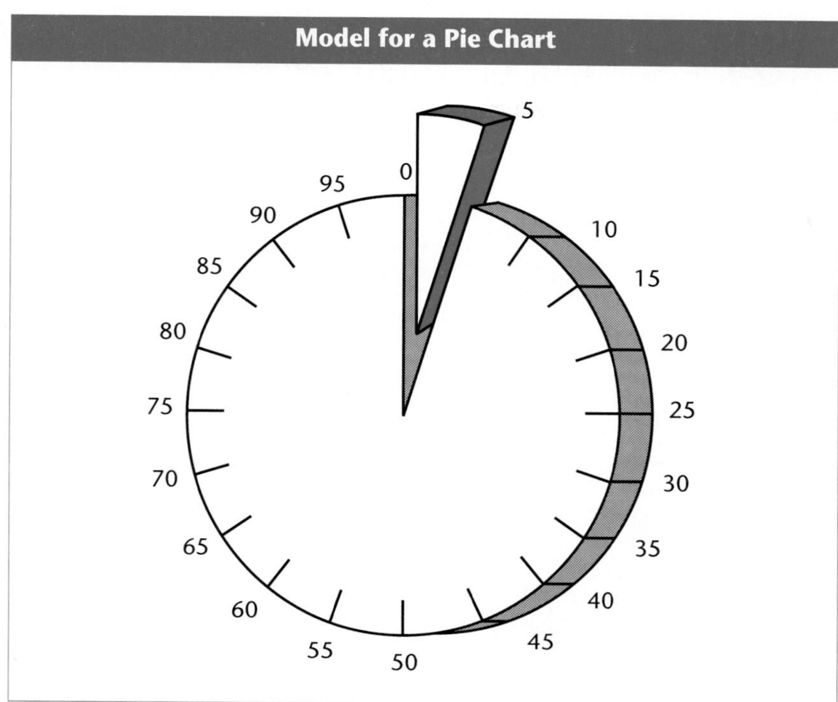

15% to the center of the pie. The area between 5% and 15% represents the typical amount spent on recreation. We continue this procedure for all remaining expenditures. The completed pie chart is shown in Figure 2–14 on page 54.

From Figure 2–14 it is readily apparent that the largest annual expenditure is for tuition, followed by the expenditure for room and board (30%), and so on.

As another example where does the federal government get its money and how does it spend it? The major source of revenue is taxes, and the major expenses are national defense and social security. The two pie charts in Figure 2–15 summarize the information.

Self-Review 2-9

The annual expenditures of Cartersville Township during last year were:

Expenditure	Amount
Roads	$1,000,000
Schools	600,000
Administration	300,000
Graft	100,000

Portray the expenditures in the form of a pie chart.

Figure 2-14

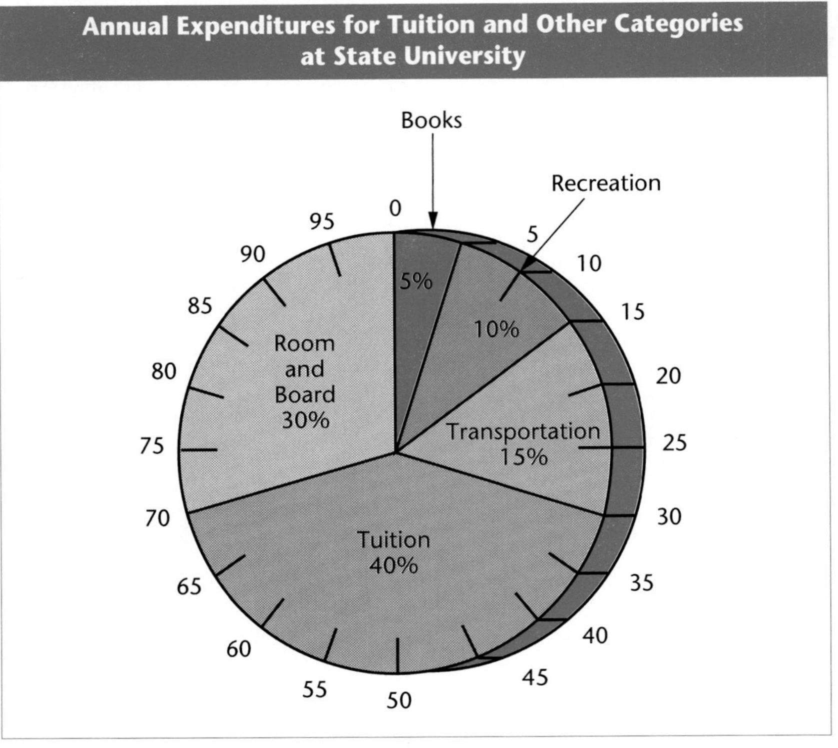

Figure 2-15

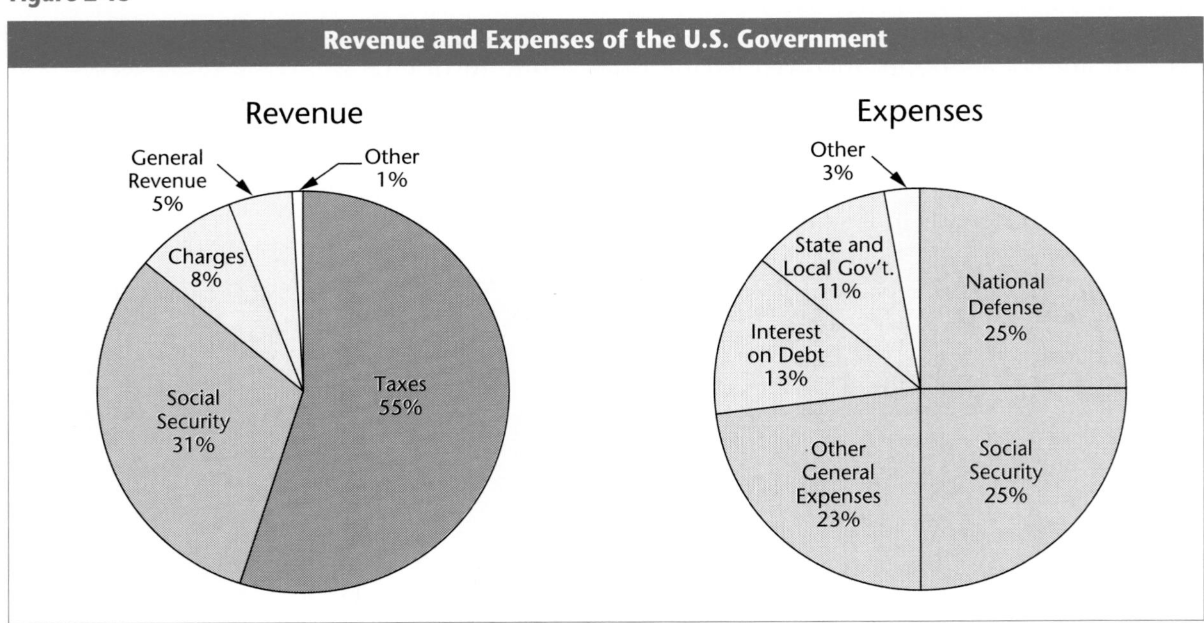

SOURCE: The World Almanac and Book of Facts, *1993 ed., 130.*

Exercises

15. A recent report showed the average value of Iowa farmland from 1986 to 1992. Plot these data in the form of a line chart.

Year	Average Value (per acre)
1986	$1,840
1987	1,999
1988	1,889
1989	1,684
1990	1,499
1991	1,064
1992	841

16. The long-term debt of Pier 1 Imports from 1984 to 1992 was as follows:

Year	Long-Term Debt (in millions)
1984	$26.9
1985	26.7
1986	26.7
1987	101.5
1988	96.5
1989	121.3
1990	92.6
1991	140.6
1992	106.8

SOURCE: *Pier 1 Imports, Inc.,* 1992 Annual Report, *21.*

Portray the trend in long-term debt in the form of a bar chart.

17. According to the Bureau of the Census, the population of the State of Arizona for the years 1910–1990 was:

Year	Number
1910	204,000
1930	436,000
1950	750,000
1970	1,792,000
1990	2,724,000

Plot the population data in the form of a bar chart.

18. In their *Motor Vehicle Facts and Figures,* the Motor Vehicle Manufacturers Association gives the following data regarding the cost of owning and operating an automobile based on 10,000 miles of travel per year:

Item	Cost per Year
Gas and oil	$520
Maintenance	190
Insurance	663
Depreciation	2,094
Finance charge	626
All other	231

Plot the data in the form of a pie chart.

19. The Council of Environmental Quality estimated the following amounts to be spent by various groups on water pollution during the most recent 9-year period:

Group	Amount (in billions of dollars)
Federal government	$ 2.5
State and local governments	39.8
Industry	57.0
Utilities	11.5

Portray the data by drawing a pie chart.

SUMMARY

A frequency distribution is a table that shows how often each value occurs. To construct a frequency distribution, we tally the raw data into predetermined classes. We then count the tallies to arrive at the frequency for each class. A histogram is often used to graphically portray a frequency distribution. For the hourly wage example, the number of employees is plotted on the vertical axis and the wages along the horizontal axis. Another useful graph is a frequency polygon. It is similar to the histogram, except the bars are replaced by lines connecting the class midpoints.

We construct a relative frequency distribution, or percent frequency distribution, by converting the class frequencies to percents. This construction is useful for comparing frequency distributions that have unequal numbers of observations. A "less-than" cumulative frequency distribution, or ogive, reports the number of observations less than the upper limit of the class. A "more-than" cumulative frequency distribution reports the number of observations greater than the lower limit of the class.

Often, several other graphic techniques are used to present statistics because their appearance has more reader appeal. These include stem-and-leaf displays, line charts, bar charts, and pie charts.

Exercises

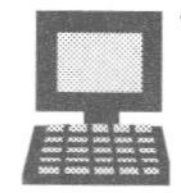

***20.** Citizens' Trust lists the following end-of-month checking-account balances for 40 of its customers:

203	37	141	43	55
303	252	758	321	123
425	27	72	87	215
358	521	863	284	279
608	302	703	68	149
327	127	125	489	234
498	968	350	57	75
503	712	440	185	404

a. Tally the data into a frequency distribution using $100 class intervals and $0 as the starting point.

b. Draw a less-than and a more-than cumulative frequency polygon.

c. The bank considers a "preferred" customer to be one with an account balance of $500 or more. Estimate the percent of all customers who are preferred customers.

d. The bank is planning to offer free checking accounts to the highest 30% of its customers. Based on the cumulative frequency polygon, what is the minimum balance for the free checking account?

21. The Quickie Change Oil Company has a number of outlets in the metropolitan area. No appointment is required. The number of oil changes at the Oak Street outlet in the past 20 days are:

65	98	55	62	79	59	51	90	72	56
70	62	66	80	94	79	63	73	71	85

The data are to be organized into a frequency distribution.

a. How many classes would you recommend?

b. What class interval would you suggest?

***Note:** This symbol indicates that a computer is recommended.

c. What lower limit would you recommend for the first class?
d. Organize the number of oil changes into a frequency distribution.
e. Comment on the shape of the frequency distribution.

22. The following is the percent of births with low weight, by state. (A low birth weight means that the child weighs 2,800 grams or less at birth.)

Area	Percent	Area	Percent
New England			
Maine	5.1	*East North Central*	
New Hampshire	5.2	Ohio	6.7
Vermont	5.2	Indiana	6.4
Massachusetts	5.8	Illinois	7.4
Rhode Island	6.4	Michigan	6.9
Connecticut	6.6	Wisconsin	5.4
Mid Atlantic		*South Atlantic*	
New York	7.3	West Virginia	7.0
New Jersey	6.5	North Carolina	7.9
Pennsylvania	6.9	South Carolina	8.6
		Georgia	8.1
West South Central		Florida	7.6
Arkansas	7.6		
Louisiana	8.6	*East South Central*	
Oklahoma	6.5	Kentucky	7.1
Texas	6.8	Tennessee	7.9
		Alabama	8.0
Pacific		Mississippi	8.7
Washington	5.2		
Oregon	5.1	*Mountains*	
California	6.0	Montana	5.9
Alaska	4.6	Idaho	5.2
Hawaii	6.9	Wyoming	6.8
		Colorado	7.7
South Atlantic		New Mexico	7.1
Delaware	7.4	Arizona	6.2
Maryland	7.7	Utah	5.4
Virginia	7.0	Nevada	7.4

a. Organize the percents in a frequency distribution, using 4.5% as the lower limit of the first class and an increment of 1%.
b. Portray the frequency distribution as a histogram.
c. Draw a more-than frequency polygon.
d. Based on the more-than frequency polygon, approximately how many states have low birth-weight rates greater than 7.0%?

23. The percent of the total number of births attributed to unmarried women for each of the 50 states (excluding the District of Columbia, which is 57.7%) is:

19.0	13.9	16.7	19.3	19.8
19.0	29.4	22.9	24.4	23.4
21.0	27.1	19.3	19.6	16.3
15.0	22.5	12.9	17.5	15.5
16.7	27.0	30.5	22.4	20.3
19.5	23.6	27.6	27.2	26.7
20.0	25.3	25.9	34.0	24.0
30.2	18.6	17.7	17.8	11.9
13.9	18.0	27.9	25.6	9.8
16.6	19.8	20.7	26.5	20.8

a. Organize the percents in a frequency distribution, using 9.0% as the lower limit of the first class. You decide on the class interval.
b. Portray the frequency distribution as a histogram.
c. Draw a less-than cumulative frequency distribution.
d. Based on the cumulative frequency distribution, approximately how many states have less than 20% of their births attributed to unmarried women?

24. The average annual percent changes in consumer prices for the past five years for selected countries are:

United States	5.5	Japan	2.7
Australia	8.3	Luxembourg	6.9
Austria	4.9	New Zealand	12.0
Belgium	7.0	Netherlands	4.2
Canada	7.4	Norway	9.0
Denmark	7.9	Portugal	23.2
Poland	8.6	Spain	12.2
France	9.8	Sweden	9.0
Germany	3.9	Switzerland	4.3
Greece	20.7	Turkey	37.8
Ireland	12.3	United Kingdom	7.2
Italy	13.7		

a. Using 0% as the lower limit of the first class and a class interval of 5%, organize the changes in consumer prices into a frequency distribution.
b. Describe the frequency distribution.
c. Organize these percents into a stem-and-leaf display. Rank the leaves in each row.
d. Draw a histogram of these data.
e. Construct a frequency polygon of these data.

25. Kolvey Travel Agency, a nationwide travel agency, offers special rates on certain Caribbean cruises to senior citizens. The president wants additional information on the ages of those people taking cruises. A random sample of 40 customers taking a cruise last year revealed these ages:

77 18 63 84 38 54 50 59 54 56 36 26 50 34 44
41 58 58 53 51 62 43 52 53 63 62 62 65 61 52
60 60 45 66 83 71 63 58 61 71

a. Organize the data into a frequency distribution, using seven classes and 15 as the lower limit of the first class.
b. Would you suggest a different interval? If so, what interval?
c. Describe the distribution. Where do the data tend to cluster?
d. Convert the distribution to a relative frequency distribution.

26. The average annual pay for selected states for 1986 is to be analyzed and later compared with the present year. For 1986, the average pay was:

Alabama	$17,638	Arizona	$18,870
Arkansas	16,162	Florida	17,679
Georgia	18,746	Hawaii	18,101
Idaho	16,602	Iowa	16,598
Kentucky	17,357	Colorado	20,275
Mississippi	15,420	Ohio	19,902
Minnesota	19,630	Rhode Island	17,733
Wyoming	18,969	West Virginia	18,405
New Mexico	17,301	Nevada	18,739
Oregon	18,322		

a. Organize this data into a frequency distribution.
b. Describe the frequency distribution.
c. Organize this data into a stem-and-leaf display. Rank the leaves in each row.
d. Draw a histogram of this data.
e. Construct a frequency polygon of this data.

27. Average per capita incomes for selected states in the East and Midwest are:

South Carolina	$12,225	Virginia	$16,408
Maine	13,794	Alabama	12,326
Rhode Island	15,580	Florida	15,646
Connecticut	20,600	Georgia	14,446
New York	18,111	Delaware	16,012
Pennsylvania	15,249	New Hampshire	16,912
Maryland	17,785	Massachusetts	18,727
New Jersey	19,726	Vermont	14,348
Michigan	15,775	District of Columbia	19,982
Ohio	14,934	Kansas	15,560
Wisconsin	14,908	Minnesota	15,993

a. Organize this data into a frequency distribution.
b. Describe the frequency distribution.
c. Organize this data into a stem-and-leaf display. Rank the leaves in each row.
d. Draw a histogram using this data.
e. Construct a frequency polygon using this data.

28. The following scores were obtained on the first statistics exam in Dr. Vernon Kiel's section:

68	52	49	56	69	74	41	59
79	81	42	57	60	88	87	47
65	55	68	65	50	78	61	90
85	65	66	72	63	95	72	74

a. Organize the data into a frequency distribution. Use an interval of 10 and use 40 as the lower limit of the first class.
b. Draw a histogram.
c. Draw a stem-and-leaf display.

29. The following data shows the amount spent on social programs in the United States from 1965 to 1990, and the projected amount for 1995. Draw a line chart.

Year	Spending (in millions of dollars)
1965	$ 25.0
1970	36.0
1975	73.0
1980	166.0
1985	307.0
1990	459.0
1995*	56.2

*Estimated

30. The expenditures on new plants and new equipment for Wong Electronics from 1984 to 1990 and projected to 1995 are shown below. Develop a line chart.

Year	Expenditure (in millions)
1984	$ 9.4
1985	10.3
1986	11.5
1987	14.6
1988	17.2
1989	20.3
1990	19.0
1995*	22.3

**Estimated*

31. The following information is a breakdown of our personal computer sales last year. Develop a pie chart.

Purchasers	Units (in thousands)
Business	3,254
Science	264
Education	679
Home	2,435

32. The following graph shows a less-than ogive of the ages of a sample group of automobiles. The plotted values are (3, 35), (6, 60), (9, 75), (12, 78), and (15, 80). Develop a table of the corresponding frequency distribution.

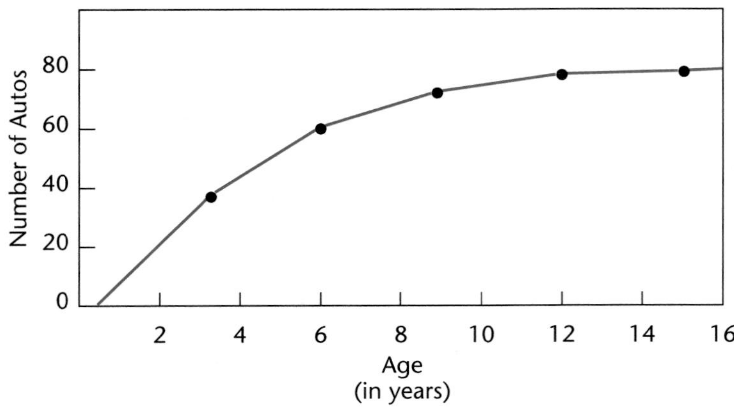

33. The sales manager for Ed Smith Auto Mart wishes to place radio ads for the super-new Belchfire 4 on one of two local stations. Belchfire 4 is a sports car aimed toward a younger buyer. Surveys of listener audiences indicate the following distribution of ages for the two stations:

	Station	
Ages	KSOP	KROK
15–24	10	58
25–34	22	52
35–44	53	35
45–54	45	30
55–64	60	25
65–74	10	0
	200	200

Which station would you recommend he select? Why?

34. The following table shows a breakdown of the time Americans spend watching TV by type of station. Develop a horizontal bar chart.

Type of Viewing	**Time per Day**
Network Affiliate	4 hr. 58 min.
Independent station	1 hr. 35 min.
Pay TV	35 min.
Cable	23 min.
Public stations	15 min.

35. The following are the number of games won by each of the 26 major league baseball teams during the 1992 season:

96	92	89	76	76	75	73	96	90	86	77	72	72
64	96	87	83	78	72	70	98	90	82	81	72	63

a. Organize the data into a frequency distribution. Use a class interval of 5 and let 55 be the lower limit of the first class.
b. How many games did the typical team win?
c. Develop a less-than cumulative frequency distribution.
d. Estimate the percent of teams that won 80 games or fewer.
e. Seventy-five percent of the teams won how many games or fewer?

36. A recent survey of prices for a gallon of unleaded regular gasoline at a self-service station in the metropolitan Little Rock area showed the following:

$1.043	$1.043	$1.029	$1.019	$1.109	$1.069	$1.079
1.034	1.054	0.999	0.989	1.019	1.079	1.039
1.019	1.015	1.029	1.009	0.999	0.979	1.079
1.095	1.019	1.099	1.024	1.035	1.043	1.019
0.998	1.012	1.076	1.056	1.038	1.044	1.059
1.109	1.018	1.036	1.049	1.034	1.078	1.041

a. Organize the data into a frequency distribution.
b. What is a typical price for the gasoline? Where does the data tend to be concentrated?
c. Develop a less-than cumulative frequency polygon.
d. Sixty percent of the stations sell their gasoline below what amount?

37. Tom Sevits plays nine holes of golf several times a week at the Tanglewood Golf Club. Last summer he maintained records of his scores on the front nine:

45	41	44	39	38	47	41	51	49
45	45	43	49	40	47	38	42	45
49	45	42	48	49	45	39	47	45
52	47	50	41	46	47	43	45	49
42	48	48	48	44	49	47	44	41

a. Organize the data into a frequency distribution.
b. What is a typical golf score? Where does the data tend to be concentrated?
c. Develop a less-than cumulative frequency polygon.
d. Tom scored what amount or less in 25% of the rounds?
e. Suppose Tom shot a 40 on his first round this season. Estimate the percent of time he shot less than 40 last summer.

38. A trade publication reported that breakfast cereals had reached a total revenue of $2.3 billion for the last 12 months. The top five brands (in millions of dollars) are: Frosted Flakes, $239.6; Cheerios, $235.3; Honey Nut Cheerios, $174.1; Kellogg's Corn Flakes, $171.8; and Rice Krispies, $170.5. Draw a histogram that shows the information regarding the top five brands.

39. In the 1992–1993 academic year, there were 565 faculty at the University of Toledo: 227 professors, 163 associate professors, 144 assistant professors, and 31 instructors. Develop a pie chart to show the relative sizes of the four academic ranks.

40. The Environmental Protection Agency recently reported the composition of household trash as follows:

Material	**Percent of Trash**
Paper and paperboard	40
Yard wastes (leaves, grass)	18
Metals	9
Glass	8
Rubber, leather, and wood	8
Food wastes	8
Plastic	7
Other	2
Total	100

Draw a pie chart to portray the results.

41. The following are the number of births per year per 1,000 population for 60 countries:

34	24	10	15	22	15
17	22	10	17	15	31
25	32	15	20	31	18
37	12	15	18	28	27
19	13	20	19	40	35
19	16	22	13	43	35
27	18	16	13	31	20
19	14	10	13	44	32
45	12	17	18	34	38
24	16	14	30	24	32

a. What level of measurement are the data? Explain.
b. Organize the data into a frequency distribution. Start with a lower limit of 10 and use an interval of 5.
c. Draw a histogram.
d. Draw a frequency polygon.
e. Draw a stem-and-leaf display.
f. Draw a more-than cumulative frequency polygon.
g. The United States has a birth rate of 15 per 1,000 population. What percent of the countries have a birth rate that is equal to or greater than the birth rate found in the United States?

42. A study of the grade point averages of a sample of 120 undergraduate students with GPAs of 1.80 or higher revealed the following information:

2.30	2.81	2.71	2.95	2.25	3.15	2.19	2.22	3.07	3.70
2.18	2.00	2.17	2.16	2.54	2.41	2.87	2.45	3.25	3.79
2.28	2.20	2.39	1.99	2.71	2.22	2.11	2.63	3.49	2.21
3.97	2.41	2.59	2.59	3.18	2.18	2.30	1.81	2.01	2.49
2.45	2.60	2.73	2.72	2.48	2.89	2.49	3.27	2.82	2.00
2.78	3.00	3.89	3.81	2.26	2.12	2.65	3.51	2.21	2.19
2.55	3.40	2.51	2.47	1.89	2.24	3.08	2.01	2.42	2.17
2.32	3.31	2.31	2.47	2.90	2.50	3.33	1.80	3.77	3.43
2.31	3.62	2.24	2.29	2.12	1.88	3.51	2.07	2.61	2.57
1.99	2.51	2.58	2.22	2.36	3.12	2.06	2.22	3.01	1.91
2.18	2.59	2.37	2.97	2.52	3.37	2.85	2.44	3.23	2.76
2.38	4.00	1.97	2.14	2.65	2.22	2.09	2.61	3.42	2.40

a. What level of measurement are the GPAs? Explain.
b. Organize the GPAs into a frequency distribution. Use an interval of 0.20 and a lower limit of 1.70 for the first class.
c. Draw a histogram.
d. Draw a frequency polygon.
e. Organize the GPA data into a less-than cumulative frequency distribution.
f. Draw a less-than cumulative frequency polygon.
g. Organize the GPA data into a more-than cumulative frequency distribution.
h. Draw a more-than cumulative frequency polygon.
i. Write a brief paragraph summarizing your findings.

43. Dean Lindsley is studying the number of humanities students per class at the University of the West. For the fall semester, he obtained the following information:

29	60	60	52	60	38	14	14	56	18	25	44	60
59	60	51	21	40	41	40	40	37	30	16	44	48
42	19	29	15	11	22	32	26	28	15	32	31	54
32	60	41	39	60	60	48	19	40	55	58	60	57
57	60	42	44	50	30	44	55	28	10	14	28	32
13	30	24	34	24	18	12	42	40	19	24	13	22
22	60	33	51	24	26	18	41	53	74	75	43	57
11	38	9	45	4	35	15	18	19	27	29	15	13
33	50	41	50	42	50	50	50	50	27	44	28	39
18	45	22	45	29	45	45	44	27	14	32	10	31
30	30	12	15	12	12	36	35	18	24	22	13	36
33	37	40	50	43	59	15	13	39	39	40	34	40
40	40	22	39	40	38	17	28	28	26	33	14	3
12	19	17	32	35	13	60	60	40	33	59	29	13
13	10	40	26	46	44	56	28	16	55	41	41	36
25	19	14	23	15	10	38	22	16	10	7	9	

a. Organize the data into a frequency distribution. An interval of 10 is suggested.
b. Develop a less-than cumulative frequency distribution.
c. Based on a cumulative frequency, or ogive, 75% of the classes have fewer than how many students?
d. What is a typical class size?

Data Exercises

44. Refer to the Real Estate Data Set in the appendix, which reports information on homes sold in Northwest Ohio during 1992.
    a. Select an appropriate class interval and organize the selling prices into a frequency distribution.
    b. What is the selling price of a typical home?
    c. Estimate the percent of homes that sell for less than $200,000.
    d. About 40% of the homes sell for less than what amount?
    e. Describe the shape of the distribution of selling prices.
    f. Write a brief summary of your findings.
45. Refer to the Salary Data Set in the appendix, which refers to a sample of middle managers in Sarasota, Florida.
    a. Select an appropriate class interval and organize the salaries into a frequency distribution.
    b. What is a typical salary?
    c. Estimate the percent of salaries less than 70.0 ($70,000).
    d. About 20% of the managers make less than what amount?
    e. Describe the shape of the distribution of salaries.
    f. Write a brief summary of your findings.

HOW HIGH IS HIGH ENOUGH?

CASE 

You are a member of the Student Housing Authority, which is designing new housing for fraternities and sororities. At today's meeting, the plumbing contractor raised the issue of the height of shower heads—that is, how far above the floor should they be located? Height is important because of the cost of the pipe that must run from the base to the shower head in each room. According to the current design, there are 300 rooms in the male complex and 500 rooms in the female complex. A 1-inch reduction in the height of each shower head would save more than 65 feet of pipe.

A rule of thumb in the building industry is to locate shower heads so they are higher than 90% of the intended users. To respond to the plumbing contractor's question, the

Student Housing Authority randomly selected 55 students and obtained the following data:

Student	Sex	Height
Abu, Rashidee	m	58
Arnold, Donna	f	59
Ballard, Andrew	m	60
Bilka, Daniel	m	61
Burke, Lori	f	62
Burkin, Valerie	f	59
Cherrill, Kate	f	63
Chieng, Siew	f	59
Chin, Christopher	f	62
Ciorach, Paulette	f	65
Cramer, Julie	f	66
Davis, Matthew	m	64
Donaldson, Scott	m	64
Eyerly, Robin	f	63
Figmaka, John	m	60
Filbrun, Melissa	f	61
Fong, Tina	f	59
Good, Todd	m	62
Grzeszczak, Denise	f	62
Hashim, Zahbedah	m	61
Herriott, Bernadine	f	63
Holben, Douglas	m	66
Horton, James	m	58
Hottmann, Tonya	f	60
Iskandarani, Miriam	f	58
Kemp, Lisa	f	68
Lim, Kuan	f	61
Mansfield, Shawn	m	66
McQueen, Andrew	m	60
McQuillian, Timothy	m	62
Midgley, Charlotte	f	62
Miller, Lisa	f	67
Mokhtar, Mohbfarid	m	69
Monato, Antoinette	f	65
Montanaro, Daniela	f	72
Moore, Sheila	f	63
Nagielski, Sara	f	69

Najjar, Mousa Mohammad	m	69
Noonan, Brenda	f	69
Omotoye, Gboyega	m	73
Onodu, Nnadi Matthew	m	71
Pfahler, Amy	f	75
Ray, Brenda	f	71
Ray, Lesa	f	71
Repp, Julie	f	64
Saba, David	m	72
Selders, Michael	m	69
Sheely, Dale	m	68
Simmons, Floyd	m	68
St. Clair, Shawn	m	69
Tan, Chong Leng	f	69
Tyrrell, Andrew	m	64
Wasnich, Chuck	m	71
Watt, Lisa	f	68
Whiteley, Gary	m	68

At what height would you recommend the shower heads be placed? Explain your reasons.

CHAPTER ACHIEVEMENT TEST

Answer all the questions. The answers are at the back of the book.

MULTIPLE-CHOICE QUESTIONS. Select the response that best answers each of the questions.

1. When arranging data into classes it is suggested you have
 a. fewer than 5 classes.
 b. between 5 and 20 classes.
 c. more than 20 classes.
 d. between 10 and 40 classes.

2. When constructing a line chart, you plot "time" along
 a. the vertical axis.
 b. the horizontal axis.
 c. either axis.
 d. neither axis.

3. A frequency distribution requires that the data be of at least what scale?
 a. Nominal
 b. Interval

c. Ordinal
d. None of these is correct.

4. The class midpoint is the
a. number of observations in a class.
b. center of the class.
c. upper limit of the class.
d. width of the class.

The following frequency distribution records the number of empty seats on flights from Cleveland to Tampa:

Number of Empty Seats	Frequency
0–4	3
5–9	8
10–14	15
15–19	18
20–24	12
25–29	6

5. The midpoint of the 0–4 class is
a. 2.
b. 4.
c. 2.5.
d. 0.

6. The true lower limit of the 0–4 class is
a. 0.
b. –0.5.
c. 2.0.
d. 2.5.

7. The size of the class interval is
a. 5.
b. 4.
c. 4.5.
d. 3.

8. About what percent of the flights had 19 or fewer empty seats?
a. 15%
b. About 29%
c. About 71%
d. Cannot be determined from grouped data

The following less-than cumulative frequency polygon is developed for the distance (in miles) commuting students travel from home to their college campus:

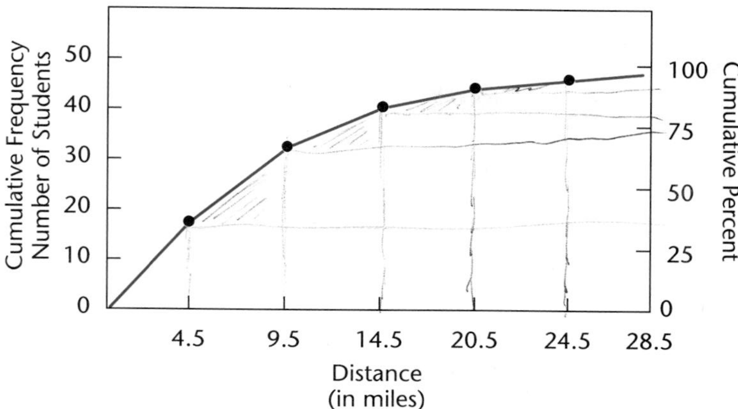

9. Approximately what percent of the students travel less than 7 miles to campus?
- **a.** About 50%
- **b.** All
- **c.** About 36%
- **d.** Cannot be determined

10. Seventy-five percent of the students travel about how many miles or less to campus?
- **a.** 5
- **b.** 7
- **c.** 12
- **d.** 20

COMPUTATION PROBLEMS

11. A hospital administrator has commissioned a study of the length of time, in minutes, a patient spends waiting in the emergency room before receiving treatment. For a typical day, the information is as follows:

12	6	5	23	10
25	19	11	25	18
10	14	12	10	16
5	19	17	17	14
2	21	9	6	21

Construct a frequency distribution using an interval of 5 and a lower-class limit for the first class of 1 minute.

12. The Bureau of Motor Vehicles of the State of Nevada made a study of the tread depth of tires. Fifty-two passenger cars that had stopped at a highway rest station were tested. The tread depth, measured to the nearest 1/32", is reported in the table below. (a) Draw a less-than cumulative frequency polygon. (b) What percent of the tires have less than 7/32" tread? (c) Forty percent of the tires have less than how much tread?

Tread Depth (1/32″)	Frequency
0–3	4
4–7	15
8–11	25
12–15	5
16–19	3

13. A family's monthly expenses are as follows: housing, $400; utilities, $140; medical, $25; food, $190; transportation, $150; clothing, $50; savings, $50; miscellaneous, $50. Draw a pie chart showing this information in terms of percents.

ANSWERS TO SELF-REVIEW PROBLEMS

2–1 a.

Age		Number
15–24	///	3
25–34	////	4
35–44	//	2
45–54	/	1
	Total	10

b. Raw data

c. Based on the frequency distribution, the ages range from 15 to 54. The largest concentration is in the 25–34 class.

2–2

Leading Digit	Trailing Digit
1	67
2	159
3	009
4	1
5	0

2–3

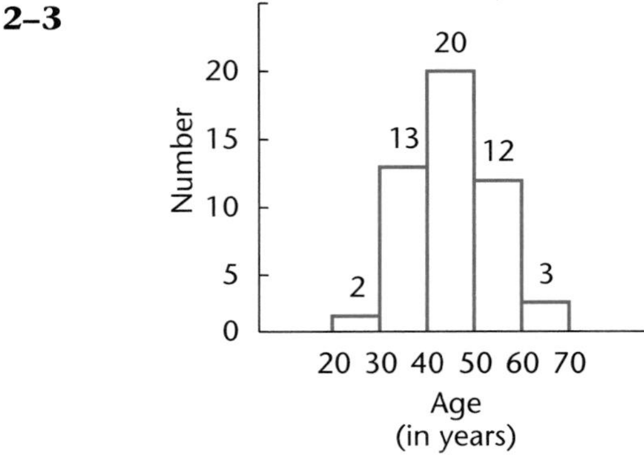

2–4

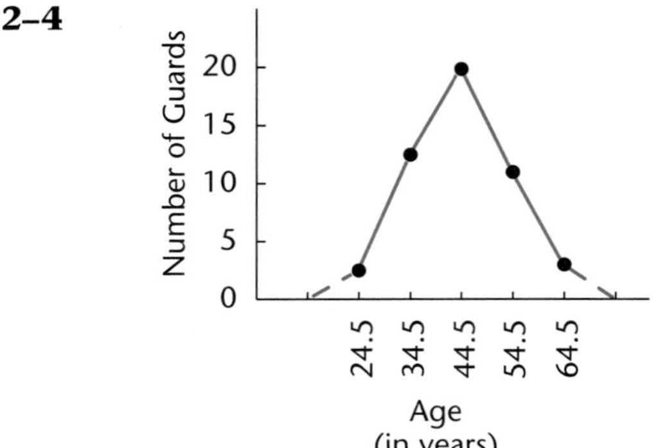

2–5 a.

Age	Cumulative Number
20–29	2
30–39	15
40–49	35
50–59	47
60–69	50

b.

Upper True Class Limits	Number Less-Than
19.5	0
29.5	2
39.5	15
49.5	35
59.5	47
69.5	50

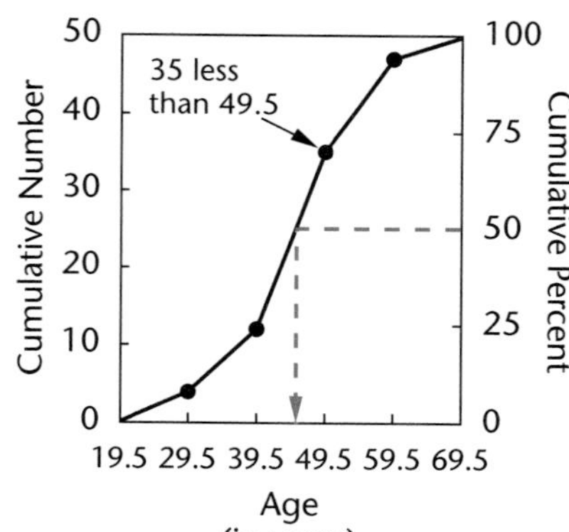

c. About 47 years

2–6

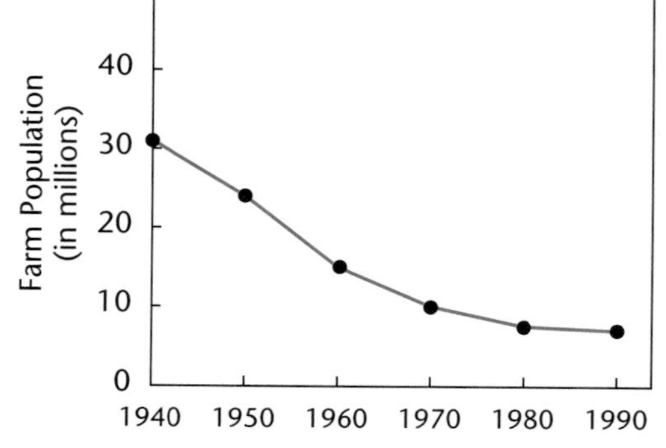

2–7 The percent below the poverty level has declined for all groups since 1959. Blacks still have the largest percent below the poverty level. The percent of whites below the poverty level is the smallest.

2–8 a.

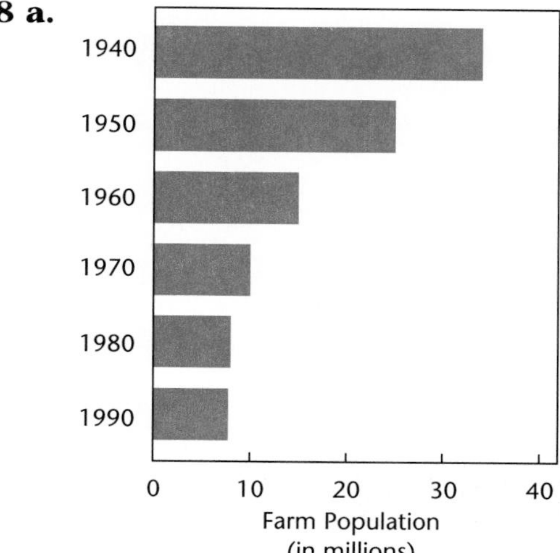

b.

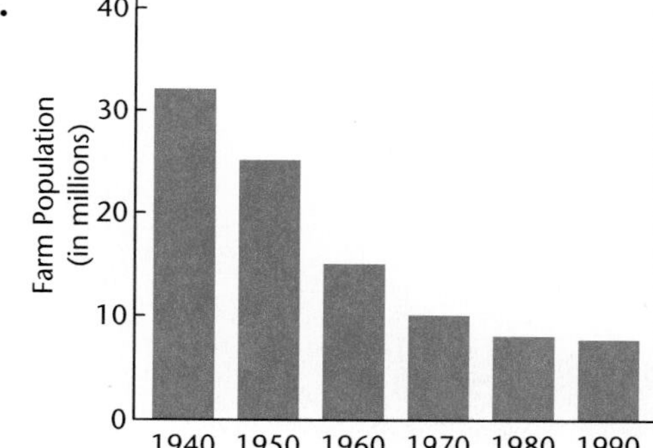

2–9

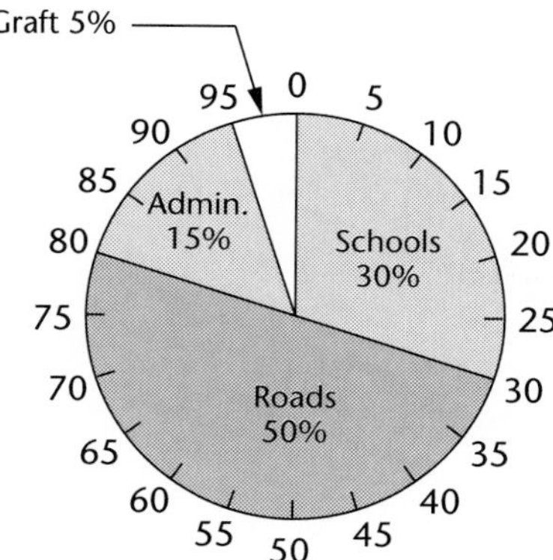

CHAPTER 3

Descriptive Statistics: Measures of Central Tendency

OBJECTIVES

When you have completed this chapter, you will be able to

- compute the mean, median, and mode;
- describe the characteristics of the mean, median, and mode;
- describe the position of the mean, median, and mode for symmetric and skewed distributions.

CHAPTER PROBLEM Pricing the Dream

RECALL FROM THE CHAPTER PROBLEM IN CHAPTER 2 THAT CHARLES REYNOLDS, OWNER OF REYNOLDS DEVELOPMENT, IS PUTTING TOGETHER A PROSPECTUS FOR NEW HOMES TO BE BUILT IN THE RAFTER J RANCH SUBDIVISION IN JACKSON HOLE, WYOMING. IN CHAPTER 2, HE FOUND THE SELLING PRICE OF EACH OF THE 80 EXISTING HOMES AND ORGANIZED THEM INTO A FREQUENCY DISTRIBUTION. NOW MR. REYNOLDS WANTS TO DETERMINE THE SELLING PRICE OF A TYPICAL HOME.

INTRODUCTION

In Chapter 2, the selling prices of homes in the Rafter J Ranch subdivision were organized into a frequency distribution. Next, a histogram or a frequency polygon was drawn to help describe the data. In this chapter, we will describe the selling prices further by computing several **measures of central tendency.** As the name implies, these measures locate the center of a set of observations. The value that either is typical of the set of data or best describes it is often referred to as an **average.**

Examples of Averages

- The average life expectancy in the United States is 74.2 years.
- In 1970, the median age of the population was 27.9 years, and in 1985 it was 31.5 years. By 1995 the median age is projected to be 35.7 years.
- The average cost per mile to operate a motor vehicle is 38.2¢, according to *Automobile Facts and Figures.*
- The Bureau of the Census reports that the median household income is $29,100.
- The Institute of International Education reports that there were an average of 6,200,000 students enrolled per year during the last five years in institutions of higher education.
- The average occupancy rate of nursing care facilities is 90.5%.
- A survey by the Hertz Corporation reports that the average annual maintenance and repair cost is $269 for a new car and $565 for a car older than one year.
- The team batting average for the Minnesota Twins during the 1992 season was .277; for the Detroit Tigers, it was .256.

As you will soon see, there are several kinds of averages, so you should develop the habit of speaking precisely regarding the various types. What we usually refer to as "averages" should be called "measures of central tendency." Each measure of central tendency has distinct characteristics, advantages, and disadvantages. For the same set of data, all measures of central tendency might have different values. In this chapter, we will examine three commonly used measures of central tendency—the **arithmetic mean,** the **median,** and the **mode.**

THE SAMPLE MEAN ($\bar{X}$)

The most frequently used measure of central tendency is the **arithmetic mean,** or simply the **mean.**

> **Arithmetic Mean** The sum of the values divided by the number of values.

In everyday language, the mean is often called the "average," however, to avoid confusion you should call it the "arithmetic mean."

Because the arithmetic mean of many different sets of data will be calculated, it is convenient to use a simple formula in which the components of the arithmetic mean are represented by a few general symbols. First, let $X_1, X_2, \ldots, X_n$ represent the values of n items or observations in a sample. The arithmetic mean of these sample items, written $\overline{X}$, is then defined as:

$$\overline{X} = \frac{X_1 + X_2 + \cdots + X_n}{n}$$

[a shorthand way of saying the sample mean of all scores ($\overline{X}$) is equal to the sum of all scores (X_1, X_2, and so on) divided by the number of scores]

For simplicity, instead of writing $X_1, X_2, \ldots, X_n$ for the observations, we use the symbol Σ (which is the Greek capital letter *sigma,* equivalent to our S) to represent a summation, or addition. Hence, the formula for the sample mean of n items is

$$\text{Mean} = \frac{\text{Sum of the values}}{\text{Number of values}}$$

or, in symbolic notation,

$$\overline{X} = \frac{\Sigma X}{n} \qquad \textbf{3-1}$$

where

$\overline{X}$ (read X-bar, or bar X) is the designation for the arithmetic mean of a sample.

Σ is the Greek capital letter sigma, the symbol for addition. In this case, it directs us to sum all the X values.

X refers to the individual values in the sample.

n is the total number of items in the sample.

Problem The estimated ages of five rocks uncovered at an excavation site were 5, 4, 9, 2, and 10 million years. What is the arithmetic-mean age of this sample of five rocks?

Solution In the above notation, n refers to the number of observations in the sample, which is 5. The values of X are $X_1 = 5$, $X_2 = 4$, $X_3 = 9$, $X_4 = 2$, and $X_5 = 10$. Hence, using formula 3 – 1, the arithmetic mean of the sample of the five rocks is computed as follows:

$$\overline{X} = \frac{\Sigma X}{n} = \frac{X_1 + X_2 + X_3 + X_4 + X_5}{n}$$

$$= \frac{5+4+9+2+10}{5} = 6 \text{ (million years)}$$

The number 6, which was computed from a sample, is called a **statistic.**

Statistic A measurable characteristic of a sample.

Problems This is the first in a series of drill problems, designed to provide you immediate feedback on your understanding of the topic just discussed. Because of the type of material covered, they appear only in the first half of the text. We suggest you work them before moving on to the next section. The solutions follow.

Determine the mean for each of the following sets of data:

a. 3, 5, 7
b. 6, 5, 8, 4
c. 3, 21, 16, 9, 13

Solutions

a. 5
b. 5.75
c. 12.4

Self-Review 3-1

Answers to all Self-Review problems are at the end of the chapter.

A sample of six school buses in the Carlton District travel the following distances each day: 14.2, 16.1, 7.9, 10.6, 11.2, and 12.0 kilometers.

a. Give the formula for the arithmetic mean of this sample.
b. Insert the appropriate figures and compute the arithmetic-mean distance traveled by the buses in the sample.

Exercises

Answers to the even-numbered exercises are at the back of the book.

1. The agility test scores for a sample of eight retarded children are 95, 62, 42, 96, 90, 70, 99, and 70. What is the arithmetic-mean score for the group?

2. The number of gallons of gasoline sold to a sample of eight customers yesterday at Frank's I-75 Marathon was 12.1, 9.7, 5.3, 6.7, 8.4, 9.2, 10.1, and 7.7 gallons. Compute the arithmetic mean.
3. According to *Motor Vehicle Facts and Figures,* the numbers of motorcycles registered in the past five years (in thousands) are: 5,444; 5,262; 4,886; 4,584; and 4,376. What is the arithmetic-mean annual registration of motorcycles?
4. The suicide rates for white and black males (per 100,000 population) for the years 1989–1991 were:

Year	White Males	Black Males
1989	18.0	8.0
1990	19.9	10.3
1991	21.7	11.5

Compare the arithmetic-mean suicide rates for white and black males for this three-year period. Comment.

THE POPULATION MEAN (μ)

Usually, a statistic is computed for the purpose of estimating an unknown population **parameter.** Hence, the sample mean $\overline{X}$ is a statistic that may be thought of as an estimate of the population mean from which the sample was drawn. The population mean is a parameter.

Parameter A measurable characteristic of a population.

The computation of the population mean is the same as that of a sample, except that the symbols are slightly different. By convention, the population mean is denoted by the Greek letter μ *(mu)*. The number of observations in a population is denoted by the capital letter N. Hence, the mean of a population is computed from the following formula:

$$\mu = \frac{X_1 + X_2 + \cdots + X_n}{N} = \frac{\Sigma X}{N} \qquad \textbf{3-2}$$

STAT BYTE

Seventy-eight million post–World War II babies are now in middle age. American retailers are selling "user friendly" products, like loose-fitting jeans, walking shoes, lineless bifocals, and lo-cal desserts to this youth-oriented generation.

Problem For the four years of Governor Lamb's administration, the numbers of pieces of legislation enacted by the state legislature were 310, 780, 960, and 842. What is the arithmetic-mean number of pieces of legislation per year?

Solution Note that this is a population because it covers all four years of the Lamb administration. The population consists of four observations and the population mean is:

$$\mu = \frac{\Sigma X}{N} = \frac{310 + 780 + 960 + 842}{4}$$
$$= \frac{2892}{4} = 723.0$$

PROPERTIES OF THE ARITHMETIC MEAN

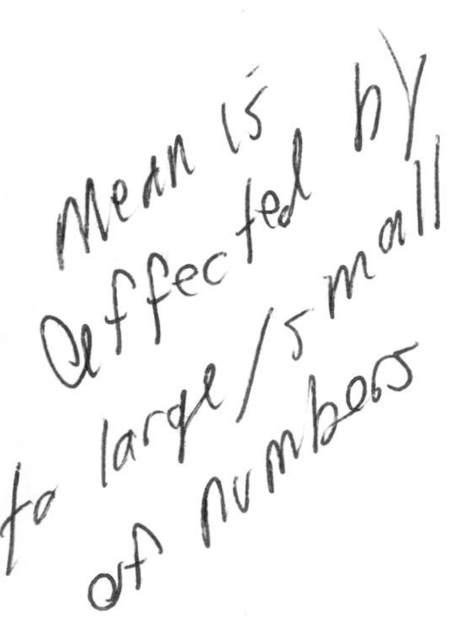

The popularity of the arithmetic mean as a measure of central tendency is not accidental. Not only is it simple, familiar, and easy to calculate, but it also has the following desirable properties:

1. It can be calculated for any set of interval-level or ratio data. Therefore, it always exists.
2. A set of data has only one arithmetic mean. Therefore, it is referred to as a "unique" value.
3. It is quite reliable. This property will be discussed in connection with sampling in Chapter 8.
4. All the data items are used in its calculation.

The arithmetic mean has one additional important characteristic, namely, *the sum of the deviations from the mean will always be zero.* This concept can be illustrated by a seesaw like those found on most playgrounds. Suppose that three children weighing 30, 40, and 90 pounds approach the seesaw with the balance point set at 60 pounds. One child stands on the seesaw at 30 (representing 30 pounds), another at 40, and the third at 90. They expect the board to balance so they can go up and down on the seesaw, but it does not. Why? Because 60 pounds is not the arithmetic mean of 30, 40, and 90 pounds (see Figure 3–1).

Suppose the child on the right side of the seesaw keeps moving toward the end of the board to try to make the seesaw balance. Finally, when he reaches 110 it balances and the children are able to seesaw up and down. Why does it balance? Because the arithmetic mean of 110, 40, and 30 is now at the balance point of 60. Thus, the mean of 60 is the center of gravity of the three weights. The sum of the deviations to the left of 60 is equal to the sum of the deviations to the right of 60 (see Figure 3–2).

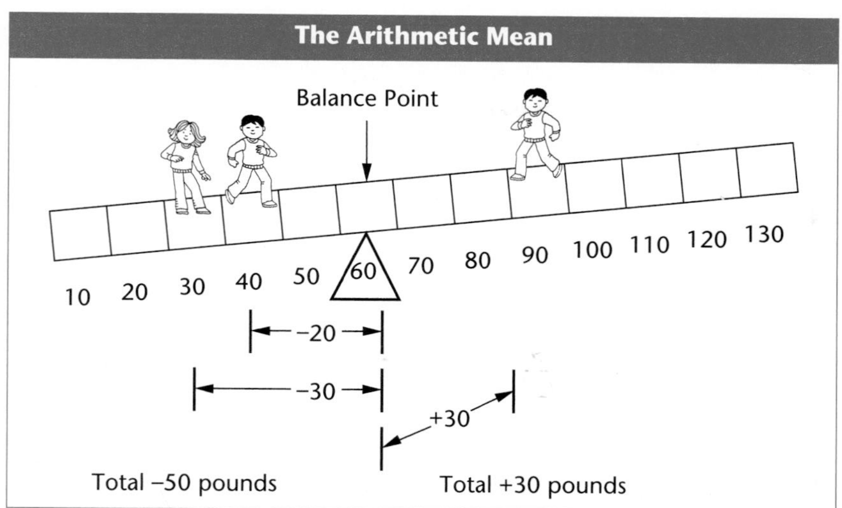

Figure 3-1

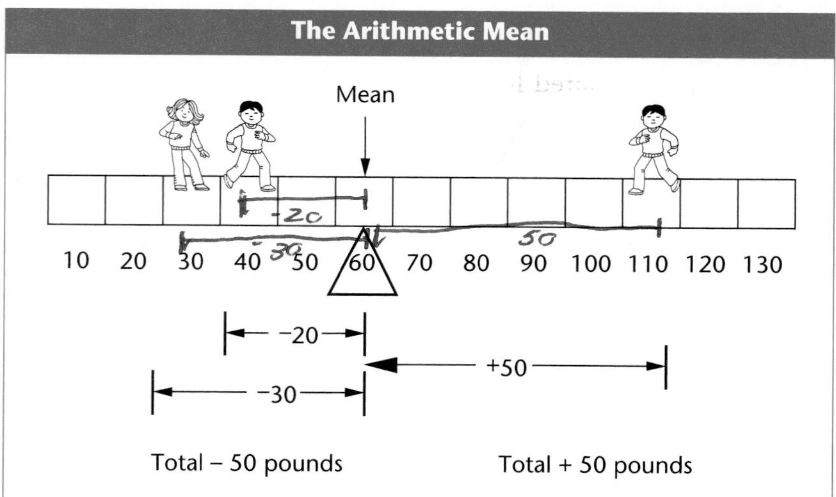

Figure 3-2

Note that in Figure 3–2 the distance (deviation) from the balance point to the child standing on the right end of the board (+50) is equal in magnitude to the sum of the deviations to the left of the balance point (–20 and –30). The balance point of 60 is the arithmetic mean of the distances (30 + 40 + 110)/3. We show this algebraically by letting each observation be represented by X and the mean by $\bar{X}$. In the seesaw problem, $\bar{X} = 60$. The sum of the deviations from the mean is zero. That is, $\Sigma(X - \bar{X}) = -20 + (-30) + 50 = 0$. The mean is the only average for which this is always true.

Self-Review 3-2

A group of football players enrolled in Statistics 166 want to check whether the sum of the deviations from the arithmetic mean is actually zero. Their weights are 200, 210, 230, and 320 pounds. They stand on the seesaw as shown below.

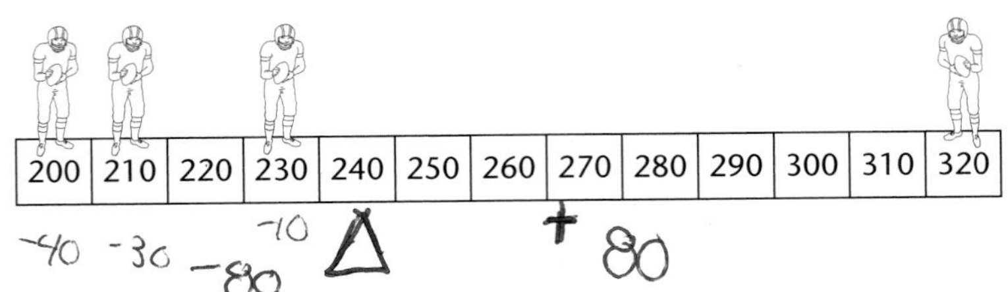

a. Where must the balance point be placed to make the seesaw balance?
b. What is this balance point called?
c. Show that the sum of the deviations from the mean equals zero.

THE WEIGHTED MEAN

Suppose it had been reported that the ages of the senior citizens on the Happy Boys slow-pitch softball team were 60, 70, 80, and 90. It might be concluded that the arithmetic-mean age is 75, found by (60 + 70 + 80 + 90)/4. This is true only if the *same* number of senior citizens are aged 60 as are aged 70, 80, or 90. However, suppose that one is 60, one 70, one 80, and nine are 90 years. To find the mean, we **weight** 60 by 1, 70 by 1, 80 by 1, and 90 by 9. The resulting average is aptly called the **weighted arithmetic mean.**

In general, the weighted mean of a group of numbers designated $X_1, X_2, X_3, \ldots, X_n$ with corresponding weights $w_1, w_2, w_3, \ldots, w_n$ is computed by

$$\overline{X}_w = \frac{w_1X_1 + w_2X_2 + w_3X_3 + \cdots + w_nX_n}{w_1 + w_2 + w_3 + \cdots + w_n}$$

or shortened to

$$\overline{X}_w = \frac{\Sigma w \cdot X}{\Sigma w} \qquad \textbf{3-3}$$

Problem Fifteen secretaries have 20 years of service each, three have 30 years of service each, and two have 50 years of service each. What is the weighted-mean length of service, using formula 3–3?

Solution

$$\overline{X}_w = \frac{15(20)+3(30)+2(50)}{15+3+2} = \frac{490}{20} = 24.5$$

The weighted arithmetic-mean length of service is 24.5 years.

Exercises

5. The monthly incomes, in thousands of dollars, for a sample of eight minority executives are \$2, \$2, \$2, \$2, \$10, \$10, \$25, and \$100.
   a. Find the arithmetic-mean income.
   b. Show that the sum of the deviations from the mean is equal to zero.

6. A sample of six students traveled 1, 4, 9, 8, 5, and 6 miles to class.
   a. Find the arithmetic-mean distance.
   b. Show that the sum of the deviations from the mean is equal to zero.

7. The following percent of flights left within 15 minutes of their scheduled departure time from selected cities:

City	Percent of Departures
Philadelphia	76.0
New York (Kennedy)	78.4
Seattle	80.2
Tampa	83.2
Miami	83.9
Los Angeles	85.9
Memphis	87.4

   a. Find the arithmetic-mean percent.
   b. Show that the sum of the deviations from the mean is equal to zero.

8. An accountant sampled the unpaid balances of C & G's customers. The results are:

\$46, \$42, \$83, \$90, \$76, \$84, \$56, \$46

a. Find the arithmetic-mean unpaid balance.
b. Show that the sum of the deviations from the mean is equal to zero.

9. A car rental agency purchased two tires at $43 each, five at $76 each, and three at $59 each. What is the weighted-mean price of a tire? Why can't the three prices be summed and the sum divided by 3 to arrive at the mean price per tire?

10. We are studying eight states with respect to various types of crime reported last year. From the FBI's annual *Uniform Crime Reports,* we discover that, per 100,000 population, four states had 28.3 rapes, three had 31.7 rapes, and one had 17.0 rapes. What is the weighted-mean number of rapes for the eight states being studied?

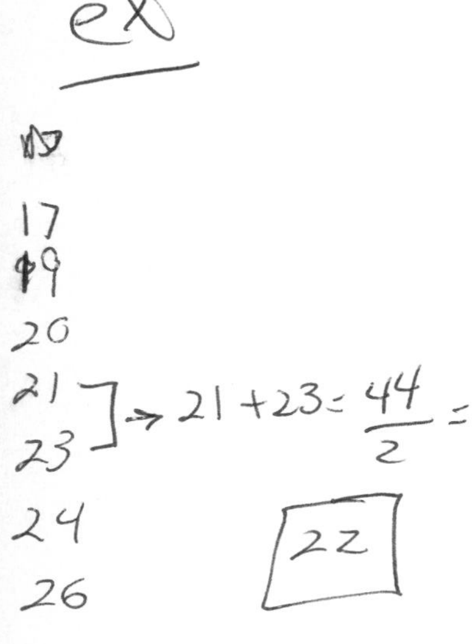

THE MEDIAN

If the data contain an observation that is either very large or very small compared with the other values, that value may make the arithmetic mean unrepresentative. To avoid this possibility, we sometimes describe the "center" of the data with other measures of central value. One measure used to designate the central value of a set of data is the **median.**

> **Median** The midpoint of the values after they have been arranged from the smallest to the largest (or the largest to the smallest). There will be as many values above the median as below the median.

Like the mean, the median is unique for any set of data. It is easy to calculate once the data have been arranged according to size. Unlike the mean, however, the median is not affected by extremely large or extremely small observations. Also, while the arithmetic mean requires the interval scale, the median merely requires that the data be at least of ordinal scale. For an *odd* number of values, the median is the middle value after the data have been ordered from low to high.

Problems

a. The test scores of a sample of five students are 92, 86, 2, 96, and 90. What is the median test score?

b. The incomes of a sample of seven local prison wardens are $41,500, $43,900, $47,100, $28,600, $32,500, $44,200, and $44,200. What is the median income?

Solutions

a. Arrange the test scores from low to high. Then select the middle value.

2
86
90 ← the median
92
96

The median score is 90.

b. Arrange the incomes of seven prison wardens from low to high. Then select the middle value.

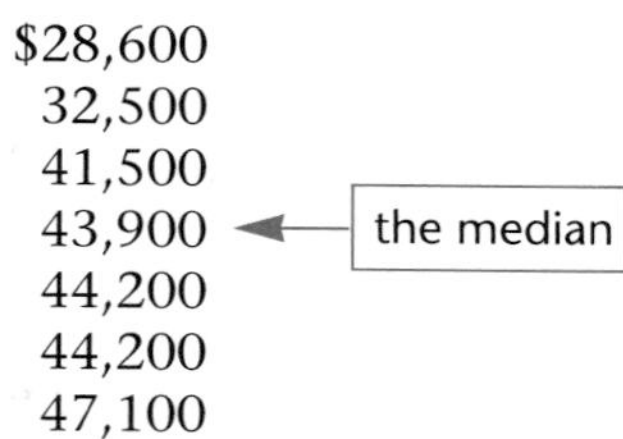

The median income is $43,900. Note that there are the same number of test scores (or incomes) below the median as above it.

Self-Review 3-3

The ages of prisoners in cell block D are 22, 75, 18, 40, and 34.

a. Determine the median age.
b. How many prisoners are above the median? Below it?

For an *even* number of values, first arrange the values from low to high. It is common practice to determine the median by finding the arithmetic mean of the two center values.

Problems

a. A park ranger recorded the lengths of a sample of several rainbow trout caught in the Yellowstone River. The lengths (in inches) were $24\frac{1}{2}$, 15, $10\frac{1}{2}$, $12\frac{1}{2}$, 21, and 26. What was the median length?

b. The weights of eight collegiate football players are 198, 240, 230, 240, 210, 250, 225, and 188 pounds. What is the median weight?

Solutions

a. Arrange the lengths from low to high:

$10\frac{1}{2}$
$12\frac{1}{2}$
15
21 ← 18
$24\frac{1}{2}$
26

The median is 18 inches, which is halfway between 15 and 21.

b. Arrange the weights from low to high:

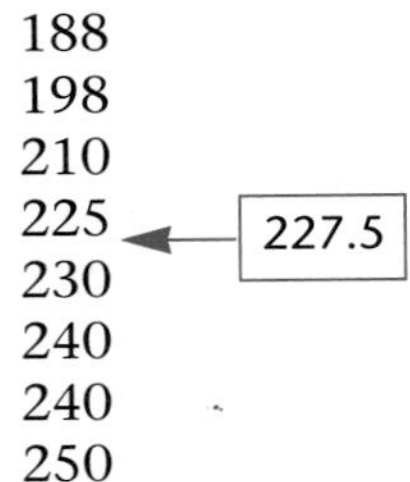

The median weight is 227.5 pounds.

Problems

Determine the median for each of the following sets of data:

a. 2, 9, 4
b. 6, 5, 8, 2
c. 16, 8, 12, 3, 1
d. 1, 2, 2, 1

Solutions

a. 4
b. 5.5
c. 8
d. 1.5

Self-Review 3-4

a. In their last eight games, the Rams football team scored the following number of points: 13, 34, 0, 0, 9, 42, 21, and 7. What is the median number of points scored?

b. The numbers of telephones in use in the United States for four recent years were 149,010,000; 138,290,000; 155,170,000; and 143,970,000. Determine the median number in use.

Exercises

11. The agility test scores for the retarded children in the exercise on page 78 are 95, 62, 42, 96, 90, 70, 99, and 70. What is the median score?

12. The monthly incomes (in thousands of dollars) for several minority executives in exercise 5 on page 83 are $2, $2, $25, $2, $10, $100, $2, and $10. What is the median income?

13. The sales for the last six customers at Churchill's Supermarket (rounded to the nearest dollar) were $12, $35, $86, $21, $135, and $72. Compute the median.

14. The statistics department at your college offers seven sections of basic statistics with 43, 41, 52, 13, 21, 39, and 46 students enrolled in each section. Compute the median section size.

15. *Modern Healthcare* reported these average patient revenues (in millions) for various types of hospitals:

Hospital Type	Patient Revenue (in millions)
Catholic	$46.6
Other church	59.1
Non-profit	71.7
Public	93.1
For-profit	32.4

What is the median patient revenue? Explain what it indicates.

16. The *Bank Rate Monitor* reported the following savings rates as of September 3, 1992:

Instrument	Savings Rate (percent)
Money market mutual fund	3.01
Bank money market account	2.96
6-month CD	3.25
1-year CD	3.51
$2\frac{1}{2}$-year CD	4.25
5-year CD	5.46

What is the median savings rate? Explain what it indicates.

THE MODE

The **mode** is yet another measure that may be used to describe the central tendency of data.

Mode The value of the observation that appears most often.

The mode has two major advantages; it requires no calculation and it can be determined for both nominal and ordinal data.

Problem The lengths of confinement (in days) for a sample of nine patients in Ward C are 17, 19, 19, 4, 19, 26, 3, 21, and 19. What is the modal length of confinement?

Solution The mode is 19 days because that value appears most frequently.

There is no mode for these hourly incomes: \$4, \$9, \$7, \$16, and \$10. There are two modes (76 and 81) for these test scores: 81, 39, 100, 81, 69, 76, 42, and 76. In the last example, the data are considered to be **bimodal.**

CHOOSING AN AVERAGE

The question often arises of what average to employ—the mean, the median, or the mode. We will now consider some of the factors influencing a decision.

The mean is probably the most frequently used measure of central tendency. It is the measure most people think of when an "average" is mentioned. The mean is considered the most reliable, or precise, average because the means of several samples taken from a population will not fluctuate as widely as either the median or the mode. (We will return to this discussion later in the book.)

The selection of an average depends in part on the level of data. As noted in Chapter 1, there are four such levels: nominal, ordinal, interval, and ratio. At the very least, an interval level of measurement is required to determine the arithmetic mean. We could, for example, compute the arithmetic-mean age of persons between 5 and 59 years old who have died of AIDS since 1982, as given in Table 3–1. (It appears that the mean age is about 32 years. The computations are presented shortly.) Incidentally, the median age and the modal age can also be determined for the AIDS data.

As another example, the arithmetic mean, median, and mode can be computed for the temperatures in Table 3–2.

The mean of ordinal-level data (see Table 3–3) or nominal-level data would be meaningless. We cannot possibly add warrant officer, lieutenant, captain, major, colonel, and general to determine the mean rank. Nor can we order the political parties in Table 3–4 from low to high to arrive at the mean or median.

Table 3-1 **AIDS Deaths Since 1982**

Age	Number
5–15	182
16–26	15,044
27–37	34,790
38–48	17,001
49–59	6,393

SOURCE: *U.S. Centers for Disease Control, unpublished data.*

Table 3-2

Interval-Level Data

Temperature (in °C)
2
1
3
12
3
3

$\bar{X}$ = 4 degrees
Median = 3 degrees
Mode = 3 degrees

Table 3-3

Ordinal-Level Data

Rank	Number (in thousands)
Warrant Officer—W–1	3.2
Chief Warrant—W–4	3.1
2d Lt.—0–1	34.0
1st Lt.—0–2	42.0
Captain—0–3	106.2
Major—0–4	53.6
Lt. Colonel—0–5	32.4
Colonel—0–6	14.0
Brig. General—0–7	0.5
Major General—0–8	0.4
Lt. General—0–9	0.1
General—0–10	(Z)

Z = Fewer than 50
Median and modal categories = Captain

SOURCE: *U.S. Dept. of Defense,* Selected Manpower Statistics, *annual, and Office of the Comptroller, unpublished data.*

ESTIMATING THE MEAN ($\bar{X}$) FROM GROUPED DATA

(not the population mean (μ))

Data on ages, incomes, education, and the like are often released by the Bureau of the Census and others in the form of a frequency distribution. In order to estimate the mean of a frequency distribution,

Table 3-4 **Nominal-Level Data**

Party	Number
Democratic	2,561
Socialist	732
Republican	1,602
Independent	1,814

Only modal category can be located.
Modal category = Democratic

it is assumed that *the values are spread evenly throughout each class.* Logically, the mean of all the values in a class is its **midpoint.** The midpoint of a class, therefore, is used to represent the class. These computations are similar to those for finding a weighted mean.

The arithmetic mean of sample data organized in a frequency distribution is estimated by

Sample

$$\overline{X} = \frac{\Sigma fX}{n} \quad \textbf{3-4}$$

where

$\overline{X}$	is the designation for the sample mean.
f	is the frequency in each class.
X	is the midpoint of each class.
fX	is the frequency of each class times its midpoint.
ΣfX	directs one to add these products.
n	is the total number of observations in the sample.

If the frequency distributions represent a population, the symbols are slightly different but the computations are the same. The formula for the arithmetic mean of a population is

Population

$$\mu = \frac{\Sigma fX}{N} \quad \textbf{3-5}$$

where

μ is the population mean.

ΣfX is the sum of the products of the class midpoints times the number of frequencies in each class.

N is the total number of observations in the population.

Problem Table 3–5 repeats the frequency distribution from Chapter 2 regarding the selling prices of a sample of 80 homes in the Rafter J Ranch subdivision. What is the arithmetic-mean selling price?

Solution It is assumed that the five selling prices in the \$79,500–\$84,500 interval averaged \$82,000 (which is the midpoint of that class). Thus, the total selling price of the five homes is approximately \$410,000, found by 5 × \$82,000. The class midpoint of the next higher class is \$87,000, used to represent the \$84,500–\$89,500 class. The total selling price of the 10 homes in that class is 10 × \$87,000 = \$870,000. This process is continued for all the classes. The total value of the 80 homes is estimated to be \$7,705,000 and the estimated arithmetic mean is \$96,313, found by \$7,705,000/80 (see Table 3–6 on page 92).

Solving for the estimated arithmetic mean using formula 3–4, we obtain

$$\overline{X} = \frac{\Sigma fX}{n} = \frac{\$7{,}705{,}000}{80} = \$96{,}313$$

Note that there is usually a slight discrepancy between the arithmetic mean estimated from data grouped in a frequency distribution and the mean of raw data. In the case of the selling prices in the

Table 3-5

Selling Prices of 80 Homes in the Rafter J Ranch Subdivision

True Limits	**Class Frequencies**
Selling Price	Number of Homes (f)
\$ 79,500–84,500	5
84,500–89,500	10
89,500–94,500	15
94,500–99,500	26
99,500–104,500	13
104,500–109,500	7
109,500–114,500	4

Table 3-6 **Calculations Needed for the Arithmetic Mean**

True Limits	**Class Frequencies**	**Midpoint**	**Frequency × Midpoint**
Selling Price	Number of Homes (f)	X	fX
\$ 79,500–84,500	5	\$ 82,000	\$ 410,000
84,500–89,500	10	87,000	870,000
89,500–94,500	15	92,000	1,380,000
94,500–99,500	26	97,000	2,522,000
99,500–104,500	13	102,000	1,326,000
104,500–109,500	7	107,000	749,000
109,500–114,500	4	112,000	448,000
	$\Sigma f = n = 80$		$\Sigma fX =$ \$7,705,000

Rafter J Ranch subdivision, the mean estimated from the grouped data is \$96,313. The exact mean of the raw data is \$96,488, or a difference of \$175. This is a difference of less than 0.2%. This small loss of accuracy is usually more than offset by the improved ease of calculation of the frequency distribution. But remember that the mean of data grouped into a frequency distribution should be viewed as an estimate of the actual mean.

Self-Review 3-5

A sample of television viewers was asked to rate a new soap opera titled "Mother Knows Best." The highest possible score is 100. A score near 0 indicates that the viewer thinks this is a very poor program. A score near 100 indicates that it is an outstanding soap opera. The scores were organized into a frequency distribution:

Rating	**Frequency**
50–59	2
60–69	6
70–79	10
80–89	8
90–99	4

Determine the arithmetic-mean rating.

Exercises

17. A sample of vehicles traveling on Interstate 75 was clocked by radar, and the following frequency distribution was constructed:

a. What is the total number of frequencies?
b. Determine the midpoint for the class 50–54.
c. Find the estimated arithmetic-mean speed by using the class midpoints.

Speeds (mph)	Number of Vehicles
40–44	9
45–49	15
50–54	30
55–59	17
60–64	17
65–69	12

18. Last month a total of 60 antique cars were sold within the Buffalo metropolitan area. The selling prices were organized into the following frequency distribution:

Selling Price (in thousands of dollars)	Frequency
$40.0–49.0	3
50.0–59.0	6
60.0–69.0	19
70.0–79.0	23
80.0–89.0	9
Total	60

Estimate the arithmetic-mean selling price.

19. A sample of light trucks using diesel fuel revealed the following miles per gallon:

Mileage	Number of Trucks
11–13	3
14–16	5
17–19	12
20–22	8
23–25	4

Estimate the arithmetic-mean mileage for the sample.

20. A sample of 50 listeners to WRQN was obtained. Data on their ages were organized into the following frequency distribution. Estimate the mean age.

Age	Frequency
15–19	10
20–24	12
25–29	14
30–34	9
35–39	5

ESTIMATING THE MEDIAN FROM GROUPED DATA

We noted previously that for some data the arithmetic mean is not a representative measure of central tendency. Usually, in these cases the median can be used to represent the typical value. How is the median determined if the data are grouped into a frequency distribution?

We can estimate the median of data organized into a frequency distribution by (1) locating the class in which the median lies and (2) interpolating within that class to arrive at the median.

Problem The selling prices of homes in the Rafter J Ranch subdivision are reintroduced to illustrate the procedure for estimating the median (see Table 3–7). The cumulative frequencies in the right-hand column will be helpful for locating the class in which the median lies. What is the median selling price?

Table 3-7 **Selling Prices of 80 Homes in the Rafter J Ranch Subdivision**

True Limits	Class Frequencies	Cumulative Frequency
Selling Price	Number of Homes *(f)*	*CF*
$ 79,500–84,500	5	5
84,500–89,500	10	15
89,500–94,500	15	30
94,500–99,500	26	56
99,500–104,500	13	69
104,500–109,500	7	76
109,500–114,500	4	80

Solution Note that there are 80 selling prices. The middle observation is determined by $n/2$, or $80/2 = 40$. [To be consistent with the way the median was located earlier in the chapter, the selling price of the $(n + 1)/2$, or 40.5th, home should be located. However, it is common practice to use the more convenient $n/2$ value. Any error involved is very slight.]

We find the class interval containing the fortieth selling price by referring to the cumulative frequency column. There are 30 selling prices less than $94,500 and 56 prices less than $99,500. Logically, the fortieth selling price is in the $94,500–$99,500 class. That particular class contains 26 selling prices.

To compute the median, we again assume the selling prices of the 26 homes in the median class are evenly distributed between the lower and upper limits of that class. Shown in a diagram, that information appears as follows:

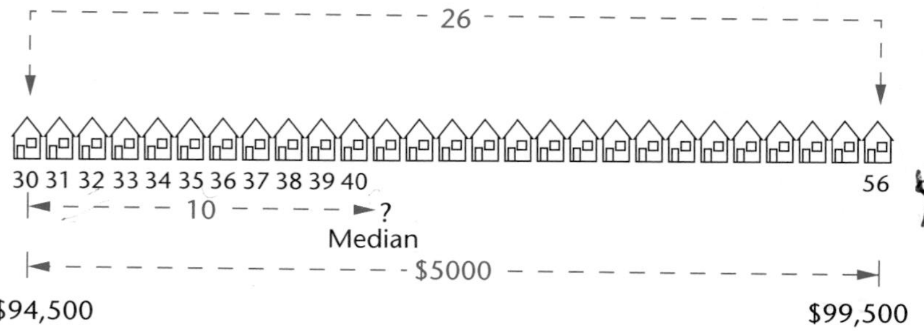

The median can be interpolated from the diagram. There are 26 prices in the class containing the median. And there are 10 selling prices between the thirtieth price and the fortieth price (40–30). Therefore, the median price is 10/26 of the amount between $94,500 and $99,500. That amount is $5,000. Thus, 10/26 of $5,000 is $1,923. Adding $1,923 to $94,500 we obtain the estimated median of $96,423. Summarizing these calculations:

$$\text{Median} = \$94,500 + \frac{10}{26}(\$5,000) = \$96,423$$

Alternatively, we can use the following formula to determine the median:

$$\text{Median} = L + \frac{\frac{n}{2} - CF}{f}(i) \qquad \textbf{3-6}$$

where

L	is the lower true limit of the class containing the median. The median is in the \$94,500–\$99,500 class and \$94,500 is the true lower limit of that class.
n	is the total number of frequencies. It is 80.
CF	is the cumulative number of frequencies in all the classes immediately preceding the class containing the median. The class containing the median is \$94,500–\$99,500, and the cumulative number of selling prices prior to that class is 30.
f	is the frequency in the class containing the median. There are 26 in that class.
i	is the width of the class in which the median lies. The width is \$5,000, found by \$99,500 – \$94,500.

Inserting these values into this formula, we obtain:

$$\text{Median} = \$94{,}500 + \frac{\frac{80}{2} - 30}{26}(\$5{,}000) = \$96{,}423$$

Again, it should be noted that a measure computed using data grouped into a frequency distribution will probably not be identical to the measure we would obtain using the raw data. Here the estimated median using the frequency distribution was \$96,423, while the median of the selling prices of Rafter J Ranch homes listed in Chapter 2 is \$97,000.

Self-Review 3-6

On arriving in Hawaii, a sample of vacationers is asked their ages by the tourist bureau. This information is organized into the following frequency distribution (the true limits and the cumulative frequencies are given for clarification):

Ages	True Limits	Number of Vacationers	Cumulative Number
20–29	19.5–29.5	4	4
30–39	29.5–39.5	9	13
40–49	39.5–49.5	20	33
50–59	49.5–59.5	8	41
60–69	59.5–69.5	5	46
70–79	69.5–79.5	4	50

Compute the median age.

THE MODE FROM GROUPED DATA

For data grouped into a frequency distribution, the mode is the *midpoint* of the class containing the largest class frequency. As an example, refer back to Table 3–7. The largest class frequency is 26 and the midpoint of that class is $97,000–the **mode.** This indicates that the largest number of Rafter J Ranch homes sold for $97,000.

Self-Review 3-7

Refer to Self-Review 3–6 for the ages of the Hawaii vacationers.

a. What is the mode of the age distribution of vacationers arriving in Hawaii? This question also could be stated as, What is the *modal* age of the vacationers arriving in Hawaii?
b. Explain what the mode indicates.

Exercises

21. Refer to the frequency distribution of speeds on I-75 in Exercise 17.
- **a.** Which class interval contains the median value?
- **b.** What is the true lower limit of the class containing the median?
- **c.** What is the class interval?
- **d.** Determine the cumulative number of frequencies in all classes smaller than the median class.
- **e.** Compute the median speed.
- **f.** What is the mode of the distribution of speeds?
- **g.** Discuss which measure of central tendency you would use as an average for the data.

22. Refer to the frequency distribution in Exercise 18.
- **a.** Which class contains the median value?
- **b.** What is the true lower limit of the class containing the median?
- **c.** What is the class interval?
- **d.** Determine the cumulative number of frequencies in all classes smaller than the median class.
- **e.** Compute the median.
- **f.** Determine the mode of the distribution.

23. A sociologist is studying the impact of television on the family. Of particular interest is the number of hours of television school-aged children watch each day. A random sample of 50 homes reveals the following number of hours watched on a Wednesday between 4 p.m. and 11 p.m.:

Number of Hours	Frequency
0	9
1	14
2	10
3	9
4	5
5 or more	3
	50

a. Compute the median number of hours of TV viewing.
b. Determine the modal number of hours of TV watched.

24. An advertising campaign is being designed for a local car dealer. One relevant piece of information, the ages of recent purchasers of cars sold by the dealership, is shown in the following frequency distribution:

Age	Frequency
Under 20	7
20 but less than 30	13
30 but less than 40	26
40 but less than 50	15
50 but less than 60	6
60 or more	3
Number of cars sold	70

a. Compute the median age.
b. What is the modal age?
c. Explain why the mean cannot be determined.

STAT BYTE

In 1967, the average salary of a baseball player in the major leagues was $19,000; by 1992, it had increased to $1,084,408. This is an average annual rate of increase of 17.5%. The increase from 1991 to 1992 alone was 27%! Last year, the New York Mets had the highest average salary, $1.7 million, while the Cleveland Indians had the lowest—$271,089.

CHOOSING AN AVERAGE FOR DATA IN A FREQUENCY DISTRIBUTION

If the data being organized into a frequency distribution contain extremely high or extremely low values, the resulting distribution is said to be **skewed.** The frequency distribution in Figure 3–3 is **positively skewed.** It can be identified by the long "tail" on the right.

Note the relative positions of the three averages. The arithmetic mean ($\overline{X}$) is being pulled upward toward the tail by a few very highly paid employees. It is the highest of the three averages. If the skewness is very pronounced, the mean is not an appropriate average to describe the central tendency of the data.

The median is the next higher measure of central tendency for a positively skewed distribution. It, of course, divides the incomes

Figure 3-3

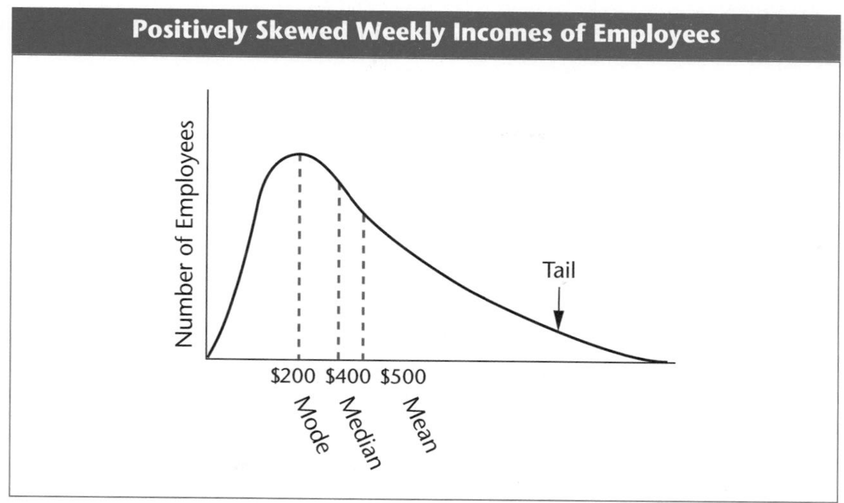

into two parts and may be more typical of the data. The mode is the smallest of the three measures. It occurs at the peak, or apex, of the curve and indicates that the largest number of employees earns $200 a week.

The distribution of the ratings in Figure 3–4 is **negatively skewed.** The arithmetic mean is being pulled down by a few very low ratings, and once again it is not the best average to use. The median is the next higher average, with the mode of 54 the highest of the three averages.

By contrast, Figure 3–5 reveals that there is no skewness in the distribution of the annual incomes of the social workers. This type of distribution is called symmetric. A **symmetric distribution**

Figure 3-4

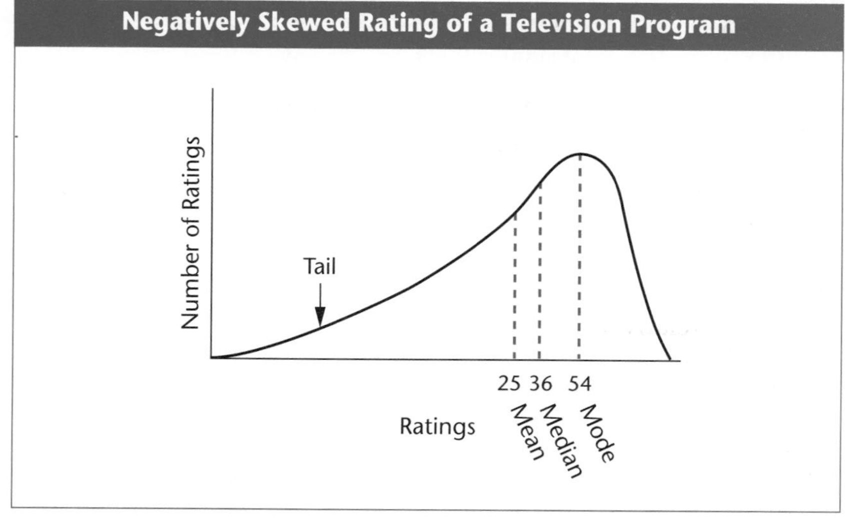

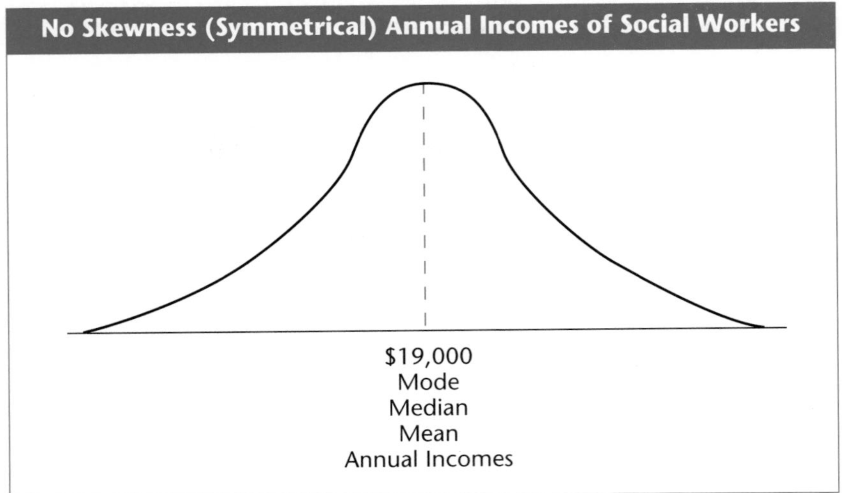

has the same shape on either side of the median. That is, if the distribution graph were folded in half at the median value, the two halves would be identical. With such a distribution all three values—mode, median, and mean—are the same. In Figure 3–5, the value for all three is $19,000.

Open-ended frequency distributions pose a problem for the computation of the arithmetic mean. Unless the midpoint of the open-ended class can be approximated, the mean cannot be computed. In the example, unless the midpoint of the "$30,000 and over" class can be approximated, the mean of this open-ended distribution cannot be calculated.

Income	Number
$10,000–19,999	42
20,000–29,999	86
30,000 and over	19

The median or the mode, however, can be used to represent the central value of the income distribution.

SUMMARY

This chapter dealt with three measures of central tendency used to describe the typical value of a set of data. The most commonly used measure of central tendency is the arithmetic mean. For raw data, we compute this mean by summing the values and dividing the total by the number of values. For data grouped into a frequency distribution, we multiply each class frequency by its midpoint and sum the products. We then divide this sum by the number of observations to obtain the arithmetic mean.

The median divides the data into two equal parts. For raw data, the values are arranged from low to high and the center value is the median. For data in a frequency distribution, we first identify the median class from the cumulative frequency distribution and then interpolate to determine the median. The mode is the value of the item that appears most often.

A number of characteristics make the mean a very reliable average. However, because all the values are included in its computation, the mean may not be representative of the data when there are extremely high or low values. The median or the mode are preferred in such highly skewed samples. The mode can be used for all levels of data, the median requires at least ordinal data, and the mean at least interval scale.

Exercises

25. The Law Enforcement Assistance Administration reported that on December 31 of each of the past five years there were 213.0, 211.0, 196.0, 241.0, and 263.0 prisoners in federal and state institutions (data reported in thousands). Consider this information to be a population.

a. Find the arithmetic-mean number of prisoners.

b. Show that the sum of the deviations from the mean is zero.

c. Find the median number of prisoners.

26. The National Oceanic and Atmospheric Administration reported the following monthly breakdown of normal daily mean temperatures for Juneau, Alaska, and San Juan, Puerto Rico:

	Temperature (°F)			
Month	Juneau	CF	San Juan	CF
January	23.5	23.5	75.4	75.4
February	28.0	51.5	75.3	150.7
March	31.9	83.4	76.3	227
April	38.9	122.3	77.5	304.5
May	46.8	169.1	79.2	383.7
June	53.2	222.3	80.5	464.2
July	55.7	278	80.9	545.1
August	54.3	332.3	81.3	626.4
September	49.2	381.5	81.1	707.5
October	41.8	423.3	80.6	788.1
November	32.5	455.8	78.7	866.8
December	27.3	483.1	76.8	943.6

a. Is the information a sample or a population?
b. Find the mean annual temperature for both Juneau and San Juan.
c. Find the median annual temperature for both Juneau and San Juan.

27. The National Science Foundation reported that the median time candidates in the social sciences take to earn a doctorate degree, after receiving a bachelor's degree, is 8.5 years. Explain what the 8.5 years indicates.

28. The Surgeon General released these figures for the United States: 54 million smokers smoked 615 billion cigarettes in one year. What is the average annual number of cigarettes smoked by persons who smoke? What is the average daily number?

29. A report on the literacy rate for a sample of countries revealed the following percents:

Country	Literacy Rate (percent)
Turkey	55
Tunisia	32
Tonga	93
Benin	20
Bangladesh	25
Colombia	47
Iceland	99
Iran	37

a. Find the mean percent.
b. Find the median percent.

30. A student is currently enrolled for a 3-credit-hour history course, a 5-credit-hour English course, and a 4-credit-hour statistics course. Grades are reported on a 4-point system where A = 4, B = 3, C = 2, D = 1, and F = 0. The student received an A in statistics, a B in English, and a C in history. What is the student's grade point average for the semester? What type of average is this?

31. At Mercy General Hospital, registered nurses (RNs) are paid $12.00 per hour, licensed practical nurses (LPNs) are paid $9.00 per hour, and nurses' aides are paid $6.00 per hour. If the Pediatrics Wing employs six RNs, four LPNs, and four aides, compute the weighted mean salary for the wing.

32. A study of the mileage per gallon of 35 compact cars of the same make and model revealed the following:

Miles per Gallon	Number of Cars
20–24	2
25–29	7
30–34	15
35–39	8
40–44	3

a. Compute the arithmetic-mean mileage.
b. Compute the median mileage.

33. Are the baseball batting averages in the American League higher than those in the National League? Consider these data to be a sample:

Batting Averages	American League	National League
.150–.199	5	4
.200–.249	34	27
.250–.299	64	52
.300–.349	27	14
.350–.399	2	0

a. Estimate the mean batting average for the two leagues.
b. Compare the median averages.
c. Determine the mode for each league.
d. Does this show there is much difference in the two leagues?

34. A sample of seven independently owned service stations in the San Francisco metropolitan area revealed the following cost per pint of transmission fluid:

Andy's Amoco	\$0.83
Rossi's Gastown	0.79
Kernie's Mobil	0.92
Mark's Exxon	0.71
Yamamoto's Chevron	0.69
Ali's Sunoco	0.83
Deckers' Union 76	0.74

a. Calculate the mean, median, and modal cost per pint.
b. Discuss which average is most representative of the information or numbers.

35. The American Association of Retired Persons is studying the 1992 voting statistics. The following distribution presents a breakdown of voters by age. Compute the median age.

Age	Frequency (in millions)
20–29	22.1
30–39	21.9
40–49	16.4
50–59	15.6
60–69	14.4
70 or more	11.5

36. In a survey of chief executive officers of large companies, the following information was obtained regarding the time they awake each weekday morning:

Time	Percent of Total
4–5 a.m.	7.0
5–6 a.m.	65.0
6–7 a.m.	24.0
7–8 a.m.	4.0

What is the typical wake-up time?

37. The 1992 season payrolls (in millions of dollars) of the 26 major league baseball teams were as follows:

\$44.4	\$43.5	\$43.3	\$42.7	\$40.2	\$35.5	\$34.7	\$34.0	\$33.1
32.5	31.8	31.8	30.5	29.2	29.1	28.7	28.5	27.8
27.6	27.5	24.3	22.4	20.5	15.5	12.7	8.1	

a. Determine the mean and the median team payroll.

b. Organize the data into a frequency distribution and compute the median team payroll. (*Hint:* You may want to use an open-ended class for the teams with the lower payrolls.)

38. The Neal Nail Company manufacturers and packages nails. For a sample of 40 packages, the following numbers of nails are found:

Number of Nails	20 or less	21	22	23	24	25 or more
Number of Packages	5	8	11	10	5	1

Determine the median number of nails per package.

39. A study of 50 small and median-size firms in a large metropolitan area reveals the following about the percent change in number of persons employed:

Percent Change	Frequency
more than –20.0	13
–20.0 up to –10.0	21
–10.0 up to 0.0	48
0.0 up to 10.0	21
10.0 up to 20.0	5
20.0 or more	1

Determine the median percent change in employment.

40. A large taxicab company has a fleet of 45 cabs. Below are the numbers of gallons of gasoline each cab used last month.

123	132	130	119	106	97	121	109	118
128	132	115	130	125	121	127	144	115
107	110	112	118	115	134	132	139	144
104	128	138	114	121	129	128	116	138
129	113	105	142	122	131	126	111	142

a. Organize the data into a frequency distribution.
b. Determine the mean number of gallons used per taxicab.
c. Determine the median number of gallons used per taxicab.

Data Exercises

41. Refer to the Real Estate Data Set in the appendix, which reports information on homes sold in Northwest Ohio during 1992.
a. Determine the mean and median selling price. Is one a better measure of central tendency than the other? Defend your choice in a short paragraph.
b. Describe a typical home. How many bedrooms does it have? How many bathrooms? How far is it from the center of the city?
c. Compare the mean selling price of a home with an attached garage to that of a home without an attached garage.
d. Write a brief summary of your findings.

42. Refer to the Salary Data Set in the appendix, which refers to a sample of middle managers in Sarasota, Florida.
a. Determine the mean and the median salary. Is one a better measure of central tendency than the other? Defend your choice in a short paragraph.
b. Describe a typical manager. How many years of experience does he or she have? How many employees does the "typical" manager supervise?
c. Compare the salaries of male and female managers.
d. Write a brief summary of your findings.

CASE: HOW MANY VIDEOS DOES A "TYPICAL" FAMILY RENT?

The number of homes with video equipment has increased markedly over the last several years. The National Video Rental Association recently reported that 67% of all households rented at least one video in the past year.

To determine if a new store would be profitable, the Video Connection commissioned Gulf Coast Marketing to survey the local area regarding the number of videos rented per household per year. The following information was taken from their report:

Number of Videos Rented	Number of Households
0	542
1–4	168
5–9	216
10–19	334
20–49	181
50 or more	59
	1500

The owner of the Video Connection is considering opening a new outlet in Ocean City, a resort community to which he often travels. Use the above information to project a typical rental rate for Ocean City.

CHAPTER ACHIEVEMENT TEST

Answer all the questions. The answers are at the back of the book.

MULTIPLE-CHOICE QUESTIONS. Select the response that best answers each of the questions.

1. The lowest scale of measurement required to calculate the mode is
   a. nominal.
   b. ordinal.
   c. interval.
   d. ratio.

2. The lowest scale of measurement required to calculate the mean is

a. nominal.
b. ordinal.
c. interval.
d. None of the above

3. In a positively skewed distribution, the mean is always
a. smaller than the median.
b. equal to the median.
c. larger than the median.
d. None of the above

4. Half the observations are always larger than the
a. mean.
b. median.
c. mode.
d. total.

5. If a frequency distribution has an open-ended class,
a. a median cannot be computed.
b. the arithmetic mean and the median will always be exactly equal.
c. the distribution is always positively skewed.
d. a mean cannot be computed.

6. Which of the following is *not* a characteristic of the arithmetic mean?
a. All the observations are used in its calculation.
b. It is influenced by extreme values.
c. The sum of the deviations from the mean is zero.
d. Fifty percent of the observations are larger than the mean.

7. Which of the following is a true statement about the median?
a. It is always one of the data values.
b. It is influenced by extreme values.
c. The sum of the deviations from the mean is zero.
d. Fifty percent of the observations are larger than the median.

8. The median is larger than the arithmetic mean when
a. the distribution is positively skewed.
b. the distribution is negatively skewed.
c. the data is organized into a frequency distribution.
d. raw data are used.

9. A sample of 10 students is obtained. The students are weighed and ranked according to their weight. The median weight is the
a. weight of the fifth student.
b. weight of the sixth student.
c. The median does not exist.
d. average weight of the fifth and sixth students.

10. The value that occurs most often in a set of data is called the
a. mode.
b. arithmetic median.
c. mean.
d. None of the above

COMPUTATION PROBLEMS

11. The weekly incomes for a sample of four registered nurses are \$420, \$445, \$600, and \$395.

a. What is the arithmetic-mean weekly income?

b. What is the median weekly income?

12. The amounts spent on mail advertising in a day by a sample of five salespersons are \$22.50, \$85.90, \$18.60, \$140.20, and \$76.10. What is the median amount spent?

Questions 13–15 are based on the following frequency distribution of a sample of the weights of king crab caught in Alaskan waters:

Weights (in pounds)	Number of Crabs
0.0 to 0.4	3
0.5 to 0.9	8
1.0 to 1.4	20
1.5 to 1.9	10
2.0 to 2.4	6
2.5 to 2.9	3

13. What is the arithmetic-mean weight (to the nearest hundredth of a pound)?

14. What is the median weight (to the nearest hundredth of a pound)?

15. What is the modal weight (to the nearest hundredth of a pound)?

ANSWERS TO SELF-REVIEW PROBLEMS

3–1 a. $\overline{X} = \frac{\Sigma X}{n}$

b. $\overline{X} = \frac{72}{6}$

$= 12.0 \text{ kilometers}$

3–2 a. $\overline{X} = \frac{960}{4} = 240.0$

b. The mean

c. $\Sigma(x - \overline{x}) = (200 - 240) + (210 - 240) + (230 - 240) + (320 - 240)$

$= (-40) + (-30) + (-10) + 80$

$= 0$

3–3 a. Arranged from low to high:

18
22
(34) median
40
75

b. 2, 2

3–4 a. 11 points—halfway between 9 and 13

b. 146,490,000 telephones—halfway between 143,970,000 and 149,010,000

3–5 a.

X	f	fX
54.5	2	109
64.5	6	387
74.5	10	745
84.5	8	676
94.5	4	378
		2,295

$$\overline{X} = \frac{\Sigma fX}{n}$$
$$= \frac{2295}{30} = 76.5$$

3–6 $\overline{X} = 39.5 + \dfrac{\frac{50}{2} - 13}{20}(10)$

$$= 39.5 + \frac{12}{20}(10)$$
$$= 45.5 \text{ years}$$

3–7 a. 44.5 years

b. The largest number of vacationers were 44.5 years of age.

CHAPTER 4

Descriptive Statistics: Measures of Dispersion and Skewness

OBJECTIVES

When you have completed this chapter, you will be able to

- describe the spread or dispersion in a set of data;
- calculate and interpret the range, interquartile range, quartile deviation, mean deviation, variance, and standard deviation;
- compute the coefficients of variation and skewness.

CHAPTER PROBLEM Price Disparities

RECALL FROM THE CHAPTER PROBLEM IN CHAPTERS 2 AND 3 THAT CHARLES REYNOLDS, OWNER OF REYNOLDS DEVELOPMENT, IS PUTTING TOGETHER A PROSPECTUS FOR NEW HOMES TO BE BUILT IN THE RAFTER J RANCH SUBDIVISION IN JACKSON HOLE, WYOMING. IN CHAPTER 2, HE FOUND THE SELLING PRICE OF EACH OF THE 80 EXISTING HOMES AND ORGANIZED THEM INTO A FREQUENCY DISTRIBUTION. IN CHAPTER 3, HE DETERMINED SEVERAL MEASURES OF CENTRAL TENDENCY TO FIND THE ESTIMATED SELLING PRICE OF A TYPICAL HOME. NOW MR. REYNOLDS WANTS TO DETERMINE THE SPREAD OF THE ESTIMATED SELLING PRICES. HOW CAN HE CHARACTERIZE THE DISPERSION IN THE DATA?

INTRODUCTION

STAT BYTE

In 1947, the price of gasoline was 23 cents per gallon; by 1991, it had increased to $1.15 per gallon. That is an increase of 400% , or an average annual increase of 3.7% over the 44-year period. However, if we adjust for inflation, the 23 cents in 1947 is equivalent to $1.49 in 1991 dollars. So the cost of gasoline has actually *decreased* by 23% since 1947.

In Chapter 2, we noted that the usual first step in summarizing raw data is to organize it into a frequency distribution. A histogram or a frequency polygon may be drawn to portray a distribution in graphic form. Chapter 3 described a summary measure called an "average." Three different measures of central tendency or averages were examined: the arithmetic mean, the median, and the mode. An average is a central value around which the data tend to cluster.

A direct comparison of the central value of two or more distributions may be misleading. For example, suppose that a sample of the lengths of imprisonment for armed robbery in Alabama revealed the arithmetic-mean length to be 10 years. Suppose further that a similar study of the lengths of imprisonment in neighboring Georgia also revealed a mean length of 10 years. Based on the two sample means, one might conclude that the distributions of the lengths of imprisonment were about the same in both states. Figure 4–1, however, reveals this conclusion to be incorrect. Despite the fact that the two means are equal, the lengths of imprisonment in Georgia are spread out, or dispersed, more than those in Alabama.

The Range

As Figure 4–1 suggests, the next step in analyzing a set of data is to develop measures that describe the spread, or **dispersion,** of that

Figure 4-1

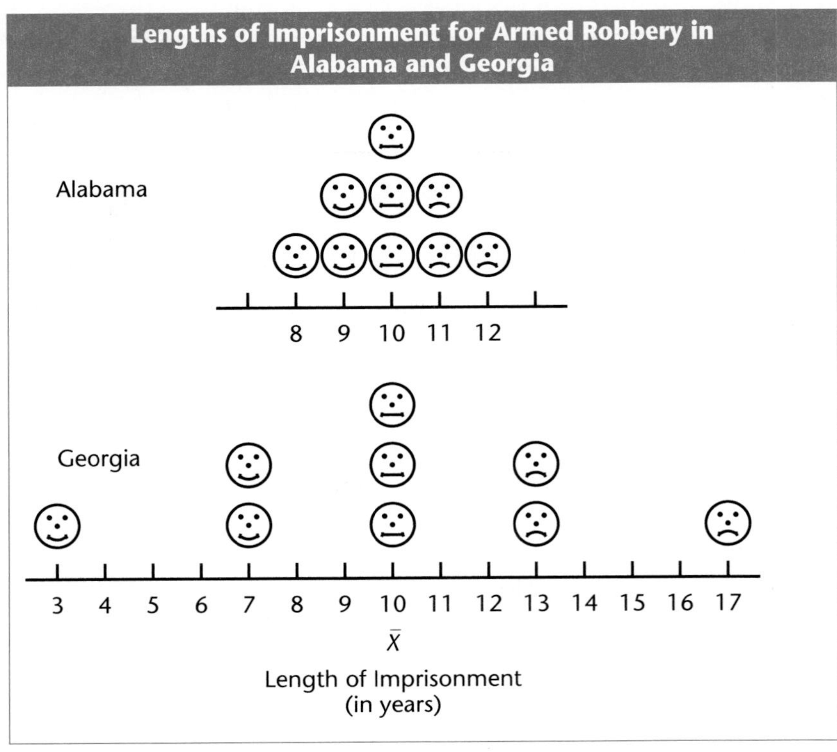

set. Perhaps the simplest measure of dispersion is the **range.** The range is the difference between the highest and lowest values in a set of data.

Range = Highest value – Lowest value 

Problem Calculate the range of the length-of-imprisonment data in Figure 4–1 for both Alabama and Georgia. Compare the two ranges.

Solution For the Alabama data, the longest stretch in prison is 12 years; the shortest is 8 years. The range is 4 years (12 – 8). The range for the Georgia prisoners is 14 years, or 17 – 3. A comparison of the two ranges indicates the following:

1. There is more spread in the Georgia data because the range of 14 years for Georgia is greater than the range of 4 years for Alabama.
2. The lengths of imprisonment for the Alabama inmates are clustered more closely about the mean of 10 years than those for the Georgia inmates, because the range of 4 years is less than the range of 14 years.

Obviously, the range is easy to compute. It does have a serious disadvantage, however. It is based on only two values—the two extreme observations. The range, therefore, does not reflect the variation in data that lie between the high and low observations. Further, one extremely high (or low) value might give a misleading picture of the dispersion. For example, suppose that the ages of a group at a U2 concert were 18, 20, 19, 20, 19, 17, 18, 18, and 78. Reporting the range of 61 years, found by subtracting 17 from 78, would give the impression that there is considerable variation in the ages of the audience. Except for one person, however, the distance between the youngest (17 years) and the oldest (20 years) is only 3 years! For that reason, the range can be considered only a rough index of the variation in a set of data.

Problems

Compute the sample range for each of the following sets of values:

a. 2, 9, 1
b. 6, 13, 10, 2
c. 0.1, 5.8, 1.2
d. $7\frac{3}{4}$, $3\frac{3}{16}$, $1\frac{1}{4}$

Solutions

a. 8
b. 11

c. 5.7
d. 3,202

Self-Review 4-1

Answers to all Self-Review problems are at the end of the chapter.

A radar check of several automobiles traveling on the Sunshine Highway recorded these speeds in kilometers per hour: 86, 70, 91, 110, and 89. Speeds recorded at about the same time on a nearby state highway were 89, 87, 89, 92, 85, and 92.

a. Compute the mean speed for each set of data.
b. Compute the range for each set of data.
c. Compare the dispersion for the two sets of speed checks.

The Interquartile Range and the Quartile Deviation

Recall that half of the ordered values are above the median and half are below it. The lower half of the ordered values can be further subdivided into two parts, so that one-fourth of all the ordered values are smaller than a particular value. The point where this subdivision occurs is called the **first quartile** and is designated Q_1. Similarly, the upper half of the ordered distribution can be divided into two equal parts. The **third quartile,** designated Q_3, is the point below which lie three-fourths of the ranked observations. The median, of course, is the same as the second quartile.

The **interquartile range** is the distance between the first quartile and the third quartile.

$$\textbf{Interquartile Range} = Q_3 - Q_1 \qquad \textbf{4-2}$$

Figure 4–2 illustrates this concept; note that the third quartile is 20 days, the first quartile 12 days. The interquartile range is 8 days, which we find by subtracting 12 from 20. Thus, the interquartile range reports the difference between the two values that bound the middle 50% of the observations.

Another measure of dispersion is the **quartile deviation,** abbreviated *QD,* which is one-half of the interquartile range.

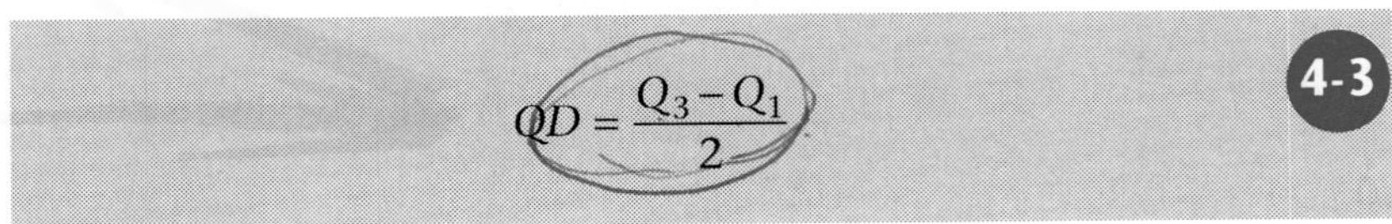

$$QD = \frac{Q_3 - Q_1}{2} \qquad \textbf{4-3}$$

Problems

Compute the interquartile range and the quartile deviation for each of the following sets of values:

	Third Quartile	First Quartile
a.	12	8
b.	27	20
c.	103	91
d.	1200	1000

Solutions

	Interquartile Range	Quartile Deviation
a.	4	2
b.	7	3.5
c.	12	6
d.	200	100

As mentioned earlier, one of the objections to the range as a measure of spread is that its value is based on the two extreme observations in a set of data. An unusually large or small extreme value might give a distorted picture of the spread in the data. The interquartile range and the quartile deviation offset this objection because they are not based on the extreme, or end, values. The first and third quartiles may be more representative of the data. Since

Figure 4-2

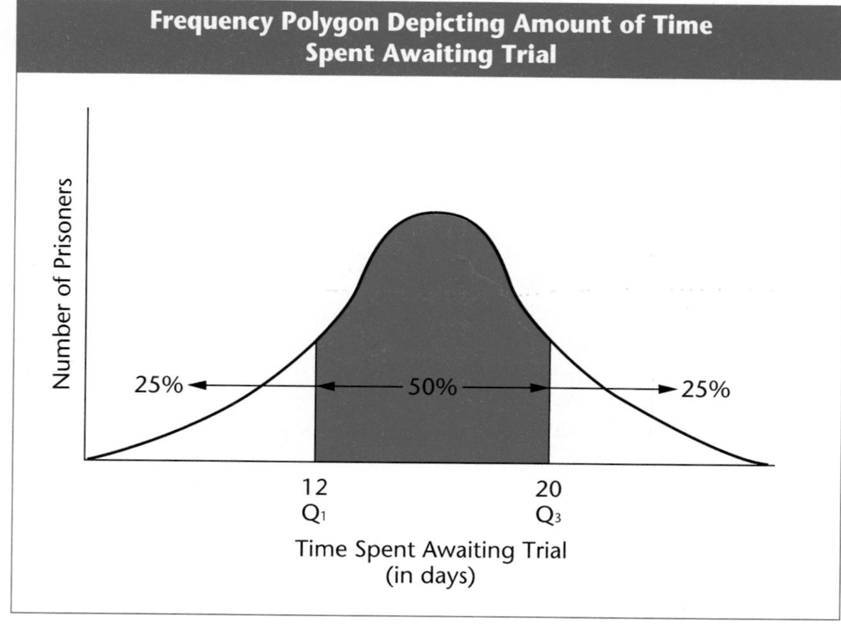

the *QD* is based on the relative positions of Q_1 and Q_3 in the frequency distribution, it may be computed not only for interval-scaled data, but also for ordinal-scaled data. Later in the chapter, we will estimate the interquartile range and the quartile deviation for data organized into a frequency distribution.

The Mean Deviation (*MD*)

One of the disadvantages of the range, the interquartile range, and the quartile deviation is that they do not make use of all the observations. However, several measures that incorporate all values do exist; they are based on the deviation of each observation from its mean. These measures include the **mean deviation** (often called **average deviation**), the **variance,** and the **standard deviation.**

We compute the **mean deviation,** abbreviated *MD,* by first finding the difference between each observation and the mean. Next, we sum these deviations using their absolute values—that is, disregarding their algebraic signs (+ and –). Finally, we divide the sum obtained by the number of observations to arrive at the mean deviation.

$$MD = \frac{\Sigma|X - \overline{X}|}{n} \qquad \textbf{4-4}$$

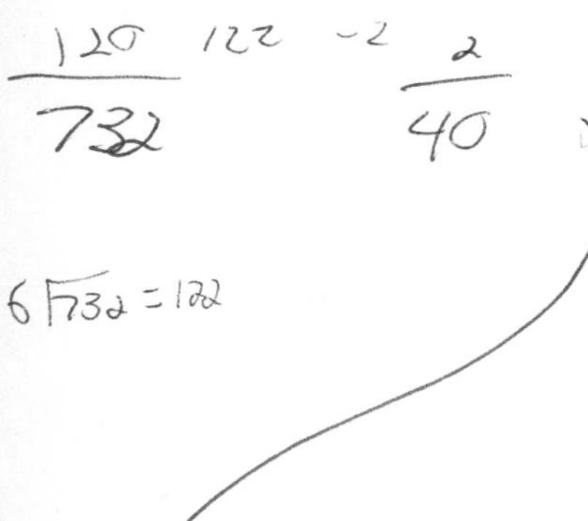

where

X is the value of each observation.

$\overline{X}$ is the mean.

n is the number of observations.

(The symbol | | indicates the absolute, or unsigned, value of a number.)

Problem The test kitchen of Cellu-Lite, a large producer of cake mixes, constantly monitors the weight, moisture content, and flavor of its cakes. The weights (in grams) of a sample of five peach upside-down cakes are 498, 500, 503, 498, and 501. What is the mean deviation? How is it interpreted?

Solution First, we find the arithmetic-mean weight of the cakes, which is 500 grams: (498 + 500 + 503 + 498 + 501)/5 = 2500/5 = 500. Next, we find the deviation from the mean for each cake. For example, the first cake selected, weighing 498 grams, deviates 2 from the mean of 500. Calculations for the mean deviation are shown in Table 4–1. Interpreting the result: On the average, the weights of the five peach upside-down cakes deviate 1.6 grams from the mean of 500 grams.

Table 4-1

Calculations Needed for the Mean Deviation

Computations for the Mean Deviation

Weight (in grams)	$X - \overline{X}$	Absolute Deviations
498	$\|-2\| =$	2
500	$\|0\| =$	0
503	$\|+3\| =$	3
498	$\|-2\| =$	2
501	$\|+1\| =$	1
2,500 ($n = 5$)		8

$$MD = \frac{\Sigma|X - \overline{X}|}{n} = \frac{8}{5} = 1.6 \text{ grams}$$

Diagram Showing Deviations from the Mean

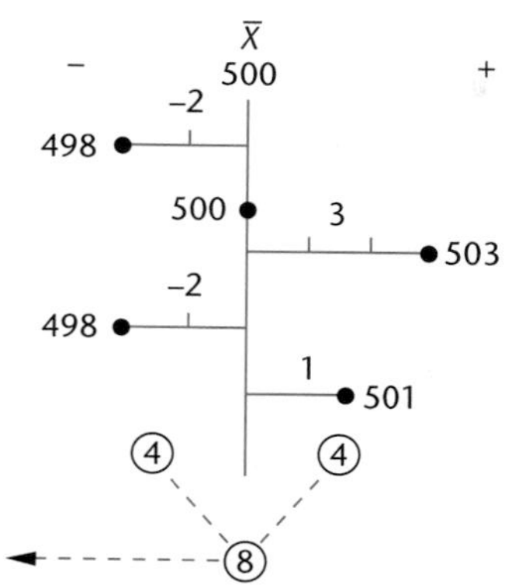

Problems Compute the mean deviations of each of the following sets of values:

a. 2, 1, 3
b. 5, 1, 4, 6
c. 10, 9, 7, 14

Solutions

a. 0.67
b. 1.5
c. 2

Self-Review 4-2

The Cellu-Lite test kitchen now wants to compare the variation in the weights of blueberry cakes with that of the peach upside-down cakes studied in the previous case. The weights of the blueberry cakes are 484, 503, 496, 510, 491, and 516 grams.

a. Compute the mean and the mean deviation.
b. Interpret the mean deviation.
c. Compare the variation in the weights of the peach upside-down cakes with the variation in the weights of the blueberry cakes.

Exercises

Answers to the even-numbered exercises are at the back of the book.

1. The amount (in gallons) of fuel oil consumed during the winter heating season for a sample of newer homes is 600, 590,

605, 600, 603, 610, 597, and 595. A second group of older homes used the following amounts: 810, 750, 790, 800, and 850.

a. What is the range for the newer homes? For the older homes?

b. What is the mean deviation for the newer homes and for the older homes?

c. Compare the dispersion in the data for the newer homes and older homes.

2. The Department of Fisheries stocked two ponds with trout fingerlings. The fish in one pond (pond A) were fed the traditional food. Those in the other pond (pond B) were fed an experimental food. Consider this as sample data. After four years, the lengths of the fish were as follows:

Length (in inches)	
Pond A	Pond B
12.5	18.0
14.0	20.0
13.5	12.0
14.5	14.5
15.0	19.0
16.5	13.5
16.0	14.5
12.0	13.0
14.0	20.5
14.0	14.0

a. Compute the arithmetic-mean lengths for both ponds.

b. Compute the range for both sets of data. Interpret.

c. Compute the mean deviation for each set. Interpret.

3. The number of customers entering Romaker's Boutique each day for a six-day period was 91, 89, 88, 58, 63, and 74.

a. What is the range?

b. What is the mean deviation?

4. Many governmental agencies are authorized to issue tax-exempt bonds (bonds whose interest earnings are tax-exempt to the owner) as a means of generating funds at a rate lower than current interest rates. A sample of 10 such bonds revealed the following interest rates: 9.75, 3.75, 5.50, 7.30, 6.00, 7.30, 7.25, 4.25, 3.30, and 6.60.

a. What is the range?

b. What is the mean deviation? Explain what this indicates.

5. The National Association of Realtors released the following median prices of existing single-family homes for selected cities in the United States:

City	Price
Akron, OH	\$ 65,800
Anaheim, CA	243,600
Chicago, IL	112,300
Honolulu, HI	290,400
Las Vegas, NE	88,600
Los Angeles, CA	211,500
Memphis, TE	77,900
New York, NY	173,700
Spokane, WA	77,700
Toledo, OH	61,300

a. What is the range in the median price of the homes in these selected cities?

b. What is the mean deviation for these prices?

6. The infant mortality rates of black and white males for the years 1983–1992 are:

Year	Black Males (percent)	White Males (percent)
1983	49.1	26.0
1984	36.2	20.0
1985	23.3	12.3
1986	21.7	11.7
1987	21.5	11.2
1988	21.1	10.8
1989	19.8	10.5
1990	19.9	10.6
1991	20.0	10.0
1992	19.6	9.6

a. Determine the range and the mean deviation for black and white males.

b. Is there a difference in the spread of the data for the two groups?

The Variance (σ^2) and the Standard Deviation (σ)

The mean deviation has two distinct advantages: It uses all available data in its computation, and it is easy to understand—that is, the mean deviation is the average amount by which the observations differ from the mean. As a measure of variation, however, it does have a major flaw: The absolute values used in its computation are difficult to work with.

Two other measures of dispersion—the **variance** and the **standard deviation**—are more versatile than the mean deviation. As a result, they will be used extensively in later chapters. Like the mean deviation, the variance and the standard deviation are based on the deviations from the mean. They differ from the mean deviation, however, in that the sign (+ or –) of the deviation from the mean is *not* ignored in their computation. To compute the variance and the standard deviation, we square the difference between each value and the mean. This squaring eliminates the possibility of negative numbers (since multiplication of two negative numbers yields a positive number). We then total the squared deviations, and divide the total by the number of values.

STAT BYTE

Where did the .400 hitters go? In 1941, Ted Williams of the Boston Red Sox batted .406 for the season. Since 1941, there have been no .400 hitters. However, the mean batting average of all players has been steady at .260 for more than 100 years. The standard deviation of the averages has declined over the period from .049 to .031. Would this explain why there are no .400 hitters today, even if players are as good or better?

Variance The arithmetic mean of the squared deviations from the mean.

Standard Deviation The square root of the variance.

Because of a difference in computation, it is necessary to distinguish between the variance of a *population,* designated by σ^2 (the lowercase Greek *sigma*), and the variance of a *sample,* which is designated s^2. Recall that in Chapter 3 we made a distinction between a parameter, which is a measure associated with a population, and a statistic, which is a measure associated with a sample. The formulas for the variance of a population and the variance of a sample are

Variance of a population **4-5**

$$\sigma^2 = \frac{\Sigma(X-\mu)^2}{N}$$

where

- μ is the population mean.
- X is the value of each observation in the population.
- N is the number of observations in the population.

Variance of a sample **4-6**

$$s^2 = \frac{\Sigma(X-\overline{X})^2}{n-1}$$

where

X	is the value of each observation in the sample.
$\overline{X}$	is the sample mean.
n	is the number of observations in the sample.

Note that we compute the sample variance by dividing the sum of the squared deviations by the quantity $n - 1$, rather than by n. The use of the sample mean $\overline{X}$ instead of the population mean μ causes the numerator to be too small. So we "underestimate" the denominator as well, to compensate. Stated differently, s^2 would underestimate σ^2 if $\Sigma(X - \overline{X})^2$ were divided by n. For a practical illustration, see the problem and solution that follow.

Problem Refer back to the weights for a sample of five peach upside-down cakes in Table 4–1. What are the sample variance and the standard deviation?

Solution The data needed to calculate the variance and the standard deviation are included in Table 4–2. The sample-mean weight, $\overline{X}$, is 500 grams. To determine the sample variance, we divided the sum of the squared deviations (18) by the number of observations minus one, using formula 4–6:

$$s^2 = \frac{\Sigma\left(X - \overline{X}\right)^2}{n-1} = \frac{18}{5-1} = 4.5$$

Table 4-2

Calculations Needed for the Variance and the Standard Deviation

Weight (in grams) X	Finding the Difference $X - \overline{X}$	Squaring the Difference $(X - \overline{X})^2$
498	−2	4
500	0	0
503	+3	9
498	−2	4
501	+1	1
	0	18

sum of the deviations from mean is always zero

The variance is 4.5. We then determine the standard deviation by taking the square root of the variance. The sample standard deviation is 2.12.

Problems Compute the variance and the standard deviation for each of the following samples:

a. 4, 2, 3
b. 9, 6, 7, 2
c. 2, 6, 1

Solutions

	Variance	Standard Deviation
a.	1.0	1.0
b.	8.67	2.94
c.	7.0	2.646

Self-Review 4-3

The test kitchen wants to compare the variation in the weights of the blueberry cakes in Self-Review 4–2 with the peach cakes—using the variance and the standard deviation. The weights of a sample of blueberry cakes were 484, 503, 496, 510, 491, and 516 grams.

a. Compute the variance and the standard deviation.
b. Compare the variation in the weights of the peach cakes and of the blueberry cakes.

Exercises

7. The amounts of fuel oil (in gallons) consumed for a sample of homes in Exercise 1 are repeated below.

Newer homes: 600, 590, 605, 600, 603, 610, 597, and 595.
Older homes: 810, 750, 790, 800, and 850.

a. What are the variance and the standard deviation for the newer homes?
b. What are the variance and the standard deviation for the older homes?
c. Compare the dispersion in the consumption of fuel oil for the newer and older homes.

8. The lengths of trout fingerlings stocked in two ponds are repeated here from Exercise 2. Consider them as sample data.

Length (in inches)	
Pond A	Pond B
12.5	18.0
14.0	20.0
13.5	12.0
14.5	14.5
15.0	19.0
16.5	13.5
16.0	14.5
12.0	13.0
14.0	20.5
14.0	14.0

a. Determine the variance for both sets of lengths.
b. Determine the standard deviation for both sets.
c. Write an analysis of your findings with respect to the variation in the lengths of the fish in the two ponds.

9. The number of customers entering Romaker's Boutique each day for a six-day period as reported in Exercise 3 was 91, 89, 88, 58, 63, and 74. Consider these to be sample data, not the entire population. Compute the variance and the standard deviation.

10. The interest rates from the tax-exempt bonds of Exercise 4 were 9.75, 3.75, 5.50, 7.30, 6.00, 7.30, 7.25, 4.25, 3.30, and 6.60. Consider these to be sample data, not the entire population. Compute the variance and the standard deviation.

11. Refer to Exercise 6. Assume that the infant mortality rates are populations—that is, all the blacks and whites born were included in the mortality rates.
a. Determine the variance and the standard deviation for each set of mortality rates.
b. Compare the dispersion of the mortality rates for the two groups.

12. Refer to Exercise 5. Assume that the selected cities refer to a population. Compute the variance and the standard deviation.

COMPUTING THE VARIANCE (σ^2)

Unless the mean is an integer—which is usually not the case—subtracting each observation from the mean and then squaring the differences to arrive at the variance tends to be rather laborious. Rounding errors can also become a problem. A more convenient computational formula for both the variance and standard deviation is based directly on the raw data. The variance formula is:

Population

$$\sigma^2 = \frac{\sum X^2 - \frac{(\sum X)^2}{N}}{N} \tag{4-7}$$

Sample

$$s^2 = \frac{\sum X^2 - \frac{(\sum X)^2}{n}}{n-1} \tag{4-8}$$

As before, we still derive the standard deviation (σ for a population, s for a sample) by computing the square root of the variance.

Note: ΣX^2 directs us to first square each observation and then sum the squares. This will not yield the same result as $(\Sigma X)^2$, which directs us to sum the numbers first and then square their total.

How do we know this computational formula will give us the same answer as the conceptual formula? Here is the proof.

$$\text{Recall that}: \mu = \frac{\sum X}{N}$$

$$\sigma^2 = \frac{\sum (X-\mu)^2}{N} = \frac{\sum \left(X - \frac{(\sum X)}{N}\right)^2}{N}$$

$$= \frac{\sum \left(X^2 - \frac{2X\sum X}{N} + \frac{(\sum X)^2}{N^2}\right)}{N}$$

$$= \frac{\sum X^2 - \frac{2(\sum X)^2}{N} + \frac{(\sum X)^2}{N}}{N}$$

$$\text{Therefore}: \sigma^2 = \frac{\sum X^2 - \frac{(\sum X)^2}{N}}{N}$$

Problem The number of days all the patients currently in St. Luke Hospital's North Wing have been confined are 6, 4, 5, 3, and 4 days. Compute the standard deviation of the lengths of confinement using the deviation method and the computational method.

Solution First, note that the five observations are a population because they include all patients currently in the North Wing. The formulas for populations, 4–5 and 4–7, will therefore be used.

Method Using Deviations from the Mean (μ = 22/5 = 4.4)

X	$X - \mu$	$(X - \mu)^2$
6	1.6	2.56
4	−0.4	0.16
5	0.6	0.36
3	−1.4	1.96
4	−0.4	0.16
22	0.0	5.20

$$\sigma^2 = \frac{\Sigma(X-\mu)^2}{N}$$

$$\sigma^2 = \frac{5.20}{5}$$

$$\sigma^2 = 1.04$$

$$\sigma = \sqrt{1.04} = 1.0198 \text{ days}$$

Method Using Raw Data Only

X	X^2
6	36
4	16
5	25
3	9
4	16
22	102

$$\sigma^2 = \frac{\Sigma X^2 - \frac{(\Sigma X)^2}{N}}{N}$$

$$\sigma^2 = \frac{102 - \frac{(22)^2}{5}}{5}$$

$$\sigma^2 = \frac{102 - 96.8}{5} = 1.04$$

$$\sigma = \sqrt{1.04} = 1.0198 \text{ days}$$

Because the variance and the standard deviation use every observation in their computation, they are more reliable measures of spread than the range. Two or more variances (or standard deviations) can be used to compare the variability around their respective means. If the observations are clustered close to the mean, the standard deviation will be small. As the observations become more dispersed from the mean, the variance and standard deviation become larger.

It should be noted that if the raw data have a few extreme values, the variance and standard deviation may not be representative of the dispersion in the data. In such cases, the deviations from the mean will be large, and squaring these deviations will result in an unusually large variance and standard deviation. The standard deviation is especially valuable in sampling theory and statistical inference, topics that will be discussed starting with Chapter 9. The variance will be used extensively in Chapter 12.

Problems Consider the following sets of data as populations. Determine the variance and the standard deviation.

a. 3, 5, 4
b. 2, 5
c. 4, 2, 1, 3

Solutions

	Variance	Standard Deviation
a.	0.67	0.82
b.	2.25	1.50
c.	1.25	1.12

Self-Review 4-4

Six families live on Merrimac Circle. The number of children in each family is 1, 2, 3, 5, 3, and 4.

a. Is this a sample or a population?
b. Apply formula 4–5, which is based on the deviations from the mean, and compute the variance and the standard deviation.
c. As a check, compute the variance and the standard deviation using formula 4–7, which is based on the raw data (the number of children in the family).

Exercises

13. A study was made of the length of time it takes ambulances to travel from the scene of the accident to the hospital when responding to emergency calls in Zone A. The times in minutes were 6.1, 5.9, 4.8, 10.2, 9.6, and 6.1.
   a. Using the raw data only (not the deviations from the mean), determine the variance. Assume the data represent a population.
   b. Find the standard deviation.

14. In Exercise 8, you computed the variance and the standard deviation of the fingerling trout data using deviations from the mean. Verify your calculations using the raw data only. Consider these to be sample data, not the entire population.

15. In Exercise 9, you computed the variance and the standard deviation using the deviation method of the number of customers entering Romaker's Boutique. Verify your calculations using the raw data. Consider these data to be a sample, not the entire population.

16. In Exercise 10, you computed the variance and standard deviation using the deviations from the mean. Consider this to be sample information, not the entire population, and compute the variance and standard deviation using the formula using the raw data.

MEASURES OF DISPERSION FOR DATA IN A FREQUENCY DISTRIBUTION

The Range

We estimate the **range** for data in a frequency distribution by finding the difference between the highest true limit and the lowest true limit. Note that if there are any open-ended classes in the frequency distribution, the range cannot be computed.

Problem Table 4–3 shows the weekly amounts that a sample of young newlyweds spend on food. What is the range?

Table 4-3

Weekly Amounts Spent on Food by a Group of Newlyweds

Amount Spent	Number
$40–44	4
45–49	11
50–54	20
55–59	31
60–64	19
65–69	11
70–74	4

Solution The highest true limit is $74.50; the lowest is $39.50. We find the range, $35, by subtracting $39.50 from $74.50.

Self-Review 4-5

A fast-food chain received several complaints about the weight of their new hamburger, "The 125 Gramburger." A check of a sample of hamburgers revealed these weights:

Weight (in grams)	Number
116–118	7
119–121	19
122–124	28
125–127	16
128–130	2

Determine the range for the weights.

The Interquartile Range and The Quartile Deviation

Recall that the **interquartile range** and the **quartile deviation** are based on the distance between the third quartile (Q_3) and the first quartile (Q_1).

$$\textbf{Interquartile range} = Q_3 - Q_1 \tag{4-2}$$

$$\textbf{Quartile deviation} = \frac{Q_3 - Q_1}{2} \tag{4-3}$$

Problem Table 4–4 reintroduces the familiar frequency distribution of the selling prices of homes in the Rafter J Ranch subdivision. Determine the interquartile range and the quartile deviation of the selling prices.

Table 4-4

Selling Prices of Homes in the Rafter J Ranch Subdivision

True Limits	**Class Frequency**	**Cumulative Frequency**
Selling Price	Number of Homes (f)	CF
$ 79,500–84,500	5	5
84,500–89,500	10	15
89,500–94,500	15	30
94,500–99,500	26	56
99,500–104,500	13	69
104,500–109,500	7	76
109,500–114,500	4	80
	Total 80	

first quartile → (89,500–94,500 row)

STAT BYTE

In 1970, women purchased 23% of all new cars and trucks. By 1991, this number had risen to 50%, and is expected to be about 60% by the year 2000. Safety and dependability are two of the most important vehicle features to women. Women also usually visit three or four dealers before making a decision, and tend to be more loyal to a particular brand. Price is a major factor in a woman's decision to purchase, but the salesperson's attitude is also important.

Solution The computations for the first and third quartiles are similar to those for the median (which is the second quartile, Q_2), as discussed in Chapter 3. The procedure for locating the first quartile, Q_1, is as follows:

Step 1. Determine how many observations are in the first quarter. Since there are 80 selling prices, one-fourth, or 20, lie in the first quarter.

Step 2. Locate the class in which the first quartile lies. Note that the cumulative column of Table 4–4 contains 15 selling prices in the first two classes and 30 in the first three classes. Clearly, the first quartile is in the third class because 20 is larger than 15 but smaller than 30.

Step 3. Using the true class limits, we have located the first quartile in the \$89,500 – \$94,500 class. We must determine how far into that class to go to find the first quartile. We move five frequencies into that class, found by 20 – 15. Since there are 15 selling prices in that class, the first quartile is five-fifteenths of the way between \$89,500 and \$94,500, or \$91,166.67. Q_1 is found as follows:

$$Q_1 = \$89{,}500 + (5/15)(\$5{,}000) = \$91{,}166.67$$

We can also find the *first quartile*, Q_1, by using the formula

$$Q_1 = L + \frac{\frac{n}{4} - CF}{f}(i) \qquad \textbf{4-9}$$

where

- L is the lower true limit of the class containing the first quartile, Q_1. It is in the \$89,500–\$94,500 class, and \$89,500 is the lower true limit of that class.
- n is the total number of frequencies. There are 80 selling prices.
- CF is the cumulative number of frequencies in all of the classes preceding the class in which the first quartile lies. There are 15 cumulative frequencies in the class preceding the \$89,500–\$94,500 class.
- f is the frequency in the first quartile class. There are 15 prices in that class.
- i is the width of the class in which the first quartile falls. The width of the \$89,500–\$94,500 class is \$5,000.

Solving for Q_1 using formula 4–9:

$$\begin{aligned} Q_1 &= L + \frac{\frac{n}{4} - CF}{f}(i) \\ &= \$89{,}500 + \frac{\frac{80}{4} - 15}{15}(\$5{,}000) \\ &= \$91{,}166.67 \end{aligned}$$

The formula for the *third quartile*, Q_3, is:

$$Q_3 = L + \frac{\frac{3n}{4} - CF}{f}(i) \qquad \text{4-10}$$

The procedure is the same as for the first quartile, except that *L, CF, f,* and *i* refer to the values needed for the third quartile.

Three-fourths of 80 is 60. The sixtieth selling price is in the \$99,500–\$104,500 class. That class contains 13 observations. A total of 56 observations (*CF*) have accrued prior to the class containing the third quartile. Hence the sixtieth observation is 4 observations into the \$99,500 – \$104,500 class, found by 60 – 56. The class interval (*i*) is \$5,000. The lower true limit of the class containing the third quartile is \$99,500. Substituting these values in formula 4 – 10, we obtain:

$$\begin{aligned} Q_3 &= L + \frac{\frac{3n}{4} - CF}{f}(i) \\ &= \$99{,}500 + \frac{\frac{3(80)}{4} - 56}{13}(\$5{,}000) \\ &= \$99{,}500 + \frac{4}{13}(\$5{,}000) \\ &= \$101{,}038.46 \end{aligned}$$

The interquartile range is \$9,871.79, found by $Q_3 - Q_1 =$ \$101,038.46 – \$91,166.67. The quartile deviation is \$4,935.90, found by $(Q_3 - Q_1)/2$.

Self-Review 4-6

In Self-Review 4–5, a fast-food chain had received complaints about the weight of "The 125-Gramburger." A check of 72 hamburgers had revealed these weights:

Weight (in grams)	Number
116–118	7
119–121	19
122–124	28
125–127	16
128–130	2

a. Determine the first quartile (Q_1).
b. Find the third quartile (Q_3).

c. What are the interquartile range and the quartile deviation (*QD*)?

Exercises

17. Using the following table on the distribution of per capita state taxes for our 50 states during a recent year, answer the following questions. (Be sure to use true limits.)
    a. Determine the first quartile. Explain what the first quartile indicates.
    b. Determine the third quartile. Explain what the third quartile indicates.
    c. What is the interquartile range?
    d. What is the quartile deviation?

Per Capita Tax	Frequency
$375–449	6
450–524	15
525–599	10
600–674	6
675–749	9
750–824	4
	50

18. A hospital administrator compared the number of emergency admissions on a given Monday with those on the Friday of the same week:

Emergency Admissions	Number	
	Monday	Friday
4–7	1	1
8–11	4	4
12–15	15	21
16–19	26	22
20–23	16	13
24–27	7	3
28–31	3	0
	72	64

    a. Calculate the quartile deviation for each set of data. (Be sure to use the true limits.)
    b. Compare the dispersion in the Monday and Friday admissions.

19. The owner of a local movie theater that shows R-rated films tabulated a sample of the ages of customers attending yesterday's showings:

Age	Frequency
18–22	15
23–27	33
28–32	45
33–37	26
38–42	13
43–47	8

a. Compute the first and third quartiles.
b. Compute the quartile deviation.

20. The owner of the Kinzua Inn is studying the occupancy rate at his motel. The Inn is near a hunting, fishing, and camping area, so the number of guests is much higher during the summer months. There are a total of 92 nights in the months of June, July, and August. The following frequency distribution shows the number of guests per night. Consider these data to be a sample, not the entire population.

Guests	Frequency
at least 50 but fewer than 60	5
at least 60 but fewer than 70	12
at least 70 but fewer than 80	23
at least 80 but fewer than 90	28
at least 90 but fewer than 100	17
at least 100 but fewer than 110	7

a. Determine the first and third quartiles.
b. Compute the quartile deviation.

The Standard Deviation

If the data are grouped into a frequency distribution, the standard deviation for both samples and populations may be computed as follows:

Population

$$\sigma = \sqrt{\frac{\Sigma f X^2 - \frac{(\Sigma f X)^2}{N}}{N}} \quad \text{(4-11)}$$

Sample

$$s = \sqrt{\frac{\Sigma f X^2 - \frac{(\Sigma f X)^2}{n}}{n-1}} \qquad \textbf{4-12}$$

where X represents the midpoint of each class and f the number of observations in each class. As usual, n is the sample size and N the population size.

Problem Compute the standard deviation and the variance for the distribution of selling prices of the homes in the Rafter J Ranch subdivision. Assume that this information is sample data.

Solution The data and the needed calculations are shown in Table 4–5.

Using Formula 4–12 to compute the standard deviation of the sample of 80 selling prices (in thousand of dollars) we obtain:

$$s = \sqrt{\frac{\Sigma f X^2 - \frac{(\Sigma f X)^2}{n}}{n-1}}$$

$$= \sqrt{\frac{746,475 - \frac{(\$7,705)^2}{80}}{80-1}} = \$7.452$$

The sample variance, s^2, is 55.534, found by $(\$7.452)^2$.

Table 4-5

Calculations Needed for the Standard Deviation

True Limits Selling Price	**Class Frequency** Number of Homes (f)	X	fX	fX^2
$ 79,500–84,500	5	$ 82	$ 410	33,620 ← $82 × $410
84,500–89,500	10	87	870	75,690
89,500–94,500	15	92	1380	126,960
94,500–99,500	26	97	2522	244,634
99,500–104,500	13	102	1326	135,252
104,500–109,500	7	107	749	80,143
109,500–114,500	4	112	448	50,176
Total	80		$7705	746,475

Self-Review 4-7

The hourly wages, excluding tips, of a sample of restaurant employees are:

Hourly Wages	Number
$ 1–3	4
4–6	10
7–9	20
10–12	11
13–15	5

a. Determine the standard deviation.
b. Compute the variance.

Exercises

21. The per capita tax for the 50 states is repeated from Exercise 17 in the following table:

Per Capita Tax	Frequency
$375–449	6
450–524	15
525–599	10
600–674	6
675–749	9
750–824	4
	50

a. Is this a sample or a population?
b. Compute the standard deviation.
c. Determine the variance.

22. The numbers of Monday and Friday emergency hospital admissions are repeated below from Exercise 18.

Emergency Admissions	Number	
	Monday	Friday
4–7	1	1
8–11	4	4
12–15	15	21
16–19	26	22
20–23	16	13
24–27	7	3
28–31	3	0
	72	64

a. Calculate both the standard deviation and the variance for each set of data.

b. Compare the dispersion in the Monday and Friday emergency admissions.

23. The ages of the customers at a local theater are repeated below from Exercise 19.

Age	Frequency
18–22	15
23–27	33
28–32	45
33–37	26
38–42	13
43–47	8

Compute the standard deviation.

24. The occupancy rates of the Kinzua Inn are repeated below from Exercise 20.

Guests	Frequency
at least 50 but fewer than 60	5
at least 60 but fewer than 70	12
at least 70 but fewer than 80	23
at least 80 but fewer than 90	28
at least 90 but fewer than 100	17
at least 100 but fewer than 110	7

Calculate the standard deviation.

RELATIVE DISPERSION

A *direct* comparison of two or more measures of dispersion may result in an incorrect conclusion regarding the variation in the data. Also, if one distribution, for example, were in dollars and the other in meters, clearly it would be impossible to compare them. Therefore, it is preferable to use a *relative* measure of dispersion when either

1. the means of the distributions being compared are far apart, or
2. the data are in different units.

One such relative measure of dispersion is known as the **coefficient of variation,** abbreviated *CV*.

Coefficient of Variation The standard deviation divided by the mean.

In order to express the coefficient of variation as a percent, we multiply $\frac{s}{\overline{X}}$ (for a sample) by 100.

$$CV = \frac{s}{\overline{X}}(100) \qquad \text{4-13}$$

Problem The mean income of a sample of homeowners in Precinct 12 is \$40,000 and the standard deviation is \$4,000. In Precinct 9, the mean income of a sample of homeowners is \$24,000, and the standard deviation is \$2,400. Note that the means are far apart and the standard deviations are considerably different. Compare and interpret the relative dispersion in the two groups of incomes.

Solution The first impulse is to say there is more dispersion in the incomes in Precinct 12, because the standard deviation associated with that group (\$4,000) is greater than the \$2,400 for the Precinct 9 incomes. However, converting the two sets of measurement to relative terms, and using the coefficient of variation, we would find that the relative dispersion is the same! Here are the calculations for the coefficients of variation:

Precinct 12

$$CV = \frac{s}{\overline{X}}(100) = \frac{\$4,000}{\$40,000}(100) = 10\%$$

Precinct 9

$$CV = \frac{s}{\overline{X}}(100) = \frac{\$2,400}{\$24,000}(100) = 10\%$$

Interpreting the results: The incomes in both precincts are spread out 10% from their respective means.

Problem To illustrate the use of the coefficient of variation when two or more distributions are in different units, we will now compare the dispersion in the ages of the homeowners in Precinct 12 with the dispersion in their incomes. The coefficient of variation allows each unlike set of data to be converted to a common denominator (a percent).

The mean age of the sample of homeowners is 40 years, the standard deviation is 10 years. Recall that for their incomes $\overline{X}$ = \$40,000 and the standard deviation is \$4,000. Compare the dispersion in their ages and incomes.

Solution

$$\text{Income}: CV = \frac{s}{\overline{X}}(100) = \frac{\$4,000}{\$40,000}(100) = 10\%$$

$$\text{Age}: CV = \frac{s}{\overline{X}}(100) = \frac{10}{40}(100) = 25\%$$

There is greater relative dispersion in the ages of the homeowners in Precinct 12 than in their incomes (because 25% is greater than 10%).

Problems Determine the coefficient of variation for each mean and standard deviation.

a. $\overline{X} = 25, s = 5$
b. $\overline{X} = 8, s = 2$
c. $\overline{X} = 50, s = 2$
d. $\overline{X} = 1650, s = 148.5$

Solutions

a. 20%
b. 25%
c. 4%
d. 9%

Self-Review 4-8

A sample of Harber High School seniors has recently completed two aptitude tests, one dealing with mechanical aptitude, the other with aptitude for social work. The results are:

a. Mechanical aptitude: mean 200, standard deviation 30.
b. Aptitude for social work: mean 500, standard deviation 40.

Compare the relative dispersion in the test results.

Exercises

25. An investor is considering the purchase of one of two stocks. The yield of Venture Electronics has averaged $105 per share over the past 10 years with a standard deviation of $15 per share. Aerospace Ltd. has yielded an average of $330 per share during the same period with a standard deviation of $40. Compare the relative dispersion of the two stocks.

26. A recent study of Ohio College of Business faculty revealed that the arithmetic mean salary for nine months is $31,000 and the standard deviation of the sample is $3,000. The study also showed that the faculty had been employed an average (arithmetic mean) of 15 years with a standard deviation of 4 years.

How does the relative dispersion in the distribution of salaries compare with that of the lengths of service?

27. Radio commercials on the local rock station WNAE average 35 seconds with a standard deviation of 8 seconds. WLQR, the local easy-listening station, has a mean commercial length of 30 seconds with a standard deviation of 5 seconds. Compare the relative dispersion of the two groups.

28. A large insurance company offers both homeowner and auto coverage. A study of last year's claims showed that the mean claim settlement for homeowner claims was $1,260 with a standard deviation of $425. The mean claim for an auto policy was $875 with a standard deviation of $300. Compare the relative distributions of the two types of claims.

THE COEFFICIENT OF SKEWNESS

An average describes the central tendency of a set of observations, while a measure of dispersion describes the variation in the data. The degree of **skewness** in a distribution can also be measured. Skewness shows the lack of symmetry in a set of data. Recall from Chapter 3 that if the mean, median, and mode are equal, there is no skewness (see Figure 4–3). If the mean is larger than the median and the mode, the distribution is said to have **positive skewness** (see Figure 4–4). If the mean is the smallest of the three averages, the distribution is **negatively skewed** (see Figure 4–5).

The degree of skewness is measured by the **coefficient of skewness,** abbreviated *sk*. It is found by:

$$sk = \frac{3(\text{Mean} - \text{Median})}{\text{Standard deviation}} \qquad \textbf{4-14}$$

Usually, its value ranges from –3 to +3.

Problem What is the coefficient of skewness for the distribution of ages in Figure 4–4?

Solution Using formula 4–14:

$$sk = \frac{3(\text{Mean} - \text{Median})}{\text{Standard deviation}} = \frac{3(22-21)}{5}$$
$$= +0.6$$

This indicates that there is a slight positive skewness in the age distribution.

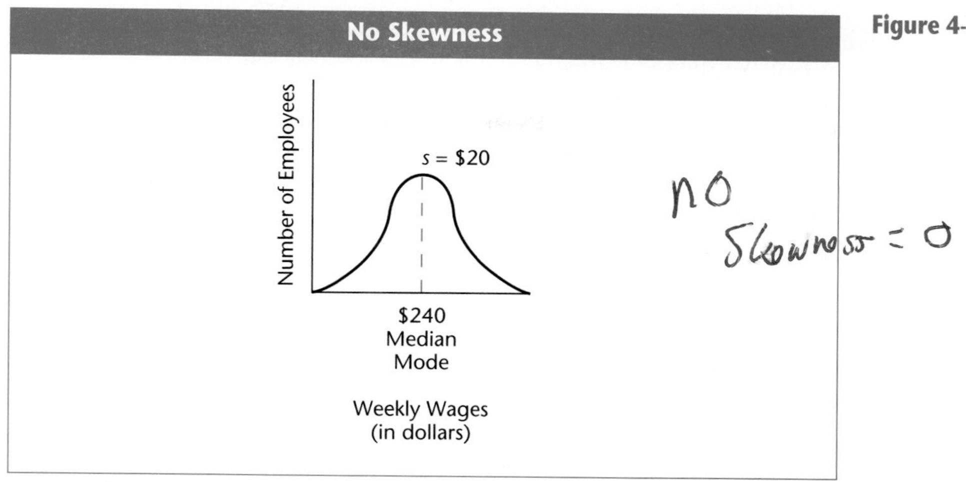

Figure 4-3

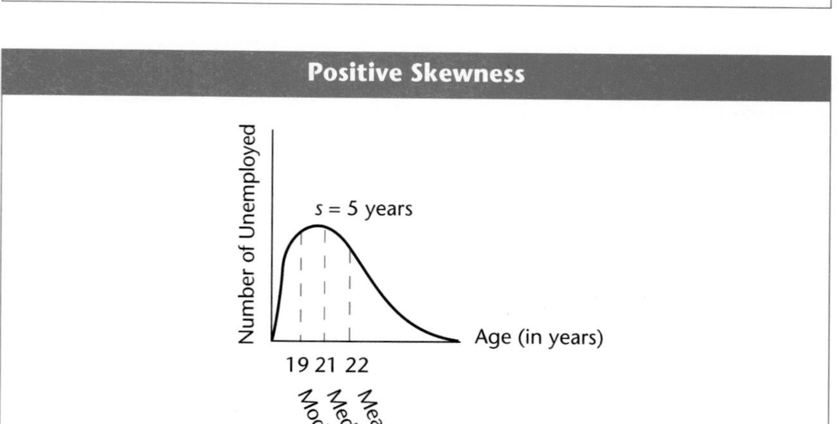

Figure 4-4

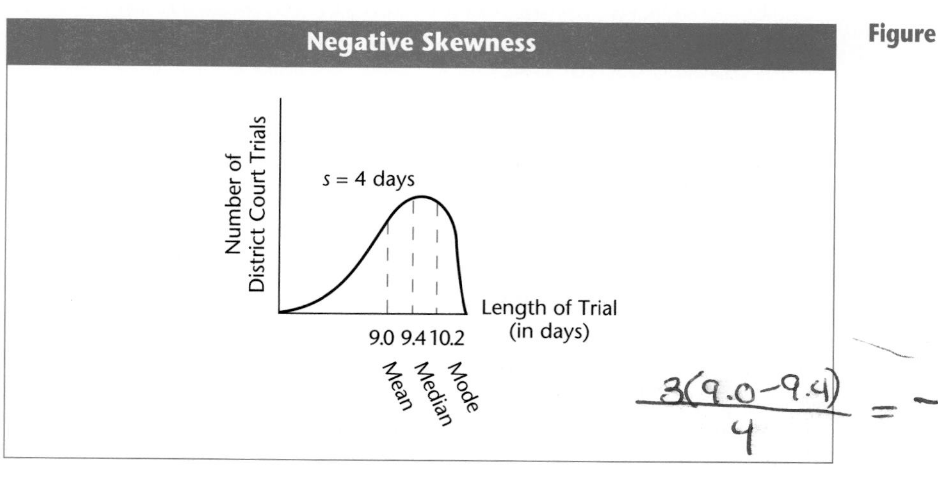

Figure 4-5

Problems Determine the coefficient of skewness for each of the following sets of data:

a. $\overline{X} = 16$, Median = 15, Mode = 13, $s = 2$
b. $\overline{X} = \$800$, Median = \$820, Mode = 830, $s = \$50$
c. $\overline{X} = 70$, Median = 70, Mode = 70, $s = 8$

Solutions

a. 1.5
b. –1.2
c. 0

Self-Review 4-9

a. Determine the coefficient of skewness for Figure 4–5.
b. Determine the coefficient of skewness for Figure 4–3.
c. Interpret the two coefficients.

Exercises

29. The Flightdeck computed the arithmetic-mean dinner check for two persons to be \$54.00, the median \$50.50. The standard deviation was \$3.75. What is the coefficient of skewness? Describe the skewness.

30. The research director of a large oil company conducted a study of consumer buying habits with respect to the amount of gasoline purchased at self-service pumps. The arithmetic-mean amount was 11.5 gallons, and the median amount was 11.95 gallons. The standard deviation of the sample was 4.5 gallons. Is this distribution negatively skewed, positively skewed, or symmetrical? What is the coefficient of skewness?

31. At Lander Community College, the mean age of the students is 19.2 years with a standard deviation of 1.2 years. The median age is 18.6 years. Compute the coefficient of skewness. Describe the skewness.

32. A survey of students at Willard University revealed that the mean checking account balance was \$376 with a standard deviation of \$120. The median balance was \$406. Compute the coefficient of skewness. Describe the skewness.

A COMPUTER EXAMPLE

The work of tallying raw data into a frequency distribution (Chapter 2), computing measures of central tendency (Chapter 3), and finding measures of dispersion (this chapter) can be accomplished quickly by a computer. As noted previously, a number of computer "packages" are used extensively in the social sciences, business, and education. Check with your computing center to determine which statistical packages are available for your use. Computer packages are generally easy to use and will save you a great deal of time.

MINITAB was used to generate the following frequency distribution for the Rafter J Ranch data. The computer has removed the tedious steps of tallying the raw data into the frequency distribution. The variable name used is 'Price.'

```
MTB  >   'Price';
SUBC >   increment = 5;
SUBC >   start 82.

Histogram of Price   N = 80

Midpoint   Count
   82.00       5   *****
   87.00      10   **********
   92.00      15   ***************
   97.00      26   **************************
  102.00      13   *************
  107.00       7   *******
  112.00       4   ****
```

MINITAB was also used to compute the following summary measures for the data that appeared originally in Table 2–1. Again, the variable name is 'Price.'

```
MTB >     describe 'Price'

               N      MEAN    MEDIAN    TRMEAN    STDEV    SEMEAN
Price         80    96.488    97.000    96.361    7.552     0.844

             MIN       MAX        Q1        Q3
Price     81.000   114.000    91.000   101.000
```

SUMMARY

A measure of central tendency pinpoints the center of a set of data. To describe the variability in the data, we can use the range, interquartile range, quartile deviation, mean deviation, variance, or standard deviation.

The range is the difference between the highest and lowest values. Caution should be exercised in applying the range, because one extremely high (or low) observation may cause it to be an unreliable measure of spread. The interquartile range and the quartile deviation are based on the first quartile and the third quartile. Hence, the middle 50% of the observations are used. One of the weaknesses of these two measures, and of the range, is that they do not include all the observations. By contrast, the mean deviation, the variance, and the standard deviation all use every available value to measure the scatter, or dispersion, about the mean. In its computation, the mean deviation ignores the algebraic signs of the deviation about the mean. Both the variance and the standard deviation use the square of each difference from the arithmetic mean in their computation. The standard deviation is especially useful in statistical inference, to be discussed in the forthcoming chapters.

A direct comparison of the dispersion in distributions that have means that are far apart, or have means that are measured in different units, may be misleading. To correct for this, we can divide the standard deviations by their respective arithmetic means, resulting in a relative measure of dispersion called the coefficient of variation. Since its outcome is in percents, even dissimilar distributions can be compared to assess the variation.

The coefficient of skewness measures the degree to which the distribution is not symmetrical. A coefficient of zero indicates that the frequency distribution is symmetrical—that is, it is neither negatively nor positively skewed. The coefficient usually varies between –3 (negative skewness) and +3 (positive skewness).

Exercises

33. The ages for a sample of 10 women who have just given birth to their first child are 18, 25, 31, 19, 22, 21, 19, 25, 16, and 27. Compute each of the following descriptive measures:

a. The range
b. The average deviation
c. The variance
d. The standard deviation
e. The coefficient of skewness

34. The number of yards gained rushing by the Redskins football team during the first seven games of the season are 210, 203, 162, 134, 390, 184, and 211. These data constitute a population. Compute the following descriptive measures:

a. The range. Interpret it.
b. The average deviation. Interpret it.
c. The variance
d. The standard deviation
e. The coefficient of skewness. Interpret it.

35. The percents of minorities enrolled in 12 New Orleans high schools are 32, 62, 58, 26, 61, 34, 46, 27, 56, 61, 64, and 24. Consider the data to be a population, and find the following descriptive measures:
a. The range. Interpret it.
b. The average deviation. Interpret it.
c. The variance
d. The standard deviation
e. The coefficient of skewness. Interpret it.

36. The distribution of the number of children under 18 years of age in a sample of 40 families selected at random is shown below:

Number of Siblings	Frequency
0	5
1	11
2	9
3	5
4	5
5	0
6	4
7	1

Find the following descriptive measures:
a. The range. Interpret it.
b. The variance
c. The standard deviation
d. The quartile deviation. Interpret it.
e. The coefficient of skewness. Interpret it.

37. The United States was divided into 25 statistical regions, and the percent of households in which a female is the head of the household was determined for each region. These data were organized into the following frequency distribution:

Percent of Households	Frequency
5–9	5
10–14	7
15–19	9
20–24	3
25–29	1

Consider these data a population, and determine the following descriptive measures:

a. The range. Interpret it.
b. The variance
c. The standard deviation
d. The quartile deviation. Interpret it.
e. The coefficient of skewness. Interpret it.

38. Two hundred physically disabled veterans are enrolled in a physical fitness program. As an initial test of strength, everyone has been asked to lift a weight. The maximum weight each person lifted has been recorded and organized into the following frequency distribution:

Weight (in pounds)	Number of Persons
0–4	10
5–9	37
10–14	50
15–19	68
20–24	30
25–29	5

a. Determine the range. Interpret it.
b. Find the interquartile range and the quartile deviation. Interpret them.
c. Find the standard deviation and the variance.

39. The pay for 1,052 top executives in 396 of the largest U.S. corporations is as follows:

12 earned less than $100,000
213 earned $100,000 to $200,000
300 earned $200,000 to $300,000
241 earned $300,000 to $400,000
135 earned $400,000 to $500,000
76 earned $500,000 to $600,000
34 earned $600,000 to $700,000
41 earned $700,000 to $1,600,000

a. What is the range?
b. What is the quartile deviation?
c. What characteristic of the income distribution makes it difficult to compute the variance, the standard deviation, and the coefficient of skewness?

40. A sample of light trucks that use diesel fuel revealed the following miles per gallon:

Mileage	Number of Trucks
11–13	3
14–16	5
17–19	12
20–22	8
23–25	4

a. Determine the first and third quartiles.
b. Compute the quartile deviation.
c. Compute the standard deviation.

41. A sample of 50 listeners to WRQN was obtained. Data on their age were organized into the following frequency distribution. Compute the standard deviation.

Age	Frequency
15–19	10
20–24	12
25–29	14
30–34	9
35–39	5

42. Last month, a total of 60 homes were sold within the Rossford city limits. The selling prices were organized into the following frequency distribution. Compute the standard deviation. Since all Rossford homes are considered, this is a population.

Selling Price (in thousands of dollars)	Frequency
$40.0–49.0	3
50.0–59.0	6
60.0–69.0	19
70.0–79.0	23
80.0–89.0	9
Total	60

43. U.S. production of lignite and anthracite coal (in millions of short tons) during the past 10 years was as follows:

Lignite: 47.2, 50.7, 52.4, 58.3, 63.1, 72.4, 76.4, 78.4, 85.1, 89.6
Anthracite: 6.1, 5.4, 4.6, 4.1, 4.2, 4.7, 4.3, 3.6, 3.8, 3.5

a. To analyze and compare the production of the two types of coals, compute such measures as the coefficient of variation, the coefficient of skewness, and so on.

b. Write a brief comparison of lignite and anthracite production during the past 10 years.

44. The Bureau of Labor Statistics gathers data monthly on the prices paid by consumers for food, entertainment, and so on. The Consumer Price Index (CPI) measures the change in prices of goods and services purchased by all urban consumers. The percent change in prices for food, new cars, and medical care during the last 12 years was as follows:

Food: 6.3, 9.9, 11.0, 8.6, 7.8, 4.1, 2.1, 2.3, 3.2, 4.1, 4.1, 5.8
New cars: 5.2, 7.7, 7.9, 8.1, 6.0, 3.9, 2.6, 3.2, 4.2, 3.6, 2.0, 2.0
Medical care: 9.6, 8.4, 9.2, 11.0, 10.7, 11.6, 8.8, 6.3, 7.5, 6.6, 6.5, 7.7

a. To analyze the changes in price for these three consumer goods, compute various measures of central tendency and measures of dispersion.

b. Write a brief analysis of your findings.

45. The Bureau of Labor Statistics publishes the average annual salary, by state. A random sample of the 50 states revealed these annual incomes:

$19,593	Alabama	$17,694	Nebraska
27,500	Connecticut	19,320	North Carolina
18,148	Idaho	22,312	Pennsylvania
19,202	Maryland	21,740	Texas
17,047	Mississippi	20,024	Wisconsin

a. What is the arithmetic-mean annual income?
b. What is the median annual income?
c. What is the standard deviation?
d. What is the coefficient of skewness?
e. What is the coefficient of variation?

Use the following data for Exercises 46 and 47. The National Institute on Drug Abuse studies the drug usage of American high school students, by class. The percent of students who used drugs is shown below for a sample of classes from 1975 to 1989.

	Percent Who Used				
Class of	Marijuana/Hashish	LSD	Cocaine	Crack	Heroin
1975	47.3	11.3	9.0	NA	2.2
1978	59.2	9.7	12.9	NA	1.6
1980	60.3	9.3	15.7	NA	1.1
1983	57.0	8.9	16.2	NA	1.2
1984	54.9	8.0	16.1	NA	1.3
1985	54.2	7.5	17.3	NA	1.2
1986	50.9	7.2	16.9	NA	1.1
1987	50.2	8.4	15.2	5.6	1.2
1988	47.2	7.7	12.2	4.8	1.1
1989	43.7	8.3	10.3	4.7	1.3

46. Compare the usage of marijuana/hashish and LSD using selected measures of central tendency and dispersion. Write a brief analysis of your findings.

47. Compare the usage of cocaine, crack, and heroin using selected measures of central tendency and dispersion. Write a brief analysis of your findings.

48. The National Center for Health Statistics cited the following number of births to unmarried women, by age of the mother, for 1992:

Age of Mother	Number
10–14	9,907
15–19	312,499
20–24	350,905
25–29	196,365
30–34	94,874
35–39	34,408
40–44	6,341

a. Determine the interquartile range.
b. What is the first quartile? Interpret it.
c. What is the third quartile? Interpret it.
d. Compute the quartile deviation. Interpret it.
e. What is the range? Interpret it.

49. The National Center for Health Statistics reported these figures on the use of family planning services:

Age	Number (in thousands) White	Black
15–19	7,313	1,409
20–24	7,401	1,364
25–29	8,672	1,459
30–34	9,010	1,406
35–39	7,936	1,170
40–44	6,745	872

a. What is the range for both groups?
b. What is the first quartile for both groups? The third quartile?
c. What is the interquartile range for both groups?

Data Exercises

50. Refer to the Real Estate Data Set in the appendix, which reports information on homes sold in Northwest Ohio during 1992.

a. What is the standard deviation of the selling price? Using the mean and median computed earlier, determine the coefficient of variation and the coefficient of skewness.

b. Determine the mean, median, and the standard deviation of the distance from the center of the city. Determine the coefficient of variation and the coefficient of skewness.

51. Refer to the Salary Data Set in the appendix, which refers to a sample of middle managers in Sarasota, Florida.

a. Determine the standard deviation of the salaries. Using the mean and median computed earlier, determine the coefficient of variation and the coefficient of skewness.

b. Find the mean, median, and the standard deviation of the number of employees supervised. Compute the coefficient of variation and the coefficient of skewness.

CASE: WHICH SUMMARY STATISTICS TO REPORT?

The local chapter of the American Association of Retired Persons (AARP) is conducting a study of monthly medication expenses for senior citizens. Helen Boggart, who lives in the Sandlewood subdivision, sampled 220 of her neighbors. Their expenses averaged \$112.50 per month with a standard deviation of \$23.00. Sam Steward, who resides in Venice Isles, reports the following results for his subdivision:

Monthly Expense	Frequency
\$ 0 up to \$ 25	5
25 up to 50	12
50 up to 75	18
75 up to 100	23
100 up to 125	17
125 up to 150	12
150 up to 175	13
175 up to 200	6

In order to report to the national office of the AARP, Helen and Sam must aggregate their reports. What numerical measures of central tendency and dispersion would you recommend they report?

CHAPTER ACHIEVEMENT TEST

Answer all the questions. The answers are at the back of the book.

MULTIPLE-CHOICE QUESTIONS. Select the response that best answers each of the questions.

1. The major weakness of the range is that
  a. it does not use all the observations in its calculation.
  b. it can be influenced by an extreme value.
  c. Both a and b are correct.
  d. None of the above

2. The major weakness of the mean deviation is that
  a. it is based on only two observations.
  b. it is influenced by a large mode.
  c. it employs absolute values, which are often difficult to use.
  d. None of the above

3. The major strength of the standard deviation is that
  a. it uses all the observations in its calculation.
  b. it is not unduly influenced by extreme values.
  c. Both a and b are correct.
  d. None of the above

4. The standard deviation is
  a. the square of the variance.
  b. two times the standard deviation.
  c. half the variance.
  d. the square root of the variance.
  e. None of the above

5. If the original data are measured in pounds, the variance is
  a. also measured in pounds.
  b. measured in pounds squared.
  c. measured in "half" pounds.
  d. None of the above

6. The median is equal to the
  a. quartile deviation.
  b. third quartile minus the first quartile.
  c. square of the standard deviation.
  d. second quartile.
  e. None of the above

7. The coefficient of variation is measured in
  a. the same units as the mean and the standard deviation.
  b. percent.
  c. squared units.
  d. None of the above

8. If the mean is larger than the median, the coefficient of skewness will be
  a. zero.
  b. positive.

c. negative.
d. None of the above

9. If the "tail" of a frequency distribution is in the positive direction (to the right), the coefficient of skewness is
a. zero.
b. positive.
c. negative.
d. None of the above.

10. The standard deviation of a frequency distribution is \$10, the mean is \$250, the median is \$250, and the mode is also \$250. The coefficient of skewness is
a. zero.
b. positive.
c. negative.
d. measured in dollars.
e. None of the above

COMPUTATION PROBLEMS

11. Several joggers ran these distances during the day:

Jogger	Distance (in miles)
Peter	2
Jan	6
Betty	3
Doug	4
Bill	5

a. What is the range? Interpret it.
b. What is the mean deviation? Interpret it.
c. What is the variance?
d. What is the standard deviation?.

12. The appraised values for a sample of single-family dwellings in the Jefferson tax district are:

Appraised Value (in thousands)	Number of Dwellings
\$20–29	8
30–39	15
40–49	20
50–59	50
60–69	18
70–79	9

a. Determine the range.

b. Determine the quartile deviation.
c. Determine the variance and the standard deviation.

13. The distribution of the appraised values of the single-family dwellings in the Sanford tax district reveals: mean, \$87,000; median, \$84,000; mode, \$78,000; standard deviation, \$9,000.
a. Compare the relative dispersion in this distribution with that in the Jefferson district in Problem 12.
b. Compare the skewness of the two distributions.

ANSWERS TO SELF-REVIEW PROBLEMS

4–1 a. Sunshine: $\overline{X} = \frac{446}{5} = 89.2$

State: $\overline{X} = \frac{534}{6} = 89.0$

b. Sunshine: 40 km per hour (110–70). State: 7 km per hour (92–85).
c. There is more dispersion in speeds on Sunshine because 40 is greater than 7. Speeds on the state highway clustered closer to the mean because 7 is less than 40.

4–2 a. $\overline{X} = \frac{3{,}000}{6} = 500$

X	$\lvert X-\overline{X} \rvert$
484	$\lvert -16 \rvert$
503	$\lvert +3 \rvert$
496	$\lvert -4 \rvert$
510	$\lvert +10 \rvert$
491	$\lvert -9 \rvert$
516	$\lvert +16 \rvert$
3,000	58

$$MD = \frac{58}{6} = 9.7 \text{ grams}$$

b. On the average, blueberry cakes deviate 9.7 grams from the mean of 500 grams.
c. There is greater variation in the weights of blueberry cakes because 9.7 > 1.6. Peach cakes clustered closer to the mean weight.

4–3 a. $\overline{X} = 3{,}000/6 = 500$ grams

X	$X-\overline{X}$	$(X-\overline{X})^2$
484	−16	256
503	+3	9
496	−4	16
510	+10	100
491	−9	81
516	+16	256
3,000	0	718

$$s^2 = \frac{718}{6-1} = 143.6$$

The variance is 143.6.

$$s = \sqrt{143.6} = 11.98$$

The standard deviation is 11.98 grams.

b. There is greater variation in blueberry cakes because the standard deviation of 11.98 > 4.5

4–4 a. Because there are only six families and all are being studied, this is a population.
b. $\overline{X} = 18/6 = 3$, $\sigma^2 = 10/6 = 1.6667$

$\sigma = \sqrt{1.6667} = 1.29$ children

c.

X	X^2
1	1
2	4
3	9
5	25
3	9
4	16
18	64

$$\sigma^2 = \frac{64 - \frac{(18)^2}{6}}{6} = \frac{64-54}{6}$$

$$= 1.6667$$

$$\sigma = \sqrt{1.6667} = 1.29 \text{ children}$$

4–5 The range is 14 grams, found by 130–116.

4–6

Cumulative Frequencies
7
26
54
70
72

a.

$$Q_1 = 118.5 + \frac{\frac{1}{4}(72)-7}{19}(3)$$

$$= 120.2 \text{ grams}$$

b.

$$Q_3 = 124.5 + \frac{\frac{3}{4}(72)-54}{16}(3)$$

$$= 124.5 \text{ grams}$$

c.

$$Q_3 - Q_1 = 4.3 \text{ grams}$$

$$QD = \frac{4.3}{2} = 2.15 \text{ grams}$$

4–7 a.

f	X	fX	fX^2
4	$ 2	$ 8	16
10	5	50	250
20	8	160	1,280
11	11	121	1,331
5	14	70	980
50		$409	3,857

$$s = \sqrt{\frac{3857 - \frac{(409)^2}{50}}{50-1}}$$

$$s = \sqrt{\frac{511.38}{49}} = \sqrt{10.44}$$

$$s = \$3.23$$

b. The variance is 10.44.

4–8 Mechanical: $\frac{30}{200}(100) = 15\%$

Social work: $\frac{40}{500}(100) = 8\%$

There is a larger relative dispersion in mechanical aptitude (15%) compared with social work aptitude (8%).

4–9 a. -0.3, found by:

$$sk = \frac{3(9.0-9.4)}{4} = -0.3$$

b. 0, found by:

$$sk = \frac{3(\$240-\$240)}{\$20} = 0$$

c. Figure 4–5: slight negative skewness
Figure 4–3: no skewness

UNIT REVIEW

In these first four chapters, you have learned the basic vocabulary of statistics and the fundamentals of descriptive statistics. Major concepts covered are listed below, and are also defined in the Glossary at the back of the book.

Key Concepts

1. **Statistics** is the collection, organization, presentation, analysis, and interpretation of data for the purpose of making better decisions.
2. Statistics may be divided into two areas, **descriptive** and **inferential.**
3. Data may be classified into four levels of measurement—**nominal, ordinal, interval,** or **ratio.**
4. A **population** is a collection or set of individuals, objects, or measurements. A **sample** is a part of the population. Calculations made from populations are **parameters,** and those made from samples are **statistics.**
5. A **frequency distribution** is a comprehensive summary of a set of observations. It separates the data into classes and shows the number of occurrences in each class.
6. **Histograms, frequency polygons,** and **stem-and-leaf charts** are graphic displays of frequency distributions.
7. A **less-than cumulative frequency distribution, or ogive,** shows the number of observations that are less than or equal to the upper limit of each class.
8. An **average** is a single representative value that is typical of the data considered as a whole. Three averages were discussed: **mean, median,** and **mode.**
9. When a data set contains extremely large or extremely small values, the resulting distribution may be **skewed.** A skewed distribution is one that is not **symmetrical.** The skewness can be measured by the **coefficient of skewness.**
10. The **dispersion** of a set of data is the amount of variability or spread in the data.
11. The **standard deviation** is the most useful and widely used measure of dispersion. It is the positive square root of the **variance.** The variance is the mean of the squared differences between the actual observation and the mean of the data set.
12. The **coefficient of variation** is a measure for comparing the relative variability of two sets of data. It is particularly useful when the data sets to be compared are in different units.

Key Terms

This list of terms is included in order for you to verify your recall of the material covered in Chapters 1–4. As you read each term, provide its definition in your own words. Then check your answers against the definitions given both in the chapter and in the Glossary at the back of the book.

Descriptive statistics
Inferential statistics
Population
Sample
Mutually exclusive
Exhaustive
Nominal scale
Ordinal scale
Interval scale
Ratio scale
Raw data
Classes
Class frequency
True limits
Stated limits
Class midpoint
Frequency distribution
Histogram
Stem-and-leaf charts
Frequency polygon
Cumulative frequency distribution (ogive)
Line chart
Bar chart
Pie chart
Mean
Median
Mode
Bimodal
Midpoint
Skewed
Symmetric distribution
Dispersion
First quartile
Third quartile
Weighted mean
Parameter
Statistic
Range
Quartile deviation
Interquartile range
Mean deviation
Variance
Standard deviation
Coefficient of variation
Coefficient of skewness

Key Symbols

$\overline{X}$	The mean of a sample
n	The number of observations in a sample
Σ	The symbol that indicates a group of values are to be added
μ	The mean of a population
N	The number of observations in a population
$\overline{X}_w$	The weighted mean
Q_3	The value of the third quartile
Q_1	The value of the first quartile
QD	The quartile deviation
MD	The mean deviation or, as it is sometimes called, the average deviation
Md	The mode
σ^2	The variance of a population
σ	The standard deviation of a population

s^2 The variance of a sample
s The standard deviation of a sample
CV The coefficient of variation
sk The coefficient of skewness

Review Problems

1. The following number of students failed freshman English in each of 10 sections of the course offered at a midwestern university during the last fall quarter: 5, 1, 1, 4, 6, 7, 9, 7, 10, and 4.
   a. Is this example a sample or a population?
   b. Compute the mean.
   c. Compute the median.
   d. What is the mode?
   e. Compute the range.
   f. Compute the standard deviation.
2. The following are the audience ratings for the first 12 weeks of a new prime-time soap opera called "Houston": 12.6, 13.9, 13.3, 15.7, 12.0, 15.1, 13.7, 16.5, 17.5, 12.5, 14.7, and 15.0.
   a. Is this an example of a sample or a population?
   b. Compute the mean rating.
   c. Compute the median.
   d. What is the mode?
   e. Compute the range.
   f. Compute the standard deviation.
3. A study is being conducted to determine the number of TV sets per household. The following sample data were obtained for eight homes: 1, 3, 3, 4, 1, 0, 4, and 3.
   a. Compute the mean.
   b. Compute the median.
   c. What is the mode?
   d. Compute the standard deviation.
4. Lander Technical College is surveying students to find out how many hours per week they work. A sample of 50 students yielded the following information:

Hours Worked	Number of Students
0–9	5
10–19	9
20–29	15
30–39	10
40–49	9
50–59	2
	50

a. Compute the mean.
b. Compute the median.
c. Compute the quartile deviation.
d. Compute the variance.

5. A study made by Canard Caribbean Cruises entails recording the number of pieces of luggage checked by passengers. A sample of 24 passengers reveals the following breakdown: 1, 9, 9, 1, 10, 1, 6, 8, 9, 4, 11, 9, 4, 5, 5, 1, 8, 4, 6, 12, 4, 7, 5, and 8. Organize the data into a frequency distribution with 0 as the lower limit of the first class.
   a. Compute the mean.
   b. Compute the median.
   c. Compute the standard deviation.
   d. What is the range, using both methods?
   e. Compute the quartile deviation.
6. The Parry-Mutual Insurance Co. is studying the number of group insurance claims submitted each week during the last year. Company records have disclosed the following information. Consider these data a population.

Number of Claims per Week	Weeks
0–2	6
3–5	15
6– 8	19
9–11	9
12–14	3
	52

   a. Compute the mean.
   b. Compute the median.
   c. Compute the standard deviation.
   d. Compute the quartile deviation.
7. At the Silver Net Seafood Restaurant, which overlooks the scenic Conewango River, the mean dinner check for two people is \$58.50, the median is \$54.00, and the standard deviation \$5.25. Compute the coefficient of variation and the coefficient of skewness.
8. The following are the weights of a sample of five Hereford steers (in pounds) after two months in a feed lot: 968; 1,014; 1,247; 959; and 642. Consider these as sample data.
   a. Compute the mean.
   b. Compute the standard deviation.
   c. What is the variance?

9. During the fall semester of 1992, there were 472 full-time faculty at the University of Western Ohio. The following table shows the number of full-time faculty in each of the six colleges. Construct a pie chart to depict these data:

College	Number of Faculty
Arts & Sciences	245
Business	45
Education	97
Engineering	45
Law	27
Pharmacy	13
	472

10. A student newspaper reported that the attendance at the five home football games for last season was 2,520; 2,140; 29,600; 2,750; and 2,280.
   a. Calculate the mean.
   b. Calculate the median.
   c. Assume that the third value was a misprint and should have been 2,960. Recalculate the mean and the median.
   d. Comment on the effect of the misprint on the mean and the median.

Using Statistics

Situation The Banks Brothers Men's Clothiers, a regional chain of stores, is considering introducing a new line of designer clothes into some or all of its five districts. The marketing research staff reports that the product has been profitable whenever it is introduced into a district containing 500,000 or more households whose gross annual purchase of men's clothing amounts to at least $350. Data on the five districts are shown below. Which, if any, of the districts meet these requirements? Which districts deserve more research?

		Annual Purchases			
District	**Number of Households**	**Mean**	**Median**	**Mode**	**Standard Deviation**
Atchafalaya	2,500,000	$450	$ 87	$ 75	$75
Columbia	1,750,000	385	109	97	52
Ohio	950,000	367	360	358	18
St. Lawrence	1,000,000	365	340	310	20
Yukon	1,350,000	353	352	348	10

Discussion The St. Lawrence district is a definite no. The median annual purchase is $340, which means that 500,000 households spend less than $340 and 500,000 spend more than $340. Hence, there are too many households below the minimum required purchase amount of $350.

By similar reasoning, the Yukon sales district definitely does qualify. There are 675,000 households spending at least the median amount of $352. The new line should also be introduced into the Atchafalaya district. The population is so large that only 20% of the households need to make the purchases. The mean purchase is more than a full standard deviation above the $350 requirement. So you are sure that more than 40% of the households will spend more than $350 annually.

The Columbia and Ohio districts are very questionable. Even though their mean values are above the critical sales level, the Columbia median is far below $350 and the Ohio population is quite small. The standard deviations are large; hence, additional information is needed.

So, it is appropriate to introduce the new line in the Yukon and Atchafalaya districts; more data is necessary on the Columbia and Ohio districts. The new line will not be introduced in the St. Lawrence district.

CHAPTER 5

An Introduction to Probability

OBJECTIVES

When you have completed this chapter, you will be able to

- define the terms *probability, experiment, outcome,* and *event;*
- describe the classical, relative frequency, and subjective concepts of probability;
- calculate probabilities using the rules of addition and multiplication;
- count the number of permutations and combinations.

CHAPTER PROBLEM Are There Enough Cherry Life Savers?

JOHNNY SMITH HAS JUST PURCHASED A ROLL OF LIFE SAVERS AT THE CANDY KOUNTER. AS HE LEAVES THE STORE, HE PEELS OFF THE OUTER WRAPPING AND COUNTS THE NUMBER OF DIFFERENT FLAVORS IN HIS PACKAGE. THREE ARE CHERRY, AND THERE ARE TWO EACH OF LEMON, LIME, PINEAPPLE, AND ORANGE. TWO OF JOHNNY'S FRIENDS, KERRY AND MARK, JOIN HIM OUTSIDE, AND HE OFFERS EACH OF THEM ONE OF HIS CANDIES. AS IT TURNS OUT, ALL THREE PREFER CHERRY, SO THERE ARE JUST ENOUGH TO GO AROUND. DID THE MANUFACTURER KNOW THIS WAS GOING TO HAPPEN? OR WAS IT JUST LUCK? CAN WE COMPUTE THE PROBABILITY OF SELECTING VARIOUS CANDY FLAVORS? WHAT ARE THESE PROBABILITIES?

INTRODUCTION

Chapters 2, 3, and 4 dealt with descriptive statistics. The emphasis in Chapter 2 was on organizing raw data into a frequency distribution. Recall, for example, the selling prices of homes in the Rafter J Ranch subdivision. When prices were organized into a frequency distribution, the lowest was about $80,000 and the highest $114,000. Furthermore, the largest concentration of selling prices was in the $95,000–$99,000 range. In Chapters 3 and 4, measures of central tendency and dispersion were introduced. It was determined, for example, that the average selling price of a Rafter J Ranch home was about $97,000. The main focus in those three chapters was to describe a set of data that had already been observed.

The purpose of this and of subsequent chapters is to examine ways of calculating the probability that some event will occur. For instance, you may be interested in determining the probability that the incumbent governor in your state will be reelected in November, or that fewer than 80% of first-time speeding offenders will be fined a second time for this offense. Whatever your purpose, the ability to calculate probability will be a valuable tool in your everyday decision making.

Why Study Probability?

One of the major purposes of the science of statistics is to make it possible to infer, from the results of studies of just one part of a group, general statements about the entire group. As defined in Chapter 1, one portion of the entire group is called a **sample,** while the entire group is called the **population.** The process of reasoning from a set of sample observations to a general conclusion about a population is **statistical inference.** Probability is the "yardstick" used to measure the reasonableness that a particular sample result could also be valid for a certain population. At some point you may decide that the observed sample result is so improbable that you reject the claim that it applies to a specific population. Hence, in the science of statistics we "prove" or "disprove" statements by calculating how probable or improbable they are.

Probability will help us make decisions with respect to statistical inference. Throughout, we will use case studies like the two that follow to illustrate the potential use of probability in decision making.

CASE 1. Senator Simeon claims that 75% of all taxpaying Americans believe additional revision of the 1987 tax "reform" is necessary. To investigate his contention, 500 taxpaying Americans were selected at random and interviewed. Of the 500 sampled, 70%, or 350, said they were in favor of additional tax reform. Can we now challenge the senator's claim as incorrect? Or could his contention still be fundamentally valid because the difference between his stated percent (75%) and that of the sample results (70%) may be attributable to chance and therefore would not be significant? In a later chapter, we will show how calculating probabilities can help us make this kind of decision.

CASE 2. Forty percent of the surnames on a very long list of eligible jury members are of Hispanic origin. Each person whose name is on the list has an equal chance of being selected for jury duty. Out of 12 jury members, how many would you expect to have Hispanic surnames? Would you be surprised if no persons with a Hispanic surname were selected for jury duty? You should be! The probability of selecting no Hispanic surname is $(1 - 0.4)^{12}$, or 0.0022. That is, the chance of selecting no Hispanic surname is only about 2/10 of 1%! In this case, probability is used to show that, with a fair method of selection, it would be quite unreasonable to expect no Hispanic-surname jury members to be selected out of 12.

CONCEPTS OF PROBABILITY

All of us have some idea of what a **probability** is—although it may be rather difficult to arrive at a precise definition. The weather forecaster notes that "there is a 30% *chance* of rain." The *likelihood* that humankind will land on Jupiter in the next four years is rather remote. Terms such as *chance* and *likelihood* are used interchangeably for the word *probability*.

What is a probability?

> **Probability** A value between 0 and 1, inclusive that measures the likelihood that a particular event will occur.

Three key words are used in the study of probability: experiment, outcome, and event. While they commonly appear in our language, in statistics these terms have specific meanings.

> **Experiment** The observation of some activity, or the act of taking some measurement.

This definition is more general than the one used in the physical sciences, where we picture researchers manipulating test tubes and microscopes. In statistics, an experiment has two or more possible results, and it is uncertain which will occur.

STAT BYTE

It is often assumed that there are an equal number of male and female births, however, national records consistently indicate that more boys than girls are born. The probability of a male birth in the United States is about 0.513. Some researchers believe that timing the act of intercourse during a woman's fertile period may explain this non-equal probability.

Outcome A particular result of an experiment.

For example, the tossing of a coin is an experiment. You may observe the coin tossing, but you will be unsure whether it will come up "heads" or "tails." Similarly, asking 500 voters whether or not they intend to vote for a $3.1 million school bond is also an experiment. If a coin is tossed, one particular outcome might be a "head." Or the outcome might be a "tail." As for the school-bond experiment, one outcome might be that 342 people favor its issuance. Another outcome might be that only 67 approve of it. Still another outcome is that 203 favor the bond's issue. When one or more of the experiment's outcomes is combined, we call this an **event.**

Event A collection of one or more possible outcomes of an experiment.

Consider, for example, the single throw of a die. There are exactly six possible outcomes for this experiment. However, many events may be associated with it—the fact that the number of spots coming face up on the toss is odd, for example, or that the number that comes up is smaller than three. With the school-bond experiment, one event might be that a majority favor the $3.1 million issue. This event would occur whether the number of people in favor is 251, 252, 253, or any number up to and including 500. Note that an event is not always simply an outcome.

Probabilities may be expressed as fractions (1/4, 5/9, 7/8), decimals (0.250, 0.556, 0.875), or percentages (25%, 56%, 88%). A probability is always between zero and one, inclusive. Zero describes the probability of something that cannot happen. If a corn seed is planted, the probability of having an elephant sprout from the seed is zero. At the other extreme, a probability of one represents a sure thing. In Portland, Maine, the mean high temperature is sure to be higher in July than in January.

Problems

Indicate whether each of the following statements is true or false:

a. If a die is rolled, one particular outcome is ⚀.
b. A probability of 1.00 indicates that something cannot happen.
c. A probability of –100.00 indicates that some event will definitely happen.

Solutions

a. True
b. False
c. False

Self-Review 5-1

Answers to all Self-Review problems are at the end of the chapter.

A politician has recently delivered a major speech. He receives six pieces of mail commenting on it. He is interested in the number of writers who agree with him.

a. What is the experiment?
b. What are the possible outcomes?
c. Describe one possible event that might occur.

Exercises

Answers to the even-numbered exercises are at the back of the book.

1. One card is drawn from a well-shuffled, standard 52-card deck. If only the four suits are of interest (spades, hearts, diamonds, and clubs), what are all the possible outcomes? Suppose that in your deck hearts and diamonds are red cards and spades and clubs are black. What is the possibility of selecting a red card called?

2. Three persons are campaigning for mayor. Schwartz is a Democrat, White is a Republican, and Rossi is an Independent. Schwartz and Rossi are men. White is a woman.
  a. What is the experiment?
  b. Describe possible outcomes with respect to political party.
  c. Describe possible outcomes with respect to the sex of the candidates.

TYPES OF PROBABILITY

There are three approaches to establishing the probability of events: **classical, relative frequency,** and **subjective.** The **classical** concept of probability is based on the assumption that several outcomes are *equally likely.*

Classical Concept of Probability The number of favorable outcomes divided by the number of possible outcomes.

Problem Recall the Chapter Problem about the roll of Life Savers purchased by Johnny Smith. If Johnny randomly selects a single candy from the package, what is the probability that it will be cherry?

Solution Of the 11 candies in the roll, three are cherry, so the probability is 3/11, or 0.273, that the Life Saver John selects will be cherry.

Self-Review 5-2

One card is selected at random from a standard deck of 52 cards.

a. What is the probability that the card will be the ace of spades?
b. What is the probability that the card will be an ace?
c. What concept of probability does this problem illustrate?

The classical approach is useful when dealing with games of chance, such as dice and card games, and situations in which the outcomes are equally likely to occur. The classical approach is similarly useful when random selection is important—that is, when every outcome has the same chance of occurring. Serious problems, however, develop when the outcomes are not equally likely. If you are an excellent student, for example, there is a good chance you will earn an A in this course. In your case, the outcome A does not have the same likelihood as the outcome F.

A second approach to probability is based on **relative frequencies.** This method uses the frequency of *past* occurrences to develop probabilities for the future.

Relative Frequency Probability Concept The number of times an event occurred in the past divided by the number of observations.

First, we compute the number of times a particular event happened in the past. Then we use this value to determine the likelihood that it will happen again. To illustrate, a mortality table revealed that out of 100,000 men aged twenty-five, 138 die within a year. Based on this experience, a life insurance company would estimate the probability of death for this age as:

$$P(A) = \frac{138}{100,000} = 0.00138$$

This type of probability is used in so-called "actuarial tables" to help insurance companies establish the premiums to be charged for various types of life insurance—term, ordinary life, and so on.

In the mortality-table example, the probability of an event is denoted by a capital *P*. An abbreviation for the event is then written in parentheses. Commonly, capital letters or numbers are used to denote an event in a concise manner. In this case, *P*(*A*) stands for the probability that a 25-year-old male will die during the year.

Self-Review 5-3

A political scientist randomly selected 200 eligible voters and determined the number of times they had voted in the last five general elections:

Number of Times Voted	Frequency of Occurrence
1	30
2	41
3	40
4	62
5	27
	200

a. What is the probability that a particular voter cast his or her ballot only once in the last five elections?
b. What is the probability that a particular voter cast his or her ballot exactly three times in the last five elections?
c. What concept of probability does this illustrate?

21. Weeks of Vacation | # of Employees
1 | 30
2 | 65
3 | 190
4 | 15
300

a. The employee had 3 wks vacation:

190/300 = 0.633 = .63

b. 3 wks or more of vacation

190 + 15 = 205

205/300 = 0.683 = .68

c. less than 4 wks vacation

30 + 65 + 190 = 285

285/300 = 0.95

For situations in which there is little or no historical information from which to determine a probability, **subjective** probabilities can be employed. Using this approach, experts given the same information often differ in their estimates of the probability. They differ with respect to the likelihood of another major earthquake occurring in California during this decade, just as they differ in their selection of the team most likely to win the National League pennant this coming season. Subjective probability can be thought of as the *probability assigned by an individual or group based on whatever evidence is available.*

Subjective Probability Concept The likelihood of an event assigned on the basis of whatever information is available.

Self-Review 5-4

a. What probability would you assign to a deep economic recession happening within a year?
b. What concept of probability does this illustrate?

In summary, there are three concepts of probability. The classical viewpoint assumes that the outcomes are equally likely. If the outcomes are not equally likely, the relative frequency viewpoint is used. If no past experience is available, subjective judgment can be used to assign the probability that an event will occur. Regardless of the viewpoint, the same laws of probability discussed in the following sections will be applied.

Self-Review 5-5

What concept of probability is used to assign the likelihood of the various events in the following experiments?

a. The probability of obtaining a when rolling a single die.
b. The probability is 0.75 that the mean high temperature in Tampa, Florida, is lower in January than in September.
c. The probability is 0.005 that a nuclear accident will occur at the Davis-Besse Plant within the next month.
d. Jones, Smith, Jackson, Keller, and Archer submit welfare claims. If only one is selected, the probability of selecting Jackson's claim is 0.20.

PROBABILITY RULES

In applications of probability, often there is a need to combine the probabilities of related events in some meaningful way. In this section, two of the fundamental methods of combining probabilities—by addition and by multiplication—are discussed. Generally, events formed using the word *or* are handled by addition; those formed using the word *and* are handled by multiplication.

Addition Rules

THE SPECIAL ADDITION RULE. We apply the **special rule of addition** to calculate the probability of an event consisting of one or more mutually exclusive outcomes. Recall from Chapter 1 that *mutually exclusive* means that when one of the events occurs, none of the other events can occur at the same time. If the result of a single roll of a die is a 4, it cannot be a 6 at the same time. Thus, the outcomes of a 4 and a 6 are mutually exclusive. Similarly, if the respondents to a questionnaire are classified as either male or female, then the events (male respondent, female respondent) are mutually exclusive. The special rule of addition for two events is:

$$P(A \text{ or } B) = P(A) + P(B) \qquad \textbf{5-1}$$

Recall that capital letters, such as A and B, refer to events. This formula can be expanded to any number of events. For example, for three events it is $P(A \text{ or } B \text{ or } C) = P(A) + P(B) + P(C)$.

Problem What is the probability that an even number will result from one roll of a single die?

Solution There are six possible outcomes:

⚀ ⚁ ⚂ ⚃ ⚄ ⚅

The event (an even number) is composed of three outcomes, namely:

The outcome of a ⚁ is event A.

The outcome of a ⚃ is event B.

The outcome of a ⚅ is event C.

The probability of each of the outcomes (2, 4, 6) is 1/6. To find the probability of the event "the outcome is an even number," we use formula 5–1 and add the three probabilities. That is, 1/6 + 1/6 + 1/6 = 3/6, or 0.50. In symbols, if A stands for the outcome of a 2, B the outcome of a 4, and C the outcome of a 6, we compute the probability of an even number appearing by:

$$\begin{aligned} P(\text{even}) &= P(A) + P(B) + P(C) \\ P(A \text{ or } B \text{ or } C) &= P(A) + P(B) + P(C) \\ &= \frac{1}{6} + \frac{1}{6} + \frac{1}{6} \\ &= 0.50 \end{aligned}$$

Recall that $P(\quad)$ denotes the probability of the event described inside the parentheses.

Addition of the probabilities of the events of a 2, a 4, and a 6 in the previous die-rolling problem is allowed because the events are mutually exclusive. This notion of mutual exclusivity can be shown in diagram form (see Figure 5–1). The six regions represent all possible outcomes of an experiment.

STAT BYTE

In the days before airport X-ray machines and metal detectors, there was once a little old lady who carried a bomb with her whenever she traveled. Since she had never heard of *two* bombs on the same plane, she reasoned that hers would be the only one on board!

Figure 5-1

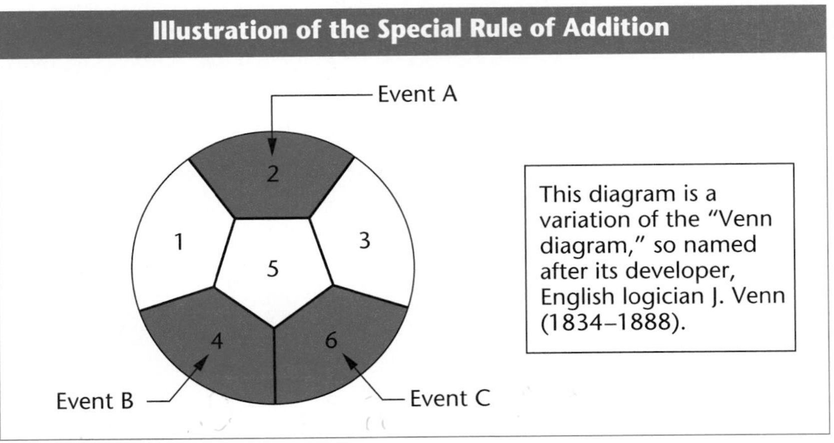

Self-Review 5-6

Two hundred randomly selected prisoners in cell block M are surveyed and classified by type of crime committed.

Type of Crime	Number
Murder	48
Armed robbery	42
Rape	101
Kidnapping	7
Other	2

a. What is the probability that a particular prisoner selected in the sample is a convicted murderer?

b. What is the probability that a particular prisoner selected is a convicted kidnapper?

c. What is the probability that a particular prisoner selected is either a convicted murderer or a convicted kidnapper? What rule of probability is employed?

Exercises

3. The events A and B are mutually exclusive. Suppose $P(A) = 0.20$ and $P(B) = 0.25$. What is the probability of the event (A or B) occurring?

4. The events X and Y are mutually exclusive. Suppose $P(X) = 0.05$ and $P(Y) = 0.10$. What is the probability of the event (X or Y) occurring?

5. A study was made of 138 children who exhibited evidence of abuse by an adult. The following table shows the abused child's position in the family:

Position	Frequency
Only child	34
Oldest	24
Youngest	50
Other	30

a. What is the probability of randomly selecting a child from the group described above and finding that the child was either the youngest or the oldest child in a family?

b. What is the probability of selecting a child who is not an only child?

6. Shown in the following table are the reported annual deaths of males aged 75 or over from one of the five leading types of cancer:

Site of Cancer	Number of Deaths
Lung	12,226
Prostate	10,835
Colon or rectum	8,426
Stomach	3,037
Pancreas	3,031
	37,555

For a randomly selected deceased male cancer victim, what is the probability that

a. he died of one of the two primary causes?

b. he died of either cancer of the stomach or of the pancreas?

c. he did not die of lung cancer?

THE GENERAL ADDITION RULE. We apply the **general rule of addition** to calculate the probability of events that are not mutually exclusive. For example, the current Social Security law has both a disability provision and a retirement provision. A welfare worker studying the residents of the Sunshine Retirement Community finds that 20% of the residents are receiving disability payments and 85% are receiving retirement income. If a retiree is randomly chosen for study, what is the probability that the person selected is receiving either disability payments or retirement income (or possibly both)?

The percents are converted to probabilities:

	Probability
Probability of receiving disability income	0.20
Probability of receiving retirement income	0.85
Total	1.05

Is the probability of 1.05 possible? It is not! A probability of more than 1.0 was ruled out in an earlier section. A probability is always 0 to 1, inclusive. What happened? Some of the people who receive payments have been "double-counted." That is, they receive both disability and retirement incomes.

To overcome this difficulty, the welfare worker must determine the percent of people who were counted twice and deduct this percent from the total. Suppose the investigator were to find that 15% of the people receive *both* disability and retirement incomes. Subtracting the corresponding probability of 0.15 from the total leaves 0.90. This is the probability of a person receiving at least one of these two types of income. The following table summarizes these calculations:

	Probability
Probability of receiving a disability benefit	0.20
Probability of receiving a retirement benefit	0.85
Probability of receiving *both* a disability benefit and a retirement benefit	−0.15
Probability of receiving one or the other benefit (or possibly both)	0.90

Symbolically, the general rule of addition is written

$$P(A \text{ or } B) = P(A) + P(B) - P(A \text{ and } B) \quad \textbf{5-2}$$

where *A* and *B* are two events. The word *or* takes into account the possibility that both *A* and *B* may occur and may need to be separated. This is sometimes called an "inclusive" *or* because it includes the possibility that both *A* and *B* happen and the possibility that either occurs separately.

Apply formula 5–2 to the welfare problem:

Let *A* stand for the event "the retiree receives disability benefits."

Let *B* stand for the event "the retiree receives retirement benefits."

Let *A* and *B* stand for the event "the retiree receives both a retirement benefit and a disability benefit."

To find the probability of receiving a disability benefit *or* a retirement benefit, use formula 5–2:

$$\begin{aligned} P(A \text{ or } B) &= P(A) + P(B) - P(A \text{ and } B) \\ &= 0.20 + 0.85 - 0.15 \\ &= 0.90 \end{aligned}$$

The general rule of addition applied to the welfare problem is illustrated by a Venn diagram (see Figure 5–2). Note the overlapping of events.

Figure 5-2

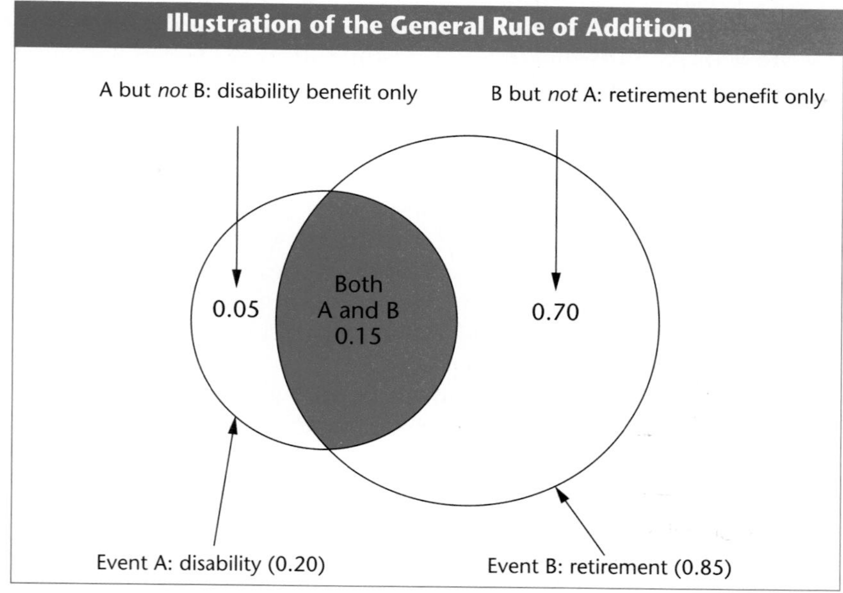

Problems

a. Let $P(X) = .50$ and $P(Y) = .20$, and $P(X \text{ and } Y) = 0$. What is the $P(X \text{ or } Y)$? Are the events mutually exclusive?

b. Let $P(D) = .35$ and $P(E) = .15$, and $P(D \text{ and } E) = .05$. What is the $P(D \text{ or } E)$? Are the events mutually exclusive?

Solutions

a. 0.70; yes

b. 0.45; no

Self-Review 5-7

An analysis of the student records at Solid State University revealed that 45% of the students have a grade point average above 3.00. Twenty-five percent of the students are employed. Ten percent of the students are employed *and* have a grade point average above 3.00. What is the probability that a student selected at random will have a grade point average above 3.00 *or* be employed?

Exercises

7. The probabilities of two events *A* and *B* are 0.20 and 0.30, respectively. The events are not mutually exclusive. The probability that they both occur is 0.15. What is the probability of either *A* or *B* occurring?

8. Let $P(X) = 0.55$ and $P(Y) = 0.35$. Assume that these events are not mutually exclusive, and that the probability that they both occur is 0.20. What is the probability of either X or Y occurring?

9. A study of patient records at a public health clinic showed that 15% of patients had a dental examination, 45% had a general physical examination, and 5% had both. If a patient's record is randomly selected, what is the probability that the patient received either a dental examination or a physical examination?

10. Twenty percent of the members of boards of directors of large corporations are women. Five percent are persons connected with a university. Two percent of large corporations have female board members with university ties. If a corporation is randomly selected, what is the probability that its board of directors will have a member who is either a woman or someone from a university?

11. A student is taking two courses, History and Math. The probability that the student will pass the math course is 0.60, and the probability of passing the history course is 0.70. The probability of passing both is 0.50. What is the probability of passing at least one of the two courses?

12. An analysis by the National Weather Service at Toledo, Ohio, showed that the low temperature in January is below freezing on 70% of the days. It snows on 30% of the January days and is below freezing *and* snows on 20% of the days. If tomorrow is January 12, what is the probability that it will snow or that the low temperature will be below freezing?

Multiplication Rules

THE GENERAL MULTIPLICATION RULE. Suppose we want to find the probability of both A and B occurring. For example, we might wish to find the probability of a student having a grade point average above 3.00 and also being employed. Such situations are different from those for which the addition rule can be used. Instead of computing the probability of one of the two outcomes occurring, we wish to find the probability that they *both* happen. This is termed a **joint probability.**

> **Joint probability** Measures the likelihood that two or more events will happen at the same time.

Another probability concept is called **conditional probability.** Recall that probability measures uncertainty. But the degree of uncertainty may change as new data become available. Conditional probability is the tool that describes the new probability corresponding to some event B after it is known that some other event A has occurred. Symbolically, it is written $P(B \mid A)$. The vertical ($\mid$) slash does not mean division. Instead, it is read "given," as in "the probability of B given A."

Conditional probability The likelihood that an event will occur given that another event has already occurred.

In the Social Security illustration, the probability that a person will receive a retirement benefit $P(B)$ is 0.85. However, if first we learn that the person is receiving a disability benefit, our estimate of the probability that he or she is also receiving a retirement benefit changes. Only 20% of the total receive disability benefits [$P(A)$ = 0.20]. Out of that 20%, 15% also receive a retirement benefit. Hence, the conditional probability is 15/20, or 0.75 [$P(B \mid A)$]. The knowledge that a person received a disability benefit reduced the likelihood from 0.85 to 0.75 that he or she would receive a retirement benefit. In symbols, this calculation is written as follows:

$$P(B|A) = \frac{P(A \text{ and } B)}{P(A)} \qquad \text{5-3}$$

This same relation gives us a way to compute P(A and B) when P(A) and P(B | A) are known. We simply multiply: $P(A) \cdot P(B \mid A)$. This is the **general rule of multiplication**, which states that if two events A and B can both occur in some experiment, then the probability that both A and B occur is $P(A) \cdot P(B \mid A)$. It is written:

$$P(A \text{ and } B) = P(A) \cdot P(B|A) \qquad \text{5-4}$$

Problem A study is conducted to determine the possible correlation between the level of education attained and a person's position on abortion. The results of a random sample of 100 individuals are shown in Table 5–1.

The question to be explored is, What is the probability of selecting an individual with fewer than four years of high school education who favors abortion? The events occurring at the same time are "fewer than four years of high school" and "favors abortion."

Table 5-1

Level of Education and Position on Abortion

Position on Abortion	Less Than Four Years High School	High School Graduate	Some College	College Graduate	Total
Favor	5	15	15	25	60
Oppose	10	10	10	10	40
Total	15	25	25	35	100

Solution Applying the general rule of multiplication gives

$$P(B \text{ and } A) = P(A) \cdot P(B|A)$$

where

A stands for the event "a person favors abortion."

B stands for a person with fewer than four years of high school education.

$P(B \mid A)$ represents the probability of selecting a person who has less than a high school education when it is known (or given) that the person favors abortion. As noted previously, the vertical line is read as "given that." The problem reads, "We find the probability that a person selected has less than four years of high school education and favors abortion by multiplying the probability that the person favors abortion by the probability that the person has less than a high school education, given that the individual favors abortion." Referring to Table 5–1 for the probabilities and using formula 5–4, we obtain:

$$\begin{aligned} P(B \text{ and } A) &= P(A) \cdot P(B|A) \\ &= 60/100 \cdot 5/60 \\ &= 0.60 \cdot 0.0833 \\ &= 0.05 \end{aligned}$$

We could also have computed this probability by determining the probability of selecting an individual with less than a high school education [$P(A)$ = 15/100 = 0.15] and then the probability that the person favors abortion, given that he or she has less than a high school education [$P(B \mid A)$ = 5/15 = 0.33]:

$$\begin{aligned} P(A \text{ and } B) &= P(A) \cdot P(B|A) \\ &= (0.15) \cdot (0.33) \\ &= 0.05 \text{ (the same as observed and calculated earlier)} \end{aligned}$$

If more than two simple events are involved, the general rule of multiplication can be extended. For example, $P(A \text{ and } B \text{ and } C) = P(A) \cdot P(B \mid A) \cdot P(C \mid A \text{ and } B)$.

In the previous example about people's attitudes toward abortion, the joint probability could have been read directly from the table. It was pointed out that the general rule of multiplication gives the same answer. Sometimes, however, the information we have is expressed in percents instead of counts as in the previous illustration. The general rule of multiplication must be used to solve such problems.

Problem The police in a small municipality know that 25% of the homeowners leave their doors unlocked. Crime records show that 4% of the homes whose doors are left unlocked are burglarized. What is the probability that a home is both left with its doors unlocked *and* burglarized?

Solution The general rule of multiplication applies—that is, formula 5–4.

Let A represent the event "left with doors unlocked."
Let B represent the event "burglarized."

We know that:

$$P(A) = 0.25$$
$$P(B|A) = 0.04$$

4% of the homes *with unlocked doors* are burglarized

Solving gives:

$$\begin{aligned} P(A \text{ and B}) &= P(A) \cdot P(B|A) \\ &= (0.25)(0.04) \\ &= 0.01 \end{aligned}$$

This indicates that 1% of *all* the homes are *both* left with their doors unlocked and burglarized.

Problems

a. For two independent events, $P(X) = .50$ and $P(Y) = .20$. Determine the probability of $P(X \text{ and } Y)$.
b. For two dependent events, $P(D) = .35$ and $P(E \mid D) = .15$. Determine the probability of $P(D \text{ and } E)$.

Solutions

a. 0.10
b. 0.0525

Self-Review 5-8

A sociologist conducted a study of a sample of 60 students to determine how many of each sex smoke marijuana. The results:

	Male	Female	Total
Smokes marijuana	15	8	23
Does not smoke marijuana	20	17	37
Total	35	25	60

Using the general rule of multiplication, determine the probability of selecting a male who smokes marijuana.

a. Letting A be the event "the student smokes marijuana" and M the event "a male is selected," supply the appropriate formula.
b. Determine the joint probability.

An interesting technique used to show probabilities, joint probabilities, and conditional probabilities is to plot them on a so-called **tree diagram.** Table 5–1 is repeated here to illustrate the tree diagram's application.

Position on Abortion	Less Than Four Years High School	High School Graduate	Some College	College Graduate	Total
Favor	5	15	15	25	60
Oppose	10	10	10	10	40
Total	15	25	25	35	100

In the tree diagram in Figure 5–3, note that there are two main branches going out from the trunk on the left. The upper branch is labeled "Favor," the lower branch "Oppose." Note that the probability written on the "favor abortion" branch is 60/100 and on the "oppose abortion" branch is 40/100.

Problem Using the tree diagram in Figure 5–3, locate the probability that a person favors abortion and has less than a high school education.

Solution

1. Find the upper branch representing the event "favor abortion." As noted before, its probability (60/100) is written on the tree branch.
2. Continuing along the same path, find the branch "Less than 4 years high school." The conditional probability of 5/60 is written on this branch.
3. Multiplying those two probabilities gives 0.05, shown at the end of the path. This is the joint probability of selecting a person who both favors abortion and has less than four years of high school education.

Figure 5-3

Tree Diagram Showing Calculations for Joint Probabilities

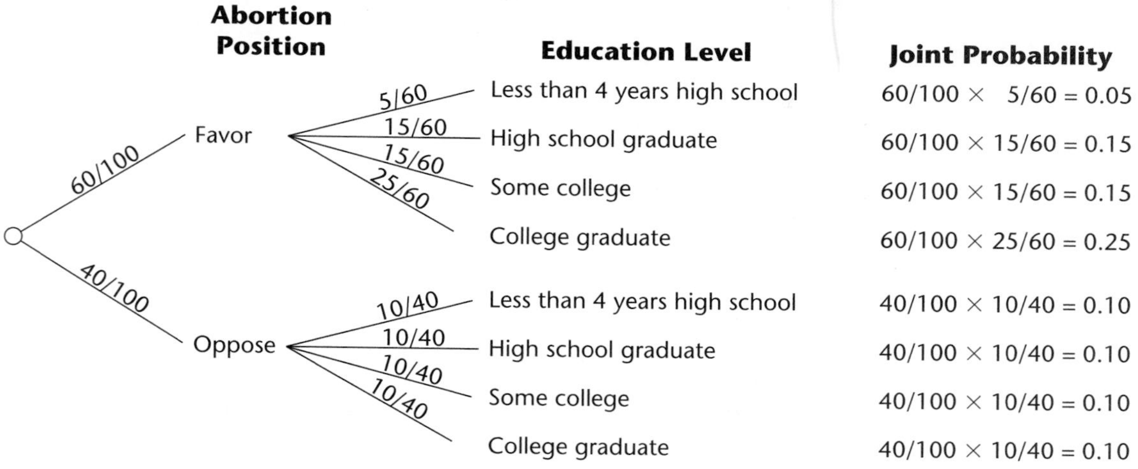

Abortion Position		Education Level	Joint Probability
60/100 Favor	5/60	Less than 4 years high school	60/100 × 5/60 = 0.05
	15/60	High school graduate	60/100 × 15/60 = 0.15
	15/60	Some college	60/100 × 15/60 = 0.15
	25/60	College graduate	60/100 × 25/60 = 0.25
40/100 Oppose	10/40	Less than 4 years high school	40/100 × 10/40 = 0.10
	10/40	High school graduate	40/100 × 10/40 = 0.10
	10/40	Some college	40/100 × 10/40 = 0.10
	10/40	College graduate	40/100 × 10/40 = 0.10

Self-Review 5-9

The study from Self-Review 5–8 is repeated below.

	Male	Female	Total
Smokes marijuana	15	8	23
Does not smoke marijuana	20	17	37
Total	35	25	60

a. What is the probability of selecting a female?
b. Assuming the person selected is a female, what is the probability that she smokes marijuana?
c. Draw a tree diagram showing all possible joint probabilities.

Exercises

13. Suppose $P(A) = 0.30$ and $P(B \mid A) = 0.60$. Determine $P(A \text{ and } B)$.
14. Refer to the following table:

	Other Events			
Events	M	N	O	**Total**
X	2	1	3	6
Y	1	2	1	4
Total	3	3	4	10

a. Compute $P(Y)$.
b. Compute $P(M \mid Y)$.
c. Compute $P(M \text{ and } Y)$.

15. A study was undertaken to correlate students' mathematical ability with their interest in statistics. The results:

	Interest in Statistics			
Math Ability	Low	Average	High	Total
Low	40	9	11	60
Average	15	16	19	50
High	6	10	25	41
	61	35	55	151

a. What is the probability of selecting a student with both a low math ability and a low interest in statistics?
b. What is the probability of selecting a student with both an average math ability and a high interest in statistics?
c. Draw a tree diagram showing all possible joint probabilities.

16. Freshmen were classified according to two traits: whether they had won a high school letter for athletics, and college grade point average.

	College GPA			
High School Athletics	Low	Average	High	Total
Letter winner	50	30	50	130
Did not win letter	20	30	20	70
	70	60	70	200

A freshman is selected.
a. What is the probability of selecting a student with a low college GPA who won a letter in high school?
b. What is the probability of selecting a student with a high college GPA who won a letter in high school?
c. Draw a tree diagram showing all possible joint probabilities.

17. Thirty-six players enter a tennis tournament. They are classified by sex and age:

	Juniors	**Seniors**
Males	6	10
Females	8	12

What is the probability that
a. two players are selected and both are senior females?
b. two players are selected and both are junior males?
c. two players are selected and both are females?

d. the first player selected is a senior male and the second is a junior female?
e. the first player selected is male and the second is female?
f. the first four players selected are senior females?

18. A hospital cafeteria menu includes the items shown below. If a patient must select one item from each group, draw a tree diagram to show the possible luncheon choices.

Entree	Salad	Fruit
Chicken	Lettuce	Apple
Tuna	Gelatin	Banana
Hot dog		Orange

a. How many different outcomes are possible?
b. What is the probability of simultaneously selecting chicken as the entree, lettuce as the salad, and a banana as the fruit?

THE SPECIAL MULTIPLICATION RULE. If there are two *independent* events A and B (meaning that the occurrence of event A does not affect the occurrence of event B), the **special rule of multiplication** is used to find the probability of A and B happening. This can be written

$$P(A \text{ and B}) = P(A) \cdot P(B) \qquad \textbf{5-5}$$

Problem There are six faces to a die: ⚀ ⚁ ⚂ ⚃ ⚄ ⚅. Suppose we have two of these dice and one is colored blue, the other white. They are rolled on the floor. What is the probability of obtaining a pair of 2s—that is, a 2 on the blue die ⚁ and a 2 on the white die ⚁?

Solution The outcome of the white die is not dependent on the outcome of the blue die; the two events are independent. Thus, the special rule of multiplication can be applied.

One of the six sides of a die is a 2-spot. The probability of a 2-spot, therefore, is 1/6. The probability of a 2-spot coming face up on the blue die is 1/6, and the probability of a 2-spot on the white die is also 1/6. The probability of both coming face up is 1/36, found by using formula 5–5:

$$\begin{aligned} P(A \text{ and B}) &= P(A) \cdot P(B) \\ &= \frac{1}{6} \times \frac{1}{6} \\ &= 0.0278 \end{aligned}$$

Figure 5–4 illustrates this concept. Note:

1. There are 36 points (dots).
2. A is the event "a 2-spot appears on the blue die."
3. B is the event "a 2-spot appears on the white die."
4. The shaded intersection of A and B is the point at which a 2-spot appears on both the blue and the white dice.

Note that while there are 36 points (dots) in Figure 5–4, the shaded intersection includes only 1 point. Hence, 1 out of 36 points meets the stipulations, and the probability is 0.0278, found by 1/36. This agrees with the result we obtained using the special rule of multiplication.

Self-Review 5-10

The probability of a newborn being a boy is 1/2; the probability of it being a girl is 1/2. Let A be the event "the first child born to Jane and Doug Schmitz is a boy." B is the event "the second child is a boy."

a. Are these events independent?
b. Supply the correct formula and determine the probability that the first two children born are boys.
c. What is the probability that the first three children are boys?

Figure 5-4

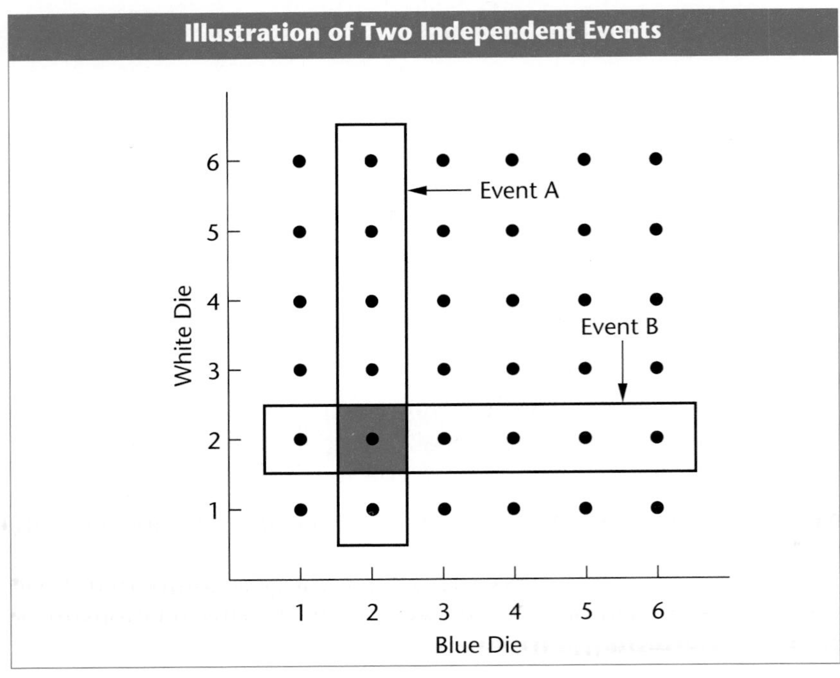

Exercises

19. We know that $P(A) = 0.30$ and $P(B) = 0.20$. If the events A and B are independent, determine the joint probability of A and B.

20. Refer to the following table:

Events	Other Events M	N	O	Total
X	2	1	3	6
Y	2	1	3	6
Z	4	2	6	12
Total	8	4	12	24

a. Compute $P(Y)$.
b. Compute $P(M \mid Y)$.
c. Compute $P(M \text{ and } Y)$.

21. What is the probability that an archer with 90% accuracy will hit a bull's-eye four times in a row?

22. Out of every 100 cars that start in the Texas Grand Prix race, only 60 finish. Two cars are entered by the Penske team in this year's race.
a. What is the probability that both will finish?
b. Are the two events independent?
c. What rule of probability does this illustrate?
d. What concept of probability does this problem illustrate?
e. What is the probability that neither of Penske's two cars will finish the Grand Prix?

23. The Penn Bank has two computers. The probability that the newer one will break down in any particular month is 0.05. The probability that the older one will break down in any particular month is 0.10.
a. Are these events independent?
b. What is the probability that both will break down during the month of October?
c. What is the probability that neither will break down during October?

24. At Mendoza University, 30% of the students live on campus.
a. If three students are selected, what is the probability that all three live on campus?
b. What is the probability that none of the three students lives on campus?

The Complement Rule

We may wish to describe a situation in which an event does *not* happen. For example, we want to estimate the probability that it will not rain next Saturday. It is easier to find the probability that it will rain and then subtract that number from 1. This relationship is termed the **complement rule.**

STAT BYTE

If you want to cause a stir at the next party you attend, announce that you believe at least two people present were born on the same date. If there are 30 people in the room and every birthdate is equally likely, the probability of this event is 0.706. (*Hint:* To verify this, use the complement rule to find the probability of no identical birthdays.) With 60 people, the probability is 0.994. With as few as 23 people in the room, there is a better than even chance that two people have identical birthdays!

$$P(\sim A) = 1 - P(A) \qquad \textbf{5-6}$$

If A is an event, we can express the complement of A in symbols as $\sim A$. The symbol ~ is called a tilde. It is written in front of the symbol for the first event to denote the second event, which is the "complement" of the first event. This rule is often very effective in reducing the number of calculations needed.

Problem A sample of 50 families revealed the following information about the number of children in each family:

Children	**Number of Families**
0	20
1	15
2	10
3 or more	5
	50

Compute the probability that a family has at least one child.

Solution These events are mutually exclusive. That is, for a particular family the number of children cannot be both 1 and 2. We can obtain the probability of selecting a family with at least one child by using the special rule of addition, formula 5–1. That is, we add the probability of having one, having two, and having three or more:

$$\begin{aligned} P(\text{at least } 1) &= P(1) + P(2) + P(3 \text{ or more}) \\ &= \frac{15}{50} + \frac{10}{50} + \frac{5}{50} \\ &= 0.60 \end{aligned}$$

The complement rule, formula 5–6, leads to a more efficient solution:

$$\begin{aligned} P(\text{at least } 1) &= 1 - P(0) \\ &= 1 - \frac{20}{50} \\ &= 1 - 0.40 \\ &= 0.60 \end{aligned}$$

Problem Recall from the Chapter Problem the roll of Life Savers purchased by Johnny Smith. If Johnny randomly selects two candies from the package, what is the probability that both will be cherry? Neither will be cherry? At least one will be cherry?

Solution The probability that both candies will be cherry is found using the general rule of multiplication. Multiply the probability that the first candy selected will be cherry (3/11) by the conditional probability that the second one chosen will be cherry after it is known that the first selection was cherry (2/10). The result is 6/110, or 0.0545, found by (3/11) (2/10).

The probability that neither will be cherry is also found using the general rule of multiplication, but the events are defined differently. First, find the probability that the first candy selected will not be cherry—that is, 8/11. Next, multiply this probability by the conditional probability that the second candy chosen will not be cherry (7/10). The result is 56/110, or 0.5091.

The complement rule is used to find the probability that at least one candy will be cherry. Use the fact that the probability *neither* will be cherry is 0.5091, and subtract this value from 1. The probability that at least one will be cherry is .4909, found by 1 – .5091.

Exercises

25. If $P(A) = 0.30$, what is $P(\sim A)$?

26. Refer to Exercise 20. Compute $P(\sim Y)$.

27. The Post-Anesthesia Care facility at St. Michael's Hospital has three beds. The probability that only one is occupied is 0.30, the probability that two are occupied is 0.25, and the probability that all three are occupied is 0.20. The probability that there are no patients in the facility is 0.25. Compute the probability that there is at least one patient in the facility.

28. According to the National Weather Service, the probability that it will rain in Pittsburgh, Pennsylvania, on a particular day in September is 0.20. Assume that each day is an independent event. What is the probability that it will not rain on the next two Saturdays?

SOME COUNTING PRINCIPLES

When the number of outcomes is small, as in the previous experiments, it is not difficult to list all the possibilities. For example, to list the outcomes for two children in a family is quite simple. The possibilities are two boys; one boy and one girl; and two girls. Suppose, however, you had to list and count the possible outcomes for a family of 15 children! Obviously, you would need a more efficient method of counting. Three principles of counting that facilitate calculations in such cases will be explored next: (1) the multiplication principle, (2) the permutation formula, and (3) the combination formula.

The Multiplication Principle

Multiplication Principle If a choice consists of two steps, the first of which can be made in m ways and the second in n ways, there are $m \cdot n$ possible choices.

Problem The Bullocks art department is designing an advertisement for the new catalog. The artist must show all possible variations on a new outfit consisting of two blouses and three skirts. The blouse options are long sleeve or short sleeve. The skirt options are solid color, plaid, or long evening. How many different interchangeable outfits are there?

Solution The letter m represents the number of blouse options. The letter n represents the number of skirt options. Using the multiplication principle, we obtain:

$$\begin{aligned}\text{Number of options} &= m \times n \\ &= 2 \times 3 \\ &= 6\end{aligned}$$

These six options are shown in Figure 5–5. This counting method can be extended to more than two events.

Problem A new-car buyer has a choice of five body styles, two engines, and eight different colors. How many different car choices does the buyer have?

Solution There are $5 \times 2 \times 8 = 80$ different choices among the cars that could be ordered.

Self-Review 5-11

An operating team generally consists of a surgeon, an anesthetist, and a nurse. A team must be scheduled for a delicate operation at Harbor Hospital. There are two surgeons capable of performing the operation. Three anesthetists are available, and five nurses are on call. How many different operating teams could be assembled?

Exercises

29. Data are grouped according to sex (male or female) and income level (low, medium, or high). How many different pairs are there?

30. A restaurant offers five sandwiches, four cold drinks, and six desserts as part of its luncheon special. How many different specials are there?

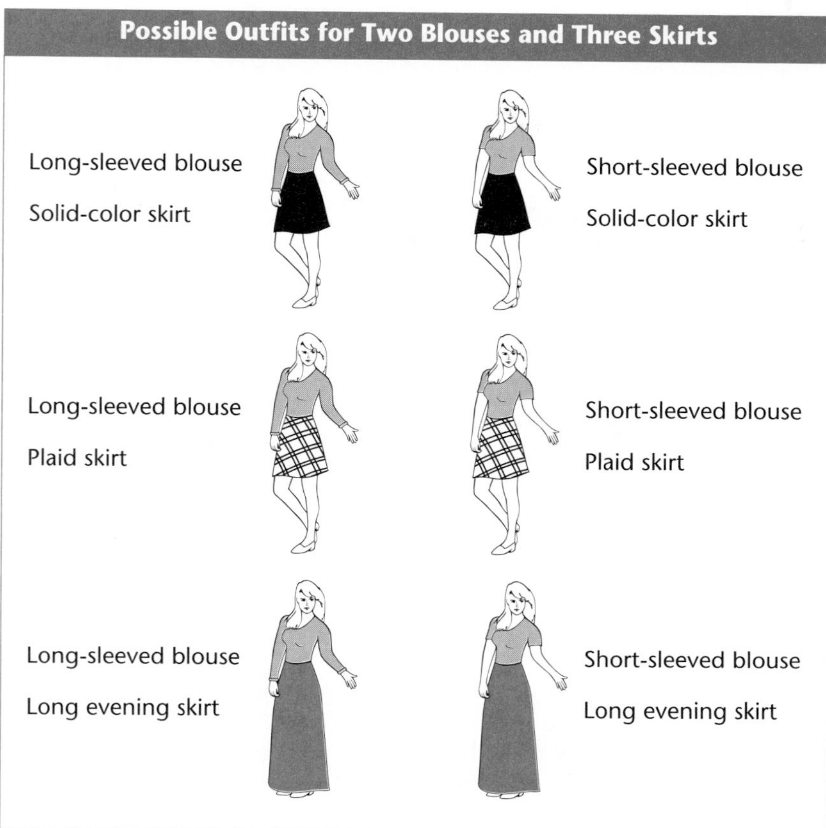

Figure 5-5

31. Tex can travel from El Paso to New Orleans by bus, train, or airplane. He can travel from New Orleans to Tampico, Mexico, by airplane or ship. How many choices for the trip from El Paso to Tampico are there? He can continue on to Veracruz by bus, airplane, or ship. How many possible travel choices are there for the trip from El Paso to Veracruz via New Orleans and Tampico?

32. One of Canada's western provinces is considering using only numbers on their automobile license plates. If only four digits are to appear on the plates (such as 8313), how many different plates are possible? (*Hint:* The first digit could be any number between 0 and 9.)

Permutations

In the previous section, one item was chosen from each of several groups (one blouse from the blouses and one skirt from the skirts, for example). In contrast, if more than one item is to be selected from the *same* group and if the **order of the selection is important,** the resulting arrangement is called a **permutation** of the items.

Permutation An ordered arrangement of a group of objects.

The permutation formula is

$$_nP_r = \frac{n!}{(n-r)!} \qquad \textbf{5-7}$$

where n represents the total number of objects in the group and r represents the number of objects actually selected. For example, if there are seven persons and three are to fill the offices of president, vice president, and secretary, $n = 7$ and $r = 3$.

The notation $n!$ is called "n factorial." It is the product of $n \cdot (n-1) \cdot (n-2) \ldots (1)$. So, 4! (four factorial) is 24, found by $(4) \cdot (3) \cdot (2) \cdot (1)$. and $5! = 120$. Zero factorial, written 0!, is treated as a special case and is defined to equal 1.

Problem Five avid football fans organize the Yellow Jacket Booster Club. The five fans are Smith, Topol, Jackson, Lopez, and McNeil. Three officers are to be selected: a president, a secretary, and a treasurer. One group of officers might consist of President Lopez, Secretary Topol, and Treasurer McNeil. How many distinct slates (permutations) are possible?

Solution

n is the number of avid boosters. (There are 5.)

r is the number of offices to be filled. (There are 3.)

$$_nP_r = \frac{n!}{(n-r)!}$$

$$_5P_3 = \frac{5!}{(5-3)!}$$

$$= \frac{5 \times 4 \times 3 \times 2 \times 1}{2 \times 1}$$

$$= 60$$

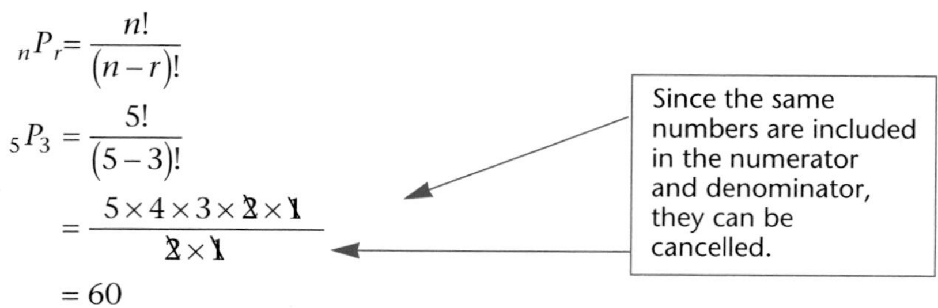

Problem How many three-letter "words" can be made from the letters A, B, C, D, E, and F? No letter may be repeated in a word (such as DBD).

Solution

$$
\begin{aligned}
{}_nP_r &= \frac{n!}{(n-r)!} \\
{}_6P_3 &= \frac{6!}{(6-3)!} \\
&= \frac{6\cdot 5\cdot 4\cdot \cancel{3}\cdot \cancel{2}\cdot \cancel{1}}{\cancel{3}\cdot \cancel{2}\cdot \cancel{1}} \\
&= 120
\end{aligned}
$$

Self-Review 5-12

From the four characters *, + , /, and #, make up computer passwords that are three symbols long, without repeating any of the symbols. For example, the three symbols * # + could make up one password. How many distinct passwords of three-unit length are possible from the four characters?

Exercises

33. There are eight different objects. How many permutations of these eight objects can be selected three at a time?

34. Moss has four bowling trophies that he plans to display on a shelf. In how many ways can they be arranged?

35. "The Triple" at the local racetrack consists of correctly picking the order of finish of the first three horses in the eighth race. Ten horses have been entered in today's race. How many "Triple" outcomes are possible?

36. A real estate developer has eight basic house designs. Zoning regulations in his community do not permit look-alike homes on the same street. The developer has five lots on Mallard Road. In how many different ways can the new homes be arranged?

Combinations

In the previous section dealing with permutations, the *order* of the outcomes was important. For example, one slate of Yellow Jacket officers might be President Jackson, Secretary Topol, and Treasurer McNeil. Another slate might have President McNeil, Secretary Jackson, and Treasurer Topol. Every time there is a change in the order, the count of possibilities is increased by one.

In another problem, the order may not be important. If, for example, Gonzales, Clay, and Higbee were selected to serve as social committee, it would be the same committee if the order were Clay, Higbee, and Gonzales. In other words, the committee of Higbee, Clay, and Gonzales is counted just *once,* regardless of the order of names. Technically, such a group is considered a **combination.**

> **Combination** One particular arrangement of a group of objects or persons selected from a larger group without regard to order.

If order is not important, then we can use the *combination formula* to count the number of combinations:

$$_nC_r = \frac{n!}{(n-r)!r!} \qquad \textbf{5-8}$$

Problem The Yellow Jacket Booster Club wants to select two persons to serve as a membership committee. How many different committees could be selected? One possible committee might consist of Lopez and McNeil. Of course, that committee would be the same as McNeil and Lopez.

Solution There are 10 possible membership committees:

Lopez, McNeil	McNeil, Smith	Topol, Jackson	Jackson, Smith
Lopez, Smith	McNeil, Topol	Topol, Smith	
Lopez, Topol	McNeil, Jackson		
Lopez, Jackson			

Using the combination formula to determine the total number of possible membership committees of two ($r = 2$) from the five-member booster club ($n = 5$), we find:

$$_nC_r = \frac{n!}{(n-r)!r!}$$

$$_5C_2 = \frac{5 \cdot 4 \cdot \cancel{3} \cdot \cancel{2} \cdot \cancel{1}}{(\cancel{3} \cdot \cancel{2} \cdot \cancel{1})(2 \cdot 1)}$$

$$= 10 \text{ (same as listed by name)}$$

Problems

a. A student must register for one of three humanities courses, one of four math courses, and one of three English courses next semester. How many different schedules are possible?

b. There are five different objects. How many permutations of these five objects can be taken two at a time?

c. There are six different objects. How many combinations of the objects can be taken two at a time?

Solutions

a. 36
b. 20
c. 15

Self-Review 5-13

A group of seven mountain climbers wishes to form a mountain-climbing team of five. How many different teams could be formed?

Exercises

37. How many different poker hands are possible from a deck of 52 cards? (Note: You are dealt five cards.)

38. The director of welfare has just received 20 new welfare cases for investigation. A caseworker will be assigned eight new cases. How many different groups of eight cases could caseworker Klein be assigned?

39. Ten employees of Aztec Industries are covered by a union contract. If three of them must be selected to form a bargaining committee, how many distinct committees are possible?

40. The new-car showroom at Berg Motors has room to display four vehicles. The company receives a trailer load of six new models. How many different groups can the manager put out on the floor?

Contrasting Permutations and Combinations

To reemphasize the difference between a permutation and a combination, recall that a **permutation** is an arrangement in which order is important. That is, a, b, c is one permutation; c, a, b is a second permutation; and b, a, c is a third. The permutation formula counts these different orders as different permutations. There are six permutations of the three letters a, b, and c taken three at a time.

If the order of selection is unimportant, the number of orders is called a **combination.** That is, if the three letters can be written a, b, c, or c, a, b, or b, a, c, the combination formula will count these as a single combination. Consequently, there is only one combination of three items taken three at a time.

Self-Review 5-14

A tuna fleet has four flags of different colors. Two flags are arranged on a mast as a signal to the ships in the fleet.

a. If you wish to count the number of different signals that can be constructed using the four flags hoisted two at a time, would you use a permutation or a combination? (Each new order has a different meaning.)

b. A blue flag on top and a yellow one below means "fish sighted." However, a yellow flag on top and a blue flag below means "stormy weather ahead." How many different signals can you construct using the four flags two at a time?

c. Suppose it has been decided that any two flags, such as a green flag on top and a red one below, *or* a red flag on top and a green one below, have the same meaning. If you wish to count the number of signals that could be used, would you apply the permutation formula or the combination formula?

d. Refer to part c above. How many different signals could you construct using the four flags two at a time?

SUMMARY

Probability measures the likelihood that a particular event will occur. A probability may range between 0 and 1, with 0 representing the likelihood that an event cannot happen and 1 representing something that is absolutely certain. A probability may be in the form of a fraction, such as 1/4; a decimal, 0.25; or a percent, 25%.

Three concepts of probability were examined: the classical viewpoint, the relative-frequency viewpoint, and the subjective viewpoint. The classical approach is based on the assumption that the outcomes are equally likely. For example, on the roll of a die, the outcomes ⚀ ⚁ ⚂ ⚃ ⚄ ⚅ are equally likely. We find the probability of any of these outcomes by dividing the number of favorable outcomes by the total number of outcomes.

The relative-frequency approach to probability is based on past experience. If, for example, past experience reveals that 38 out of every 1,249 prisoners in honor farms walk off the farm and escape, the probability of an escape is 38/1,249 = 0.0304.

If no past experience is available to determine a probability, subjective judgment is used. This probability is based on whatever evidence is available. It is not surprising, therefore, that 10 football "experts" disagree with respect to the probability that the Dallas Cowboys, the New York Giants, or the Denver Broncos will go to the Super Bowl next year.

Four rules of probability were examined. The special rule of addition requires that the outcomes of an experiment be mutually exclusive. For two events, $P(A \text{ or } B) = P(A) + P(B)$. The general rule of addition is appropriate when the events are not mutually exclusive. For two events, $P(A \text{ or } B) = P(A) + P(B) - P(A \text{ and } B)$.

The general rule of multiplication allows us to determine the probability that several events may occur at the same time. For two events, $P(A \text{ and } B) = P(A) \cdot P(B \mid A)$. If one event is not dependent on another, then we apply the special rule of multiplication to find the probability that both event A and event B will occur. The rule for two independent events is $P(A \text{ and } B) = P(A) \cdot P(B)$.

Conditional probability reports the likelihood that an event will occur given that another event has already occurred. For example, suppose there are 60 students in a statistics class, 40 men and 20 women. Of the 10 students who earned an A, six were women. The likelihood of selecting a student who received an A given she is a woman is 6/20, where P (female and A)/P (female) = (6/60)/(20/60).

The complement rule is used to compute the likelihood of an event happening, by subtracting the likelihood of the event not happening from 1. For example, if 8% of a case of tennis balls are defective, the probability of selecting a good tennis ball is 92%, found by 1 – 0.08.

Three counting formulas were introduced. The multiplication formula states that if there are m distinct results from doing one thing and n distinct results from doing a second thing, there is a total of $m \times n$ possible results. We use the permutation formula to count if the order of an arrangement is important. The formula is:

$$_nP_r = \frac{n!}{(n-r)!}$$

If the order of an arrangement is not important, we use the combination formula to count the number of possible combinations. The formula is:

$$_nC_r = \frac{n!}{(n-r)!r!}$$

Exercises

41. A dealer receives a shipment of four TV sets, one of which is known to be defective. If one of the sets is sold to a customer, what are the possible outcomes?

42. A car dealer has ten new cars in stock. Four are subcompact, four are compact, and two are luxury models. A car is sold. What is the experiment? What are the possible events regarding the type of car? If the luxury models are four-door sedans and all the others are two-door, describe the possible outcomes with respect to the number of doors.

43. What concept of probability is used in each of the following events to assign the likelihood of the outcomes?

a. Ford decides to market a new subcompact car. The company has never offered a car in this competitive field before, so no comparable data are available. The probability that Ford will sell 2,000,000 cars next year is 0.70.

b. An outfielder for the Toronto Blue Jays, has a season's batting average of .285. What is the probability that he will get a base hit the next time he comes to bat?

c. A student is randomly selected from your class. The probability that his or her birthday is in July is 1/12 = 0.083.

44. Mr. Blume and Ms. White are planning a date on Saturday night. The probability that they will see a basketball game, a movie, or a horse race is 0.25, 0.20, and 0.10, respectively. They will not do more than one of these activities. Of course, they may decide to do something else entirely. What is the probability that they will do one of these things? What is the probability that they will choose one of the two sporting events?

45. A newly married couple plans to have two children and is thinking about the possible sexes of their offspring.

a. What is the experiment?

b. What are the possible outcomes for the sex of their children?

46. A positive integer is picked at random—for example, the last digit of the nine digits in your Social Security number. What is the probability that

a. it is larger than 4?

b. it is an odd number?

c. it is an odd number larger than 4?

47. A review of the records at Lander Community College revealed the following ethnic breakdown of the student population:

Caucasian	1,200
Black	640
Hispanic	280
Asian	80
Native American	60
Other	20

a. What is the probability that a randomly selected student is either Caucasian or Black?

b. What is the probability that the selected person is not a member of either of the two largest categories?

48. A study by the Florida Tourist Commission revealed that 70% of tourists entering the state visit Disney World, 50% visit Busch Gardens, and 40% visit both. What is the probability that a particular tourist visits at least one of the attractions?

49. A certain large city has a morning newspaper, *The Mirror,* and an afternoon paper, *The Observer.* A study shows that 30% of households subscribe to the morning paper and 40% to the afternoon paper. A total of 20% of households subscribe to both. What percent of households subscribe to at least one of the papers?

50. The owner of a ski resort has been informed by a private weather service that the probability of an abundant snow base is 0.80. If there is an abundant snow base, the probability that the owner will make more than a normal profit is 0.85. What is the probability that there will be an abundant snow base and that the owner will make more than a normal profit?

51. A survey of students at Pemberville Tech revealed the following employment breakdown:

	Full-Time	Part-Time	Unemployed	Total
Male	75	75	50	200
Female	125	75	100	300
	200	150	150	500

a. What is the probability of selecting a female student who works full-time?
b. What is the probability of selecting a male student who works?
c. Given that a male student is selected, what is the probability that he is not employed?

52. The Department of Labor conducted a study of 29.3 million families with children under 18 years old. Two traits were noted for each family, namely, whether the family was headed by a female and whether the mother worked. These were the results:

Status of Mother	Male Family Head (in millions)	Female Family Head (in millions)
In labor force	10.9	2.4
Not in labor force	14.4	1.6

If a family is selected at random from this population,
a. what is the probability that the mother is in the labor force?
b. what is the probability that the family is headed by a female?
c. what is the probability that a selected family is headed by a female, or that the mother is in the labor force?
d. given that the mother is in the labor force, what is the probability that she heads the family?

53. A recent newspaper article stated that the probability of downing an attacking airplane at each of four independent missile stations was 0.25. If an attacking plane had to pass over all four stations and the four stations were independent, what is the probability that the plane would reach the target?

54. Mr. Benzy just retired at age 65. His wife is 62 years old. If the probability is 0.50 that a man aged 65 will live another 10 years, and the probability is 0.70 that a woman aged 62 will live another 10 years, what is the probability that both Mr. and Mrs. Benzy will live another 10 years? (Assume that the husband's life expectancy is independent of the wife's life expectancy.)

55. Titusville Oil Company is currently drilling at two new sites. Based on the available geological evidence, the probability of striking oil at Site I (in Oklahoma) is 0.40, and at Site II (in Pennsylvania) is 0.50.

a. Are these events independent?

b. What is the probability that both drilling operations will be successful?

c. What is the probability that neither will be successful?

d. Use the complement rule to compute the probability that at least one is successful.

56. Consider the generating of telephone numbers. Within a given exchange, say 874, how many different numbers are possible? (Assume that the numbers 874-0000 and 874-9999 are acceptable.)

57. Three scholarships are available to students at Salem College. The values are $500, $600, and $700. If 15 qualified students applied, in how many different ways could these scholarships be awarded? (No student may receive more than one scholarship.)

58. A mail order company sells eight different cheeses. As part of a special Christmas holiday package, customers may select three different cheeses for their package. How many different gift packages are possible?

59. A hospital has two independent energy sources. Historical records show there is a 0.98 chance that the primary source will operate during severe weather conditions. The probability that the backup power source will operate under severe weather conditions is 0.95.

a. What is the probability that both will fail?

b. What is the probability that the primary source will fail and the backup source will operate properly?

60. Tests employed in the detection of lung cancer are 90% effective—that is, they fail to detect the disease correctly 10% of the time. If three persons, all known to have cancer, are tested, what is the probability that the disease will be detected in all three cases? What is the probability that two cases will be detected and one case will not?

61. A zoologist has five male and three female guinea pigs. She randomly selects two for an experiment.

a. What is the probability that both are males?

b. What is the probability that both are females?

c. What is the probability that the two are not of the same sex?
d. Are these events independent?

62. The combinations on the gym lockers at Hammerville High have three numbers from 0 to 39, inclusive. For example, one combination is 13–36–27.
a. What is the maximum number of lockers possible in the school so that no two students have the same combination?
If a combination is chosen at random from all possible combinations, what is the probability that
b. it will open a particular lock?
c. each number contains the digit 3?
d. the first number is even, the second number is odd, and the third number is a single digit?
e. the first two numbers are multiples of 11 and the third number is even?

63. A new soft drink will be introduced in three different bottle sizes (6 oz., 8 oz., and 12 oz.) and four different flavors (lemon, lime, grape, and cherry). How many different outcomes are possible?

64. A sports newsletter estimates that the probability that Nick Faldo will win the British Open golf tournament next year is 1/9 and the probability that Seve Ballesteros will win is 1/11. If these estimates are correct, what is the probability that
a. neither of them will win the tournament?
b. one of them will win the tournament?

65. In the Washington School District, there are 100 voters. Of these voters, 45 are male and 62 are in favor of annexation by the city of Greenville. Of those in favor of annexation, 22 are male. Find the probability of selecting one of these voters and the probability of that voter being in favor of annexation and female.

66. A recent study randomly selected 670 Americans and asked them how they get to work. The results:

Type of Transporation	**Type of Worker**		
	Urban	Rural	Total
Automobile	400	200	600
Public transportation	50	20	70
Total	450	220	670

If a worker is selected at random, what is the probability that the worker
a. is an urban worker?
b. uses public transportation?
c. is an urban worker or uses public transportation?
d. is an urban worker and uses public transportation?

e. is an urban worker given that the worker uses public transportation?

67. A 1991 report on drunk drivers in Ohio indicated the age and number of Ohio first-time and repeat offenders.

Age	First-Time	Repeat	Total
16–20	5,311	519	5,830
21–25	10,713	4,104	14,817
26–30	10,301	5,719	16,020
31–35	8,246	4,344	12,590
36–40	5,442	2,596	8,038
41–45	3,474	1,719	5,193
Total	43,487	19,001	62,488

A driver is selected at random.

a. What is the probability that the driver is a repeat offender?

b. What is the probability that the driver is a repeat offender or under 21?

c. What is the probability that the driver is a repeat offender and under 21?

d. What is the probability that the driver is a repeat offender given that the driver is under 21?

68. A recent study revealed that 20% of adult males and 17% of adult females cannot read or write. Assume the population is 49% male and 51% female.

a. In the table below, determine the joint probabilities by indicating the fraction of the adult population that falls in each of the categories.

	Illiterate	Not Illiterate	Total
Male			.49
Female			.51
Total			1.00

b. What is the probability that a randomly selected individual cannot read or write?

c. A person who is illiterate is selected at random. What is the probability that the person is male?

Data Exercises

69. Refer to the Real Estate Data Set, which reports information on homes sold in Northwest Ohio during 1992.

a. Arrange the selling prices into three classes: \$0 to \$179,900; \$180,000 to \$189,900; and \$190,000 or more. Then create a table that shows the grouped selling prices and whether the home had an attached garage.

(1) What percent of the homes sold for less than $180,000?

(2) What percent of the homes sold for less than $180,000 and did not have an attached garage?

(3) Given that a home did not have an attached garage, what percent of the homes sold for less than $180,000?

(4) What percent of the homes did not have an attached garage or sold for less than $180,000?

b. Using the grouped selling prices from part a, create a table that shows the grouped selling prices and the number of bedrooms.

(1) What percent of the homes sold for $190,000 or more?

(2) Given that a home sold for at least $190,000, what percent had five bedrooms?

(3) What percent of the homes sold for at least $190,000 and had five bedrooms?

(4) What percent of the homes sold for at least $190,000 or had five bedrooms?

(5) Suppose that two of the new homeowners are to be contacted by the real estate company. What is the likelihood that both of their homes have five bedrooms?

70. Refer to the Salary Data Set, which refers to a sample of middle managers in Sarasota, Florida.

a. Arrange the salaries into three groups: $0 to $59,900; $60,000 to $64,900; and $65,000 or more. Then create a table that shows both the grouped salaries and the sex of the manager.

(1) What percent of the managers are female?

(2) Given that the manager is female, what percent earn at least $65,000?

(3) What percent of the managers are female and earn at least $65,000?

(4) What percent of the managers are female or earn at least $65,000?

b. Using the grouped salaries from part a, create a table that shows the salaries by region.

(1) What percent of the managers are from Region 3?

(2) If a manager is from Region 3, what percent earned at least $65,000?

(3) What percent of the managers are from Region 3 and earned at least $65,000?

(4) What percent of the managers are from Region 3 or earned at least $65,000?

(5) Suppose that three managers are selected at random. What is the likelihood that all three earn less than $65,000?

CASE: IS A SECOND ONE SAFER?

On January 28, 1986, the space shuttle *Challenger* exploded at an altitude of 46,000 feet, resulting in the deaths of all seven astronauts aboard. The Air Force had studied the possibility of such a disaster and had estimated the probability to be 1 in 34. The *Challenger* crashed on the twenty-fifth mission in the shuttle program.

NASA engineers isolated a particularly vulnerable part in the shuttle design as the cause of the disaster. To decrease the likelihood of this kind of failure in the future, they introduced a second or "redundant" part of this subsystem. The shuttle will fly if either of the parts operates.

Determine the probability that the shuttle program could have at least one disaster in its next 25 flights. Estimate the probability of failure for this new mechanical arrangement. Justify any assumptions you make.

CHAPTER ACHIEVEMENT TEST

Answer all the questions. The answers are at the back of the book.

MULTIPLE-CHOICE QUESTIONS. Select the response that best answers each of the questions.

1. Which definition of probability includes the condition that the events be equally likely?
 a. Relative frequency
 b. Subjective
 c. Classical
 d. None of the above

2. Someone's "educated guess" fits which of the following definitions of probability?
 a. Classical
 b. Relative frequency
 c. Subjective
 d. None of the above

3. The result of an experiment is called
 a. an event.
 b. a subjective probability.
 c. a probability.
 d. an outcome.

4. An experiment has
 a. only one result.
 b. only two results.
 c. two or more results.
 d. None of the above

5. The combination of two or more outcomes results in
 a. an event.

b. a subjective probability.
c. an experiment.
d. a multiplicative rule.

6. For the special rule of addition, the events must be
a. independent.
b. mutually exclusive.
c. equally likely.
d. not equally likely.

7. If the occurrence of one event does not affect the occurrence of another event, the two events are
a. mutually exclusive.
b. independent.
c. always equal.
d. conditional.

8. Two events are mutually exclusive if
a. they overlap on a Venn diagram.
b. when one event occurs, the other cannot occur.
c. the probability of one event occurring does not alter the probability that the other will occur.
d. Both a and b

9. If the order of a set of objects matters, then the set is
a. a combination.
b. a permutation.
c. a sample.
d. an independent event.

10. Selecting a red ace from a deck of cards could be described as
a. an outcome.
b. a sample.
c. an event.
d. a complement.

11. The combined events A and B are shown in a Venn diagram by the area
a. inside both regions A and B.
b. included in region A or region B.
c. outside the two regions.
d. None of the above

12. If two events are mutually exclusive, the probability that both will occur
a. is 0.
b. is 1.
c. is 0.5.
d. cannot be determined.

13. If the outcome of event A is not affected by event B, the two events are
a. mutually exclusive.
b. exhaustive.
c. independent.
d. None of the above

14. We find the number of ways four workers can be chosen for a trip from 50 employees by calculating

a. 4^{50}.

b. 50!/46!.

c. 4×50.

d. 50!/4!46!.

15. We find the number of distinct ways six people can be seated at a table by calculating

a. 6^6.

b. 6!.

c. 6×6.

d. 6!/3!3!.

COMPUTATION PROBLEMS

16. Of 500 items produced at Magnum Manufacturing, 200 are produced on the first shift, 175 on the second, and 125 on the third. What is the probability that a randomly selected item is produced on either the first or the third shift?

17. An English class, Introduction to American Poetry, includes 30 students—10 men and 20 women. Five of the men and two of the women are out-of-state students.

a. Summarize this information in a table.

b. What is the probability of selecting a male student who is from out of state?

c. What is the probability of selecting a male student, given that the student is from out of state?

d. What is the probability of selecting a male student or an out-of-state student?

18. A box contains six ham and four cheese sandwiches. Two sandwiches are randomly selected and eaten. What is the probability that both were ham sandwiches?

19. If 35% of the visitors to St. Croix tour Long John Silver's Castle, 25% shop at Cadbury's Department Store, and these events are independent, what proportion of tourists visit *both* attractions?

20. Forty percent of the residents of LaCasa visited the zoo at least once during the last year. What proportion did *not* go to the zoo last year?

21. You have seen three of eight movies showing at the local cinema. You and a friend randomly select a movie to attend tonight. What is the probability that you haven't seen the movie?

22. For two events A and B, $P(A) = 0.67$, $P(B) = 0.23$, and $P(A \text{ and } B) = 0.12$. Find each of the following values:

a. $P(\sim A)$ **b.** $P(A \text{ or } B)$

23. A contest judge must select a winner, a first runner-up, and a second runner-up from 30 entrants. How many distinct ways can the judge do it?

24. A pollster will select three states from a region consisting of 15 states. How many different samples are there?

25. You must match five numbers selected at random from the numbers 1 through 30, inclusive, to win the Big Lotto. What is the probability that your ticket will win?

ANSWERS TO SELF-REVIEW PROBLEMS

5–1 **a.** The count of favorable responses
b. Any integer between 0 and 6, inclusive
c. Here are three possible events:
1. The majority of comments (4, 5, or 6) are favorable.
2. Nobody agrees (0).
3. Everybody liked it (6).

5–2 **a.** 1/52, or 0.0192
b. 4/52, or 0.0769
c. The classical concept

5–3 **a.** $P(1) = 30/200 = 0.15$
b. $P(3) = 40/200 = 0.20$
c. The relative frequency concept

5–4 **a.** Our estimate is 0.10; your estimate will no doubt be different.
b. The subjective concept

5–5 **a.** Classical
b. Relative frequency
c. Subjective
d. Classical

5–6 **a.** $48/200 = 0.24$
b. $7/200 = 0.035$
c. $55/200 = 0.275$, using the special rule of addition

5–7 "Over 3.00" is event A, "employed" is event B.

$$P(A \text{ or } B) = P(A) + P(B) - P(A \text{ and } B)$$
$$= 0.45 + 0.25 - 0.10$$
$$= 0.60$$

5–8 **a.** $P(A \text{ and } M) = P(A) \cdot P(M \mid A)$
b. 0.25, found by $23/60 \times 15/23$

5–9 **a.** 25/60
b. 8/25
c.

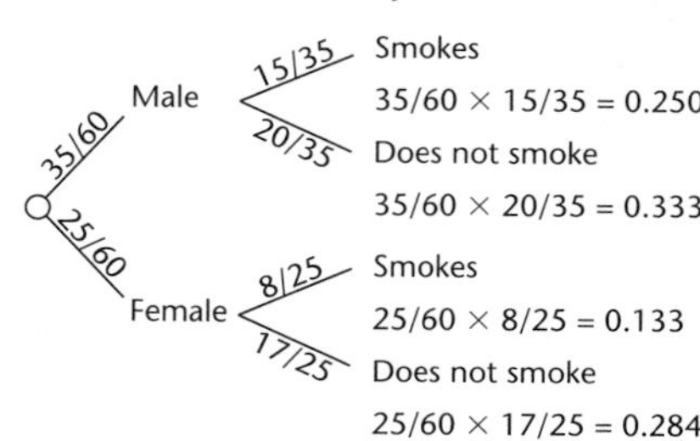

5–10 **a.** Yes
b. $P(A \text{ and } B) = P(A) \cdot P(B) = 1/2 \times 1/2 = 1/4 = 0.25$
c. $1/2 \times 1/2 \times 1/2 = 1/8 = 0.1250$

5–11 $2 \times 3 \times 5 = 30$

5–12

$$n = 4$$
$$r = 3$$
$$_4P_3 = \frac{4!}{(4-3)!}$$
$$= 4 \times 3 \times 2 = 24$$

5–13 $_7C_5 = \dfrac{7!}{(7-5)!5!} = \dfrac{7 \cdot 6}{2 \cdot 1} = 21$

5–14 **a.** Permutation
b. $_nP_r = \dfrac{4!}{(4-2)!} = 12$
c. Combination
d. $_4C_2 = \dfrac{4!}{(4-2)!2!} = 6$

CHAPTER 6

Probability Distributions

OBJECTIVES

When you have completed this chapter, you will be able to

- define a random variable;
- compute the mean and variance of a random variable;
- define a probability distribution;
- distinguish between discrete and continuous probability distributions;
- list the characteristics of the binomial probability distribution and compute probabilities with it;
- list the characteristics of the Poisson probability distribution and compute probabilities with it.

CHAPTER PROBLEM A Taxing Concern

LEWIS GIBSON IS THE CONEWANGO COUNTY REPRESENTATIVE TO THE PENNSYLVANIA STATE LEGISLATURE. HE PRIDES HIMSELF ON HIS ABILITY TO KEEP IN TOUCH WITH HIS CONSTITUENTS. OF PARTICULAR CONCERN AT THIS TIME IS STATE INCOME TAX REFORM. MR. GIBSON ESTIMATES THAT 80% OF VOTERS IN HIS COUNTY FAVOR SOME TYPE OF REFORM, BUT DECIDES TO RANDOMLY CONTACT SIX VOTERS TO SEE HOW MANY ACTUALLY SUPPORT IT. WHAT IS THE LIKELIHOOD THAT EXACTLY FIVE OF THE SIX FAVOR INCOME TAX REFORM? WHAT IS THE LIKELIHOOD THAT A MAJORITY OF THE SIX FAVOR REFORM?

INTRODUCTION

With Chapter 5, we started our investigation of statistical inference. The main objective in statistical inference is to be able to make decisions about an entire population based on a study of just part—that is, a sample—of that population. We illustrated that calculating the probability of an event happening is a very useful technique in this type of decision making. Recall that probability is the likelihood that a particular outcome will happen.

In the previous chapter, we also calculated the probability of specific events. For example, we determined the probability of selecting no persons with a Hispanic surname for a jury of twelve. Now we will study the entire range of events that may result from an experiment. To describe the likelihood of each outcome of this range of events, we use a probability distribution.

WHAT IS A PROBABILITY DISTRIBUTION?

To draw accurate conclusions about the population from which a sample is taken, we need to know the probability of the various outcomes when the sample is selected.

A **probability distribution** describes all the possible outcomes of some random event and their corresponding likelihoods. It is similar to the percent frequency distribution (described in Chapter 2), which reports the percent of observations occurring in each class. However, instead of being merely descriptive of what *has* occurred, a probability distribution is used to project into the future and describe what *will* probably occur.

STAT BYTE

Many professional sports leagues balance the quality of teams with a multi-round draft of college players. Usually the team with the worst record in the previous season picks first, the team with the next-to-worst record picks second, and so on. However, the National Basketball Association has a novel lottery in which only the 11 non-playoff teams participate. The team with the best record has just one chance to pick first, the team with the second-best record has two chances, and so on, with the team having the eleventh-best record (worst in the league) having 11 chances.

Probability Distribution A listing of the outcomes of an experiment that may occur and their corresponding probabilities.

For example, if a coin is tossed twice, the possible outcomes are

First Toss	Second Toss
H	H
H	T
T	H
T	T

where H represents a "head" and T represents a "tail." Table 6–1 describes a situation in which only the *number* of tails that will occur

Table 6-1

Probability Distribution for the Number of Tails Resulting from Two Tosses of a Coin

Number of Tails X		Probability $P(X)$
0		1/4 = 0.25
1		1/2 = 0.50
2		1/4 = 0.25
	Total	4/4 = 1.00

is of interest, and not the order in which they occur. The probability distribution for the number of tails that may occur on two tosses of a coin is shown graphically in Figure 6–1. The number of tails (X) is shown on the horizontal axis and the probability of X on the vertical axis.

Figure 6-1

Graphic Portrayal of the Probability Distribution for the Number of Tails on Two Tosses of a Coin

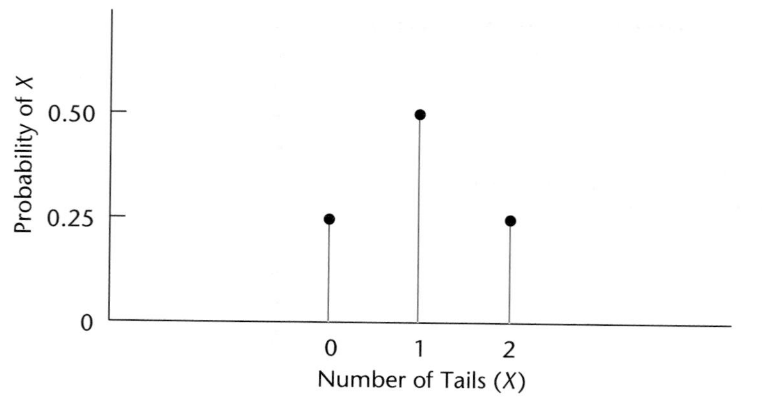

As stated, then, a probability distribution lists the values that may occur (0, 1, or 2 on the X-axis) and their corresponding probabilities (0.25 or 1/4, 0.50 or 1/2, and 0.25 or 1/4 on the Y-axis). As we shall see in later chapters, probability distributions provide the theory upon which is built statistical inference, that is, reasoning from a sampled portion of a population to characteristics of the entire population. Two important features of a probability distribution are as follows:

1. The probability of an outcome X must always be between 0 and 1, inclusive.

2. The sum of the probabilities of all possible mutually exclusive outcomes is 1.

In the coin-tossing experiment, each outcome had a probability between 0 and 1, inclusive, and the sum of the probabilities of 1/4, 1/2, and 1/4 is 4/4, or 1.00.

Self-Review 6-1

Answers to all Self-Review problems are at the end of the chapter.

As an experiment, a coin is to be tossed three times. Some of the possible outcomes are:

Toss		
1	2	3
H	H	H
H	H	T
H	T	H

a. Complete the list of possible outcomes.
b. Develop a probability distribution showing the number of tails possible.
c. Portray the probability distribution in a graph.

DISCRETE AND CONTINUOUS RANDOM VARIABLES

The number of tails that may occur when a coin is tossed twice is also an example of a **random variable.** The random variable, in that instance, was designated by the letter X and could assume any one of the three numbers (0, 1, or 2) assigned to it.

Random Variable A quantity that assumes a numerical value as a result of an experiment.

A random variable may take two forms: discrete or continuous. A **discrete random variable** can assume only certain distinct values. The number of male children in a family with five children can be only 0, 1, 2, 3, 4, or 5; likewise, the number of highway deaths in Kansas during the Thanksgiving weekend can be counted only in increments of 1. Both are examples of discrete random variables. Note that there can be 0, 1, 2, 3, . . . , but there cannot be $2\frac{1}{4}$ children or 17.375 deaths. This does not rule out the possibility that a discrete random variable would assume fractional values. However, the fractional values must have distance between them. For example, mortgage interest rates and stock prices are discrete random variables that are commonly expressed in dollar fractions. The price of IBM stock, for example, could be 98, or $98\frac{1}{4}$, or $98\frac{3}{4}$, but not 98.52174.

Continuous random variables will be discussed in detail in Chapter 7. One example of them would be the weight of young men in your class. A weight might be 176 pounds, or 176.1 pounds, or 176.13 pounds, or 176.134 pounds, and so on, depending on the accuracy of the scale.

How do you tell the difference between a discrete and a continuous random variable? Discrete random variables generally result from counting, whereas continuous random variables are the result of some type of measurement.

A **discrete probability distribution** is based on a discrete random variable. The difference between a random variable and a probability distribution is that the random variable lists only the possible outcomes, whereas *the probability distribution includes both the list of possible outcomes and the probability of each outcome.*

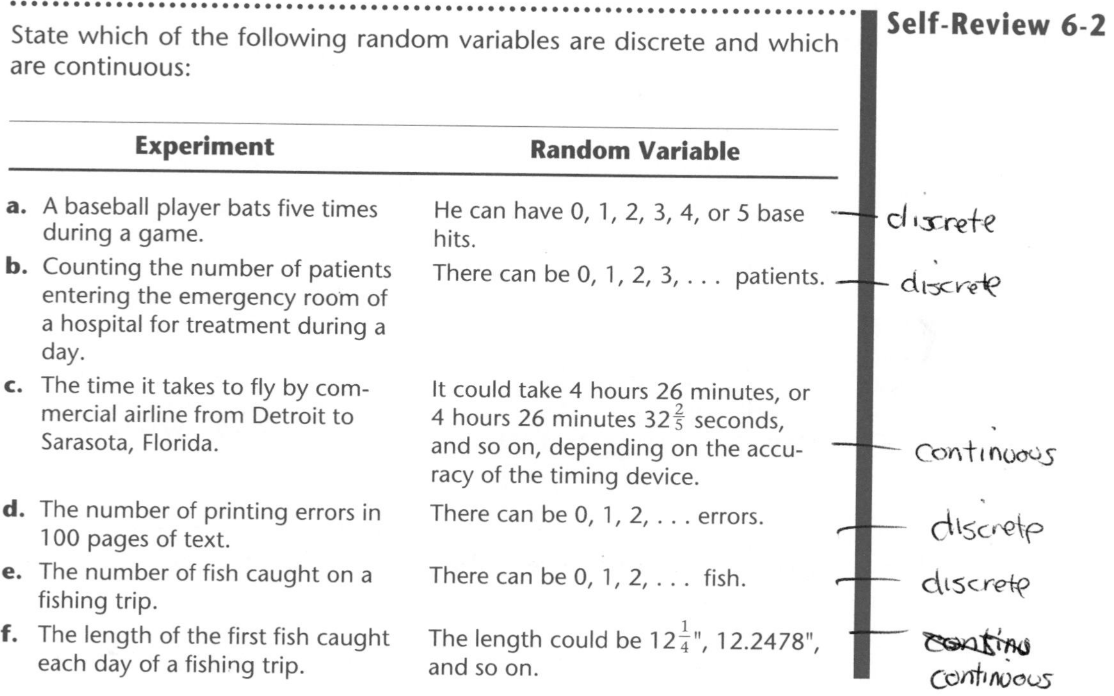

Self-Review 6-2

State which of the following random variables are discrete and which are continuous:

Experiment	Random Variable
a. A baseball player bats five times during a game.	He can have 0, 1, 2, 3, 4, or 5 base hits.
b. Counting the number of patients entering the emergency room of a hospital for treatment during a day.	There can be 0, 1, 2, 3, . . . patients.
c. The time it takes to fly by commercial airline from Detroit to Sarasota, Florida.	It could take 4 hours 26 minutes, or 4 hours 26 minutes $32\frac{2}{5}$ seconds, and so on, depending on the accuracy of the timing device.
d. The number of printing errors in 100 pages of text.	There can be 0, 1, 2, . . . errors.
e. The number of fish caught on a fishing trip.	There can be 0, 1, 2, . . . fish.
f. The length of the first fish caught each day of a fishing trip.	The length could be $12\frac{1}{4}$", 12.2478", and so on.

Why Study Probability Distributions?

We need to calculate the likelihood of a particular sample outcome so that we can draw conclusions about the entire population. The probability of each possible outcome of some experiment is organized into a probability distribution. The following example gives a hint of the manner in which a probability distribution can be used to make an inference about, or "test" for, racial bias. Consider a community composed of 70% white families and 30% black families. Suppose 10 families had been randomly selected to be interviewed on the advisability of rezoning a school district. The probability distribution for the number of black families among the 10 selected is shown in Table 6–2.

Table 6-2 **Number of Black Families in a Sample of 10**

Number X	$P(X)$	
0	0.028	
1	0.121	
2	0.233	
3	0.267	Very probable
4	0.200	
5	0.103	
6	0.037	
7	0.009	
8	0.001	Almost impossible
9	0.000	
10	0.000	

Observe that the most probable event is either 2, 3, or 4 black families. In other words, most of the time the selection process will result in 2, 3, or 4 black families being included. The event that 7, 8, 9, or 10 black families out of the 10 will be selected is quite improbable. Its occurrence would cause you to suspect racial bias in the selection of the sample.

STAT BYTE

Abraham Lincoln once observed that "you can't please all of the people all of the time." If you can please someone 9 out of 10 times and are meeting with 6 people, what is the likelihood that you can please them all? How many of them could you expect to please?

THE MEAN (μ) AND VARIANCE (σ^2) OF A PROBABILITY DISTRIBUTION

In Chapters 3 and 4, measures of location and spread were discussed for a frequency distribution. The mean is a measure of location, and the variance is a measure of spread. In an analogous fashion, a probability distribution can be summarized by its mean, denoted by the Greek letter mu (μ), and its variance, denoted by the Greek letter sigma (σ^2).

The Mean

The mean is a value that is typical of the probability distribution, and it also represents the long-run average value of the random variable. The mean for a probability distribution is also called the **expected value.** The expected value is denoted $E(X)$ and is a "weighted" average. The possible values of the probability distribution are weighted by their corresponding probabilities.

The symbol $P(X)$ is the probability of a particular value X. The mean (μ or expected value) of the probability distribution is computed by:

$$\mu = E(X) = \Sigma[X \cdot P(X)] \qquad \textbf{6-1}$$

In words, the formula directs you to multiply each value X by its probability $P(X)$ and then add these products.

Problem The Pizza Palace offers three sizes of cola—small, medium, and large—to go with its pizza. The colas are sold for \$0.60, \$0.75, and \$0.85, respectively. Forty percent of the orders are for the small cola, 40% for the medium, and 20% for the large. What is the mean amount charged for cola?

Solution This is an example of a discrete probability distribution, because the outcomes are mutually exclusive. A particular cola cannot be both medium *and* large. A customer can purchase only certain sizes of cola. (It is not possible to purchase a \$0.63 cola.)

We determine the mean amount charged by weighting the amount charged by the corresponding percent of the time each sized cola is purchased. The mean for this probability distribution is \$0.71, found by using formula 6–1:

$$\begin{aligned}\mu &= E(X) = \Sigma[X \cdot P(X)]\\ &= (\$0.60)(0.40) + (\$0.75)(0.40) + (\$0.85)(0.20)\\ &= \$0.71\end{aligned}$$

The calculations above could be shown in a table as follows:

Cola Size	Amount Charged	Probability	$X \cdot P(X)$
Small	\$0.60	0.40	\$0.24
Medium	0.75	0.40	0.30
Large	0.85	0.20	0.17
		1.00	$E(X)$ = \$0.71

The Variance

We noted in Chapter 3 that the arithmetic mean is a typical value for a distribution. However, as explained in Chapter 4, it does not describe the amount of spread (variation). The variance measures the spread. It allows you to compare two distributions with the same mean, but with possibly different amounts of spread.

The formula for the variance (σ^2) is:

$$\sigma^2 = \Sigma\left[(X-\mu)^2 \cdot P(X)\right] \qquad \textbf{6-2}$$

To determine the variance:

1. Subtract the mean from each value.
2. Square these differences.
3. Multiply each squared difference by its probability.
4. Sum the resulting products to arrive at the variance.

As noted in Chapter 4, the square root of the variance is the standard deviation.

Problem Compute the variance and the standard deviation of the amount charged for cola at The Pizza Palace. Interpret the result.

Solution The information is repeated below.

Size	*X* Amount Charged	*P*(*X*) Probability	$(X-\mu)^2$	$(X-\mu)^2 \cdot P(X)$
Small	\$0.60	0.40	$(\$0.60 - \$0.71)^2$	(0.0121)(0.40)
Medium	0.75	0.40	$(\$0.75 - \$0.71)^2$	(0.0016)(0.40)
Large	0.85	0.20	$(\$0.85 - \$0.71)^2$	(0.0196)(0.20)
				0.0094

Using formula 6–2, the variance is 0.0094, found by:

$$\begin{aligned}\sigma^2 &= \Sigma\left[(X-\mu)^2 \cdot P(X)\right]\\ &= (0.0121)(0.40) + (0.0016)(0.40) + (0.0196)(0.20)\\ &= 0.0094\end{aligned}$$

The standard deviation is \$0.097, found by $\sqrt{0.0094}$. Both the variance (0.0094) and the standard deviation (\$0.097) are measures of variation in the price of cola at The Pizza Palace. Note that the standard deviation is measured in the same units as the original values.

Self-Review 6-3

The Pizza Palace is located next to The Big Burger. The Big Burger offers four sizes of cola at the following prices:

Size	Price	Probability
Small	$0.50	0.40
Medium	0.60	0.20
Large	0.90	0.20
Extra Large	1.10	0.20

a. Compute the mean price.
b. Compute the standard deviation.
c. Compare the means and the standard deviations for colas at The Pizza Palace and The Big Burger.

Exercises

Answers to the even-numbered exercises are at the back of the book.

1. Compute the mean and the standard deviation of the following probability distribution:

Random Variable	Probability
10	0.60
20	0.30
30	0.10

2. Consider the following probability distribution:

Random Variable	Probability
25	0.10
35	0.25
45	0.40
55	0.20
65	0.05

Compute the mean and the standard deviation.

3. The Perrysburg City Ambulance Service is requesting additional funds from the city council. In support of its request, the following probability distribution was presented by Dirk Plessner, director of Ambulance Services:

Number of Trips per Day	Fraction of Days
0	0.10
1	0.40
2	0.30
3	0.10
4	0.10

According to Mr. Plessner, this indicates that on 30% of the days, two trips, or ambulance runs, were made.

Compute the mean number of trips per day and the standard deviation.

4. Mr. Tim Waltzer, Principal of Home Street School, holds three open houses each school year. Records show the following probabilities that a child's parents (one or both) will attend from 0 to 3 of the open houses:

Number of Open Houses Attended	Probability
0	0.20
1	0.20
2	0.30
3	0.30

Compute the mean and the standard deviation.

THE BINOMIAL PROBABILITY DISTRIBUTION

One widely used discrete probability distribution is the **binomial distribution.** The binomial distribution describes many experiments that require the probability of the number of successes or failures in a sample. The following characteristics identify the binomial probability distribution:

1. Each outcome is classified into one of two mutually exclusive categories. An outcome is either a "success" or a "failure." For example, a lottery ticket is either a winning number (a success), or not a winning number (a failure). The outcomes are mutually exclusive, meaning that a ticket cannot be both a success and a failure at the same time. We do not record the amount of winnings, only whether we won or lost.

2. The binomial distribution is a count of the number of successes. It is a discrete distribution. X can assume only certain integer values, namely, 0, 1, 2, 3, . . . successes. There cannot be, for example, $2\frac{1}{2}$ successes.

3. Each "trial" is *independent.* This means that the outcome of one trial does not affect the outcome of any other trial. It is common to use the word "trial" when discussing the binomial probability distribution. Usually, the total number of trials is the same as the size of the sample (n).

4. The probability of success π on each trial is the same from trial to trial. For a lottery ticket, the probability of having a winning number on one ticket is the same as the probability of having a winning number on another ticket. The Greek lowercase π

STAT BYTE

The *Annals of Internal Medicine* reports that a blood test on a normal person has a 5% chance of being read as abnormal by a laboratory computer. A 12-test profile of a healthy person will therefore appear normal only 54% of the time!

(pi) is often used in statistics to represent a proportion of "successes" in a population. It should not be confused with the mathematical constant 3.14159.

Self-Review 6-4

The probability that expectant parents will have a girl is 1/2, or 0.50. Likewise, the probability they will have a boy is 0.50. The probability of having 0, 1, 2, 3, 4, and 5 girls in a family of five children is:

Number of Girls X	Probability $P(X)$
0	0.0312
1	0.1562
2	0.3125
3	0.3125
4	0.1562
5	0.0312

Explain why this distribution qualifies as a binomial probability distribution.

Constructing a Binomial Probability Distribution

To construct a binomial probability distribution, we must know (1) the total number of trials and (2) the probability of success on each trial.

Problem Recall from the Chapter Problem that Lewis Gibson estimates 80% of the voters in Conewango County to favor some type of income tax reform in Pennsylvania. To make sure, he randomly selects six voters from his county. Construct a binomial probability distribution for the probability that exactly 0, 1, 2, 3, 4, 5, and 6 out of the six sampled voters will favor state income tax reform. In this problem, the number of trials is six and the probability of success (will vote for tax reform) is 0.80.

Solution To illustrate, let us first compute the probability that exactly five of the six voters selected in the sample will favor tax reform. If F represents a person who favors tax reform and O represents someone who opposes it, one outcome is that the first voter contacted opposes tax reform and the rest favor it. In symbols, this outcome could be written O, F, F, F, F, F. The voter preferences are presumed independent, so the probability of this particular joint occurrence is the product of the individual probabilities. Hence, the likelihood of finding a voter who is opposed to tax reform followed by five who favor tax reform is:

$$(0.2)(0.8)(0.8)(0.8)(0.8)(0.8) = 0.0655$$

However, the problem does not require that the first voter be the one who is opposed to tax reform. The voter opposing tax reform could be any one of the six people sampled. Any of the six outcomes listed below results in exactly five people favoring tax reform and one opposing it.

Order of Occurrence	Probability of Occurrence
O, F, F, F, F, F	(0.2)(0.8)(0.8)(0.8)(0.8)(0.8) = 0.0655
F, O, F, F, F, F	(0.8)(0.2)(0.8)(0.8)(0.8)(0.8) = 0.0655
F, F, O, F, F, F	(0.8)(0.8)(0.2)(0.8)(0.8)(0.8) = 0.0655
F, F, F, O, F, F	(0.8)(0.8)(0.8)(0.2)(0.8)(0.8) = 0.0655
F, F, F, F, O, F	(0.8)(0.8)(0.8)(0.8)(0.2)(0.8) = 0.0655
F, F, F, F, F, O	(0.8)(0.8)(0.8)(0.8)(0.8)(0.2) = 0.0655
	0.3930

Therefore, the probability that exactly five of the six voters sampled will favor tax reform is equal to the sum of the six possibilities, or 0.3930.

The probabilities of the other outcomes, such as 0, 1, 2, and so on, of the six sampled voters favoring tax reform could be computed in a similar fashion. A more direct method, however, is to use the **binomial formula**

$$P(X) = \frac{n!}{X!(n-X)!}\pi^X(1-\pi)^{n-X} \qquad \textbf{6-3}$$

where

n is the total number of trials. It is 6 in this problem.
π is the probability of success on each trial. It is 0.80.
X is the number of observed successes. Here it is the number favoring reform.

Substituting these values in formula 6–3, the probability that exactly five will favor tax reform is:

$$P(X) = \frac{n!}{X!(n-X)!}\pi^X(1-\pi)^{n-X}$$

$$P(5) = \frac{6!}{5!(6-5)!}(0.80)^5(1-0.80)^{6-5}$$

Recall from your algebra class that 6! (six factorial) means $6 \times 5 \times 4 \times 3 \times 2 \times 1$. When the same number is included in both the numerator and the denominator, it can be cancelled. Continuing:

$$P(5) = \frac{6 \cdot \cancel{5} \cdot \cancel{4} \cdot \cancel{3} \cdot \cancel{2} \cdot \cancel{1}}{\cancel{5} \cdot \cancel{4} \cdot \cancel{3} \cdot \cancel{2} \cdot \cancel{1} \cdot \cancel{1}}(0.80)^5(0.20)^1$$
$$= (6)(0.80)^5(0.20)^1$$
$$= 0.3930$$

This is the same probability obtained earlier. The remaining probabilities for 0, 1, 2, . . . were computed in similar fashion and are shown in Table 6–3.

Problems Let X follow a binomial probability distribution. For each X compute its probability.

a. $n = 4$, $\pi = 0.20$, $X = 2$
b. $n = 7$, $\pi = 0.70$, $X = 5$
c. $n = 5$, $\pi = 0.60$, $X = 3$

Solutions

a. 0.154
b. 0.318
c. 0.346

Binomial Probability Distribution for an n of 6 and a π of 0.80 — **Table 6-3**

Number in Favor of Tax Reform X	Probability of Occurrence $P(X)$
0	0.000
1	0.002
2	0.015
3	0.082
4	0.246
5	0.393
6	0.262
Total	1.000

Self-Review 6-5

A fire chief estimates that the probability an arsonist will be arrested is 30%. What is the probability that exactly three arsonists will be arrested in the next five deliberately set fires?

Exercises

5. An insurance representative has appointments with four prospective clients tomorrow. From past experience, she knows that the probability of making a sale on any appointment is

0.20. What is the probability that she will sell a policy to three of the four prospective clients?

6. It is estimated that 40% of those 65 or older did some type of volunteer work last year. If five individuals are randomly selected from this age group, what is the probability that exactly three did some type of volunteer work last year?

7. Sixty percent of the applicants to Gator Tech are accepted. What is the probability that exactly four of the next five applicants will be accepted?

8. One-fourth of a certain breed of rabbits are born with long hair. What is the probability that, in a litter of five rabbits, none has long hair?

Using Binomial Tables

Another way to find the probabilities needed to construct a binomial distribution is to use tables that give the probabilities for various values of n and π. Such a table is given in Appendix A; a small portion of it, for the case where $n = 6$, is duplicated in Table 6–4.

Table 6-4 **Binomial Probability Distribution, $n = 6$**

N = 6
Probability

X	.05	.1	.2	.3	.4	.5	.6	.7	.8	.9	.95
0	.735	.531	.262	.118	.047	.016	.004	.001	.000	.000	.000
1	.232	.354	.393	.303	.187	.094	.037	.010	.002	.000	.000
2	.031	.098	.246	.324	.311	.234	.138	.060	.015	.001	.000
3	.002	.015	.082	.185	.276	.313	.276	.185	.082	.015	.002
4	.000	.001	.015	.060	.138	.234	.311	.324	.246	.098	.031
5	.000	.000	.002	.010	.037	.094	.187	.303	.393	.354	.232
6	.000	.000	.000	.001	.004	.016	.047	.118	.262	.531	.735

In the previous problem involving a sample of six voters, a π of 0.80 was used to illustrate the use of formula 6–3. We find the probability of exactly five out of six voters being in favor of tax reform by the following procedure, which relies on Appendix A. The steps are:

1. Find the section where the sample size or the number of trials is given under the first heading n. In this case it is 6.
2. Within that block, locate the row headed by the number of successes under the column headed X. It is 5.
3. Move across the row until you reach the column headed by the desired probability, or value of π. It is 0.80.
4. The desired probability is where the row and column of interest intersect. In Table 6–4, for $n = 6$, the value 0.393 is located

at the intersection of row 5 and column 0.8. This result agrees with the earlier calculation.

Problems Use Appendix A to find each of the following probabilities of X:

a. $n = 6$, $\pi = .20$, $X = 1$
b. $n = 10$, $\pi = .40$, $X = 3$
c. $n = 7$, $\pi = .90$, $X = 6$

Solutions

a. .393
b. .215
c. .372

Self-Review 6-6

In Self-Review 6–5, the fire chief estimated the probability of an arsonist being arrested at 0.3. The question asked was, What is the probability that, for the next five arson-related fires, three arsonists will be arrested? The probability was computed to be 0.132. Using Appendix A, complete the binomial probability distribution for this problem.

Problem From Table 6–4, what is the probability that five or more voters favor tax reform? What is the probability that fewer than six favor tax reform?

Solution We can answer the first question by using the special rule of addition, formula 5–1, which tells us to add the probabilities for five and six: $P(5) + P(6) = 0.393 + 0.262 = 0.655$.

We determine the probability of fewer than six either by $P(0) + P(1) + P(2) + P(3) + P(4) + P(5) = 0.000 + 0.002 + 0.015 + 0.082 + 0.246 + 0.393 = 0.738$ or by applying the complement rule, formula 5–6: $P(X < 6) = 1 - P(6) = 1 - 0.262 = 0.738$.

Self-Review 6-7

The binomial probability distribution developed for arsonists in Self Review 6–6 is:

X	*P(X)*
0	0.168
1	0.360
2	0.309
3	0.132
4	0.028
5	0.002

a. What is the probability that three or more arsonists will be arrested for the next five fires?
b. What is the probability that one or more arsonists will be arrested for the next five fires?

Exercises

9. Let X follow a binomial probability distribution. Find each of the following probabilities using the binomial tables:

a. $n = 4, \pi = 0.20, X = 2$
b. $n = 7, \pi = 0.70, X = 5$
c. $n = 5, \pi = 0.60, X = 3$

10. Let X follow a binomial probability distribution. Find each of the following probabilities using the binomial tables:

a. $n = 5, \pi = 0.10, X = 2$
b. $n = 10, \pi = 0.30, X = 4$
c. $n = 8, \pi = 0.40, X = 6$

11. The instructor in Political Science 101 gives a weekly five-question multiple-choice quiz. For each question there are five choices, but only one of them is the correct answer. A student did not attend class or read the text assignments (a common occurrence). He did, however, take the weekly quiz.

a. Using the binomial table in Appendix A, develop a binomial probability distribution for the number of correct answers.
b. The instructor had announced that students who score three or more correct out of five pass the quiz. What is the probability that a student who neither attended class nor read the assignments will pass the test?

12. As noted earlier, half of the newborn babies are girls. For families with five children, what is the probability

a. of having three girls and two boys?
b. that all the children are girls?
c. of having at least one girl?

13. Thirty percent of all automobiles sold in the United States are foreign-made. Four new automobiles are randomly selected.

a. Referring to Appendix A, what is the probability that none of the four is foreign-made?
b. Construct a binomial probability distribution showing the probabilities of 0, 1, 2, 3, and 4 out of four being foreign-made.
c. What is the probability that at least one is foreign made?

14. Forty percent of the residents of North Toledo are opposed to the widening of Berdan Avenue. In a random sample of ten residents, what is the probability that the number opposed to the widening is

a. exactly five?
b. at least five?
c. between seven and nine, inclusive?
d. fewer than five?

We can also solve the tax reform problem using statistical software, such as MINITAB. The MINITAB procedure, for example, is

PDF, which stands for probability density function. A subcommand BINOMIAL is used with n, the number of trials or the sample size, and π, the probability of a success. The output is as follows:

```
MTB  >  pdf;
SUBC >  binomial  n = 6  pi = 0.8.

     BINOMIAL WITH N = 6 P = 0.800000
           K              P(X = K)
           0              0.0001
           1              0.0015
           2              0.0154
           3              0.0819
           4              0.2458
           5              0.3932
           6              0.2621
```

The output is the same as Table 6–3 except there is an additional decimal place.

Appendix A is limited in that it gives probabilities for an n of 1 to 20 and π values of 0.05, 0.10, 0.20, . . . , 0.90, 0.95. There are two methods for arriving at binomial probabilities not found in Appendix A. The normal approximation to the binomial may be used, a procedure we will describe in Chapter 7. Or a software package can generate the binomial probability distribution, given n and π. The following is the binomial probability distribution for an n of 22 and $\pi = 0.237$:

```
MTB  >  pdf;
SUBC >  binomial  n = 22  pi = 0.237.

     BINOMIAL WITH N = 22 P = 0.237000
           K              P(X = K)
           0              0.0026
           1              0.0178
           2              0.0580
           3              0.1202
           4              0.1773
           5              0.1982
```

6	0.1745
7	0.1239
8	0.0721
9	0.0349
10	0.0141
11	0.0048
12	0.0014
13	0.0003
14	0.0001
15	0.0000

Several other points should be made regarding the binomial distribution.

1. If n remains the same but π becomes closer to 0.50, the shape of the binomial probability distribution becomes more symmetrical.
2. If π, the probability of a success, remains the same but n becomes larger and larger, the shape of the distribution also becomes more symmetrical.
3. The mean (μ) and the variance (σ^2) of a binomial distribution can be computed by:

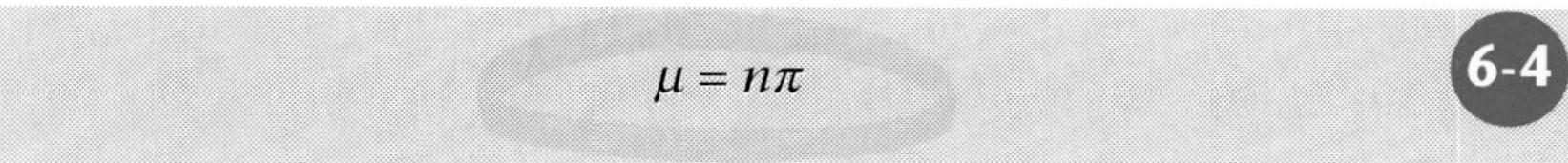

$$\mu = n\pi \quad \text{(6-4)}$$

$$\sigma^2 = n\pi(1-\pi) \quad \text{(6-5)}$$

For the example regarding the number of voters favoring tax reform, where $n = 6$ and $\pi = 0.80$, the mean and variance are computed as follows:

$$\mu = n\pi = 6(0.80) = 4.80$$
$$\sigma^2 = n\pi(1-\pi) = 6(0.80)(1-0.80) = 0.96$$

The mean and variance can also be calculated by formula 6-1 and formula 6-2.

Number in Favor X	P(X)	X · P(X)	$(X - \mu)^2 \cdot P(X)$
0	0.000	0	0
1	0.002	0.002	0.02888
2	0.015	0.030	0.11760
3	0.082	0.246	0.26568
4	0.246	0.984	0.15744
5	0.393	1.965	0.01572
6	0.262	1.572	0.37728
		4.799	0.96260

Notice that the values computed for μ and σ^2 in the above table are the same as those found by 6-4 and 6-5, accept for rounding.

THE POISSON PROBABILITY DISTRIBUTION

Another important discrete probability distribution is the **Poisson distribution.** The Poisson distribution counts the number of successes in a fixed interval of time or within a specified region. Some examples where the Poisson distribution could be applied:

- The number of calls arriving per hour at the Nawash Fire Department
- The number of murders reported in a month in Detroit
- The number of orders received by a mail order firm per day
- The number of raisins in a slice of raisin bread

The "fixed interval of time" is an hour in the case of calls to the Nawash Fire Department, and a month in the case of reported murders in Detroit. In the example of the slice of raisin bread, we are speaking of a "region." The distribution of the number of raisins in a "thin" slice will be different than in a "thick" slice.

To apply the Poisson distribution, two conditions must be met:

1. The number of successes that occur in any interval is independent of those that occur in other non-overlapping intervals.
2. The probability of a success in an interval is proportional to the size of the interval.

A *success* refers to the event of interest, such as the Nawash Fire Department receiving a call. An *interval* refers to the period of time involved, such as an hour. In short, the two important traits of the Poisson distribution are independence and proportionality.

How do these conditions relate to the Nawash Fire Department? If the Department receives three calls between 8 a.m. and 9 a.m., that outcome does not have any effect on the number they will receive between 10 a.m. and 11 a.m. The probability of a call in the two-hour interval between 8 a.m. and 10 a.m. is twice as likely as the one-hour interval between 3 p.m. and 4 p.m., for example.

The Poisson Distribution is described mathematically by the formula

$$P(X) = \frac{\mu^X e^{-\mu}}{X!} \qquad \textbf{6-6}$$

where

μ	is the mean number of successes.
e	is the mathematical constant 2.7183.
X	is the number of successes in the interval.
$P(X)$	is the probability of X successes in an interval.

The mean of the Poisson distribution is μ and the standard deviation is $\sqrt{\mu}$. It is positively skewed.

Problem The number of cars arriving at a gasoline station follows a Poisson distribution with a mean of three cars every 10 minutes. What is the probability that exactly two cars will arrive in the next 10 minutes?

Solution To evaluate this form, we enter 2 for X and 3 for μ in the formula:

$$P(X) = \frac{3^2 e^{-3}}{2!} = \frac{9(.0498)}{2} = 0.224$$

We find that the probability exactly two cars will arrive in the next 10 minutes is 0.224.

The Poisson distribution can be used to approximate the binomial distribution when the probability of a success (π) is small and the number of trials (n) is very large.

Problem Records show that the probability a tire will suffer a blowout in the first year is 0.005. If 100 tires are used for a year, what is the probability of exactly two blowouts?

Solution Note that the binomial requirements are satisfied. That is, there is a constant probability of success (0.005), a fixed number of trials (100), only two outcomes (the tire blows out or it does not), and independent trials (if the fifth tire blows out, that does not mean that the eighteenth will). The mean number of blowouts in 100 trials is 0.5, found by formula 6–4, where $\mu = n\pi = 100(0.005)$. The probability of exactly two blowouts is:

$$P(2) = \frac{(0.5)^2 e^{-.5}}{2!} = \frac{(.25)(.6065)}{2} = 0.0758$$

Appendix B is used to obtain the probability for selected values of μ. To use the Appendix for this example, find the column headed 0.5. Move down that column to the row where $X = 2$ and read the probability. It is 0.0758, the same as computed previously. This is also given in Table 6–5.

Table 6–5

Poisson Probability Distribution

	μ								
X	0.1	0.2	0.3	0.4	0.5	0.6	0.7	0.8	0.9
0	0.9048	0.8187	0.7408	0.6703	0.6065	0.5488	0.4966	0.4493	0.4066
1	0.0905	0.1637	0.2222	0.2681	0.3033	0.3293	0.3476	0.3595	0.3659
2	0.0045	0.0164	0.0333	0.0536	0.0758	0.0988	0.1217	0.1438	0.1647
3	0.0002	0.0011	0.0033	0.0072	0.0126	0.0198	0.0284	0.0383	0.0494
4		0.0001	0.0003	0.0007	0.0016	0.0030	0.0050	0.0077	0.0111
5					0.0002	0.0004	0.0007	0.0012	0.0020
6							0.0001	0.0002	0.0003

Problems Use Appendix B to find each of the following probabilities:

a. $\mu = 2.0, X = 6$
b. $\mu = 7.0, X = 9$
c. $\mu = 0.9, X = 1$

Solutions

a. .0120
b. .1014
c. .3659

Self-Review 6-8

In the Problem and Solution above, the probability of a tire blowing out in the first year is 0.005, and a set of 100 tires was obtained. Determine the probability of exactly one blowout. Then find the probability of at least one blowout. Use Appendix B.

Exercises

15. The mean number of accidents in a shoe factory is 0.10 per day. What is the probability that during a randomly selected day

a. there will be no accidents?
b. there will be exactly one accident?
c. there will be at least one accident?

16. Five percent of the graduates of the Modern Method Driving School fail to obtain the driver's license on their first attempt. If there are 60 graduates this month, what is the probability that they will all pass?

17. Lloyd's of London has determined that each year one in a thousand people is injured while driving a stunt car for a movie. If Lloyd's insures 400 stunt drivers, what is the probability that in a particular year

a. none of their policyholders will be injured?
b. at least two of their policyholders will be injured?
c. no more than two will be injured?

18. The mean number of arrests for shoplifting is 0.20 per week at Hidleman's Department Store. What is the probability that in a particular week

a. there is one arrest?
b. there are two or more arrests?

SUMMARY

A probability distribution is a listing of the outcomes of an experiment and of their corresponding probabilities. For example, the following probability distribution gives the likelihoods of having 0, 1, and 2 girls for a family with two children:

Number of Girls	Likelihood (probability)
0	1/4 = 0.25
1	2/4 = 0.50
2	1/4 = 0.25
Total	4/4 1.00

A discrete probability distribution has a countable number of outcomes. For example, the probability distribution just cited can have only three outcomes: 0, 1, or 2 girls. A continuous probability distribution is based on a continuous random variable. That is, the variable may take on an infinite number of values (within a given range), depending on the accuracy of the measuring device. Temperatures in Maine, for example, take on values ranging from 50°F to 95°F during the summer months. A temperature reading may be recorded as 58°, 58.2°, 58.23°, or 58.227°F.

The mean or expected value of a discrete probability distribution is a weighted average. The values of the random variable are multiplied by the probability, and these products are added. Thus,

$$\mu = E(X) = \Sigma[X \cdot P(X)]$$

where $P(X)$ is the probability of particular values of X. The variance of a probability distribution is computed as follows:

$$\sigma^2 = \Sigma\left[(X - \mu)^2 \cdot P(X)\right]$$

The binomial probability distribution is a discrete probability distribution. Its characteristics are as follows: (1) There are only two possible outcomes to a trial—success or failure; (2) the outcomes of a trial are mutually exclusive; (3) each trial is independent; and (4) the probability of a success remains the same from trial to trial. The probability of having a girl, for example, is the same for each child born.

To construct a binomial probability distribution, we need the total number of trials and the probability of success on each trial. For example, in developing a probability distribution for the number of girls likely to be born to a family with two children, the total number of trials is 2, and the probability of having a girl each trial is 1/2. The list of probabilities can be determined by the binomial formula:

$$P(X) = \frac{n!}{X!(n - X)!}\pi^X(1 - \pi)^{n-X}$$

As an alternative to calculating all the probabilities needed to construct a binomial probability distribution, we can use a table containing the binomial probabilities (see Appendix A). Many computer programs are also available to do the routine calculations.

Another discrete distribution discussed was the Poisson. Its formula is

$$P(X) = \frac{\mu^X e^{-\mu}}{X!}$$

where e is the constant 2.7183.

Exercises

19. Sam sells used cars. Last month he sold one car on 30% of the days, two cars on 40% of the days, three cars on 20% of the days, and no cars on 10% of the days. What is the mean number of cars sold per day? What is the standard deviation?

20. The school band members are selling boxes of candy door to door in an effort to raise money to purchase new band uniforms. Mr. Sprague, the band director, has found the following probability distribution for the number of boxes sold per house:

Number of Boxes Sold *X*	Probability *P(X)*
0	0.50
1	0.30
2	0.20

a. How many boxes does a student expect to sell at a particular house?
b. What is the standard deviation of the distribution?
c. What is the probability that a band member will not sell any candy at two consecutive houses?

21. Ten percent of the students in Professor Green's biology class earn the grade of A. In a sample of seven students who have taken this class, compute the probability that the number of A grades is
a. exactly two.
b. at least one.
c. none.
d. seven.
e. How many students out of the seven would you expect to get an A?

22. The Tameron Apartments is a rather small complex. The owner is studying the number of unoccupied apartments and has gathered the following information:

Number of Vacant Units	Probability
0	0.30
1	0.30
2	0.20
3	0.20

Compute the mean and the standard deviation for the number of vacant units.

23. A new pilot TV show has been created. The network president thinks 60% of the population will like the new show. If the pilot is shown to 15 people, and the president is correct, what is the probability that 10 or more of the 15 will indicate they liked it? How many of the 15 should the network president expect to like the show?

24. List the characteristics of the binomial distribution.

25. A poplar tree, if less than three feet high and transplanted in the spring, has a 40% chance of survival. If six such trees are transplanted, what is the probability that exactly five will survive? What is the probability that 2, 3, 4, or 5 will survive? Compute the mean and the variance.

26. A particular type of birth-control device is effective 90% of the time when used as directed. If a sample of 15 couples use the device, what is the probability that none of the devices will fail?

27. If a baseball player with a batting average of .300 comes to bat five times in a game, what is the probability he will get
- **a.** exactly two hits?
- **b.** fewer than two hits?
- **c.** at least one hit?
- **d.** Compute the mean and the variance of the number of hits.

28. It is estimated that AT&T has a 70% share of the long-distance telephone market. Fifteen long-distance calls are selected at random.
- **a.** What is the expected number of AT&T calls?
- **b.** What is the probability that 10 are AT&T calls?
- **c.** What is the probability that 10 or less are AT&T calls?

29. Researchers have defined leading a "sedentary lifestyle" as getting less than 20 minutes of exercise three times a week. It is claimed that 60% of all Americans lead a sedentary lifestyle. If you randomly select eight people, what is the probability that
- **a.** exactly six lead a sedentary lifestyle?
- **b.** none of them leads a sedentary lifestyle?
- **c.** at least five lead a sedentary lifestyle?

30. Binge drinking (consuming five or more alcoholic drinks on a single occasion) is a habit of 20% of the national population. If you go to a party with six people, what is the probability that
- **a.** none of them engages in binge drinking?
- **b.** exactly one does?
- **c.** more than one does?

31. National studies reveal that 40% of adult drivers do not use seatbelts. If nine adult drivers are stopped at random on the Florida Turnpike, what is the probability that
- **a.** exactly three are not wearing seatbelts?
- **b.** exactly three are wearing seatbelts?
- **c.** at least four are not wearing seatbelts?
- **d.** at least four are wearing seatbelts?

32. Twenty percent of all marriages fail in the first year. What is the probability that the marriage of four or fewer of the 15 couples married in Findlay, Ohio last weekend will have their marriage fail within a year?

33. Thirty percent of all elderly people have a flu shot each year. Suppose Dr. Lowe had 12 elderly patients. What is the probability that

a. none of them receives a flu shot?

b. at least one receives a flu shot?

c. most of them receive a flu shot?

34. A large department store finds that 1% of its accounts receivable will not be collected. A sample of 100 accounts is selected for audit.

a. What is the probability that all accounts will be collected?

b. What is the probability that at least one will not be collected?

35. Customers arrive at the drive-in windows of Bank One at a rate of two per minute.

a. What is the probability that exactly two customers arrive in a particular minute?

b. What is the probability of no arrivals in a particular minute?

c. What is the mean number of arrivals?

36. Health records indicate that 70% of all first-graders are fully immunized by the time they reach age six. A class of 14 first-grade students is being studied. What is the probability that more than 10 are fully immunized?

37. A bank processes 10 personal loan applications per day. The probability that a particular application will be denied is 0.30. What is the probability that more than two will be denied on the same day?

38. A certain medicine is 70% effective; that is, out of every 100 patients who take it, 70 are cured. A group of 12 patients is given the medicine. Find the probability that

a. eight are cured.

b. fewer than five are cured.

c. ten or more are cured.

39. It is known that 30% of all high-school graduates go to a four-year college. If a very small graduating class of 20 is considered, find the probability that

a. more than eight go to a four-year college.

b. fewer than six go to a four-year college.

c. more than nine go to a four-year college.

40. A telephone salesperson expects to make a sale on 8% of her calls. If 50 calls are placed, what is the probability that

a. two will result in a sale?

b. at least one will result in a sale?

41. A recent study showed that 60% of airline flights from Chicago to Atlanta are on time. A sample of 12 flights is selected.

a. How many flights would you expect to be on time?

b. What is the likelihood that more than seven flights are on time?
c. What is the likelihood that exactly seven flights are on time?

42. For the hours 4 p.m.–10 p.m., an average of five people arrive per hour needing treatment at St. Luke's Hospital Emergency Room.
a. What is the likelihood that no people will arrive at a particular hour?
b. What is the likelihood that less than five will arrive at a particular hour?
c. What is the likelihood that less than five will arrive in each of two consecutive hours?

43. From past experience, 30% of the students at Aloha University graduate within five years. In a freshman writing class, there are 15 students.
a. How many of these students would you expect to graduate in five years? What is the standard deviation?
b. What is the probability that at least four of the students will graduate?
c. What is the probability that less than four will graduate?

44. A student accuses an instructor of being chauvinistic and of harassing female students. As evidence, the accuser points to the fact that the instructor has called on a male student on only four of the last 10 occasions. Thirty percent of the large class are women. What is the probability that the result cited might have occurred by chance alone?

45. Describe the differences between the binomial probability distribution and the Poisson probability distribution.

46. In a certain manufacturing process, 5% of the product does not meet specification. A group of 20 parts is selected. Determine the probability that exactly one part is defective using both the binomial and the Poisson probability distributions.

Data Exercises

47. Refer to the Real Estate Data Set, which reports information on homes sold in Northwest Ohio during 1992.
a. Create a probability distribution for the number of bedrooms. Compute the mean and standard deviation of this distribution.
b. Develop the probability distribution of the number of bathrooms. Compute the mean and standard deviation of this distribution.

48. Refer to the Salary Data Set, which refers to a sample of middle managers in Sarasota, Florida. Create a probability distribution for the number of employees supervised. Compute the mean and standard deviation of this distribution.

CASE: WILL SAMPLING SAVE SARA TIME?

An item such as an increase in taxes, a recall of elected officials, or an expansion of public services can be placed on the ballot if a required number of valid signatures are collected on a petition. Unfortunately, many people will sign a petition even though they are not registered to vote in that particular district, or they will sign the petition more than once.

Sara Fergerson, the elections auditor, must check the validity of each signature after the petition is officially presented. Not surprisingly, her staff is overloaded, so she is considering using statistical methods to validate each page of 200 signatures, instead of validating each individual signature. At a recent professional meeting, she found that in some other communities election officials were checking only five signatures on each page and rejecting the entire page if two or more of the five signatures were invalid. She is concerned that five signatures are not enough to check. As an intermediate plan she is considering checking 10 signatures per page and rejecting the page if three or more are invalid.

Calculate the probability of rejecting each page under the two plans if the actual percent of invalid signatures on the entire petition is 5%. Repeat the calculations for invalid signatures of 10% and 20%. Does Sara's plan appear superior to that of the other election officials? Discuss the merits of the two plans.

CHAPTER ACHIEVEMENT TEST

Answer all the questions. The answers are at the back of the book.

MULTIPLE-CHOICE QUESTIONS. Select the response that best answers each of the questions.

1. A listing of all outcomes of an experiment and their corresponding probabilities is called a
  a. probability distribution.
  b. frequency distribution.
  c. random variable.
  d. subjective probability.

2. In every probability distribution, the probabilities must
  a. be increasing.
  b. add up to one.
  c. be constant.
  d. be decreasing.

3. Which of the following is *not* required for a binomial distribution?

a. Only two outcomes
b. At least 50 observations
c. Independent trials
d. Constant probability of success

4. The binomial distribution is symmetric when
a. n is small.
b. $n\pi$ is 0.5.
c. π is 0.5.
d. π is large.

5. A binomial distribution must have
a. the same probability of success from trial to trial.
b. a value of π equal to 1.34.
c. three or more possible outcomes.
d. None of the above

6. In a Poisson distribution, the
a. variance equals the median.
b. standard deviation equals the variance.
c. mean equals the variance.
d. standard deviation equals the mean.

7. A Poisson distribution
a. is negatively skewed.
b. is positively skewed.
c. is symmetric.
d. None of the above

8. Within a specific range, a continuous random variable may assume
a. an infinite number of values.
b. only certain values.
c. a countable number of values.
d. None of the above

9. If a binomial distribution has $n = 10$ and $\pi = 0.4$, then the mean is
a. 2.4.
b. 4.0.
c. 10.0.
d. None of the above

10. Which are discrete distributions?
a. The binomial
b. The Poisson
c. Both the binomial and the Poisson
d. Neither the binomial nor the Poisson

COMPUTATION PROBLEMS

11. Suppose that 90% of Toro lawn mowers will start on the first pull. Eight Toro mowers are chosen and tried once. Find the following probabilities:

a. Exactly six will start.
b. At least six will start.
c. Three, four, or five will start.

12. The Consumer Protection Division of Casino City receives a complaint on 1 out of 50 calls. If 150 calls were received today, determine the probability that
a. no complaints were received.
b. exactly two complaints were received.
c. fewer than four complaints were received.
d. at least one complaint was received.

13. Pizza Palace recorded the number of extra toppings added to each pizza. Compute the mean and the standard deviation for each number of toppings.

Number of Toppings	Percent of Purchases
0	5
1	15
2	30
3	40
4	10

ANSWERS TO SELF-REVIEW PROBLEMS

6-1 a. In any order, the complete list is:

H	H	H
H	H	T
H	T	H
H	T	T
T	H	H
T	T	H
T	H	T
T	T	T

b.

Number of Tails X	Probability of X
0	1/8 = 0.125
1	3/8 = 0.375
2	3/8 = 0.375
3	1/8 = 0.125
Total	8/8 1.000

c.

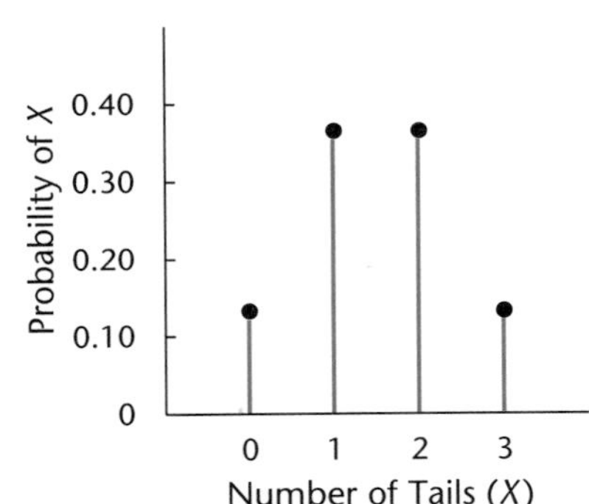

6-2 a. Discrete
b. Discrete
c. Continuous
d. Discrete
e. Discrete
f. Continuous

6-3 a. $\mu = \$0.50(0.40) + \$0.60(0.20) + \$0.90(0.20) + \$1.10(0.20) = \$0.72$

b. $\sigma^2 = (0.50 - 0.72)^2\,(0.40) + (0.60 - 0.72)^2(0.20)$

$+ (0.90 - 0.72)^2(0.20)$
$+ (1.10 - 0.72)^2(0.20)$
$= 0.0576$
$\sigma = \sqrt{0.0576} = 0.24$

c. The cola prices are nearly the same ($0.71 or $0.72), but there is more variation at Big Burger ($0.24 versus $0.097).

6-4 **1.** Mutually exclusive: A baby cannot be a girl and a boy at the same time.
2. There are only two possible outcomes: boy or girl.
3. Trials are independent. If a boy is born, this does not affect the sex of the next child.
4. Probabilities remain the same for each birth: 0.50 and 0.50.

6-5 $P(3) = \frac{5!}{3!2!}(0.3)^3(0.7)^2$
$= (10)(0.027)(0.49)$
$= 0.132$

6-6 $n = 5$, $\pi = 0.3$

X	*P*(*X*)
0	0.168
1	0.360
2	0.309
3	0.132
4	0.028
5	0.002

6-7 **a.** 0.162, found by $P(3) + P(4) + P(5) = 0.132 + 0.028 + 0.002$
b. 0.832, found by either $1 - P(0) = 1 - 0.168$ or $P(1) + P(2) + P(3) + P(4) + P(5)$ (slight discrepancy due to rounding)

6-8 $P(1) = 0.3033$
$P(X \geq 1) = 1 - P(0) = 1 - 0.6065$
$= 0.3935$

CHAPTER 7

The Normal Probability Distribution

OBJECTIVES

When you have completed this chapter, you will be able to

- list the characteristics of a normal probability distribution;
- calculate a standardized value or *z*-value;
- calculate the probability that an observation will occur between two particular points of interest on a normal distribution;
- determine a point beyond which a specific percent of the observations will occur, or a point beyond which there is a specific probability of occurrence on a normal distribution.

CHAPTER PROBLEM How Fast Was I Going, Officer?

THE FEDERAL HIGHWAY ADMINISTRATION REGULATES FEDERAL FUNDING FOR ROADWAY CONSTRUCTION AND IMPROVEMENTS. TO ENCOURAGE INDIVIDUAL STATES TO OBSERVE NATIONAL STANDARDS FOR SPEED AND SAFETY, IT WITHHOLDS FUNDS FROM THOSE STATES THAT DO NOT MAKE SERIOUS EFFORTS TO ENFORCE SPEED LIMITS. IN CONNECTION WITH A CERTIFICATION STUDY, THE OHIO DEPARTMENT OF TRANSPORTATION CONDUCTED A TRAFFIC SURVEY ON A SECTION OF I-75 NEAR LIMA. THE MEAN VEHICLE SPEED WAS 66.7 MILES PER HOUR, WITH A STANDARD DEVIATION OF 3.5 MILES PER HOUR. WHAT PERCENT OF THE VEHICLES TRAVELED AT THE 65 MILE-PER-HOUR SPEED LIMIT? WHAT PERCENT TRAVELED MORE THAN 70 MILES PER HOUR?

INTRODUCTION

In Chapter 2, raw data were organized into a frequency distribution. Then, frequency polygons and other graphs were constructed to describe the location and shape of the frequency distribution. In Chapters 3 and 4, we introduced various measures that describe the central tendency and dispersion in the data. The material in these three chapters is referred to as "descriptive statistics." With Chapter 5, we began our exploration of the idea of probability, which measures the likelihood that some uncertain event will occur. The concept of a probability distribution was developed in Chapter 6. A probability distribution extends the notion of a frequency distribution to include a description of some experiment and of all of its possible outcomes. Chapter 6 described two important *discrete* distributions—the binomial and the Poisson.

In this chapter, we will examine a very important *continuous* probability distribution known as the **normal probability distribution.** Its importance is due to the fact that, in practice, experimental results very often seem to follow this mounded, "bell-shaped" pattern. The normal probability distribution is also important because most of the sampling methods developed later in this book are based on it. Incidentally, this distribution came to be known as the "normal" probability distribution because around 1800 absolutely every set of data had to follow it, or else statisticians would think something was wrong—not "normal"—with the data. The normal distribution requires that the data be of at least interval scale. The data in Table 7–1 and Figure 7–1 are examples of this type of interval-level measurements. Other types of data that frequently array themselves into a bell-shaped pattern are:

- weights of seven-year-old girls
- heights of adult males
- IQ scores *(continued next page)*

Table 7-1 **Distribution of Ages of Surgical Patients**

Age of Patient	Percent of Total
0–10	9
10–20	10
20–30	19
30–40	20
40–50	17
50–60	14
60–70	11

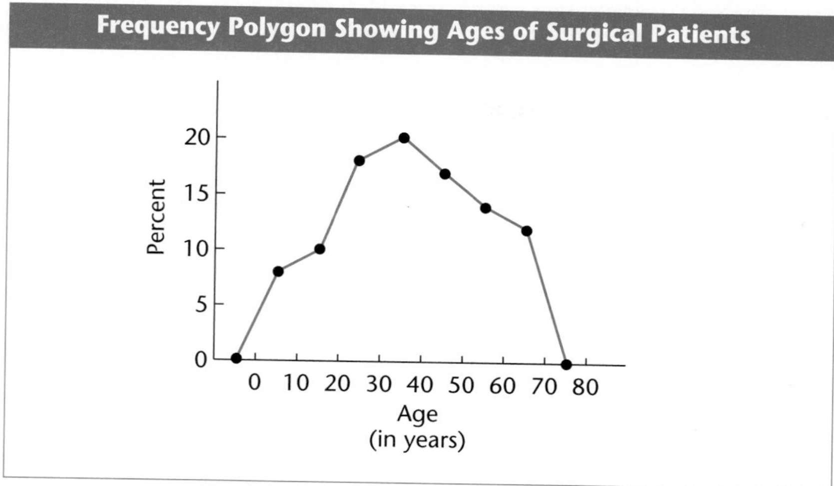

Figure 7-1

- blood-pressure readings
- scores on the first statistics test

CHARACTERISTICS OF A NORMAL PROBABILITY DISTRIBUTION

The normal probability distribution has the following characteristics:

1. The graph of the normal probability distribution has a single peak at the center of the distribution. The mean, median, and mode—which in a normal distribution are equal—are all located at the peak. Therefore, exactly one-half, or 50%, of the area is to the left of the center of the distribution and exactly one-half of the area is to the right of it.
2. A normal probability distribution is *symmetrical* about its mean. If you were to "fold" the probability distribution along its central value, the two halves would be identical.
3. The normal curve falls off smoothly in a "bell-shape" and the two tails of the probability distribution extend indefinitely in either direction. In theory, the curve never actually touches the *X*-axis (see Figure 7–2 on page 238).

THE "FAMILY" OF NORMAL DISTRIBUTIONS

There is not just one normal probability distribution. Rather, there is an entire "family" of related normal probability distributions. If you are studying the annual incomes of a group of male employees, that probability distribution might follow a normal distribution

Figure 7-2

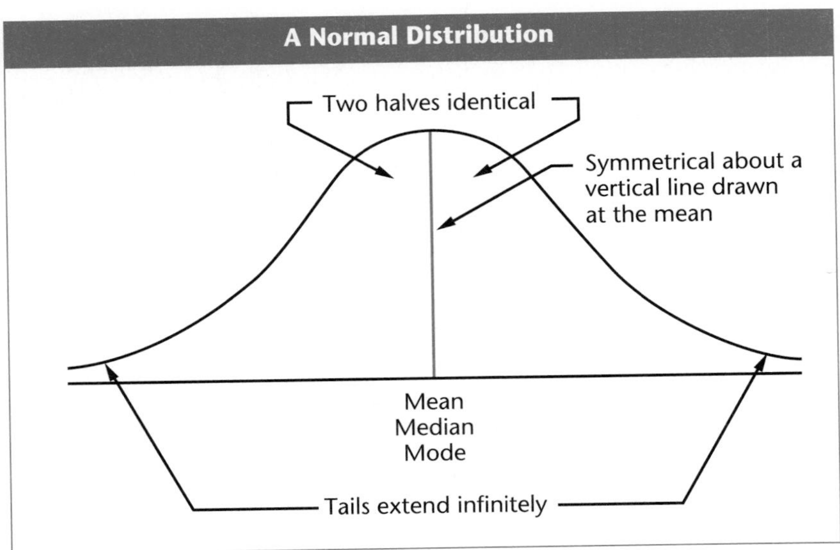

with a mean of $32,394 and a standard deviation of $842. The annual incomes of a group of female employees could follow a normal distribution with a mean of $27,652 and a standard deviation of $797. For each value of a mean and a standard deviation, there is a particular normal probability distribution. When either the mean or the standard deviation changes, a new normal probability distribution is formed.

Figure 7–3 illustrates three normal probability distributions. Each of them has a mean of 30 but a different standard deviation. Figure 7–4 shows a set of normal probability distributions that have different means but the same standard deviation of 2.

Figure 7-3

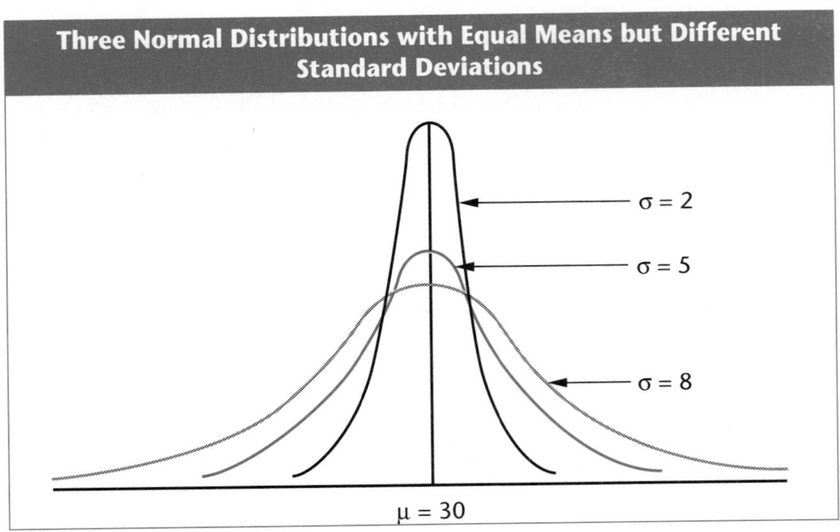

Figure 7-4

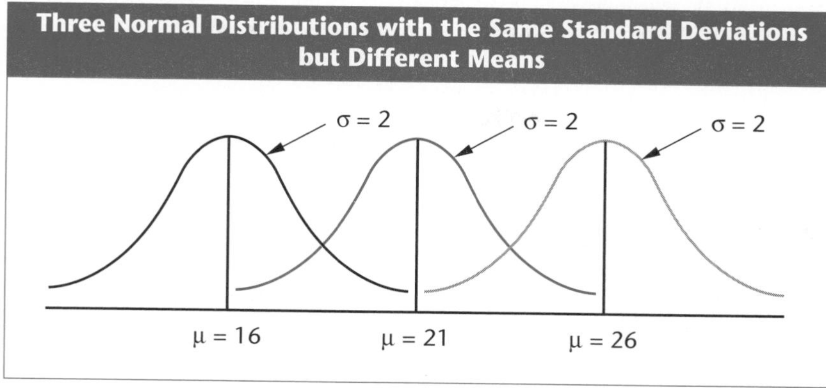

Figure 7–5 portrays two normal probability distributions for which both the means and the standard deviations are different. Although each of the normal probability distributions displayed in Figures 7–3, 7–4, and 7–5 differs somewhat in appearance, each is still a member of the "family" of normal probability distributions.

In dealing with normal probability distributions, we use three relationships extensively:

1. About 68% of the distribution is within one standard deviation of the mean.
2. About 95% of the observations are within two standard deviations of the mean.
3. Virtually all (99.73%) of the area is within three standard deviations of the mean.

For example, if a normal probability distribution has a mean of 20 and a standard deviation of 4, then

Figure 7-5

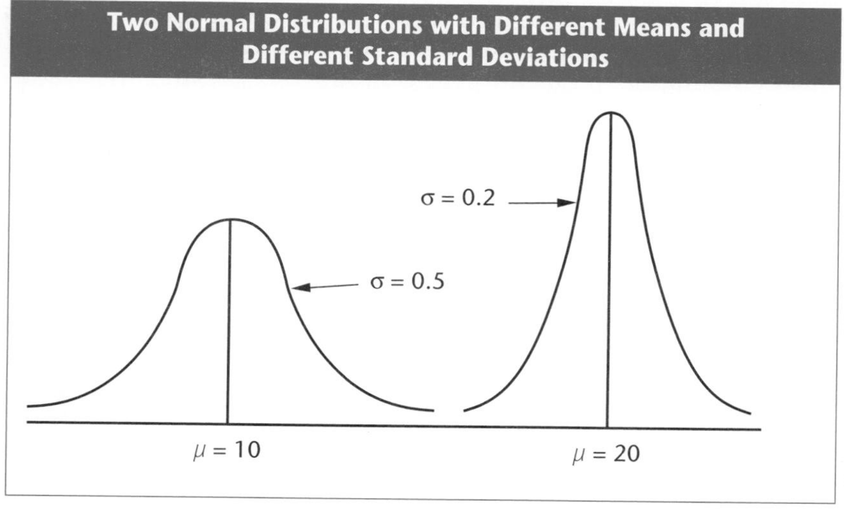

1. about 68% of the values are between 16 and 24, found by $20 \pm 1(4)$.
2. about 95% of the values are between 12 and 28, found by $20 \pm 2(4)$.
3. virtually all the values are between 8 and 32, found by $20 \pm 3(4)$.

Problem The daily high temperature for a city in Hawaii is normally distributed and the mean is 21°C. The standard deviation is 2.5°C. What percent of the temperatures are within 1, 2, and 3 standard deviations of the mean?

Solution

1. About 68% of the mean daily temperatures are between 18.5° and 23.5° C, found by $\mu \pm 1(\sigma) = 21 \pm 1(2.5)$.
2. About 95% of the mean daily temperatures are between 16° and 26° C, found by $\mu \pm 2(\sigma) = 21 \pm 2(2.5)$
3. Virtually all of the mean daily temperatures are between 13.5° and 28.5° C, found by $\mu \pm 3(\sigma) = 21 \pm 3(2.5)$.

Self-Review 7-1

Answers to all Self-Review problems are at the end of the chapter.

A test to measure anxiety is given to males between the ages of 17 and 19. The mean score is computed to be 50, the standard deviation 6. About 95% of the males have an anxiety score between what two values?

STAT BYTE

An individual's skills depend on a combination of hereditary and environmental factors. This suggests that many skills follow the standard normal distribution. For example, scores on the Scholastic Aptitude Test (SAT) are normally distributed with a mean of 1000 and a standard deviation of 140. The NCAA currently requires a student athlete to achieve a score of 700 to compete during his or her freshman year. In other words, fewer than 5% of those taking the SAT should fall below this standard.

THE *STANDARD* NORMAL PROBABILITY DISTRIBUTION

As noted, there are many normal probability distributions—one for each pair of values for a mean and a standard deviation. While this makes the normal probability distribution very versatile in describing many different real-world situations, it would be very awkward to provide tables of areas for each such normal probability distribution. An efficient method for overcoming this difficulty is available. This method calls for *standardizing* the distribution. To find the area between a value of interest (X) and the mean (μ), we first compute the difference between the value (X) and the mean (μ); then we express that difference in units of standard deviation. In other words, we compute the value

$$z = \frac{X - \mu}{\sigma} \qquad \textbf{7-1}$$

where

z is the *standardized* value, or z-value.
X is any observation of interest.
μ is the mean of the normal distribution.
σ is the standard deviation of the normal distribution.

Finally, we find the desired area under the curve, or the probability, by referring to a table whose entry corresponds to the calculated value of z. How the probability is calculated will be discussed shortly.

The value of z actually follows a normal probability distribution with a mean of zero and a standard deviation of one unit. This probability distribution is known as the **standard normal probability distribution.** Thus, we can convert *any* normal distribution to the standard normal distribution by using formula 7–1.

Problem The ages of patients admitted to the coronary care unit of a hospital are normally distributed with a mean of 60 years and a standard deviation of 12 years. What is the computed z-value (standardized value) for a patient (a) aged 78? (b) aged 45?

Solution The z-values are computed as follows using formula 7–1:

(a) $X = 78$ Computing z: $z = \dfrac{X - \mu}{\sigma}$

$$= \frac{78 - 60}{12}$$

$$= 1.5$$

(b) $X = 45$ Computing z: $z = \dfrac{X - \mu}{\sigma}$

$$= \frac{45 - 60}{12}$$

$$= -1.25$$

Note that a z-value merely transforms a selected value to a deviation from the mean expressed in standard deviation units. The location of the two ages (78 and 45) is shown in Figure 7–6 on page 242.

Problems A normal distribution has a mean of 50 and a standard deviation of 8. Compute the z-value when

a. $X = 56$.
b. $X = 43$.
c. $X = 68$.

Solutions

a. 0.75
b. – 0.875
c. 2.25

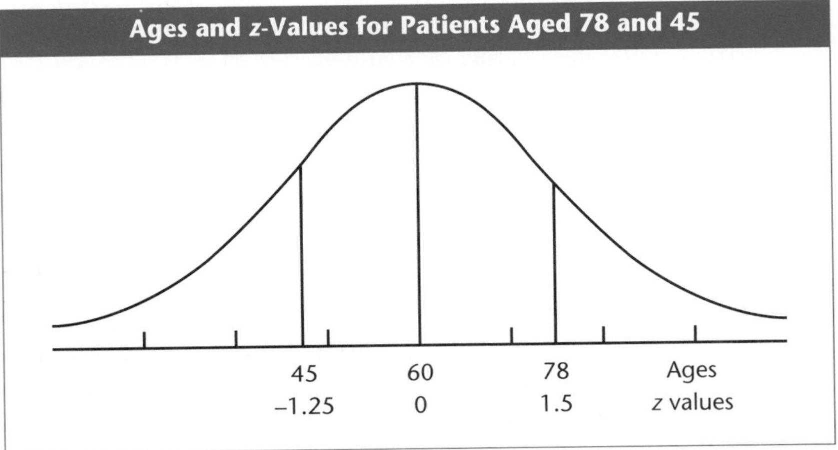

Self-Review 7-2

Referring to the ages of patients, what is the z-value ($\mu = 60$ years, $\sigma = 12$ years) for a patient

a. aged 68?
b. aged 33?

Exercises

Answers to the even-numbered exercises are at the back of the book.

1. The monthly food expenditures of families of five on welfare were studied. The mean amount spent was \$125 and the standard deviation \$20. Assuming that the monthly expenditures are normally distributed,
   a. standardize the expenditure of \$105. That is, convert the expenditure of \$105 to a z-value.
   b. standardize the expenditure of \$145.
   c. What percent of the welfare families will spend between \$105 and \$145 a month?

2. A mathematics instructor studied the lengths of time required for students to complete the final examination. She found that the mean time was 90 minutes and the standard deviation 10 minutes. If the lengths of time are normally distributed.
   a. 95% of the lengths of time will fall between what two times?
   b. 99.7% of the students will complete the final examination between what two times?

3. The life of an automatic dryer at a commercial laundry is normally distributed with a mean of 8.5 years and a standard deviation of 0.75 years.
   a. Standardize the value of 8.0 years.
   b. About what percent of the dryers will last between 7.0 and 10.0 years?

4. A dexterity test is given to young children. The mean score is 20 and the standard deviation is 5.
a. Standardize a score of 12.
b. About what percent of the scores are between 5 and 35?

Comparing Scores on Different Scales

Numbers that are of different scales or are in different units can be compared if we convert them to *z*-values. This can be explained best by an illustration.

Problem A mentally disabled person scores 84 on a special anxiety test. The scores for this test are normally distributed, with a mean of 80 and a standard deviation of 8. He also scores 28 on a mechanical aptitude test designed especially for the disabled. The scores of this test are normally distributed about a mean of 20. The standard deviation is 6. Transform the scores to *z*-values to compare the performance of this handicapped person on the two tests.

Solution The *z*-value for the anxiety score is 0.5, found by formula 7–1:

$$z = \frac{X - \mu}{\sigma} = \frac{84 - 80}{8} = 0.5$$

The *z*-value for the mechanical aptitude test score is 1.33, found by:

$$z = \frac{X - \mu}{\sigma} = \frac{28 - 20}{6} = 1.33$$

Interpretation: The handicapped person's performance on the anxiety test is slightly above average. This means that relative to other handicapped persons who took the test, this person's score is 0.5 standard deviations above the arithmetic mean. His performance on the mechanical aptitude test is well above the mean. That is, relative to other handicapped persons taking the test, this person's score is 1.33 standard deviations above the mean.

It should be noted that a negative *z*-value would indicate a below-average performance:

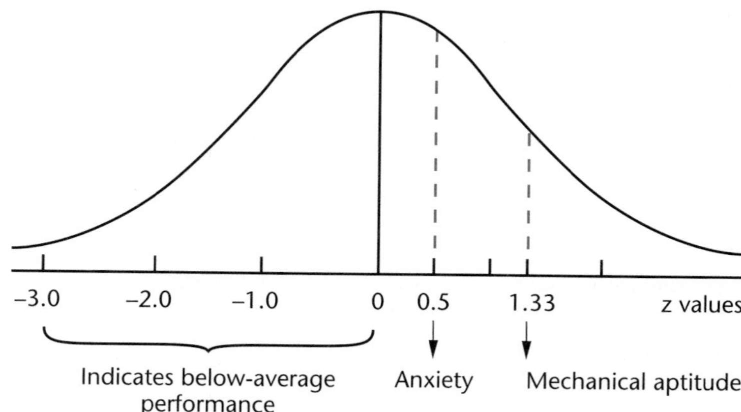

Self-Review 7-3

The mean age of prisoners in a state prison is 40 years and the standard deviation 10 years. The ages are normally distributed. The scores on a test measuring the social consciousness of the prisoners are also normally distributed, with a mean of 500 and a standard deviation of 100.

A prisoner, age 38, scored 750 on the test. Compare his relative position within the two distributions.

Exercises

5. Martha and Frank Clark are newlyweds. Both are under 25 years of age and both are college graduates. Their combined income is $65,000 a year. The combined mean income of all newlyweds who are under 25 and are college graduates is $62,000. The standard deviation of that distribution is $5,000 and it is normal. Comment on the relative position of the Clark's income.

6. The hourly wages of two people working in the trades are to be compared. Neil Holzmann, a carpenter, earns $12.00 per hour. Joe Bevilacqua, a plumber, earns $14.00 per hour. A survey of both trades in the same city reveals the following information (assume both distributions are normal):

	Plumber	Carpenter
Mean	$15.00	$10.00
Standard deviation	1.75	1.25

Compare the relative position of the two tradesmen within their respective trades.

7. According to a trade publication, the mean flying time from Detroit to Atlanta is 2 hours, with a standard deviation of 0.15 hours. Yesterday's 10 a.m. flight on Crash Airlines took 2.10 hours and the 2 p.m. flight took 1.90 hours. The evening flight at 7 p.m. took 2.25 hours. Comment on the relative position of the three flights.

8. The mean time spent in a grocery store is 10 minutes, with a standard deviation of 1.5 minutes. Mrs. Delwhiler shopped this morning and spent 9 minutes in the store. Mrs. Stevens went this afternoon and spent 12 minutes. Comment on the relative amount of time spent shopping.

Finding Probabilities Between Two Selected Values

Another use of z-values is to determine what percent of a group of observations will be located *between* two values, or to determine the probability that an observation will occur between two values. Again, we will use a problem to illustrate.

Problem A research study revealed that the amount spent by persons seeking a seat on the city council in medium-sized cities is normally distributed, with a mean of \$6,000 and a standard deviation of \$1,000. The question to be explored is, What percent of the candidates seeking office spend between \$6,000 and \$7,250?

Solution The *z*-value corresponding to \$7,250 is 1.25, found by formula 7–1:

$$z = \frac{X - \mu}{\sigma} = \frac{\$7,250 - \$6,000}{\$1,000} = 1.25$$

The actual amounts spent, and their corresponding *z*-values, are depicted on the following graph. The upper scale shows the actual amounts spent and the lower scale shows the *z*-values. Note that the mean converts to 0 on the *z*-scale.

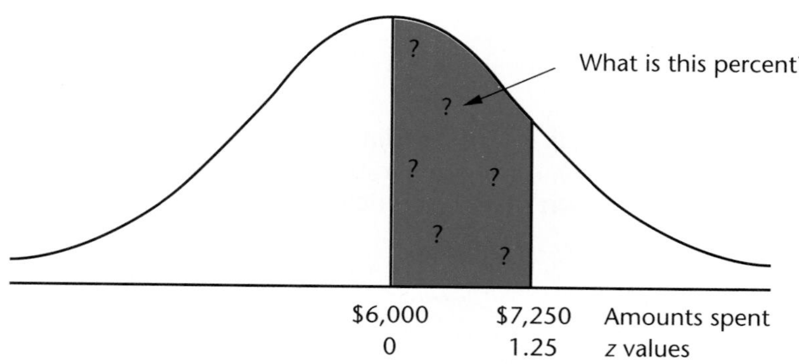

With that information, we can now determine what percent of candidates spend between \$6,000 and \$7,250. Percents derived from *z*-values have already been computed and organized into standard tables (see Appendix C, "Normal Probability Distribution"). The values found in such tables are actually areas under the standard normal curve, or probabilities. A portion of that appendix is shown in Table 7–2 on the following page.

To find the percent under the normal curve corresponding to a *z*-value of 1.25, first go down the left column of the table to a *z* of 1.2. Then move horizontally to the column headed 5 and read the probability. It is 0.3944. To convert to percents, we multiply the number in the table by 100, or, more simply, we move the decimal point two places to the right. Converted to a percent, then, 0.3944 is 39.44%. Interpreting this outcome, we find that 39.44% of the candidates for city council in medium-sized cities spend between \$6,000 and \$7,250 on their campaigns.

Recall that one of the characteristics of a normal distribution is its symmetry—that is, the left half of the curve is identical to the right half.

Table 7-2

A Portion of a z-Table (The Normal Probability Distribution, Appendix C)

	Second Decimal Place of z									
z	0	1	2	3	4	5	6	7	8	9
⋮	⋮	⋮	⋮	⋮	⋮	⋮	⋮	⋮	⋮	⋮
1.0	.3413	.3438	.3461	.3485	.3508	.3531	.3554	.3577	.3599	.3621
1.1	.3643	.3665	.3686	.3708	.3729	.3749	.3770	.3790	.3810	.3830
1.2	.3849	.3869	.3888	.3907	.3925	.3944	.3962	.3980	.3997	.4015
1.3	.4032	.4049	.4066	.4082	.4099	.4115	.4131	.4147	.4162	.4177
1.4	.4192	.4207	.4222	.4236	.4251	.4265	.4279	.4292	.4306	.4319
⋮	⋮	⋮	⋮	⋮	⋮	⋮	⋮	⋮	⋮	⋮

Thus, the percent of total observations between the z-values of 0 and + 1.00 (34.13%), is the same as the percent between 0 and −1.00 (also 34.13%). Shown diagrammatically:

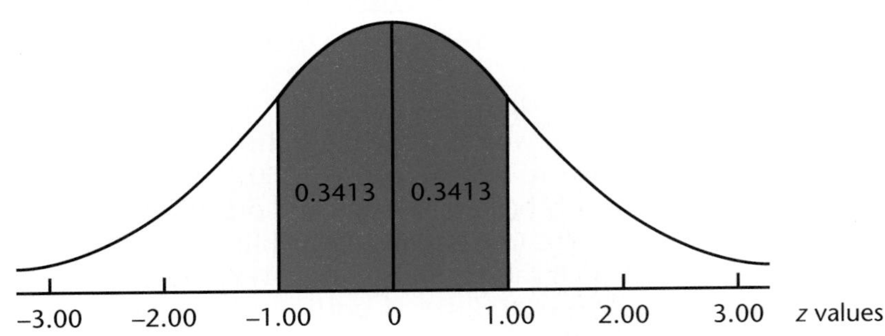

What percent of the z-values are between −1.00 and + 1.00? The answer: 0.3413 + 0.3413 = 0.6826 = 68.26%.

Self-Review 7-4

The weights of boxes of Crunchy breakfast cereal are normally distributed with a mean of 450 grams and a standard deviation of 2 grams. What percent of the boxes weigh between

a. 450 and 454 grams?
b. 447.3 and 450 grams?

Problem To illustrate further the procedure for finding the probability between two selected values, suppose Cardy Halzey decides to run for city council. What is the likelihood (probability) that she will spend between \$5,000 and \$8,000? The mean expenditure for all candidates is \$6,000 and the standard deviation is \$1,000.

Solution The problem can be divided into two parts.

1. What is the probability of a campaign expenditure between \$5,000 and \$6,000 (\$6,000 is the mean)? Using formula 7–1, the z-value is – 1.00, found by:

$$z = \frac{X - \mu}{\sigma} = \frac{\$5{,}000 - \$6{,}000}{\$1{,}000} = -1.00$$

The probability for – 1.00 (from Appendix C) is 0.3413.

2. What is the probability of a campaign expenditure between \$6,000 (the mean) and \$8,000?

$$z = \frac{X - \mu}{\sigma} = \frac{\$8{,}000 - \$6{,}000}{\$1{,}000} = 2.00$$

The probability for 2.00 (from Appendix C) is 0.4772.

Adding 0.3413 and 0.4772 gives 0.8185. Thus, the probability that Halzey will spend between \$5,000 and \$8,000 is 0.8185. The various components of this problem are shown in the following diagram:

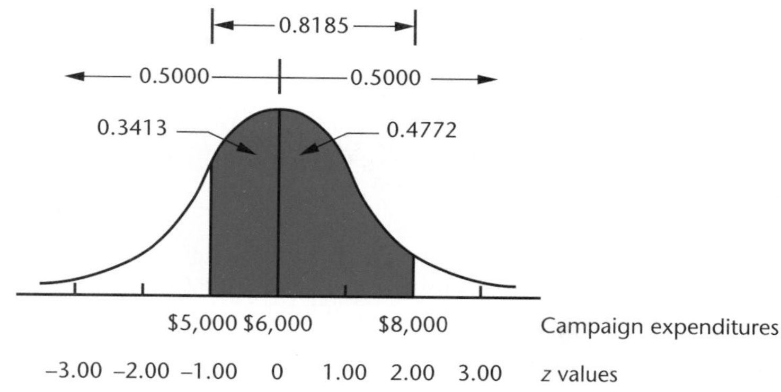

Determining Percent Above or Below a Selected Value

In the previous section, we computed the percent of observations occurring between two values, or—stated a different way—the probability that a particular observation would occur between two points. In this section, the process is reversed. We will be interested in the percent of the observations above or below a given point.

$z = \frac{X - \mu}{\sigma}$

Problem What percent of the candidates spend \$8,000 *or more* on their campaign for city council?

Conewango Valley Nursing Home mean is \$500 day, standard deviation is \$75.
What proportion of patients spend more than \$600 per day? $\frac{600-500}{75} = \frac{100}{75} = \frac{4}{3} = 1.33$
1.33 = 0.4082 0.5000

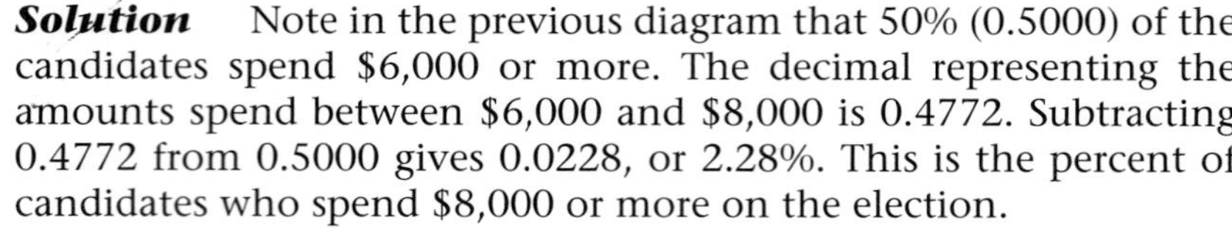

Solution Note in the previous diagram that 50% (0.5000) of the candidates spend $6,000 or more. The decimal representing the amounts spend between $6,000 and $8,000 is 0.4772. Subtracting 0.4772 from 0.5000 gives 0.0228, or 2.28%. This is the percent of candidates who spend $8,000 or more on the election.

STAT BYTE

Many processes, such as filling soda bottles and canning fruit, are normally distributed. Manufacturers must guard against both over- and under-filling. If they put in too much, they will give their product away. If they put in too little, the customer will feel cheated. "Control" charts, with limits drawn three standard deviations above and below the mean, are routinely used to monitor these kind of production processes.

Problems A normal distribution has a mean of 50 and a standard deviation of 8. What is the probability that a selected value is

a. between 50 and 62?
b. between 44 and 50?
c. between 44 and 62?
d. more than 60?

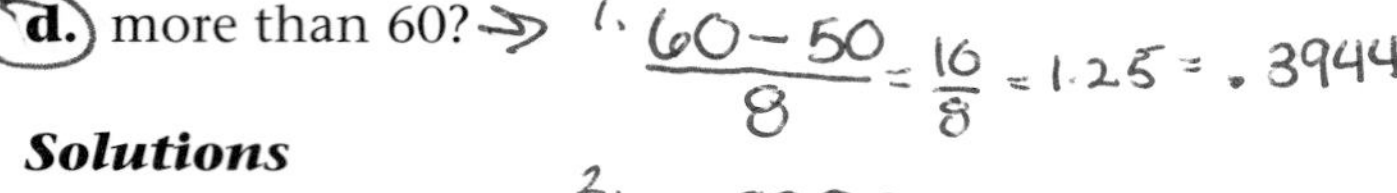

Solutions

a. 0.4332
b. 0.2734
c. 0.7066
d. 0.1056

Problem Recall from the Chapter Problem the Ohio Department of Transportation study. It found that the mean speed of vehicles on a section of I-75 near Lima was 66.7 miles per hour, with a standard deviation of 3.5 miles per hour. What percent of the vehicles exceeded the 65 miles-per-hour speed limit? What percent exceeded 70 miles per hour?

Solution First, we assume that the distribution of speeds is normal, with a mean of 66.7 miles per hour and a standard deviation of 3.5 miles per hour. To find the proportion of vehicles that exceeded 65 miles per hour, we use formula 7–1:

$$z = \frac{X - \mu}{\sigma} = \frac{65.0 - 66.7}{3.5} = -0.49$$

Next, we refer to Appendix C to find the z-value corresponding to 0.49. It is 0.1879. To determine the proportion of observations that exceeded –0.49, we add 0.1879 and 0.5000. The result is 0.6879. We therefore conclude that 0.6879 of the observations exceeded a z-value of –0.49. In other words, about 69% of the vehicles traveled faster than the posted speed limit of 65 miles per hour.

To find the proportion of vehicles that exceeded 70 miles per hour, we compute the z-value associated with 70. It is 0.94, found by (70 – 66.7)/3.5. Appendix C gives us the probability between 0.0 and 0.94 as 0.3264, but we are interested in the probability that z is greater than 0.94. This is found by 0.5000 – 0.3264 = 0.1736. That is, 17.36% of the vehicles traveled more than 70 miles per hour.

Self-Review 7-5

The grade point averages at La Siesta University are normally distributed, with a mean of 2.5 and a standard deviation of 0.75.

a. What percent of the students have a grade point average between 2.35 and 2.5?
b. What percent of the students have a grade point average between 2.45 and 2.63?
c. What percent of the students have a grade average of 3.5 or above?

Exercises

9. The nicotine content of a certain brand of king-sized cigarettes is normally distributed, with a mean of 2.0 mg and a standard deviation of 0.25 mg. What is the probability that a cigarette has a nicotine content
   a. of 1.6 mg or less?
   b. between 1.6 and 2.1 mg?
   c. of 2.1 mg or more?

10. The mean yearly amount of sap collected for making maple syrup is 10 gallons per tree. The distribution of the amounts collected per tree is normal, with a standard deviation of 2.0 gallons.
   a. What is the probability that a particular tree produces 13.0 gallons of syrup or more a year?
   b. What is the probability that a particular tree produces between 9.0 and 12.5 gallons of syrup a year?

11. A cola-dispensing machine dispenses 7.00 ounces of cola per cup. The standard deviation is 0.10 ounces. What is the probability that the machine will dispense
   a. between 6.8 and 7.00 ounces of cola?
   b. between 6.8 and 7.15 ounces of cola?
   c. more than 7.15 ounces of cola?

12. The mean starting salary for last year's accounting graduates is $23,000, with a standard deviation of $1,500. What percent of the graduates
   a. earned between $21,000 and $22,000?
   b. earned less than $20,000?

Determining *X* Values for a Given Probability

In our previous work on the normal curve, we computed the probability of an observation occurring between two values. For example, we determined the probability that Cardy Halzey would spend between $6,000 and $8,000 on her campaign. Now the process is changed. We are given an area beyond X and are asked to determine X.

Problem Return once more to the expenditures of the candidates for city council, where $\mu = \$6,000$ and $\sigma = \$1,000$. Ten percent of the candidates spent what amount or more on their campaigns? The components of this problem are shown graphically:

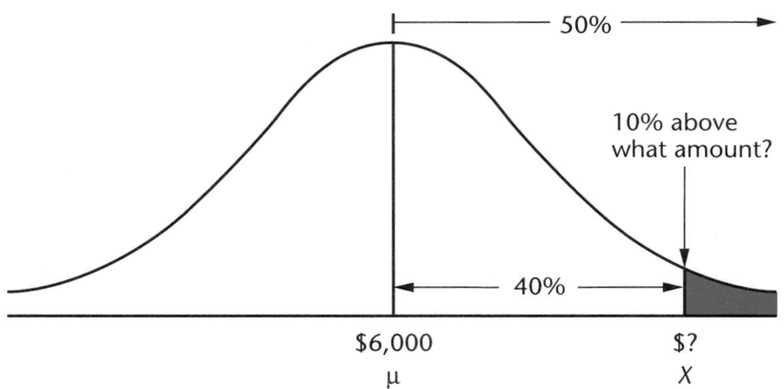

Solution As noted, 50% of the expenses are $6,000 or more, and 10% are some unknown amount or more. The unknown amount is designated by X. Logically, 40% of the campaign expenditures are between $6,000, or μ, and X. We obtain the z-value corresponding to 0.4000 by searching in Appendix C. The closest probability is 0.3997. This corresponds to a z-value of 1.28.

Inserting the z-value of 1.28 and solving for X using formula 7–1, we find:

$$z = \frac{X - \mu}{\sigma}$$

$$1.28 = \frac{X - \$6,000}{\$1,000}$$

Then:

$$\$1,280 = X - \$6,000$$
$$X = \$7,280$$

Thus, about 10% of the candidates spend $7,280 or more on their bids for election to the city council.

Self-Review 7-6

From Self-Review 7–5: The mean grade point average of students at La Siesta University was 2.5, the standard deviation 0.75. The top 3% among students are to be given special recognition. What grade point average (or above) does a student need in order to receive special recognition?

Exercises

13. The heights of adult males follow a normal distribution, with a mean of 70 inches and a standard deviation of 2.6 inches. How tall should a doorway be so that 98% of all men can pass through it without having to stoop?
14. A certain brand of passenger-car tire has a mean tread life of 40,000 miles. The tread life is normally distributed, with a standard deviation of 3,000 miles. Five percent of the tires will lose all their tread before they are driven what distance?
15. The times taken by applicants for data entry positions to type a standard passage were normally distributed, with a mean of 90 seconds and a standard deviation of 12 seconds. If it is company policy to consider only the top 10% of the applicants, what is the cutoff point between those considered by the company and those not considered?
16. A machine is set to fill milk bottles up to 1.0 liters. Most of the bottles contain 1.0 liters, but some of the bottles are slightly underweight or overweight. The fills are normally distributed, with a standard deviation of 0.05 liters. Find the value below which the smallest 10% of the fills occur.

THE NORMAL APPROXIMATION TO THE BINOMIAL

Recall from Chapter 6 that the binomial distribution is a discrete probability distribution. It is characterized by π, the probability of a success, and n, the number of trials. Figure 7–7 (on the following page) shows four binomial distributions, where π is 0.50 and n is 2, 4, 8, or 16. Note that as the size of n increases, the distribution begins to approximate the normal probability distribution.

Thus, the normal distribution may be used to estimate binomial probabilities. As a general rule of thumb, when $n\pi$ and $n(1 - \pi)$ are both at least five, the normal probability distribution is a very good approximation for the binomial distribution.

Problem The city's legal affairs director reports that, based on past experience, 60% of automobiles reported stolen are recovered and returned to their owners. In a month in which 40 automobiles are stolen, what is the probability that 28 or more will be recovered and returned to their owners?

Solution For this binomial probability distribution, the total number of trials is 40 and the probability of success on each trial is 0.60. The normal probability distribution may be used to approximate the binomial distribution because both $n\pi$ and $n(1 - \pi)$ exceed 5. (Solution continues under Figure 7–7 on the following page.)

Figure 7-7

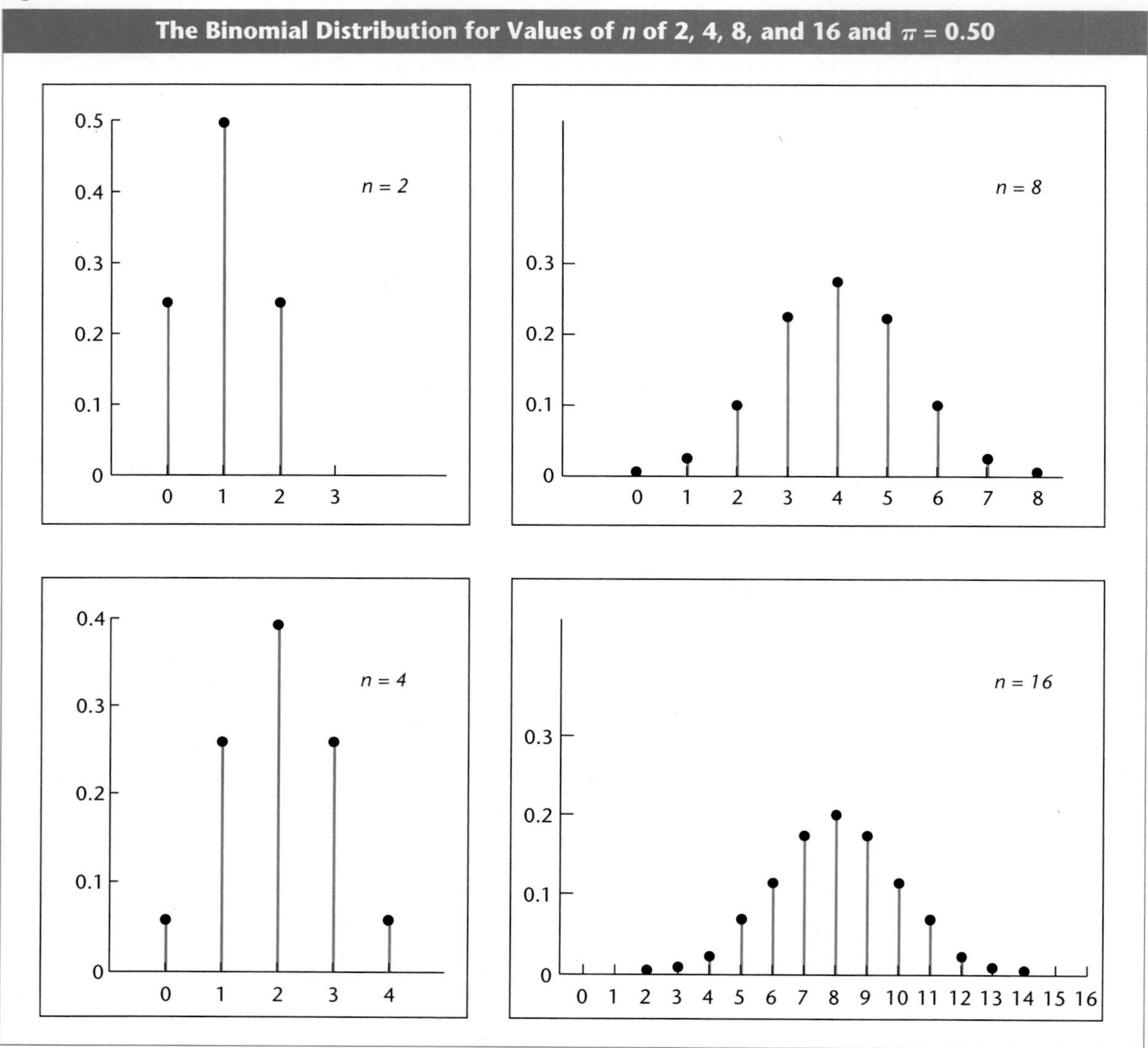

$$n\pi = 40(0.60) = 24$$
$$n(1-\pi) = 40(1-0.60) = 16$$

The steps needed to determine the probability that 28 or more (actually 27.5, as we will see shortly) cars will be recovered and returned are these:

Step 1. Compute the mean and the variance of the binomial distribution. They are computed as follows, using formulas 6–4 and 6–5 from Chapter 6 (μ is used as the designation of the mean, σ^2 as the designation for the variance):

$$\mu = n\pi \quad \text{(6-4)}$$

$$\sigma^2 = n\pi(1-\pi) \quad \text{(6-5)}$$

The mean is 24 and the variance 9.6, found by:

$$\mu = 40(0.60) = 24$$
$$\sigma^2 = (40)(0.60)(0.40) = 9.6$$

Step 2: Determine the standard normal (*z*) value corresponding to 28. A discrete distribution can be only a number of distinct or separate values. Between these values, there are "gaps" that have no probability. A continuous distribution, on the other hand, takes on any value in a range. The situation is similar to "rounding," where 6.7 is rounded up to 7 and 7.3 is rounded down to 7. In fact, all numbers between 6.5 and 7.5 are *rounded* to 7.

To adjust for a difference between a binomial (discrete) and a normal (continuous) distribution, consider the binomial probability as the area over an interval centered at the discrete value and extending one-half of a unit in both directions. This is called a **correction for continuity.** In this example, the exact binomial probability of recovering 28 cars is approximated by the normal area between 27.5 and 28.5. We find the probability of 28 or more cars returned, then, by computing the probability of 27.5 or more. The standard deviation of the distribution is found by $\sqrt{\sigma^2}$: The square root of the variance of 9.6 is 3.10. Computing z using formula 7–1 gives:

$$z = \frac{X-\mu}{\sigma} = \frac{27.5-24}{3.10} = 1.13$$

Step 3. Find the probability of a *z*-value of 1.13 or greater occurring. We determine the area, or probability, between 0 and 1.13 standard normal deviates by referring to Appendix C. Go down the left column to 1.1 and then read the probability headed by the column marked 3. The probability is 0.3708. Thus, the area beyond 1.13 standard normal deviates is 0.1292, found by 0.5000 – 0.3708. Shown graphically:

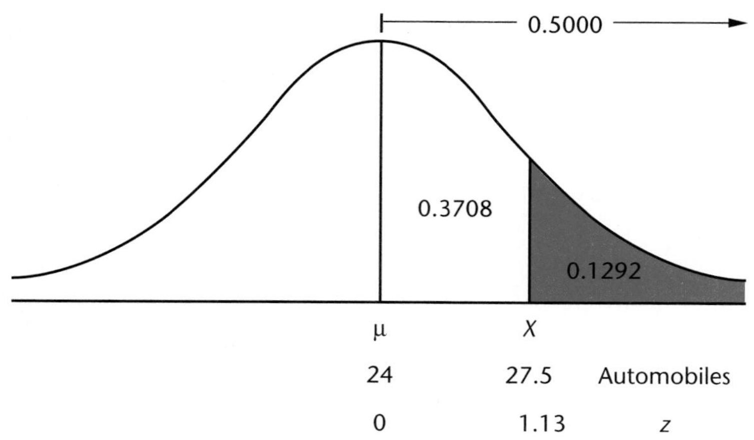

We therefore determine that the probability is 0.1292 that in a month in which 40 automobiles are stolen, 28 or more will be recovered and returned. To put it another way, 28 or more automobiles will be recovered and returned 12.92% of the months in which 40 automobiles are stolen.

The complete distribution of the probabilities of recovery for 0, 1, 2, 3, . . . ,40 automobiles stolen during the month is shown in the MINITAB on page 255. The computer output for a binomial distribution with $n = 40$ and $\pi = 0.60$ shows the probability of 28 or more successes to be 0.128512. The estimate using the normal approximation is 0.1292, a very small difference.

```
MTB   >  set c1.
DATA  >  1:40
DATA  >  end
MTB   >  pdf c1 c2;
SUBC  >  binomial 40 .60.
MTB   >  print c1 c2
```

Problems X is distributed as a binomial probability distribution with $n = 30$, $\pi = 0.30$. Compute each of the following:

a. The mean and variance of the random variable
b. The probability that X is greater than 12
c. The probability that X is greater than 5

Binomial Probability Distribution (n = 40, π = 0.60)

Cars	Probability	Cars	Probability	
1	0.000000	21	0.079163	
2	0.000000	22	0.102552	
3	0.000000	23	0.120387	
4	0.000000	24	0.127912	
5	0.000000	25	0.122795	
6	0.000000	26	0.106265	
7	0.000000	27	0.082651	
8	0.000001	28	0.057560	
9	0.000001	29	0.035727	
10	0.000006	30	0.019650	
11	0.000024	31	0.009508	
12	0.000088	32	0.004011	
13	0.000283	33	0.001459	
14	0.000819	34	0.000450	0.128512
15	0.002129	35	0.000116	
16	0.004991	36	0.000024	
17	0.010569	37	0.000004	
18	0.020257	38	0.000000	
19	0.035184	39	0.000000	
20	0.055414	40	0.000000	

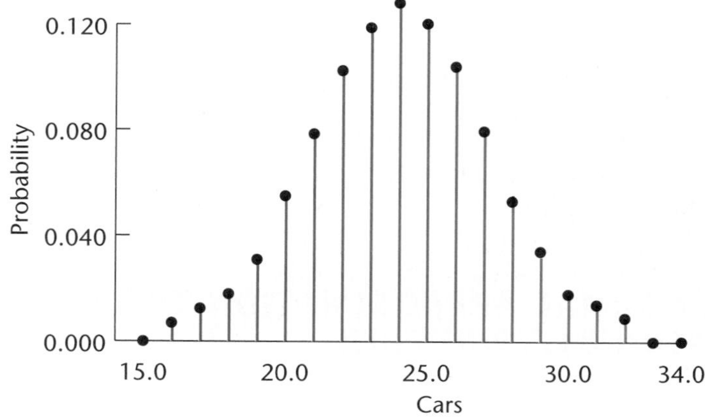

Solutions

a. $\mu = 9$, $\sigma^2 = 6.3$
b. $z = 1.39$; probability is 0.0823, found by 0.5000 – 0.4177.
c. $z = -1.39$; probability is 0.9177, found by 0.5000 + 0.4177.

Self-Review 7-7

The instructor in Geology 115 gives only a final examination. It consists of 100 multiple-choice questions with five possible answers for each. The instructor announces that at least 30 correct answers will be required to pass the course.

Assume that you are enrolled in the course but have never attended class or read any of the assignments. You decide, however, to take the final examination.

a. Can the normal approximation to the binomial be used? Why?
b. Determine the mean and the standard deviation.
c. Calculate the probability that you will pass Geology 115.
d. Portray the probabilities and other parts of the problem graphically.

Exercises

17. The Wayward Inn, a 300-room resort hotel, experiences an 85% occupancy rate, on the average, in January. Use the normal approximation to the binomial to find
  a. the probability that at least 260 rooms are occupied in January.
  b. the probability that fewer than 240 rooms are occupied in January.

18. An airline manager estimated that, on the average, 8% of the passengers flying across the Atlantic experience some airsickness. What is the probability that on a transatlantic flight of 150 passengers, at least 5 will experience some airsickness?

19. Suppose that 1 out of 10 people default on their car loans. Last month the Penn Bank approved 50 car loans. What is the probability that at least 1 borrower will default?

20. A mail order company specializes in selling men's slacks. The probability that a customer will respond to a mailing with an order is 0.05. Today's mailing includes 500 people. What is the probability that fewer than 20 people will respond with an order?

THE NORMAL APPROXIMATION TO THE POISSON DISTRIBUTION

The normal distribution was used to approximate the binomial distribution. In a similar fashion, the normal distribution is used to approximate a Poisson distribution. The approximation improves as the mean (μ) increases. A widely followed guideline is to use the normal approximation when the Poisson mean exceeds 3.

Problem The number of people using the emergency room of Addison Community Hospital during a day shift approximates a

Poisson distribution with a mean of three people. Compute the probability that between two and five people will arrive during the day shift today. Use both the exact Poisson probabilities and the normal approximation to the Poisson distribution. Compare the results.

Solution First, to find the probability that between two and five people will arrive, using the Poisson distribution where $\mu = 3$, we add the probabilities of 2, 3, 4, and 5. From the Poisson table (Appendix B) we find that:

$$P(2 \le X \le 5) = P(2) + P(3) + P(4) + P(5)$$
$$= 0.2240 + 0.2240 + 0.1680 + 0.1008 = 0.7168$$

The probability is 0.7168. As an alternative, we determine an area between 1.5 and 5.5 for a normal probability distribution with a mean and variance of 3. Recall that the standard deviation is the square root of the variance, and that the variance of a Poisson distribution equals its mean. In this problem, $\sqrt{3} = 1.732$. Substituting 1.732 into formula 7–1 for z, we find that:

$$z = \frac{X - \mu}{\sigma} = \frac{1.5 - 3.0}{1.732} = -0.87 \qquad z = \frac{X - \mu}{\sigma} = \frac{5.5 - 3.0}{1.732} = 1.44$$

Thus, the standardized values for 1.5 and 5.5 are – 0.87 and 1.44.

The area under the normal curve between the mean of 3.0 and X of 1.5, represented by a z of – 0.87, is 0.3078 (from Appendix C). The area between 3.0 and 5.5 is represented by the z of 1.44, which is 0.4251. Adding the probabilities from the normal approximation, 0.3078 + 0.4251, we obtain a total of 0.7329, shown graphically below.

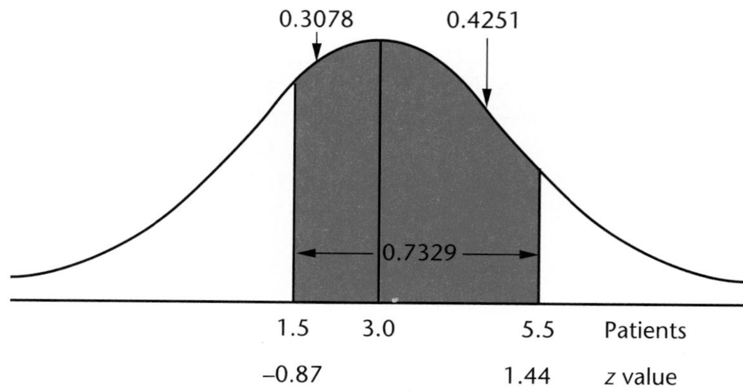

Note that the difference between the probability using the exact Poisson distribution (0.7168) and the probability using the normal approximation (0.7329) is only 0.0161. Again, this emphasizes that under certain circumstances the normal approximation can be used to approximate the Poisson.

Problems The mean of a Poisson probability distribution is 9. Use the normal approximation to the Poisson to compute the probability (don't forget the correction for continuity) that the value is

a. greater than 12.
b. between 7 and 12.
c. less than 7.

Solutions

a. $z = 1.17$; probability is 0.1210, found by 0.5000 – 0.3790.
b. $z = -0.83$ and $z = 1.17$; probability is 0.6757.
c. 0.2033, found by 0.5000 – 0.2967.

Self-Review 7-8

A pair of fair die are rolled 360 times. What is the probability that "snake eyes," a one on each dice, will appear between 5 and 10 times, inclusively? Remember that the two die are independent; hence, the probability of "snake eyes" on a single roll is (1/6)(1/6) = 1/36.

Exercises

21. The mean of a Poisson probability distribution is 5. Use the normal approximation to the Poisson to compute the probability that the value is
   a. less than 2.
   b. between 2 and 10.
   c. less than 7.

22. Sam's Carpet Shop receives an average of four orders per day. The number of orders approximate a Poisson probability distribution. Use both the Poisson and the normal distribution to approximate the Poisson to find the probability that, on a given day,
   a. two orders are received.
   b. more than seven orders are received.

23. The number of telephone calls received at a switchboard approximates a Poisson probability distribution. The mean number of calls per quarter hour is nine. Compute the probability that fewer than seven calls are received in a particular quarter hour.

SUMMARY

The normal distribution often describes an observed frequency distribution such as weights, heights, ages, wages earned, and test scores. A normal distribution is completely described by the mean μ and the standard deviation σ. There are many normal distributions—a different one for each combination of a mean and standard deviation.

A normal distribution has the following characteristics: It has a single peak, it is bell-shaped and symmetrical, and the tails drop off indefinitely. The mean, median, and mode are equal.

Computations involving a normal distribution employ the standard normal distribution. We can standardize an observation by subtracting the mean from the observation and dividing the difference by the standard deviation. In symbols, this is

$$z = \frac{X - \mu}{\sigma}$$

where

- z is the standardized normal value.
- X is the selected observation.
- μ is the mean of the normal distribution.
- σ is the standard deviation of the normal distribution.

We discussed several uses of the standard normal distribution:

1. We used z-values to compare observations on different scales or in different units (such as income and years of education).
2. We determined the probability of an observation falling between two points.
3. We computed the percent of the observations above or below a certain value. Example: What percent of National League players batted over .300?
4. We found the value of an observation designated X, given the area beyond X. Example: The top 5% of college professors have incomes above what amount?

The normal probability distribution can be used to approximate both the binomial and the Poisson distributions. If we want to approximate the binomial distribution, both $n\pi$ and $n(1 - \pi)$ should exceed 5. The mean μ of a binomial is $n\pi$, and the variance σ^2 is $n\pi(1 - \pi)$. We find the probability for each outcome, designated by X, by using the standard normal distribution. The normal distribution is used to estimate the Poisson distribution when μ is at least 3.0.

Exercises

24. An applicant for a position with the Norton Corporation earned scores of 60 on the Sales Aptitude Test, 150 on the

Personnel Human Factors Evaluation Test, and 90 on the Financial Management Test. The scores of each of these tests approximate a normal distribution. The means and standard deviations of the tests are:

Test	Mean	Standard Deviation
Sales Aptitude Test	50	7
Personnel Human Factors Test	120	25
Financial Management Test	85	5

a. Compute a *z*-value for each of the applicant's test scores.
b. On which of the tests did the applicant do the best relative to the entire group?
c. Regarding the Sales Aptitude Test, what percent scored higher than the applicant?
d. Based on the applicant's performance on the three tests, to what area would you assign her—sales, personnel, or finance? Why?

25. The number of hours per week a college student devotes to study is normally distributed, with a mean of 30 hours and a standard deviation of 8 hours.
a. What percent of students will study less than 20 hours?
b. What percent will study more than 35 hours?
c. Out of a class of 200 students, how many will study between 25 and 35 hours?

26. Memorial Hall is used as a site for both student-sponsored concerts and intercollegiate basketball games. Attendance figures for both the concerts and the basketball games are normally distributed. The means and standard deviations are:

	Concerts	Basketball
Mean attendance	8,600	7,200
Standard deviation	560	600

a. What percent of the basketball games have an attendance of 8,000 or more?
b. Tickets to concerts are priced so that, if 8,000 tickets are sold, expenses are covered. If fewer than 8,000 are sold the students lose money, and if more than 8,000 are sold they make a profit. What percent of the time will the students lose money?

27. A juice dispenser is set to fill cups with an average of 7.5 ounces of fruit juice. The standard deviation of the process is 0.3 ounces.

a. If 8-ounce cups are used, what percent of them will overflow?
b. What percent of the cups will have less than 6.8 ounces of juice in them?

28. Southern Airways is studying its service from Chicago to Atlanta. Historical data show that the mean number of passengers per flight is 235.6, and the standard deviation of the normal distribution is 36.3 passengers.
a. What is the probability that a particular flight will carry more than 260 passengers?
b. What is the probability that a particular flight will have fewer than 180 passengers?
c. What is the probability that a particular flight will have between 240 and 250 passengers?
d. What is the probability that a particular flight will have fewer than 240 passengers?

29. The owner of a fast food restaurant keeps records of the daily hamburger demand, which is normally distributed with a mean of 260 pounds and a standard deviation of 20 pounds.
a. What percent of days will the owner need more than 310 pounds of hamburger?
b. The owner does not want to run out of meat more than 1% of days. How many pounds should she order every day?

30. A researcher reports that the mean heart rate of rats is 120 beats per minute, and that 45% of all rats tested had heart rates in the range 120 to 140. Assume that these rates are normally distributed.
a. What standard deviation is implied by these data? (*Hint:* Use the formula for z to compute the standard deviation.)
b. What percent of the animals have heart rates in the range 100 to 120?
c. What percent of the rats have heart rates in excess of 150?
d. What are the standard scores corresponding to 120 and 140?

31. If an elm tree is less than three feet high and is transplanted in the spring, it has a 40% chance of survival. If 50 such trees are transplanted, what is the probability that 25 or more will survive? What is the probability that between 18 and 23 will survive?

32. A new drug is developed to treat a certain disease. It is found to be effective in 90% of patients. If the drug is administered to 80 patients having the disease, what is the probability that it is effective in at least 70 cases?

33. A particular type of birth-control device is effective 90% of the time when used correctly. If the device is employed 300 times, how many times would you expect the device to fail? What is the probability that it would fail 35 or more times?

34. The number of daily admissions at Riverside Hospital followed a normal distribution with a mean of 39.52 and a standard deviation of 6.29.

a. What is the standardized *z*-value corresponding to 50 admissions?

b. What is the standardized *z*-value corresponding to 25 admissions?

c. What percent of days had more than 50 admissions?

d. What percent of days had fewer than 25 admissions?

e. On the busiest 10% of the days, what was the minimum number of admissions?

35. Suppose the U.S. Postal Service were to claim that 80% of letters mailed in New York City destined for Los Angeles are delivered within three working days. To verify this claim, you mail 200 letters from New York to various destinations in the Los Angeles area. Compute the following probabilities.

a. What is the probability that more than 150 of the letters will be delivered within three working days?

b. Fewer than 148?

c. Between 150 and 160?

d. The probability is 10% that what number or more will be delivered within a minimum of three working days?

36. Cars arrive at a car wash at a mean rate of 12 per hour. Compute the probability that more than 15 cars will arrive today between 3 p.m. and 4 p.m.

37. During rush hour, accidents occur, according to a Poisson distribution, at an average rate of 2.5 per hour. Compute the probability that no accidents occur today between 4 p.m. and 5 p.m.

38. The First National Bank of Sylvania recently opened an automatic teller machine (ATM) at the Southview Mall. The number of customers arriving per hour at this ATM follows a Poisson distribution, with a mean of 15. Use the normal approximation to the Poisson to estimate the probability that more than 12 customers will use the machine between two and three o'clock this afternoon.

39. The high-school grade point averages of students applying to Brownlee University is normally distributed, with a mean of 2.80 and a standard deviation of 0.50. If a high-school grade point average of 3.00 is required for admission to Brownlee, what percent of the students applying meet the requirement?

40. The law firm of Tybo and Associates is concerned with the length of time prospective clients spend on hold when calling one of the firm's attorneys. A study revealed the waiting times to be normally distributed, with a mean of 80 seconds and a standard deviation of 12 seconds. What percent of the clients wait at least a minute?

41. Hannah Simpson is a commuter student at Lourdes College who drives her jeep to class each day. She finds that the driving times are normally distributed, with a mean of 20 minutes and a standard deviation of 3.6 minutes. Explain to Hannah why theoretically the probability is zero that she takes exactly 19.0 minutes to get to class. Using the correction for continuity, develop an estimate of the probability of a particular trip taking 19 minutes.

42. The Heritage Family Restaurants is studying the length of time customers spend in their establishments. The times are normally distributed, with a mean of 18 minutes and a standard deviation of 5.5 minutes.

- **a.** What percent of the customers spend less than 30 minutes in a given restaurant?
- **b.** What is the probability of a customer spending exactly 30 minutes in the restaurant?
- **c.** How would you estimate the probability of a customer spending 30 minutes in the restaurant?

43. According to a 1992 study by the American Medical Association, 40% of first-year medical students are women. The same study reports that 5.8% are black. Suppose a group of 60 medical students is selected at random.

- **a.** What is the likelihood that at least 18 are women?
- **b.** What is the likelihood that more than five of the students selected to be in the study are black? (*Hint:* Does this meet the binomial conditions?)

44. A recent study of home sales in suburban Williston revealed that the mean selling price was $107,800, with a standard deviation of $20,000. The mean number of days on the market was 43, with a standard deviation of 14.5 days. After 50 days on the market, Jeff and Mandy Hall's home sold for $89,550. Assuming that both distributions are approximately normal, determine the percent of homes that sold for less than the Halls' and the percent that were on the market for a shorter period of time.

45. Theresa's Tax Service specializes in the preparation of federal tax returns. A recent audit of her returns indicates that an error was made on 6% of the returns her service prepared last year. Assuming that rate continues this year and 100 returns are prepared, what is the probability that her service will make eight or more errors? Determine the probability three ways—normal approximation to the binomial, normal approximation to the Poisson, and the Poisson distribution—and compare the results.

46. A racing expert from the Dickey-Bend International Speedway reports that the mean length of time required to complete a routine pit stop is 15 seconds. The distribution of times is normally distributed, with a standard deviation of one second.

   a. The probability is 0.8 that a pit stop will take at least how many seconds?
   b. The probability is 0.9 that a pit stop will take at most how many seconds?

47. A management consultant is studying the daily work habits of senior-level executives. The consultant finds that a mean of 2.50 hours per day is spent performing tasks that could be handled by subordinates. The distribution of hours is approximately normal. It is also found that 10% of the executives spend more than 3.25 hours on such tasks. Estimate the standard deviation of the distribution.

Data Exercises

48. Refer to the Real Estate Data Set, which reports information on homes sold in Northwest Ohio during 1992.
   a. The mean list price of the 50 homes is $192,180, with a standard deviation of $8,250. Use the normal distribution to estimate the number of homes that would list for more than $200,000. Compare this with the actual number. Is the normal distribution a good approximation of the actual results? Why or why not?
   b. The mean area of the 50 homes is 2,214.6 square feet, with a standard deviation of 435 square feet. Use the normal distribution to estimate the number of homes with more than 2,100 square feet. Compare this with the actual number. Is the normal distribution a good approximation of the actual results? Why or why not?

49. Refer to the Salary Data Set, which refers to a sample of middle managers in Sarasota, Florida.
   a. The mean salary of the 75 middle managers is $64,548, with a standard deviation of $4,427. Use the normal distribution to estimate the number of middle managers with a salary of less than $63,000. Compare this with the actual number. Is the normal distribution a good approximation of the actual results? Why or why not?
   b. The mean age of the 75 middle managers is 47.83 years, with a standard deviation of 8.96 years. Use the normal distribution to estimate the number of middle managers more than 50 years old. Compare this with the actual number. Is the normal distribution a good approximation of the actual results? Why or why not?

CASE

WHAT ARE THE PARAMETERS OF THE DISTRIBUTION?

Invitations to join Delta Gamma Delta, an exclusive sorority at Jefferson State College, are extended dependent upon a student's cumulative grade point average. Lower-division (freshmen and sophomore) women must be in the top 1% of their class. The criterion is a bit easier for upper-division (junior and senior) women. They only need to be in the top 5% of their class! This year lower-division women with grade point averages as low as 3.67 were allowed to join the sorority, and upper-division women with grade point averages as low as 3.32 were permitted to pledge. If the distribution of grades for all women at Jefferson is normally distributed, what are the mean and the standard deviation of that distribution?

CHAPTER ACHIEVEMENT TEST

Answer all the questions. The answers are at the back of the book.

MULTIPLE-CHOICE QUESTIONS. Select the response that best answers each of the questions.

1. The normal distribution is a
  a. continuous distribution.
  b. discrete distribution.
  c. subjective distribution.
  d. historical distribution.

2. For the normal distribution, which of the following statements is *not* always true?
  a. It is symmetric.
  b. It is unimodal.
  c. The mean, median, and mode are equal.
  d. It requires a continuity correction.

3. What percentage of a normal distribution is within two standard deviations of the mean? (Select the closest answer.)
  a. 50%
  b. 95%
  c. 75%
  d. 100%

4. Which of the following statements is true for a normal distribution?
  a. It has two parameters: a mean and a standard deviation.
  b. Half of the values are greater than the mean.
  c. It goes to infinity in both directions.
  d. All of the above.

5. A normal distribution is standardized by the formula

a. $n\pi$

b. $n\pi(1 - \pi)$.

c. $(X - \mu)/\sigma$.

d. None of the above

6. The normal distribution is very close to the binomial distribution when both $n\pi$ and $n(1 - \pi)$ are greater than

a. 1.

b. 5.

c. 30.

d. The normal distribution is never very close to the binomial distribution.

7. A standardized *z*-value

a. expresses distance from the mean in units of the standard deviation.

b. applies only to the Poisson distribution.

c. must be positive.

d. All of the above.

8. The probability that a standard normal deviate (*z*-value) is greater than 1 is

a. 0.1587.

b. 0.3413.

c. 0.8413.

d. None of the above

9. If we use the standard normal distribution, the probability of obtaining a *z*-value between 1 and 2 is

a. 0.0228.

b. 0.1359.

c. 0.3413.

d. 0.4772.

10. The area under the standard normal distribution between −1.5 and 1.5 is

a. 0.1338.

b. 0.1668.

c. 0.4332.

d. None of the above

COMPUTATION PROBLEMS

11. The length of time bank customers must wait for a teller is normally distributed, with a mean of 3 minutes and a standard deviation of 1 minute.

a. What proportion of bank customers wait between 3 and 4.5 minutes?

b. What percent wait more than 4 minutes?

c. What proportion waits between 2 and 3.5 minutes?

d. What percent wait less than 1 minute?

e. Ninety percent of the customers spend less than what amount of time waiting for a teller?

12. Forty percent of a population has blood type O. Find the following probabilities for a random sample of 80 people.
   a. More than 25 have blood type O.
   b. More than 40 have blood type O.
   c. Fewer than 35 have blood type O.
   d. Between 30 and 36 have blood type O.

13. An average of six customers arrives at the Commodore II Barber Shop every half hour. Compute the following probabilities.
   a. Ten or more customers arrive within a half hour.
   b. More than four customers arrive within a half hour.

ANSWERS TO SELF-REVIEW PROBLEMS

7–1 38 and 62: $\mu \pm 2(\sigma)$
$= 50 \pm 2(6)$
$= 50 \pm 12$

7–2 a. $\dfrac{68-60}{12} = 0.67$

b. $\dfrac{33-60}{12} = \dfrac{-27}{12}$
$= -2.25$

7–3 Age: The z-value is –0.2, slightly below average age.
Social consciousness: The z-value is 2.50, indicating well above average.

7–4 a. $z = \dfrac{X-\mu}{\sigma}$
$= \dfrac{454-450}{2}$
$= 2.00$

47.72%, from Appendix C

b. $\dfrac{447.3-450.0}{2} = -1.35$

41.15%

7–5 a. 7.93%

$z = \dfrac{2.35-2.50}{0.75}$
$= -0.20$

Refer to Appendix C for 0.0793.

b. 9.54%, found by:

$z = \dfrac{2.45-2.50}{0.75}$
$= -0.07$

$z = \dfrac{2.63-2.50}{0.75}$
$= 0.17$

The probabilities from Appendix C are: 0.0279 + 0.0675 = 0.0954.

c. 9.18%, found by:

$z = \dfrac{3.5-2.5}{0.75} = 1.33$

Then

0.5000 – 0.4082 = 0.0918 = 9.18%.

7–6 A grade point average of 3.91

$$1.88 = \frac{X-2.50}{0.75}$$
$$1.88(0.75) = X - 2.50$$
$$1.41 = X - 2.50$$
$$X = 3.91$$

7–7 a. Yes. $n = 100$, $\pi = 1/5 = 0.20$. Both $n\pi$ and $n(1-\pi)$ are greater than 5.
$n\pi = 100(0.20) = 20$
$n(1-\pi) = 100(1-0.20) = 80$

b. $\mu = n\pi(100)(0.20) = 20$
$\sigma^2 = n\pi(1-\pi)$
$= (100)(0.20)(0.80)$
$= 16$
$\sigma = \sqrt{16} = 4$

c. $$z = \frac{X - \mu}{\sigma} = \frac{29.5 - 20}{4} = 2.38$$

From Appendix C, the probability for z of 2.38 = 0.4913. Then 0.5000 – 0.4913 = 0.0087. Your chance of passing is less than 1%.

d.

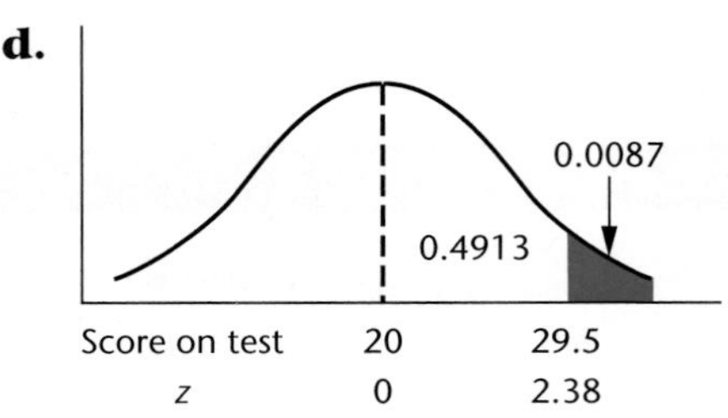

7–8 The mean and variance are 10, found by $\mu = n\pi = (360)(1/36) = 10$, so the standard deviation is $\sqrt{10}$, or 3.16. The discrete values from 5 to 10, inclusive, are included in the continuous range from 4.5 to 10.5. When 4.5 and 10.5 are standardized, we have

$$z = (4.5 - 10)/3.16 \text{ and}$$
$$z = (10.5 - 10)/3.16$$

which equal –1.74 and 0.16, respectively. So the normal probability of rolling a "snake eyes" is 0.4591 + 0.0636 = 0.5227.

UNIT REVIEW

In the last three chapters, you were introduced to the fundamental concepts of probability and to discrete and continuous probability distributions. The various methods for determining a probability, the rules for combining several probabilities, and three different probability distributions were described, discussed, and illustrated.

Key Concepts

1. A **probability** is a number that expresses the likelihood that a particular event will occur. There are three types, or definitions, of probability:
   a. **Classical probability:** Each of the possible outcomes is equally likely. If there are n outcomes, the probability of a particular outcome is $1/n$.
   b. **Relative frequency:** The total number of times the event has occurred in the past is divided by the total number of observations.
   c. **Subjective:** The assignment of probability is based on whatever information is available—personal opinion, hunches, and so on.
2. The fundamental rules of probability are the **rule of addition** and the **rule of multiplication.**
   a. *Rule of addition.* If two events A and B are mutually exclusive, the probability that one or the other of the events will occur is:

$$P(A \text{ and B}) = P(A) + P(B)$$

   This is called the **special rule of addition.** If the events are *not* mutually exclusive, the probability that one or the other will occur is

$$P(A \text{ and B}) = P(A) + P(B) - P(A \text{ and B})$$

   where $P(A$ and $B)$ is the probability of the joint occurrence of the two events. This is called the **general rule of addition.**
   b. *Rule of multiplication.* If two events A and B are unrelated (independent), the probability of their joint occurrence is the product of the two probabilities:

$$P(A \text{ and B}) = P(A) \cdot P(B)$$

This is called the **special rule of multiplication.** If the two events are related (not independent), the probability of their joint occurrence is

$$P(A \text{ and } B) = P(A) \cdot P(B|A)$$

where $P(B \mid A)$ refers to the probability that the event B occurs, given that A has already happened. This is called the **general rule of multiplication.**

3. A **probability distribution** is a listing of the outcomes of an experiment that may occur and the corresponding probability associated with each of the outcomes.
4. A probability distribution has two main features:
   a. The likelihood of a particular outcome must be between 0 and 1.0.
   b. The sum of all possible mutually exclusive outcomes must total 1.0.
5. There are two types of **probability distributions**—discrete and continuous. A **discrete probability distribution** can assume only certain distinct values and is usually the result of counting. A **continuous probability distribution** may assume an infinite number of values within a given range.
6. For a discrete probability distribution, the mean is computed by $\mu = \Sigma[X \cdot P(X)]$ and the variance by $\sigma^2 = \Sigma[(X - \mu)^2 \cdot P(X)]$.
7. The **binomial distribution** is an example of a discrete random variable, where π is the probability of a success and n is the number of successes. The mean is found by $n\pi$ and the variance by $n\pi(1 - \pi)$. A binomial distribution has the following characteristics:
   a. Each outcome is classified in one of two mutually exclusive categories.
   b. Each trial is independent.
   c. The probability of a success remains the same from trial to trial.
   d. It results from counting the number of successes in the total number of trials.
8. The **Poisson distribution** is another discrete distribution. It has the same characteristics as the binomial, but, in addition, n is usually large and π is small. The mean μ is also computed by $n\pi$. This distribution depends on only one value, namely μ.
9. The **normal distribution** is an example of a continuous probability distribution. It has the following main characteristics:
   a. It has a single peak.
   b. It is symmetrical about the mean.
   c. It is bell-shaped and the two tails extend indefinitely in both directions.

d. There is a family of normal distributions—a different one for each different mean and standard deviation.

10. The normal distribution is used to approximate the binomial if $n\pi$ and $n(1 - \pi)$ both exceed 5.0.

11. The normal distribution is used to approximate the Poisson if the mean is at least 3.0.

Key Terms

Probability
Experiment
Outcome
Event
Classical probability
Relative frequency
Subjective probability
Special rule of addition
General rule of addition
Mutually exclusive events
Joint probability
Conditional probability
General rule of multiplication
Special rule of multiplication
Independent events
Complement rule
Permutations
Combinations
Probability distribution
Random variable
Discrete probability distribution
Binomial distribution
Binomial formula
Poisson probability distribution
Continuous probability distribution
Expected value
z-value
Standard normal distribution
Correction for continuity

Key Symbols

$P(A)$	The probability of the event A happening.
$P(B \mid A)$	The conditional probability the event B will occur, given that A has already happened.
${}_nP_r$	The number of permutations of n objects taken r at a time.
${}_nC_r$	The number of combinations of n objects taken r at a time.
z	The value of the standard normal distribution.

Problems

1. A study of individuals from 10 to 18 years of age regarding their understanding of television commercials was conducted. First, 200 of them were shown various television commercials. Then they were questioned about each commercial, and it was determined whether they understood it. The results are as follows:

	Age			
	10–12	13–15	16–18	Total
Understood	30	40	50	120
Did not understand	40	30	10	80
Totals	70	70	60	200

If an individual is randomly selected,

a. what is the probability that he or she understood the commercial?

b. what is the probability that he or she is 10–12 years old and also understood the commercial?

c. given that the individual did not understand the commercial, what is the probability that he or she was 10–12 years old?

d. what is the probability that an individual either did not understand the commercial or is 10–12 years old?

2. A social club at a large university has 300 members who are registered in one of three different colleges. Their colleges of registration and their grade point averages, on a 4-point scale, are summarized in the following table:

	Grade Point Average (GPA)			
College	Greater Than 3.0	Between 2.0 and 3.0	Lower Than 2.0	Total
Arts and Sciences	20	40	30	90
Business	60	50	10	120
Education	20	60	10	90
Total	100	150	50	300

A student is selected at random from the list of club members.

a. What is the probability that the student is registered in the College of Business?

b. What is the probability that the student has a GPA greater than 3.0 *and* is in the College of Education?

c. What is the probability that the student has a GPA lower than 2.0 *or* is in the College of Education?

3. The small town of Sugar Grove has two ambulances. Records show that the first ambulance is in service 80% of the time and the second 70% of the time.

a. What is the probability that both are in service when needed?

b. What is the probability that at least one is in service when needed?

4. The security manager of a large building reports that the probability is 0.05 that a fire alarm will not operate when needed. If there are three alarms in the building, what is the probability that none of the alarms will operate during a particular fire? What is the probability that at least one will operate during a particular fire?
5. An appliance dealer sponsors advertisements on both radio and TV. A study of 200 customers revealed that 80 had seen the advertisement on TV, 120 had heard the radio advertisement, and 40 had both seen the TV ad and heard the radio ad.
   a. What is the probability that a customer heard both the radio ad and the TV ad?
   b. What is the probability that a customer heard the advertisement *either* on the radio *or* on TV?
6. A sheriff needs new tires for his road patrol cars. The probability that he will purchase Michelin, Goodyear, or Uniroyal tires is, respectively, 0.20, 0.30, and 0.40. What is the probability that he will not purchase any of these brands?
7. According to police records, 70% of reckless drivers are fined, 50% have their driver's licenses revoked, and 40% are both fined and have their licenses revoked. What is the probability that a particular reckless driver will either have her license revoked or be fined?
8. The Clegg Truck Stop offers free refills to those ordering coffee. Bill Clegg, the owner, gathered the following information on the number of refills:

Refills	Percent
1	40.0
2	30.0
3	20.0
4	10.0

   Compute the mean number and the standard deviation of the number of refills.
9. The Spoon Appliance Store ran an advertisement stating "Come save an additional \$10–\$25–\$50–\$100 off the already low price of our TVs. . . . Bust a balloon with a money-saving coupon inside and have that amount taken off of your purchase price." There is a total of 50 balloons with 35, 10, 4, and 1 containing coupons for \$10, \$25, \$50, and \$100, respectively. What is the expected savings for each customer? What is the standard deviation?
10. The Utah Department of Natural Resources reports that the mean number of fish caught per hour in the Conewango River

is 0.40. What is the probability that if Victor Anderson fishes for an hour, he will catch three or four fish? At least one fish?

11. The suicide-prevention unit in a particular city estimates that 20% of the callers are serious about taking their lives. If on a particular day the unit received 10 calls, what is the probability that none of the callers was serious? What is the probability that at least two were serious?

12. The area a painter covers with one gallon of paint is normally distributed, with a mean of 400 square feet and a standard deviation of 60 square feet. If the manufacturer specifies that one gallon should cover between 375 and 450 square feet, what percent of the time will the painter exceed the upper limit of the manufacturer's specification? What percent of the time will he be within the manufacturer's limit?

13. Patients arrive at the emergency room of St. Charles Hospital at the rate of four per hour. Assume that the arrival distribution approximates the Poisson distribution. What is the probability of no arrivals during a given hour? Use the normal approximation to the Poisson and compare your results.

14. It is estimated that 70% of the law school graduates in a particular state pass the state bar examination on the first try. What is the probability that in a group of 12 students nobody passes? At least one student passes? More than half pass?

15. It is estimated that 80% of all household plants are overwatered by their owners. In a group of 50 plants, what is the probability that more than 35 were overwatered? That up to 45 were overwatered?

16. The mean life of socks used by the Army is 60 days with a standard deviation of 12 days. Assume that the life of the socks is normally distributed. If one million pairs are issued, how many would need replacement after 50 days? After 70 days?

17. Past records indicate that 30% of the students enrolling at a particular university graduate within five years of their entrance. In a group of 14 newly enrolled students, what is the probability that fewer than half will graduate within the five years? More than eight?

18. The number of sandwiches sold daily by the deli bar in the student union is normally distributed, with a mean of 215 and a standard deviation of 20. On what percent of days does the deli bar sell more than 200 sandwiches? Fewer than how many sandwiches are sold 10% of the days?

19. According to recent newspaper reports, the typical player in the National Football League plays for 3.2 years. If the career is normally distributed and the standard deviation is 1.80 years, compute the probability that a randomly selected first-year player will play more than five years.

20. Refer to Problem 19. Assume that the length of the career is a Poisson distribution, with a mean of 3.2 years. Compute the probability that the player will play more than five years.

Using Statistics

Situation Professional baseball has a playoff system in which two teams play several games and the team that wins the majority is declared the champion. The winner of the World Series is the first team to win four games out of a possible seven. Suppose two teams that are equally matched—that is, $\pi_1 = \pi_2 = 0.50$—meet in the World Series.

a. What is the theoretical distribution of games that could be played in a seven-game series?
b. How does this theoretical distribution compare with modern World Series history?
c. Suppose the teams are not equally matched. Would you expect the series to last an average of more or less games than the case where $\pi_1 = \pi_2$?

Discussion The following table shows the number of games played in every World Series from 1945 until 1992.

Number of Games in World Series	Frequency of Occurrence, 1945–1992	Relative Frequency
4	7	.1458
5	7	.1458
6	9	.1875
7	25	.5208

In the 48 World Series' played between 1945 and 1992, it has gone seven games 25 times, or 52% of the time. The mean number of games played was 6.0828, with a variance of 1.2428, found by using formulas 6–1 and 6–2. So we conclude from the evidence that the "typical" World Series took a little more than six games on average to complete, and that more than half the time the Series went the full seven games.

Now let's compare the actual result with a theoretical result in which the teams are equally matched. To estimate the length of the World Series, assume that each game is independent of previous games and that the probability of winning is 0.50. Suppose that the Los Angeles Dodgers and the Cleveland Indians are playing in the Series. The chance that the Indians will win a particular game is 0.50, and if they win the first game the likelihood that they will win the second game is still 0.50. Given these reasonable assumptions, we can develop a probability distribution for the number of games in a World Series:

Length of Series	Probability
4	$2 \cdot {}_3C_3(0.50)^4 = 0.1250$
5	$2 \cdot {}_4C_3(0.50)^5 = 0.2500$
6	$2 \cdot {}_5C_3(0.50)^6 = 0.3125$
7	$2 \cdot {}_6C_3(0.50)^7 = 0.3125$

How did we compute these probabilities? Let's study a six-game series as an example. A six-game series means that one team wins its fourth game in the sixth game of the series. Suppose the Indians hold a three-games-to-two lead after five games. This means that they could have won any combination of three out of five games (a total of 10 possibilities, found by ${}_5C_3 = 5!/2!3! = 10$). Hence, there are 10 different ways the Indians could have won three of the first five games. There are six games, and the Indians have a 0.50 chance of winning each game. Now we must also consider the possibility that the Dodgers win their fourth game in the sixth game of the series, so we need to multiply the probability by 2. This probability is 0.3125. The probabilities for the other series lengths are computed in a similar way.

The mean of the theoretical distribution is 5.8125 and the variance 1.0273, computed by using formulas 6–1 and 6–2, respectively. In comparing the means of the actual and the theoretical distributions, the mean of the actual distribution is 0.2703 games larger, found by 6.0828 – 5.8125. That is, in the actual cases over the 48 years, the World Series ran an average of about 0.30 games longer than expected. The variance of the actual distribution is also larger. This indicates there was more variation than expected in the length of the World Series. As a final comparison, the theoretical distribution suggests that the World Series should go seven games 31.25% of the time. It actually went seven games 52.08% of the time.

Is it reasonable to conclude that the actual data contradicts the theoretical calculations? Why? Apparently, the games are not independent trials, even though that seems reasonable, and the constant assumption of $\pi_1 = \pi_2$ is not true. Teams apparently win the "must" games to extend the series. Perhaps the so-called "home-field advantage" is a factor, with such positive influences as the support of fans and familiarity with the playing field. In any case, it seems likely that what happens in one game is a factor, either positive or negative, in the next one.

CHAPTER 8

Sampling Methods and Sampling Distributions

OBJECTIVES

When you have completed this chapter, you will be able to

- design a statistical study;
- list and describe three methods of collecting data;
- describe the various types of probability sampling;
- describe what is meant by sampling error;
- develop and describe the sampling distribution of the sample means.

CHAPTER PROBLEM **Should It Be "Last Hired, First Fired"?**

THE MAYOR OF LAS PALMAS IS REVIEWING ALL CITY SERVICES IN AN EFFORT TO CUT THE GROWING BUDGET DEFICIT. IT MAY BE NECESSARY TO REDUCE THE NUMBER OF PERSONNEL IN THE DEPARTMENT OF SOCIAL SERVICES FROM SIX TO FOUR. THE LENGTH OF SERVICE FOR EACH OF THE SIX EMPLOYEES IS: 4, 3, 3, 2, 4, AND 3 YEARS. IF SENIORITY IS THE DETERMINING FACTOR, WHICH TWO WORKERS SHOULD BE LET GO?

INTRODUCTION

What do the following four problems have in common? (1) The president of the United States wants to know the proportion of voters in Texas who will vote to reelect him in November; (2) Revlon wants to know how many women will buy newly developed Siren Red lipstick; (3) the General Services Administration needs to know how many government-owned computers are not fully utilized; and (4) the National Organization for Women (NOW) wishes to know what percent of all banks in the United States have at least one female director.

These problems have one thing in common: The information is very difficult to obtain. It would be almost impossible (and very expensive) for the president to contact every potential voter in Texas by November. Likewise, it would be impossible and, again, very expensive, for Revlon to contact all women in the world (or even just in the United States) and ask each one to try a complimentary stick of Siren Red. NOW would find it difficult to contact every bank in the United States.

The usual, less costly, way to obtain this kind of information is to take a **sample.** As defined in Chapter 1, a sample is a smaller group selected from the population of interest. The objective of studying the smaller group is to obtain information about the whole **population.** The president may hire a polling service, which in turn may sample 2,000 Texans on their political preference in the forthcoming election. NOW might select, say, 50 banks at random and determine what percent have at least one female director. If 15 out of the 50, or 30%, have at least one female director, NOW might reasonably conclude that about 30% of *all* banks in the United States have at least one female director.

These four situations illustrate how information obtained from a sample can be used to say something about the entire population. This is the process called **statistical inference.** Recall from Chapter 1 that statistical inference is the process of reasoning from specific instances or data to general conclusions about the entire group or population. In this chapter and those that follow, you will be introduced to statistical techniques that are based on **probability sampling.**

> **Probability Sampling** Each member of the population of interest has a known likelihood of being included in the sample.

If probability sampling is not used, sample results may not be representative of the entire population. In such cases, it is said that the results are **biased.** To illustrate, Revlon might contact 400 women in New York City about Siren Red mainly because the group's location is convenient to Revlon's New York office. The results, however, might not be representative of all women in the United States; the color red may suggest "warmth" to women in New Mexico and "aggressiveness" to those in New York, or vice versa.

In this chapter, we will first discuss several scientific methods of design and selection of a sample and situations in which each method might be used. Then the sampling distribution of a widely used statistic—namely, the sample mean—will be examined.

DESIGNING THE SAMPLE SURVEY OR EXPERIMENT

Nowhere in statistics is the expression, "well begun is half done," more applicable than to the design of a sample survey or experiment. If a study is not carefully planned and tested, any calculations that follow from it may be useless. Further, collecting data and conducting a controlled experiment are expensive and time-consuming, so it is important to begin with clearly defined objectives and procedures. There are four questions we can ask ourselves to help us plan a sample survey or experiment.

WHAT DO WE WISH TO FIND OUT? Answer this question precisely! Now is the time to clearly focus the study by avoiding poorly worded goals and emotionally biased statements. Consider the following question sent to all office managers during a national survey: "If modern office equipment was installed, how much would productivity improve in the office?" What exactly is meant by "productivity in the office"? Could managers calculate an exact figure? If so, would it be in terms of dollars or percents? To some, "modern office equipment" may be a FAX machine or a single computer; to others, it may refer to a network of computers or even furniture.

Self-Review 8-1

Answers to all Self-Review problems are at the end of the chapter.

The following question appeared in a survey on smoking, alcohol, and drug abuse given to a group of elementary school pupils: "How often in the last month did you use an inhalant?" Evaluate this question. Is it reasonable or misleading? Why?

WHAT POPULATION ARE WE INTERESTED IN STUDYING? "All residents of the United States" is a different group than "all eligible voters" or "all registered voters." Since conclusions about one group may not apply to others, it is useful to list every member of the population being studied without duplication. This list is called the **sample frame.**

Sample frame A complete list of the population without duplication.

An alphabetical listing of all students enrolled at Kankakee Community College this semester is an example of a sample frame.

HOW WILL THE DATA BE COLLECTED? The procedure for selecting the actual sample is called the **sample design.**

Sample Design The procedure for selecting the sample.

This is the step when cost is considered. There are three basic methods for collecting data: mail surveys, personal interviews, and telephone surveys.

Mail surveys are relatively cheap and easy to administer, but a typical response rate is less than 25%. A recent survey sent to 500 MBA graduates resulted in only 89 responses, many of which were unusable! On the other hand, a higher response rate usually occurs when there is an issue involving the respondents. For example, a female student sent a survey to 62 women who were their Schools' Directors of Intercollegiate Athletics. She received responses from 37, well above the typical rate. The difference was probably due to the respondents' interest in women's issues.

Probably the worst prediction blunder ever made was the result of a mail survey. In 1936, *Literary Digest* published a survey that predicted Alf Landon would defeat Franklin D. Roosevelt in the upcoming presidential election. Among other flaws, this survey had what is termed "self-selection bias"—respondents were supposed to voluntarily send back sample ballots with their "votes" on them. Out of 10 million sample ballots sent out, however, only 2.3 million were returned. Those who returned the ballots clearly did not represent the voting public.

Only people with strong opinions bother to return questionnaires or wait to speak on a call-in radio or TV show. One recent variation of this phenomenon has been to prompt viewers to call, for a charge, one number if they support a particular position and another if they oppose it. For example, baseball fans were asked to call 1–900–283–4545 if they supported the job Fay Vincent was doing as commissioner, and 1–900–283–4546 if they did not. Do you think a cross-section of baseball fans bothered to pay 50 cents and voice their opinions? While very committed to causes, those who write letters to editors or call "Larry King Live" usually do not represent the population at large.

Personal interviews are expensive, but get fairly thorough results. In a situation where the interviewer can see the answer, instead of asking the question, the results tend to be more accurate. For example, if the researcher needs to know whether the respondent lives in an apartment, the interview setting will answer the question. Of course, respondents will try to please the interviewer or are otherwise influenced by the situation. To offset this, for questions where the respondent may give answers he or she feels are "right" rather than true, "check" questions should be included later in the interview. This helps to verify the consistency of the respondents' answers. If not included, the questions may be poorly worded or ones the person may not be qualified to answer. In either case, they should be dropped from the study.

Telephone interviews are a common compromise between mail surveys and personal interviews. Most Gallup polls or *USA Today* surveys are done by phone. With a carefully constructed sample, useful results can be obtained by contacting as few as 1,200 people. Here are some recent examples:

- While 89% of Americans could identify Shakespeare, only 47% could identify Freud.
- Sixty-nine percent felt the Postal Service was doing a good job and 64% approved the work done by the Defense Department, but only 45% gave the Justice Department good marks.
- Only 61% of American adults are married, down from a peak of 74% in 1960, reflecting the growing social acceptability of remaining unmarried and the increasing financial independence of women.

Unlisted phone numbers were a problem for pollsters until the use of "random-digit dialing." With this procedure, the first three digits of a telephone number (called an exchange) are identified, and random numbers are generated for the last four digits. However, as many as 10% of American households do not have telephones! This could clearly lead to a biased sample if those households are never contacted.

HOW WILL THE RESPONSES BE RECORDED AND PROCESSED? Trade-offs between effort and accuracy must be made. If, for example, age is a variable, tallying respondents' ages is very tedious, but grouping people into wide ranges, such as 10–20 years, leads to less accuracy. Techniques that can improve studies include "pretesting" questions on a small group to clear up language, and "follow-up" procedures to encourage those who initially do not respond.

STAT BYTE

Television shows are "rated" each week to determine the amount charged for advertising and whether the show will continue. The A. C. Nielsen Company reports statistics on each show. A ratings "point" is 931,000 households. The percent of sets in use is the "share." Part of the October 27, 1992 report is shown below. You can see that 24.9 million people watched "Full House." That was 16.3 rating points and 26% of the sets in use.

	Viewers (in millions)	Rating	Share
8:00			
Full House (ABC)	24.9	16.3	26
Rescue 911 (CBS)	21.7	14.2	23
Quantum Leap (NBC)	11.5	8.0	13
9:00			
Roseanne (ABC)	37.0	24.1	37
The President's Child (CBS)	13.9	10.0	17
Reasonable Doubts (NBC)	11.0	8.1	13

Pretesting is sometimes called a "pilot study." Including return postage and offering a copy of the study's results are good ways to encourage cooperation.

Thus far, we have only discussed ideas and language related to a **sample survey.**

> **Sample Survey** A study in which how a treatment has already affected the experimental objects is observed.

It is sometimes possible, however, to include *every* member of the population of interest in a sample. In that special case, we have conducted a **census.**

> **Census** An observation of the entire population.

In a study on absenteeism, for example, if a company has only 200 employees it is reasonable to include all of them rather than take a sample. All things equal, a census is better than a sample because there is no sampling error. However, sometimes less experienced personnel are used for census work and errors result. This occurred in the 1990 national census and led to several lawsuits. In one case, the city of Detroit claimed that much of its inner-city population was not counted, resulting in the loss of federal funding.

In some situations, we can do more than simply record information. If we modify some aspect of the group under study and then see the result, we are conducting an "experiment." In Chapter 5, we defined an experiment as the observation of some activity or the act of taking some type of measurement. In most experiments, two groups are formed. In the **control group,** no real change is introduced.

> **Control Group** A set of experimental objects that *does not* undergo change or receive a treatment during a test.

The other group is the **experimental group.**

> **Experimental Group** A set of experimental objects that *does* undergo change or receive a treatment during a test.

Consider, for example, a study to determine the effect of alcohol on the driving ability of young adults. A driving simulator reports the number of errors made during a five-minute interval. A sample of 50 young adults is obtained, each of whom is randomly assigned to either the control group or the experimental group. Each member of the control group takes the driving simulator test and the mean number of errors for the group is determined. Each member of the experimental group drinks two 12-ounce cans of beer in a half-hour period before taking the driving simulator test to determine the mean number of errors for that group. Note that the control group receives no modification, or "treatment." However, the experimental group does receive a treatment—consuming two cans of beer in a half-hour period. The effect of the modification is measured by the difference between the two groups.

As another illustration, we might form work teams in a study on absenteeism. For half of the employees, no change is made in the work environment; for the other half, employees are allowed to work in an area where music of their choice is played. Then, using employee records, we can see if there is any difference in attendance between the two groups.

As with surveys, it is important to eliminate possible bias. It is common practice to match the two groups as closely as possible with respect to age, sex, health condition, type of work, and so on, so that these differences cancel each other out. Another way to avoid biasing results in a direction one thinks they "ought" to go is to conduct "double blind" tests, in which the person doing the experiment doesn't know which group is the control group.

Self-Review 8-2

In a *USA Today*/CNN/Gallup poll of 602 registered voters taken March 22, 1992, President Bush led Governor Clinton, 52% to 43%. A July 16 poll by the same organizations showed Clinton leading 56% to 33%. How do you explain the difference in the two surveys? Was the sample too small or biased in some fashion?

Exercises

Answers to the even-numbered exercises are at the back of the book.

1. Identify the "population" for each of the following samples:
   a. A sample of 5,000 readers of *Sports Illustrated*
   b. A sample of 400 students living in southern California
   c. A sample of 30 public high school students in New Jersey

2. You roll a die five times with the following results: 5, 3, 2, 3, and 6. What is the population under study? What do you wish to learn about it? What data would you collect?

METHODS OF PROBABILITY SAMPLING

There are several types of probability sampling. We will study simple random sampling, systematic random sampling, and stratified random sampling. Each of these types of sampling has a similar goal: *to allow chance to determine the items or persons that make up the sample.* While each sample outcome may not be predictable when taken alone, groups of samples are quite predictable. The population of interest can consist either of items, such as all cassette decks produced by Pioneer during the past month, or of persons, such as all the registered voters in Precinct 9. Other examples of a population might be all the banks in the United States, all the fish in a pond, or all the students now attending Yale University.

Random Sampling

The most widely used method of probability sampling is **simple random sampling** or, as it is often called, random sampling.

Random Sample A sample chosen so that each member of the population has the same chance of being selected for the sample.

A random sample does not just happen, nor should just any collection of objects carelessly be called a random sample. The selection of a random sample must be planned with care. One method we can use is similar to a lottery. First, we write the name of each member of the population, or an identifying number—such as a Social Security number—on a slip of paper. Then the slips of paper are thoroughly mixed. Finally, the desired number of slips of paper are drawn. As an illustration, let us look at NOW's interest in the percent of all banks having at least one female director. The group could obtain a list of banks in the United States and assign a number to each of them. If number 121 were drawn first, that bank (say, the First National Bank of Arizona) would be contacted. This procedure would be repeated until the desired sample size had been selected. It is very important that the population of interest be precisely defined, NOW, for example, would have to decide whether branch banks with their own board of directors should be included on the list; a member of the population should be listed only once.

An easier way to select a random sample employs a **table of random digits.** A portion of such a table, found in Appendix H, is:

71529	51996	99289	44268	42759	72434	54402
11776	17395	61317	63290	17067	18408	08992
82437	75248	23715	61194	62175	11149	44793
14997	08398	37662	90175	65331	02562	38020
55317	50018	64380	49047	57111	41641	25427
47422	53721	11419	38616	72171	21523	80967
09540	89442	52381	35035	15884	64273	96028

The table is generated in a random fashion; that is, each of the 10 digits has the same chance of being included. Hence, blind chance determines the outcome of the selection process and bias does not enter the procedure.

The entire population from which the sample will be drawn is arranged in some systematic fashion (perhaps alphabetically); next, each item is assigned a number. Let us assume our research involves the response of psychiatric patients at Palm Hospital to a new drug. The population consists of 70 psychiatric patients who receive the drug. An identification number is assigned to each patient, starting with 00 and ending with 69. First, a starting point in the table is randomly selected. You could close your eyes and place a pencil down on the page. Suppose, for example, that number were 14 (see the table below).

71529	51996	99289	44268	42759	72434	54402
11776	17395	61317	63290	17067	18408	08992
82437	75248	23715	61194	62175	11149	44793
→ 14997	→ 08398	→ 37662	90175	65331	02562	38020
55317	50018	64380	49047	57111	41641	25427
47422	53721	11419	38616	72171	21523	80967
09540	89442	52381	35035	15884	64273	96028
Starting point	Second patient	Third patient				

The patient identified by number 14 becomes part of the sample. To select the next patient in the sample, you can move in any direction you choose. Perhaps you could look at the second hand of a clock and move in the direction in which it is pointing. Let's move horizontally to the right. The first two digits in the next column are 08, so patient 08 is also part of the sample. The next patient in the sample is number 37. The following random number is 90. Since no patient is assigned that number, it is omitted. This procedure is continued until the desired number of patients is obtained for the sample.

Self-Review 8-3

Answers are at the end of the chapter.

Here is a class roll for a beginning course in social science. Three students are to be randomly selected to investigate and report on a new community program for the mentally retarded. Suppose you had written numbers 1 through 38 on slips of paper and then had randomly selected numbers 29, 5, and 11. Which students would be included in the sample?

Winter Quarter Preliminary Class Roster

SS 101 03 INTRO TO SOC SC
2:00 P.M. 3:40 P.M. MW UH 422 W Marchal

Name	Rank	Name	Rank
1. August, Nancy M.	FR	20. McFarlin, Ireatha	FR
2. Benner, Robert A.	JR	21. Meinke, Denise M.	JR
3. Brenner, Susan M.	SO	22. Morrison, David D.	JR
4. Clark, Richard C.	FR	23. Navarre, Garry G.	JR
5. Cowan, Timothy J.	JR	24. Oyer, David	SO
6. Cross, Jill M.	SO	25. Pastor, Virginia Marie	SO
7. Daschner, John H.	SO	26. Pickens, Mitch A.	SO
8. Figliomeni, Michael A.	JR	27. Price, Doug C.	SO
9. Grady, Walter P.	JR	28. Rawson, Jeryl L.	FR
10. Heinrichs, James M.	SO	29. Rista, Vicki A.	SO
11. House, James D.	JR	30. Schmidt, Randy F.	SO
12. James, Phyllis E.	JR	31. Sherman, Mike J.	SO
13. Kimmel, Kurt D.	SO	32. Shull, Karen A.	SO
14. Lach, Jerry William	JR	33. Snow, Sue A.	FR
15. Lehman, Tim J.	SO	34. Straub, Jeff J.	SO
16. Lenz, Matthew H.	SO	35. Turco, Greg W.	SO
17. Martin, Diane M.	SO	36. Von Hertsenberg, Kevin	JR
18. Mason, Craig D.	SO	37. Wagner, Holly S.	SO
19. McCullough, Randy N.	SO	38. Yamada, Jay A.	JR

Exercises

3. The following is a list of the McDonald's Restaurants in a large city. Four locations are to be randomly selected and inspected for cleanliness, safety, customer convenience, and other features. The 28 locations have been coded from 00 to 27. Also noted is whether the location has a play area (P) or not (N).

Number	Location	Play Area	Number	Location	Play Area
00	1560 E Alexis Rd	P	14	5855 Lewis	N
01	835 Lime City Rd	P	15	90 Main	N
02	343 New Towne Square Dr	N	16	567 E Manhattan	N
03	10471 Fremont Pke	P	17	4948 Monroe St	P
04	6555 Airport Hwy	N	18	3345 Monroe St	P
05	04225 Airport Hwy	N	19	2908 Navarre	P
06	5810 W Alexis	N	20	805 N Reynolds	P
07	1736 Broadway	N	21	3138 Secor	P
08	2259 S Byrne	P	22	853 Southwyck Shp Cntr	P
09	3158 Cherry	P	23	3350 W Stern Rd	N
10	1016 Conant	P	24	3740 N Summit	P
11	3240 Dorr St	N	25	1205 W Sylvania Av	P
12	3015 N Holland	P	26	2325 Woodville	P
13	2112 W Laskey	P	27	22201 Woodville Rd	N

a. Suppose the random numbers 11, 17, 61, 03, 93, and 22 are obtained from Appendix H. Which locations would be selected?

b. Using Appendix H, select a random sample of five locations.

4. The following is a list of family practice physicians. Three physicians are to be randomly selected and contacted regarding their fees. The 39 physicians have been coded from 00 to 38. Also noted is whether they are in practice by themselves (S), have a single partner (P), or are in a group practice (G).

Random Number	Physician	Type of Practice	Random Number	Physician	Type of Practice
00	R. E. Scherbarth, M.D.	S	11	Wendy Martin, M.D.	S
01	Crystal R. Goveia, M.D.	P	12	Denny Mauricio, M.D.	P
02	Mark D. Hillard, M.D.	P	13	Hasmukh Parmar, M.D.	P
03	Jeanine S. Huttner, M.D.	P	14	Ricardo Pena, M.D.	P
04	Francis Aona, M.D.	P	15	David Reames, M.D.	P
05	Janet Arrowsmith, M.D.	P	16	Ronald Reynolds, M.D.	G
06	David DeFrance, M.D.	S	17	Mark Steinmetz, M.D.	G
07	Judith Furlong, M.D.	S	18	Geza Torok, M.D.	S
08	Leslie Jackson, M.D.	G	19	Mark Young, M.D.	P
09	Paul Langenkamp, M.D.	S	20	Gregory Yost, M.D.	P
10	Philip Lepkowski, M.D.	S	21	J. Christian Zona, M.D.	P

(continued)

Random Number	Physician	Type of Practice	Random Number	Physician	Type of Practice
22	Larry Johnson, M.D.	P	31	Jeanne Fiorito, M.D.	P
23	Sanford Kimmel, M.D.	P	32	Michael Fitzpatrick, M.D.	P
24	Harry Mayhew, M.D.	S	33	Charles Holt, D.O.	P
25	Leroy Rodgers, M.D.	S	34	Richard Koby, M.D.	P
26	Thomas Tafelski, M.D.	S	35	John Meier, M.D.	P
27	Mark Zilkoski, M.D.	G	36	Douglas Smucker, M.D.	S
28	Ken Bertka, M.D.	G	37	David Weldy, M.D.	P
29	Mark DeMichiei, M.D.	G	38	Cheryl Zaborowski, M.D.	P
30	John Eggert, M.D.	P			

a. If the random numbers 54, 08, 44, 38, and 25 are obtained, which physicians would be contacted?
b. Select a sample of four physicians, using Appendix H.

Systematic Random Sampling

If the population is large—say, the 15,599 students enrolled at the University of Utah—a list of all students could be obtained and a random starting point selected. Suppose the twenty-eighth student were chosen. To save us the effort of selecting additional random numbers, a constant number could be added to the starting number. If we decided to add 100 to the starting number, the students identified by numbers 28, 128, 228, 328, and so on would become members of the sample. This systematic procedure is aptly called **systematic random sampling.**

The results of a systematic sample will be just as representative as those from a simple random sample. It should not be used, however, if there is any possibility of bias in the ordered list. For example, if you were doing a study on absenteeism, it would be unwise to take a systematic sample of every seventh day. The results would be unduly affected by the starting day. If a Monday had been selected as the starting day, then all the other days selected would also be Mondays, and it is well known that absences are higher on Mondays.

Systematic Random Sample The members of the population are arranged in some fashion. (They may be numbered 1, 2, 3, . . . , listed alphabetically, or ordered by some other method.) A random starting point is selected. Then every *k*th element is chosen for the sample.

Self-Review 8-4

Refer to Self-Review 8–3. Suppose that this sample is to consist of every ninth student enrolled in the class after a starting student has been randomly selected from students numbered 1 through 9. Suppose this starting point is the fourth student. Which students will be in the sample?

Exercises

5. Refer to Exercise 3. A sample is to consist of every seventh location. The number 02 is selected as the starting point. Which restaurants will be contacted?
6. Refer to Exercise 4. A sample is to consist of every fifth physician. A random starting point of physician number 03 is selected. Which physicians will be included in the sample?

Stratified Random Sampling

In planning some types of surveys, it is desirable that the sample be representative not just of the population as a whole, but of certain subdivisions or groups within it. For example, Revlon might want to ensure that women in each of 10 regions of the United States be included in their research project to determine the market potential of Siren Red lipstick. To accomplish this goal, we would divide the country into 10 geographical regions and randomly select a sample of women within each region. Each woman selected would be asked to try Siren Red and report her reaction. Dividing the country into regions is called **stratifying the population.** Other traits commonly used to form strata are age, income level, and political-party affiliation. The manner in which the sample is gathered may be either nonproportional or proportional to the total number of members in each stratum.

> **Stratified Random Sample** After the population of interest is divided into logical strata, a sample is drawn from each stratum or subgroup.

We form a **stratified random sample** by identifying the natural subgroupings in the population and selecting an independent random sample from each stratum or subgroup. Obviously, it is important that we define the strata carefully to ensure that each member belongs to only one subgroup.

The advantage of stratified sampling is that one member of a subgroup is usually quite similar to other members of that subgroup while being quite different from the members of other subgroups. If, for example, the research project involved surveying executives on the role of government in business, the population could be stratified into bankers, executives of large firms, executives of small firms, and so on. Bankers tend to think alike about the role of government in business, but their opinions might differ drastically from those of small business executives. Unless the subgroups differ significantly from each other, nothing is gained by stratification. For example, if our research were concerned with the attitude of college students about compulsory military training, a stratification of the population (all college students) into those from the Far West, the East, and so on would not seem justified unless the researcher believes that geographic location affects attitude toward military service.

Self-Review 8-5

Refer to Self-Review 8–3. Separate the population into three strata: freshmen (FR), sophomores (SO), and juniors (JR). Suppose it had been decided that one freshman, five sophomores, and four juniors would constitute the sample. Randomly select the required number from each stratum.

Exercises

7. Refer to Exercise 3. Suppose that the sample is to consist of four locations, three with play areas and one without. Select a sample accordingly.

8. Refer to Exercise 4. Suppose that a sample is to consist of three physicians in group practice (G), two in solo practice (S), and two with a single partner (P). Select a sample accordingly.

STAT BYTE

In the 1992 presidential election, national polls conducted in November and the actual vote were virtually identical:

Poll	Clinton	Bush	Perot
The Harris Poll	44	38	17
ABC News	44	37	16
NBC/*Wall Street Journal*	44	36	15
CBS/*New York Times*	44	35	15
Washington Post	43	35	16
Actual Vote	43	38	19

Makes you wonder why we have elections, doesn't it?

THE SAMPLING ERROR

The previous discussion of various scientific sampling methods emphasized the importance of trying to choose a sample so that every member of the population has a known chance of being selected. In other words, the sample should be representative of the population. It would be unreasonable, however, to expect the sample characteristics to match the population *exactly*. The mean of the sample might be different from the population mean by *chance alone*. The standard deviation of the sample will probably be different from the population standard deviation. We can therefore expect some difference between the **sample statistics** (such as the mean and the standard deviation) and the corresponding population values, known as **parameters.** This difference is known as the **sampling error.**

Sampling Error The difference between the value of the population parameter and its corresponding sample statistic.

The idea of sampling error can be illustrated with a very simple example. Suppose your five grades (the population) to date in this course were 69, 86, 82, 70, and 98. A sample of two grades is selected at random from this population of grades to estimate your mean grade. They are, let us say, 70 and 86. The mean of this sample is 78. The mean of another sample of two grades (69 and 98) is 83.5. The mean of all five grades (the population) is 81. Notice that sampling errors of –3 and 2.5 are made in estimating the population mean.

Given this potential sampling error, how can political polls, for example, make accurate predictions about the behavior of the voting population based only on sample results? How can a quality control inspector in a manufacturing plant make a decision about the quality of a product after inspecting only a sample of 10 parts? We will answer such questions by developing a sampling distribution for the sample means.

THE SAMPLING DISTRIBUTION OF THE SAMPLE MEANS

As noted in the previous section, the means of samples of a specified size selected from a population vary somewhat from sample to sample. When all the sample means that might occur are organized into a probability distribution, the resulting distribution is called the **sampling distribution of the sample mean.**

> **Sampling Distribution of the Sample Mean** The probability distribution of the sample means of all possible samples of a given size selected from a population.

The problem that follows illustrates how such a distribution is constructed.

Problem Recall from the Chapter Problem that the mayor of Las Palmas is concerned about the city's loss of revenue. To make up for the loss, she is reducing the size of the Department of Social Services from six to four employees. The information in Table 8–1 lists the six employees and their years of service. Note that these data constitute a population, and that the mean of this population is 3.16667 years. Suppose the mayor decides, because the lengths of service vary from only two to four years, that it would be more equitable to randomly select the four workers who are to remain rather than to dismiss the two with the fewest years of service. How many different groups of four workers are there? How would the sampling distribution of the mean lengths of service appear? What are the possible mean lengths of service for the department when there are four employees?

Table 8-1 **Lengths of Service of Six Social Workers**

Social Worker	Length of Service (in years)
Don	4
Gary	3
Sue	3
Bob	2
Kirk	4
Sandra	3

The population mean μ is 3.16667, found by:

$$\mu = \frac{4+3+3+2+4+3}{6} = 3.16667$$

Solution The population consists of six employees; four are to be randomly selected to remain. We can use the idea of a combination from Chapter 5 to determine the total number of possibilities. Let N be the size of the population—in this case, 6—and n the size of the sample, which is 4. So the number of possible samples (groups of remaining employees) is:

$$_NC_n = \frac{N!}{(N-n)!\ n!} = \frac{6!}{(6-4)!\ 4!} = 15$$

There are 15 different possible combinations of four employees who will remain in the Social Services Department after the cutback. These possible outcomes are listed below, along with the lengths of service for each retained employee, and the mean for each sample (group of employees).

Sample Names				Length of Service (in years)	Sample Mean $\overline{X}$
Don	Gary	Sue	Bob	4, 3, 3, 2	12/4 = 3.00
Don	Gary	Sue	Kirk	4, 3, 3, 4	14/4 = 3.50
Don	Gary	Sue	Sandra	4, 3, 3, 3	13/4 = 3.25
Don	Gary	Bob	Kirk	4, 3, 2, 4	13/4 = 3.25
Don	Gary	Bob	Sandra	4, 3, 2, 3	12/4 = 3.00
Don	Gary	Kirk	Sandra	4, 3, 4, 3	14/4 = 3.50
Don	Sue	Bob	Kirk	4, 3, 2, 4	13/4 = 3.25
Don	Sue	Bob	Sandra	4, 3, 2, 3	12/4 = 3.00
Don	Sue	Kirk	Sandra	4, 3, 4, 3	14/4 = 3.50
Don	Bob	Kirk	Sandra	4, 2, 4, 3	13/4 = 3.25

Sample Names				**Length of Service (in years)**	**Sample Mean $\overline{X}$**
Gary	Sue	Bob	Kirk	3, 3, 2, 4	12/4 = 3.00
Gary	Sue	Bob	Sandra	3, 3, 2, 3	11/4 = 2.75
Gary	Sue	Kirk	Sandra	3, 3, 4, 3	13/4 = 3.25
Gary	Bob	Kirk	Sandra	3, 2, 4, 3	12/4 = 3.00
Sue	Bob	Kirk	Sandra	3, 2, 4, 3	12/4 = 3.00

The sample means for the retained employees are presented in the form of a probability distribution (see Table 8–2). Logically, it is called the **sampling distribution of the sample means.**

The distribution of the sample means in Table 8–2 gives the probability of the sample means for all possible samples of size four taken from the population of six social workers. The distribution of the sample means in Table 8–2 is portrayed graphically in Figure 8–1. The top chart depicts the sampling distribution and the bottom chart the population values.

Note these predictable patterns from Figure 8–1 on page 294:

1. The mean of the sampling distribution and the mean of the population are equal (3.16667 in this case), found by:

$$\frac{3.00+3.50+3.25+\cdots+3.00}{15}=\frac{47.50}{15}=3.16667$$

   The mean of the sample means is the same as the population mean computed in Table 8–1. This is not a coincidence! The mean of the sample means will always equal the population mean.

2. The spread in the distribution of the sample means is smaller than the spread in the population values; the sample means range from 2.75 to 3.50, while the population ranges from a low of 2 to a high of 4. In fact, the standard deviation of the

Table 8-2

Sampling Distribution of the Sample Mean

Sample Mean	Frequency	Probability
2.75	1	1/15 = 0.0667
3.00	6	6/15 = 0.4000
3.25	5	5/15 = 0.3333
3.50	3	3/15 = 0.2000
	15	1.0000

Figure 8-1

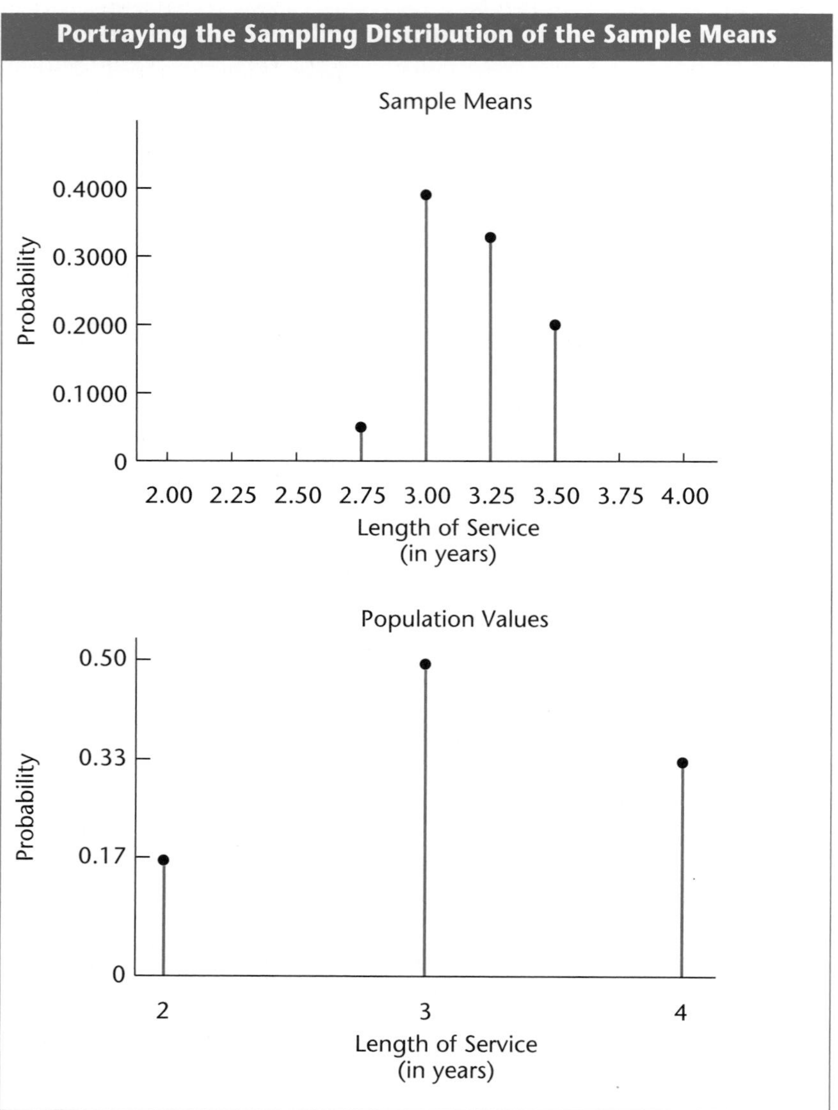

distribution of the sample means will always equal the population standard deviation divided by the square root of the sample size. So, as the sample size gets larger, the spread in the distribution of sample means becomes smaller.

3. The shape of the sampling distribution of the means and the shape of the frequency distribution of the population values are different. The distribution of sample means tends to be bell-shaped and to approximate the normal probability distribution.

In summary, we took random samples from a population and for each sample calculated a sample statistic (the mean length of service). Because each possible sample has a known chance of selection, the probability that the mean length of service will be 2.75 years, 3.0 years, and so on can be determined. The distribution of these mean lengths of service is aptly called the sampling distribution of the sample means.

Even though in practice we see only one particular random sample, in theory any sample could arise. Consequently, we view the sampling process as repeated sampling of the statistic from its sampling distribution. This sampling distribution is then employed to measure how reasonable or likely a particular outcome might be.

Self-Review 8-6

A population consists of five prisoners in cell block M. The length of time each has spent in prison is shown below.

Name	Years
Dow	5
Smith	3
Artz	6
Kim	2
Batt	4

a. Compute the mean length of imprisonment for the population.
b. Select all possible samples of two prisoners from the population. Compute the mean of each sample.
c. Does the mean of the sample means equal the population mean?
d. Give the sampling distribution of the means.
e. Plot the sampling distribution of the means and the population.
f. Does the sampling distribution tend to be bell-shaped, and does it begin to approximate a normal distribution?
g. Cite evidence to show that there is less spread in the sampling distribution compared with the population values.
h. Is the population normally distributed or nonnormal?

Exercises

9. A police training class consists of four recruits. They have had 2, 3, 4, and 4 years of education beyond high school.
   a. Determine the population mean.
   b. List all possible samples of two recruits.
   c. Compute the mean of each sample. (*Hint:* There should be six samples.)

d. Develop a sampling distribution of the means.
e. Portray the sampling distribution of the means and the population values in a histogram.
f. Compare the two distributions in terms of means, shapes, and ranges.

10. A population consists of the hourly wages of six employees. The wages are \$10, \$4, \$12, \$11, \$9, and \$8.
a. Determine the population mean.
b. A sample of four wages is to be selected at random from the population. List all the possible samples of four wages. Then compute the mean of each sample.
c. Develop a distribution of the sample means and compute the mean.
d. Portray the distribution of the sample means in the form of a histogram. Immediately below the histogram show the probability distribution for the population values.
e. Draw conclusions regarding the two means (the population mean and the mean of the sample means). Also, make an observation with respect to the spread of the two probability distributions.

11. Five children are in a preschool play group. Their heights are 39, 38, 36, 39, and 38 inches.
a. Determine the mean height of this population of five children.
b. Compute the standard deviation of their heights.
c. List all possible samples of three children.
d. Compute the mean of each sample found in part c.
e. Draw histograms of the original population and of the sampling distribution of the means on the same scale.
f. Calculate the mean and the standard deviation of the sampling distribution of the means.

SUMMARY

Research in such areas as criminology, medicine, aging, mental retardation, resource management, business, and education often involves taking a sample. Rarely does the investigation encompass a study of all the criminals, all the mentally retarded, all the fish in the lake, or all senior citizens, collectively referred to as the population. Studying an entire population is too time-consuming, and the expense is usually prohibitive.

A small number (the sample) selected from the population (the entire group) is used to reason from specific instances to generalizations about the entire population. Care must be taken that the members of the sample are representative of the entire population. Chance must govern the selection.

One method of ensuring that each member of the population has the same chance of being included in the sample is called random sampling. Another way of ensuring a representative sample is called systematic random sampling. As an example, if the population is defined as all inmates in Rahway Prison, a list of all the prisoners is secured. Then a starting point (say, the twelfth prisoner on the list) is chosen at random. Starting with the twelfth prisoner, every tenth prisoner on the list might constitute the sample. Prisoners numbered 12, 22, 32, 42, and so on would be interviewed. For certain research problems, a stratified random sample might be the most appropriate sampling method. If the population of interest contains a number of distinct groups, it may be desirable to stratify the population into subgroups.

It was pointed out that even though a sample is carefully selected from the population, by chance, the mean of the sample will probably not be exactly the same as the population mean. To examine further the concept of this sampling error, all possible samples of a given size were selected from the population and the mean of each sample was computed. When all the sample means are organized into a frequency distribution, the distribution is called the sampling distribution of the mean. It was noted that (1) the mean of the sampling distribution of the mean is the same as the population mean and (2) there is less spread in the distribution of sample means than in the distribution of the population values.

Exercises

12. A study of hotel accommodations in a metropolitan area reveals the existence of 30 such facilities. The city's convention and visitors bureau is surveying charges per day for single-occupancy rooms. The daily rates for this statistical population are $25, $35, $22, $25, $30, $24, $25, $20, $25, $24, $28, $24, $28, $25, $27, $25, $35, $25, $30, $25, $17, $25, $21, $18, $16, $24, $21, $21, $13, $19.

- **a.** Using the random numbers in Appendix H, draw a simple random sample of six daily rates from this population.
- **b.** Select a systematic random sample by randomly choosing a starting point among the first five hotels and then including every fifth observation.
- **c.** If the last 10 hotels on the list are all "cut-rate" hotels, describe how you could select a sample of four regular hotels and two "cut-rate" hotels.

13. Since 1935, the Social Security Administration has issued numbers, such as 123–45–6789, to 305 million people.

- **a.** Describe how you would select a random sample of 20 individuals from the agency's files.
- **b.** The leading digit indicates which of the 10 service centers issued the number. Describe how you would create a samsple which includes exactly two individuals from each service center.

14. You are studying voters' reactions in your state to a piece of statewide legislation. Describe how you would take a sample of these voters.

15. You wish to study the birth weights of newborn infants in your city. How would you go about obtaining the sample?

16. If you wished to estimate the percent of total working time a secretary spends on specific tasks—typing, answering the telephone, making copies, and so forth—how would you sample his or her activities?

17. A certain variety of flower grows to only three different heights: two inches, four inches, or six inches. If each of these heights is equally likely, draw a histogram of the probability distribution of this population. Find the mean and the standard deviation. List all possible samples of two flowers that could be drawn from this population and calculate the corresponding sample average. Draw a histogram of the probability distribution of sample averages. Find its mean and standard deviation.

Data Exercises

18. Refer to the Real Estate Data Set, which reports information on homes sold in Northwest Ohio during 1992. Assume that these data are a population. Use Appendix H to select five random samples of 10 homes. Compute the mean number of bedrooms in these five samples. Compare the mean of the population (3.58 bedrooms) with the sample means. Compare the shape of the population with the shape of the sampling distribution. Write a brief report comparing your findings with what you expected to find.

19. Refer to the Salary Data Set, which refers to a sample of middle managers in Sarasota, Florida. Assume that these data are a population. Use Appendix H to select five random samples of 10 managers. Compute the mean number of years of experience of the managers in these five samples. Compare the mean of the population (14.707 years of experience) with the sample means. Compare the shape of the population with the shape of the sampling distribution. Write a brief report comparing your findings with what you expected to find.

IS THIS A "SUCKER BET" OR WHAT?

A friend of yours, who will gamble on virtually anything, points out that every dollar bill printed by the federal government has an eight-digit serial number on it, such as J86581144B. He further claims that he can guess the arithmetic mean of the eight digits on any dollar bill in your pocket. He tells you, before you pull a bill out of your pocket, that the mean of the eight digits will be between 4 and 6. He challenges you to empty your pockets and give him every bill with a mean between 4 and 6. As incentive, he offers to match any bill whose arithmetic mean doesn't fall between 4 and 6 and to sweeten the payoff with an extra quarter for every such bill. Would you accept this challenge? Why or why not?

CHAPTER ACHIEVEMENT TEST

Answer all the questions. The answers are at the back of the book

MULTIPLE-CHOICE QUESTIONS. Select the response that best answers each of the questions.

1. A list of every member of the population without duplication is called a
   a. sample frame.
   b. sample design.
   c. survey.
   d. census.
2. In a study of poverty, all families receiving welfare benefits are selected. This group represents the
   a. sample design.
   b. census.
   c. survey.
   d. population of interest.
3. A "100%" survey could also be called a(n)
   a. sample frame.
   b. sample design.
   c. census.
   d. unbiased study.
4. In general, the least expensive way to obtain a sample is
   a. an experiment.
   b. a mail survey.
   c. a personal interview.
   d. a telephone survey.
5. The most accurate survey results are generally achieved by
   a. a controlled experiment.
   b. telephone surveys.
   c. personal interviews.
   d. mail surveys.
6. Each item in a population has a chance of being selected. Sampling done under these conditions is called
   a. systematic sampling.
   b. convenience sampling.

c. confidence sampling.
d. probability sampling.

7. Each item in a population has the *same* chance to be selected for a sample. What is this process called?
a. Nonprobability sampling
b. Random sampling
c. Judgment sampling
d. None of the above

8. If you select every tenth item on a list, which method of sampling are you using?
a. Stratified sampling
b. Random sampling
c. Systematic sampling
d. None of the above

9. The difference between the population mean (μ) and a sample mean ($\overline{X}$) is the
a. interval estimate.
b. point estimate.
c. sampling error.
d. standard error.

10. When all possible samples are selected and their means found, the mean of these sample means
a. is smaller than the population mean.
b. is larger than the population mean.
c. is equal to the population mean.
d. cannot be predicted.

COMPUTATION PROBLEMS

11. A population of five geriatric patients indicate that the number of care-giving relatives are 6, 4, 7, 3, and 2.
a. Compute the mean and standard deviation of this population.
b. Select all possible samples of three patients from this population and compute the mean of each sample.
c. Draw a histogram of the sampling distribution of means found in part b.
d. Compute the mean and standard deviation of the sampling distribution of the means.
e. Compare the location and spread in the original population and the sampling distribution of the sample means.

ANSWERS TO SELF-REVIEW PROBLEMS

8–1 The researcher thought it was reasonable. "Inhalant" meant the concentration and breathing of fumes from glue, liquid "white out," or similar products that can yield a "high." Unfortunately, younger students may not understand the term or, even worse, may be using an inhalant under a doctor's direction. They may give a literally correct response that, in fact, misleads the researcher.

8–2 The sample was probably valid. During the time between the two surveys, Ross Perot abruptly decided not to run! So his withdrawal dramatically changed the race.

8–3

29	Vicki Rista
5	Timothy Cowan
11	James House

8–4

4	Richard Clark
13	Kurt Kimmel
22	David Morrison
31	Mike Sherman

8–5 There are many possibilities. One of them is:

FR	5	Snow
SO	9	Mason
	3	Daschner
	11	Oyer
	12	Pastor
	8	Martin
JR	7	Lach
	10	Navarre
	4	Grady
	3	Figliomeni

No doubt the composition of your sample will be different.

8–6 **a.** $20/5 = 4.0$

b.

	$\bar{X}$
Dow, Smith	4.0
Dow, Artz	5.5
Dow, Kim	3.5
Dow, Batt	4.5
Smith, Artz	4.5
Smith, Kim	2.5
Smith, Batt	3.5
Artz, Kim	4.0
Artz, Batt	5.0
Kim, Batt	3.0

c. Yes, $40/10 = 4.0$, same as $\mu = 20/5 = 4.0$.

d.

$\bar{X}$	f	Probability
2.5	1	1/10 = 0.1000
3.0	1	1/10 = 0.1000
3.5	2	2/10 = 0.2000
4.0	2	2/10 = 0.2000
4.5	2	2/10 = 0.2000
5.0	1	1/10 = 0.1000
5.5	1	1/10 = 0.1000
	10	1.0000

e.

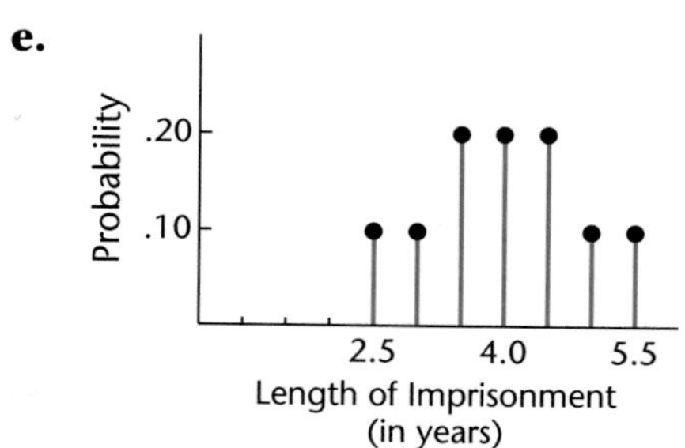

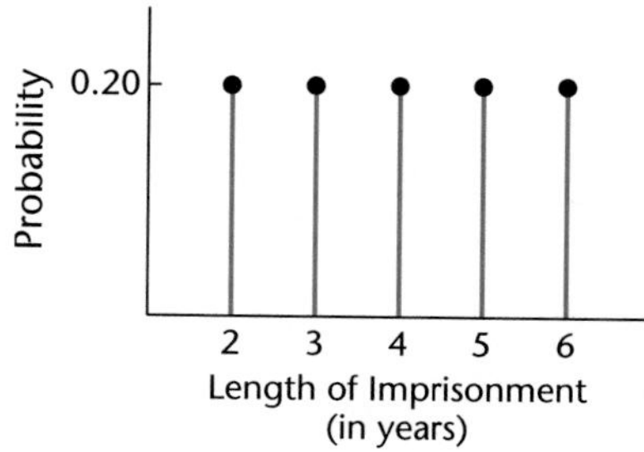

f. Yes

g. The range for the sampling distribution is $5.5 - 2.5 = 3.0$. The range for the population is $6 - 2 = 4.0$.

h. Nonnormal

CHAPTER 9

The Central Limit Theorem and Confidence Intervals

OBJECTIVES

When you have completed this chapter you will be able to

- describe the central limit theorem;
- develop and describe the distribution of sample means;
- develop and describe the distribution of sample proportions;
- compute a confidence interval for means and proportions;
- understand the need for a population correction factor.

CHAPTER PROBLEM ¿Habla Ud. Español?

A STUDY WAS UNDERTAKEN TO CHARACTERIZE THE HISPANIC POPULATION OF SUBURBAN DALLAS. A TOTAL OF 160 HISPANIC FAMILIES WERE CONTACTED AND INFORMATION WAS OBTAINED ON THE PROPORTION OF BILINGUAL RESIDENTS, THE MEAN NUMBER AND AGE OF CHILDREN, THE UNEMPLOYMENT RATE, AVERAGE ANNUAL INCOME, AND SO ON. ONE OF THE QUESTIONS ASKED WAS, "ARE YOU FLUENT IN BOTH ENGLISH AND SPANISH?" NINETY-SIX RESPONDED "YES." THE SAMPLE PROPORTION IS 96/160, OR .60. CONSTRUCT A 95% CONFIDENCE INTERVAL FOR THE PROPORTION OF BILINGUAL HISPANIC FAMILIES IN THE DALLAS AREA.

INTRODUCTION

In this chapter, we will examine one of the most important theorems in statistics—the **central limit theorem.** Its application to the sampling distribution of the sample means, introduced in Chapter 8, allows us to use the normal probability distribution to create confidence intervals for the population mean.

THE CENTRAL LIMIT THEOREM

The central limit theorem states that, for sufficiently large random samples, the shape of the sampling distribution of the sample means is very close to a normal probability distribution. The approximation is more accurate for large sample sizes than for small ones. This is one of the most useful facts in statistics. We can reason about the sample means with absolutely no information about the shape of the original distribution from which our sample is taken. In other words, this is true for all distributions!

The proof of this theorem is beyond the scope of this text, but you can visualize the result. Identify *any* population you might like to study, such as a set of telephone numbers or social security numbers. Select several samples of two items from your population and calculate their means. Do this enough times so that you can draw a reasonably accurate histogram or frequency polygon of the sample means. Now go back and repeat the process with a somewhat larger sample—say, four items. Again, draw a histogram or frequency polygon of the resulting distribution of the sample means.

Repeat this procedure for an even larger sample. You will see that the sampling distribution of the sample means begins to look more and more like a normal probability distribution. That's the point of the central limit theorem! When the size of the sample is large enough, the distribution of the sample means is normal, regardless of the shape of the population from which the sample is drawn. This is true for *any* population from which you randomly sample. Generally, if the sampled distribution is symmetrical, you will see normality in samples as small as 10. On the other hand, if you start with a distribution that is skewed or has very thick tails, it may require samples of at least 30 to note the normality. Most statisticians consider a sample of 30 or more "large enough" for the central limit theorem to be employed.

Central Limit Theorem If all samples of a fixed size are selected from any population, the sampling distribution of the sample means is approximately a normal probability distribution. This approximation improves with samples of larger size.

This concept is illustrated in Figure 9–1 on page 306. In diagram a, the population is a discrete distribution with a positive skewness. Diagram b shows five of the possible samples of size n that might be obtained from the original population. Note that the sample means ($\overline{X}_1, \overline{X}_2$, and so on) vary and that each sample differs slightly from the original shapes. In diagram c, only the sample means are shown. These sample means approximate the normal distribution as n increases. In other words, the central limit theorem tells us that no matter what the shape of the population, the shape of the sampling distribution will approach normal.

An Illustration of the Central Limit Theorem

The identification numbers of 30 students are listed in Table 9–1 on page 307. Treating the eight digits separately as sample observations, we can illustrate the central limit theorem. The table shows 30 random samples of eight digits each. We are not sure of the shape of the population distribution. The right column of the table also contains the corresponding sample means. For example, the mean of the first student's ID number, 699–14–085, is 5.250.

A histogram of the initial 240 digits from Table 9–1 is shown in Figure 9–2 on the top of page 308. Each "*" represents four observations. Note that the histogram is fairly flat or uniform and includes each of the digits from 0 to 9.

Figure 9–3 shows the histogram of the sample means found in the last column of Table 9-1. Note that this frequency distribution is much more bell-shaped, or normally distributed.

We began with a population that was unknown in shape. We took 30 random samples of eight numbers and computed the sample mean of each. The resultant distribution of the sample means approached a normal distribution. Now try repeating this sampling experiment with the telephone numbers listed in your local directory as your population!

STAT BYTE

During World War II, Allied military planners needed estimates of the number of tanks Germany was manufacturing. The information provided by traditional spying methods was unreliable, but statistics proved to be of considerable benefit. For example, espionage and reconnaissance led intelligence analysts to estimate originally that 1,550 tanks were produced in June 1941. Using the serial numbers of captured tanks, however, statisticians revised that figure to 244 tanks, only 27 fewer than the actual number manufactured. A similar problem was encountered in estimating the number of Iraqi tanks destroyed in Desert Storm.

CONFIDENCE INTERVALS

The information just developed about the shape of the sampling distribution of the sample means ($\overline{X}$) allows us to locate an interval that has a high probability of containing the population mean μ. For reasonably large samples, we can assume the following:

1. Ninety-five percent of the sample means selected from a population will lie within 1.96 standard deviations of the population mean μ.
2. Ninety-nine percent of the sample means will lie within 2.58 standard deviations of the population mean.

Intervals computed in this fashion are called the **95% confidence intervals** and the **99% confidence intervals.**

Figure 9-1

Population and Sampling Distributions

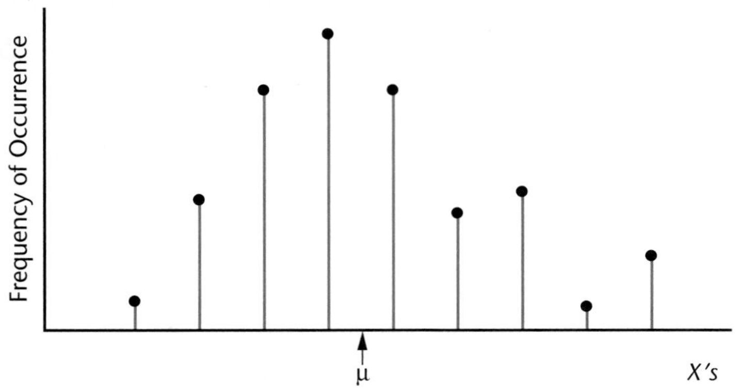

Frequency of Occurrence

μ

X's

μ = Population mean

a. Population Distribution

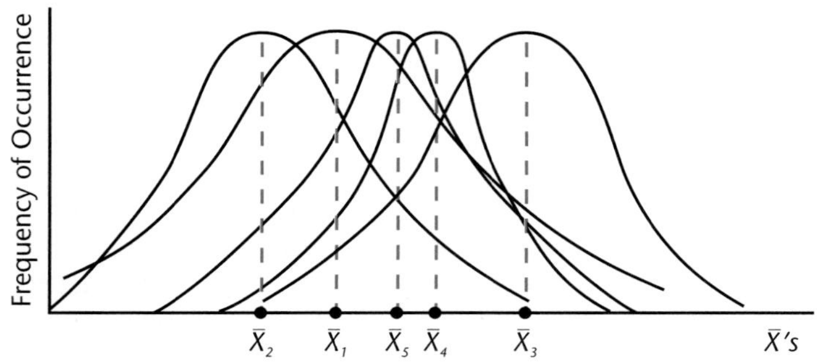

Frequency of Occurrence

$\bar{X}_2$ $\bar{X}_1$ $\bar{X}_5$ $\bar{X}_4$ $\bar{X}_3$

$\bar{X}$'s

b. Distributions of Various Samples

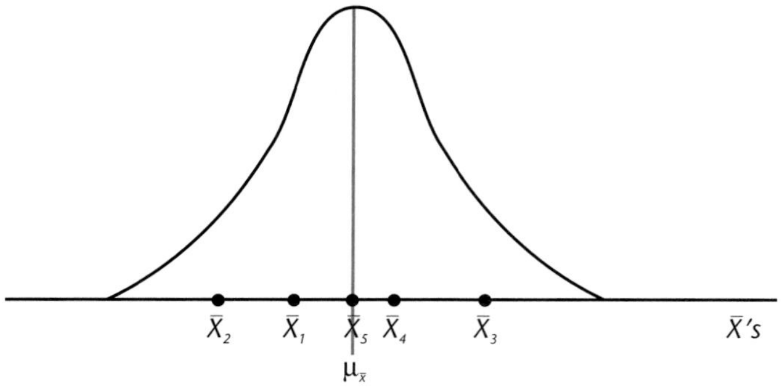

$\bar{X}_2$ $\bar{X}_1$ $\bar{X}_5$ $\bar{X}_4$ $\bar{X}_3$

$\bar{X}$'s

$\mu_{\bar{x}}$

$\mu_{\bar{x}}$ = Mean of the sampling distribution of sample means

c. Distribution of Sample Means

Random Samples and Sample Means of 30 Student ID Numbers **Table 9-1**

Trial Sample	Sample Data (Identification Number)	Sample Mean $\bar{X}$
1	6 9 9 1 4 0 8 5	5.250
2	3 4 8 0 7 5 3 2	4.000
3	6 0 1 7 4 5 4 3	3.750
4	5 7 8 0 5 7 2 2	4.500
5	3 7 1 8 5 6 0 1	3.875
6	1 9 9 1 2 2 7 1	4.000
7	9 2 5 5 5 4 1 5	4.500
8	5 2 4 2 0 0 1 7	2.625
9	4 6 5 7 6 3 5 6	5.250
10	8 9 8 3 7 5 9 9	7.250
11	4 5 3 1 7 7 2 5	4.250
12	5 2 0 7 4 5 9 5	4.625
13	4 5 8 4 4 2 8 1	4.500
14	8 1 5 5 4 3 7 6	4.875
15	8 4 4 8 4 2 8 9	5.875
16	9 9 2 6 3 7 3 4	5.375
17	3 7 0 8 0 9 3 4	4.250
18	1 9 5 2 8 7 8 6	5.750
19	4 2 0 5 9 7 3 3	4.125
20	2 3 2 2 0 8 0 6	2.875
21	8 6 7 5 9 7 2 5	6.125
22	8 3 0 7 0 2 8 9	4.625
23	2 1 6 4 9 3 6 4	4.375
24	0 8 1 7 1 7 2 6	4.000
25	2 9 1 8 6 1 1 8	4.500
26	3 7 8 6 2 7 2 2	4.625
27	9 2 8 7 3 4 3 2	4.750
28	7 1 9 9 5 1 6 1	4.875
29	4 3 9 8 8 7 9 6	6.750
30	9 2 9 0 7 3 0 5	4.375

Figure 9-2

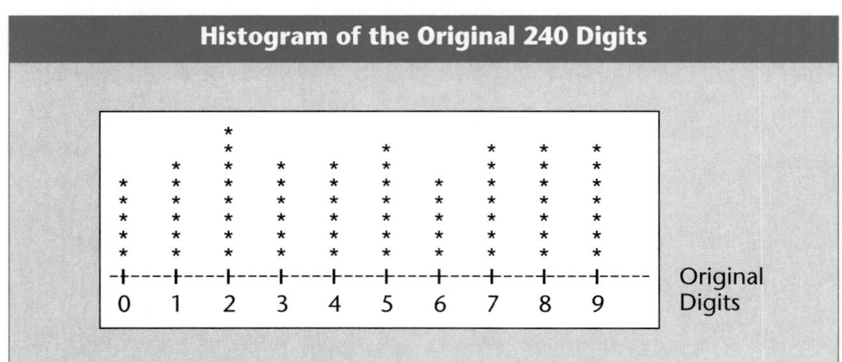

Figure 9-3

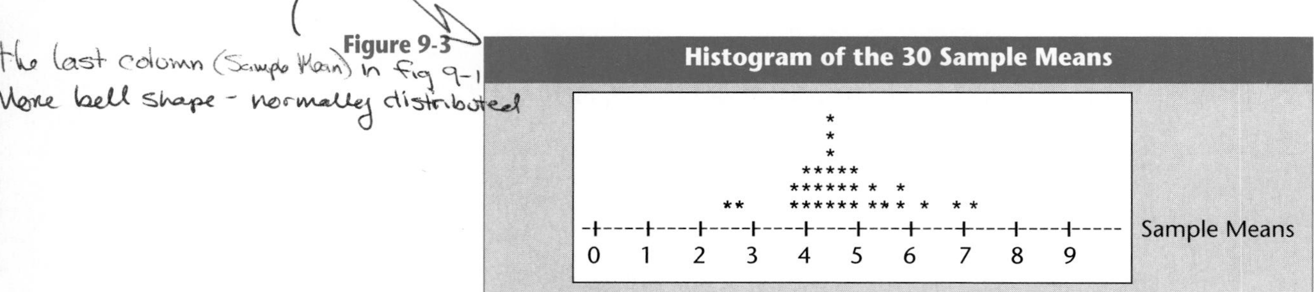

Confidence Interval A range of values constructed from sample data so that a parameter occurs within that range at a preselected probability. The preselected probability is termed the "level of confidence."

How are the values of 1.96 and 2.58 obtained? The 95% and 99% refer to the approximate percent of the time that similarly constructed intervals would include the parameter that is being estimated. The 95%, for example, refers to the middle 95% of the observations. Therefore, the remaining 5% is equally divided between the two tails. See the following diagram. The central limit theorem states that the distribution of the sample means will be approximately normal; therefore, Appendix C may be used to find the appropriate z-values. Locate 0.4750 in the body of the table, then read the corresponding row and column value. It is 1.96; that is, the probability of being in the interval between $z = 0$ and $z = 1.96$ is 0.4750. Likewise, the probability of being in the interval between -1.96 and 0 is also 0.4750. When we combine these two probabilities, the probability of being in the interval -1.96 to 1.96 is 0.95. The z-value corresponding to 0.99 is determined in a similar way.

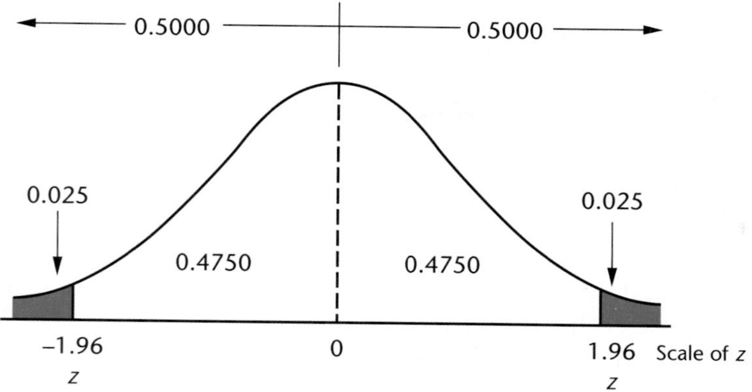

How do we construct the 95% confidence interval? To illustrate, assume our research involves the annual starting salary of graduates with a master's degree. We have computed the mean of the sample to be $32,500 and the standard deviation of the sample means to be $200. The 95% confidence interval is from $32,108 to $32,892, found by $32,500 ± 1.96($200). If 100 samples of the same size were selected from the population of interest and the corresponding 100 confidence intervals determined, one could expect to find the population mean in about 95 out of the 100 confidence intervals.

THE STANDARD ERROR OF THE SAMPLE MEAN ($\sigma_{\bar{X}}$)

In the previous section, the standard deviation of the sampling distribution was given as $200. It is called the **standard error of the sample mean** and denoted by the symbol $\sigma_{\bar{X}}$, read "sigma sub x bar," often shortened to the **standard error.**

> **Standard Error of the Sample Mean** The standard deviation of the distribution of the sample means.

The standard error is a measure of the variability of the sampling distribution. It is computed by

$$\sigma_{\bar{X}} = \frac{\sigma}{\sqrt{n}} \qquad \textbf{9-1}$$

where

$\sigma_{\overline{X}}$ is the symbol for the standard error of the mean.
σ is the population standard deviation.
n is the sample size.

In most "real-world" cases, the population standard deviation is not known. However, if the sample size is 30 or more, the sample standard deviation s will closely approximate the population standard deviation σ. The formula for this approximation of the standard error of the mean then becomes

$$\sigma_{\overline{X}} = \frac{s}{\sqrt{n}} \qquad \text{(9-2)}$$

where s is the sample standard deviation.

Note that the standard error is affected by the size of the sample, which is recorded in the denominator. As the sample size n increases, the variability of the sample means decreases. This outcome is logical, because an estimate made with a larger sample should be subject to less variability.

POINT AND INTERVAL ESTIMATES

The data in Table 8–1 (see p. 292) represented a population—the length of service of six social workers. The mean of this population was easily computed. However, in most cases the very thing we are trying to estimate is a population parameter—the mean length of service of the social workers, for example. This parameter is unknown in practice, and we are trying to find its value. The single number with which we estimate a population parameter is called a **point estimate.**

Point Estimate The value, computed from a sample, that is used to estimate a population parameter.

A sample mean $\overline{X}$ is a point estimate of the population mean μ. To estimate the mean age of purchasers, a distributor of stereo equipment records the age of a sample of 50 customers. The mean age of the sample is a point estimate of the mean age of the population of all purchasers.

However, a point estimate tells only part of the story. While we expect the point estimate to be close to the population parameter, we would like some way to measure how close it is. The **interval estimate** serves this purpose.

Interval Estimate A range of values within which we have some confidence that the population parameter lies.

For example, we estimate the mean yearly income of a group of farmers to be $35,000. The range of that estimate might be from $34,000 to $36,000. We can describe how confident we are that the population parameter is in that interval by making a probability statement. The resulting confidence interval is an interval estimate of the population parameter. Confidence levels such as 95% and 99% are often used to indicate the degree of belief or credibility to be placed on a particular interval estimate of a population parameter. We might say, for example, that we are 90% sure that the mean yearly income is between $34,000 and $36,000.

Constructing Confidence Intervals

A confidence interval for the population mean is constructed by

$$\overline{X} \pm z\frac{s}{\sqrt{n}} \quad \text{(9-3)}$$

where $\overline{X}$ is the sample mean, s is the sample standard deviation, z is the standard normal value corresponding to the desired level of confidence, and n is the sample size. The 95% confidence interval is computed by:

$$\overline{X} \pm 1.96\frac{s}{\sqrt{n}}$$

The 99% confidence interval for the mean is computed by:

$$\overline{X} \pm 2.58\frac{s}{\sqrt{n}}$$

Problem Construct a 95% confidence interval for the mean hourly wages of apprentice geologists employed by the top five oil companies. For a sample of 50 apprentice geologists, $\overline{X}$ = $14.75 and s = $3.

Solution The standard error of the sample means is estimated to be \$0.42, using formula 9–2:

$$\sigma_{\bar{X}} = \frac{s}{\sqrt{n}} = \frac{\$3}{\sqrt{50}} = \$0.42$$

Thus, the 95% confidence interval for μ goes from \$13.93 to \$15.57, found by formula 9–3:

$$\$14.75 \pm 1.96(\$0.42)$$
$$\$14.75 \pm \$0.82$$

If we repeat the sampling process for the 50 apprentice geologists, would we expect to see a sample mean of \$14.75 each time? Probably not! The central limit theorem tells us that if the *population* mean of *all* apprentice geologist's hourly wages is \$15.00 and we repeat the process of sampling and computing the corresponding 95% confidence intervals, a table similar to Table 9–2 will result.

Note from Table 9–2 that the population mean value of \$15 per hour is included in every interval except sample number 19. Therefore, 19 out of 20, or 95%, of the intervals include the population mean value. Figure 9–4 portrays this concept graphically.

Problem Recall from the Chapter Problem that 160 Hispanic families in the Dallas area were surveyed with respect to the proportion of bilingual families, average annual income, the mean number and age of children, and so on. "What is the age of your youngest child?" was one of the questions asked. The sample mean was computed to be 6.7 years, the sample standard deviation 2.5

Table 9-2 **Twenty Samples (n = 50) and Confidence Intervals**

Sample Number	Sample Mean $\bar{X}$	Standard Deviation s	95% Confidence Interval
1	\$14.99	\$3.01	\$14.13–15.84
2	15.61	3.42	14.64–16.58
3	15.12	3.25	14.19–16.04
.	.	.	
.	.	.	
.	.	.	
18	15.01	2.92	14.18–15.84
19	15.92	2.99	15.07–16.77
20	15.34	3.25	14.41–16.26

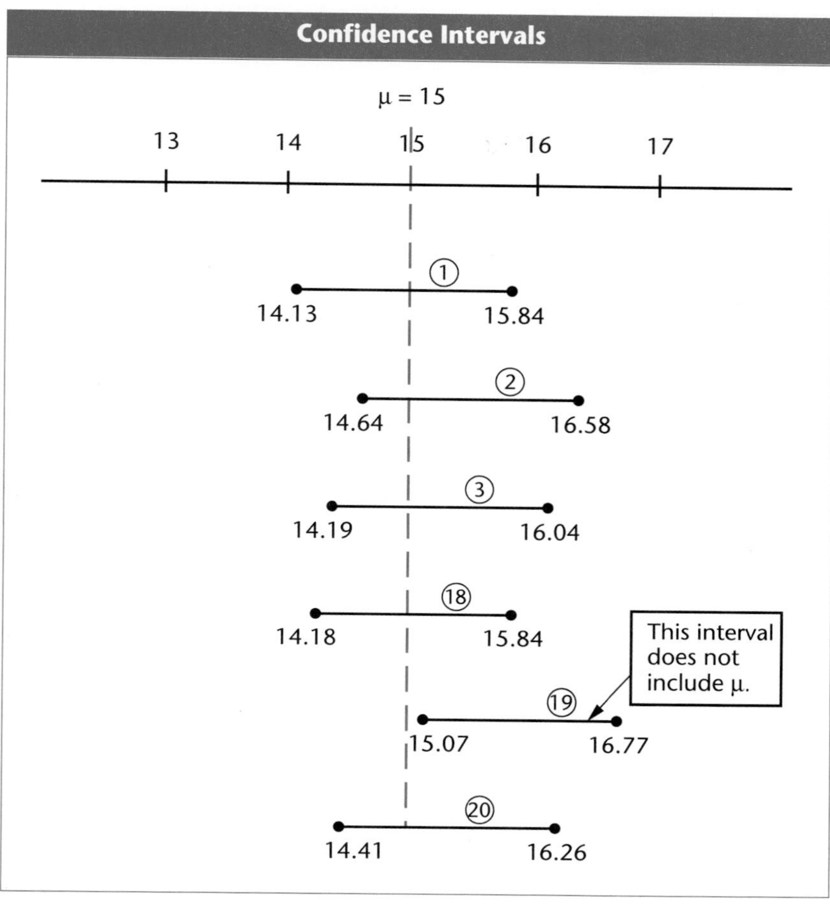

Figure 9-4

years. Construct a 99% confidence interval for the mean age of the youngest child in all Hispanic families in the Dallas area.

Solution To determine the 99% confidence interval, we need to know the standard error of the mean. It is approximately 0.2 years, found by using formula 9–2:

$$\sigma_{\overline{X}} = \frac{s}{\sqrt{n}} = \frac{2.5}{\sqrt{160}} = 0.2 \text{ years}$$

Recall that 99% of the normal distribution is within 2.58 standard deviations of the population mean. Hence, the 99% confidence interval is

$$6.7 \pm 2.58(0.2)$$
$$6.7 \pm 0.5$$
$$6.2 \text{ years up to } 7.2 \text{ years}$$

Thus, we are 99% confident that the population mean age is between 6.2 years and 7.2 years. In other words, our maximum error is 0.5 years, which is one-half of the width of the interval.

In summary, the researcher selects the desired degree of confidence (such as 95% or 99%) and then proceeds to construct an interval corresponding to that percent. Note that only one confidence interval is calculated from a particular sample. The confidence interval either includes the population mean μ or it does not. Thus, probability statements can be made only before the sample is taken.

We should be careful when interpreting a confidence interval. The level of confidence does *not* specify the probability that a population mean is included in the interval, but rather the percent of times that similarly constructed intervals could be expected to include the population mean. Referring to the problem above, we can say that about 99% of the similarly constructed confidence intervals for the youngest Hispanic child would bracket the population mean. We cannot say that the probability is 0.99 that the population mean is in that interval.

Self-Review 9-1

Answers to all Self-Review problems are at the end of the chapter.

1. A sample of 900 registered voters in the state was surveyed about age, political party, and so on. The mean age of the sample was computed to be 42 years, the standard deviation 12 years. What is the 95% confidence interval for the population mean μ?
2. A sample of 312 grades on a mathematics test given nationwide reveals that the sample mean is 560 and the sample standard deviation is 120. What is the 99% confidence interval for the population mean?

Exercises

Answers to the even-numbered exercises are at the back of the book.

1. A sample of 45 truck drivers was given the Army General Classification Test. The sample mean is 96.2, with a sample standard deviation of 9.7. Construct a 95% confidence interval for the population mean score of all truck drivers.
2. A study is made of how long inner-city families have lived at their current address. A random sample of 40 families revealed a mean of 35 months, with a sample standard deviation of 6.3 months. Construct a 99% confidence interval for the mean time that inner-city families have lived at their current address.
3. The Central Ohio Paper Company wants to estimate the mean time required for a new machine to produce a ream of paper,

wrap it, and put in a box ready for shipment. A random sample of 36 reams required a mean machine time of 1.5 minutes. Assuming a standard deviation of 0.3 minutes, construct an interval estimate with a confidence level of 95%.

4. A simple random sample of 80 schoolchildren enrolled at Home Street School was asked the distance they travel to school. The sample mean and the standard deviation are found to be 7.52 km and 1.32 km, respectively. What is the 95% confidence interval for the population mean distance traveled to school?

5. A sample of 400 fuses had a mean breaking point (in amperes of current) of 7.5, with a sample standard deviation of 1.0. Construct a 95% confidence interval for the population mean breaking point.

6. A study of the time required between landing clearance and physical touchdown for 49 airplanes at City Airport shows that the sample mean is 280 seconds, with a standard deviation of 21 seconds. Find a 99% confidence interval for the population mean landing time.

7. A sample of 100 high school students revealed that the mean time spent working at an outside job each week is 10.7 hours, with a standard deviation of 11.6 hours. What is the 99% confidence level for the mean number of hours that the population of high school students spent working each week?

8. Sixty-four teenagers who received their driver's licenses during the last month were given a test for reaction times. The mean is 8 seconds, with a standard deviation of 2 seconds. Find a 95% confidence interval for the mean reaction time of the entire population.

THE STANDARD ERROR OF A SAMPLE PROPORTION (σ_p)

Sometimes you are interested not in a mean value, but simply in the fraction or proportion of a population that has some characteristic. For example, you might wish to determine what proportion of the U.S. population owns a video cassette recorder. The symbol π (Greek letter *pi*) is used to show that fraction of the population that has the characteristic. (Do not confuse this π with the mathematical constant.) If 73% of the U.S. population has a cassette recorder, then $\pi = 0.73$. The symbol *p,* on the other hand, is used to indicate a *sample* proportion. What is a proportion?

Proportion A fraction, ratio, percent, or probability that indicates what part of a sample or population has a particular trait.

STAT BYTE

Women's opinion of men has taken a nosedive in the last 20 years, according to one recent national survey. The Roper Organization found that a growing number of women are sensitive to sexism and unhappy with men on a number of issues. It compared responses with those given to identical questions asked in 1970. Some of the changes were sizable. For example, in 1970, 67% of the respondents felt that men are "basically kind, gentle, and thoughtful." In 1990, only 51% agreed with that statement. Interestingly, in the same survey more than 9 out of 10 women reported that they preferred being married to living by themselves.

Strictly speaking, the distribution of the number of items in a sample (n) that has a given trait follows a binomial distribution. However, n is frequently large, so we may use the normal approximation to the binomial. This approximation was discussed in Chapter 7. The common rule for employing this approximation is that $n\pi$ and $n(1 - \pi)$ are at least five. Recall that for the binomial, $\sigma = \sqrt{n\pi(1-\pi)}$. The **standard error of the proportion** is the standard deviation of the sampling distribution of the sample proportion. The symbol σ_p is read as *sigma sub p*. It is found by

$$\sigma_p = \sqrt{\frac{\pi(1-\pi)}{n}} \qquad \textbf{9-4}$$

where π is the population proportion that has the trait and n is the size of the sample.

Problem Twenty-five thousand people live in a wealthy Chicago suburb. A travel agent wishes to estimate the proportion of people who have never visited Hawaii. Unknown to the travel agent, 75% have never been there. If the agent selects a random sample of 100, what is the standard error of this sample proportion?

Solution The standard error (σ_p) is found by using formula 9–4:

$$\sigma_p = \sqrt{\frac{\pi(1-\pi)}{n}} = \sqrt{\frac{(0.75)(0.25)}{100}} = 0.0433$$

CONFIDENCE INTERVALS FOR PROPORTIONS

In general, a confidence interval for a proportion is found by

$$p \pm z\sqrt{\frac{p(1-p)}{n}} \qquad \textbf{9-5}$$

where p is the sample proportion, z is the standard normal value corresponding to the desired level of confidence, and n is the number in the sample.

The 95% and 99% confidence intervals for a sample proportion are estimated by

$$p \pm 1.96\sqrt{\frac{p(1-p)}{n}}$$

and

$$p \pm 2.58\sqrt{\frac{p(1-p)}{n}}$$

respectively.

Problem Recall from the Chapter Problem the survey of 160 Hispanic families in the Dallas area. One of the questions asked was, "Are you fluent in both English and Spanish?" Ninety-six responded "yes." The sample proportion is 96/160, or 0.60. Construct a 95% confidence interval for the proportion of bilingual Hispanic families in the Dallas area.

Solution First, find the standard error of the sample proportion by:

$$\sqrt{\frac{(0.6)(0.4)}{160}} = 0.04$$

Because 95% of a normal distribution is less than 1.96 standard deviations away from the mean, the confidence interval is determined using formula 9–5.

$$0.6 \pm 1.96(0.04)$$

$$0.6 \pm 0.08$$

$$0.52 \text{ up to } 0.68$$

About 95% of similarly constructed intervals will contain the population proportion.

Self-Review 9-2

The Greenville Lodge inspected 50 of its air conditioners. Ten were not operating. Develop a 95% confidence interval for the proportion of air conditioners not operating.

Exercises

9. In an "exit" poll of 2,200 randomly selected voters conducted in Precinct 29, only 250 said they voted for Governor Long.
- **a.** Find the standard error of the proportion.
- **b.** What is a 95% confidence interval for the population proportion that voted for Governor Long?

10. A manufacturing process produced 15 defective parts in a sample of 100.
  a. Determine the standard error of the sample proportion.
  b. Develop a 90% confidence interval for the proportion defective in the population.

11. A sample of 250 urban workers revealed that 25 work with a video display terminal (VDT). What is the 95% confidence interval for the proportion of all urban workers who work with a VDT?

12. A sample of 500 purchasers of a new Mercedes-Benz shows that 325 order two-door models. What is the 99% confidence interval for the population proportion who order two-door models?

13. Of 250 college freshmen polled, 150 said they exercise regularly.
  a. Find a 95% confidence interval for the population proportion of freshmen who exercise regularly.
  b. Find a 99% confidence interval for the same population proportion.

14. Of 40 patients in a survey of current wheelchair users, 32 judged a new, medically approved wheelchair to be an improvement over a standard wheelchair.
  a. Find a 95% confidence interval for the proportion of the population already using wheelchairs for whom the new wheelchair would be an improvement.
  b. Suppose 320 out of 400 wheelchair users judged the new chair to be an improvement. Find a 95% confidence interval for the proportion in the population.

15. An audit of 840 randomly selected tax returns shows that 30% require payment of additional taxes. Find the 95% and 99% confidence intervals for the proportion of all tax returns requiring additional payments.

16. In a national magazine, a poll reported that 36% of the 980 voters surveyed approved of immigration quotas. The article also reported a "margin of error" of plus or minus 3 percentage points. What confidence interval was the pollster using?

THE FINITE POPULATION CORRECTION FACTOR

If you are sampling without replacement from a small population, the standard error is reduced. In other words, if a sample is a relatively large fraction of the population, it yields a better estimate of the population parameter. If a sample (n) is equal to the population's size (N)—that is, $n = N$—we would not expect any error.

Does this make sense? It should, because if we know the entire population of values, we will make *no error* when estimating the mean or proportion! At the other extreme, when the population size (N) is quite large compared to the sample size, there is no need for the correction. (We could never catch and weigh all the fish in Lake Erie, because there are so many fish. Therefore, n could not equal N.)

The standard error of the mean or the standard error of the proportion is corrected by the term

$$\sqrt{\frac{N-n}{N-1}} \quad \text{9-6}$$

where N is the number in the population and n is the number in the sample.

This is called the **finite population correction factor.** The standard error of the sample mean is computed by the following formula, which combines formulas 9–1 and 9–6:

$$\sigma_{\overline{X}} = \frac{\sigma}{\sqrt{n}}\sqrt{\frac{N-n}{N-1}}$$

The standard error of the sample proportion is:

$$\sigma_p = \sqrt{\frac{\pi(1-\pi)}{n}}\sqrt{\frac{N-n}{N-1}} \quad \text{9-7}$$

Statisticians usually ignore the correction factor when the sample is less than 5% of the population.

Problem In the small town of Oxford, there are 200 families. A poll of 30 families revealed a mean annual church contribution of \$470, with a standard deviation of \$130. Construct a 95% confidence interval for the mean annual contribution.

Solution Note that the sample constitutes more than 5% of the population (30/200 = 15%), and the sampling is done without replacement (no family is included twice). The 95% confidence interval is constructed as follows, using formulas 9–3 and 9–6:

$$= \overline{X} \pm z \cdot \frac{s}{\sqrt{n}}\left(\sqrt{\frac{N-n}{N-1}}\right)$$

$$= 470 \pm 1.96\left(\frac{130}{\sqrt{30}}\right)\left(\sqrt{\frac{200-30}{200-1}}\right)$$

$$= 470 \pm 46.519(0.924)$$
$$= 470 \pm 43$$
$$= 427 \text{ and } 513$$

Self-Review 9-3

The same study of church contributions in Oxford revealed that 12 of the 30 persons sampled attended church regularly. Construct a 95% confidence interval for the proportion attending church regularly.

Exercises

17. A hospital employs 250 nurses. A sample of 50 nurses revealed that 30 were graduates of a diploma school. Develop a 90% confidence interval for the proportion of diploma graduates at the hospital.

18. In River City, a total of 300 traffic citations was issued. A sample of 40 of these tickets showed that the mean amount of the citation was $29 and the standard deviation $5. Construct a 98% confidence interval for the population mean.

19. There are 257 pilots in a secret military training program. In a study of the use of hypnosis to enhance performance, ratings for a sample of 32 pilots led to a mean rating increase of 8.4, with a standard deviation of 2.1. Construct the 95% confidence interval for the population mean.

20. Five hundred students take physical education classes at Wilson High. In a sample of 50 of these students, 10 were able to run two miles in less than 14 minutes. Use this sample data to construct the 99% confidence interval for the population proportion of students who can run two miles in less than 14 minutes.

CHOOSING AN APPROPRIATE SAMPLE SIZE

A concern that usually arises when a statistical study is being designed is, "How many items should be in the sample?" If a sample is too large, money and effort are wasted collecting the data. Similarly, if the sample is too small, the resulting conclusions will be uncertain. The correct sample size depends upon three factors:

1. the level of confidence desired,
2. the variability in the population being studied, and
3. the maximum allowable error.

You, the researcher, select the level of confidence. As noted in the previous sections, confidence intervals of 95% and 99% are selected most often. A 99% confidence level corresponds to a standard normal (z) value of ± 2.58 and a 95% confidence level yields $z = \pm 1.96$.

If the population is widely dispersed, a large sample is required. On the other hand, a small standard deviation (a homogeneous population) would not require as large a sample. The population variability is measured by the sample standard deviation s, and the standard error of the sample mean is, of course, $s/\sqrt{n}$. Often a small "pilot study" is conducted to estimate s before the major study is undertaken.

Finally, the maximum allowable error (E) is the amount that was added and subtracted from the sample mean to obtain the limits of the confidence interval in previous sections. It is the amount of error the researcher is willing to tolerate. In general, it is one-half of the width of the corresponding confidence interval. A small allowable error will require a larger sample size, whereas a large allowable error will permit smaller sample sizes.

We can express this interaction among these three factors and the sample size in the following formula:

$$z = \frac{E}{s/\sqrt{n}}$$

Solving this equation for n, we obtain the required sample size

$$n = \left[\frac{z \cdot s}{E}\right]^2 \qquad \textbf{9-8}$$

where

E is the maximum allowable error.

s is the estimate of the population standard deviation.

n is the size of the sample.

z is the standard normal value corresponding to the desired level of confidence.

Because the result of this computation is not always a whole number, the usual conservative practice is to round up any fractional result. For example, 72.1 would be "rounded up" to 73.

Problem A study is to be conducted on the mean salary of mayors of cities with a population of fewer than 100,000. The error in estimating the mean is to be less than \$100, and a confidence level of 95% is desired. Suppose that the standard deviation of the population is estimated to be \$1,000. What is the required sample size?

STAT BYTE

The Career Planning and Placement Center at a western state university surveyed their May 1990 graduates. The survey was conducted in the fall of 1990 and completed by 60% of the 1,512 bachelor's degree recipients. Here are some of the survey's findings:

- Ninety-five percent of the graduates available for employment were employed.
- The mean starting salary of graduates in new full-time positions was \$2,077 per month.
- Seventy-one percent of the employed graduates were employed in business and industry.

Solution The allowable error, E, is \$100. The value of z for a 95% level of confidence is 1.96. When we substitute the values into the formula 9–8, the required sample size is determined to be:

$$n = \left[\frac{(1.96)(\$1{,}000)}{\$100}\right]^2 = (19.6)^2 = 385$$

Thus, a sample of 385 is required. If a higher level of confidence were desired, say 99%, then a larger sample would also be required:

$$n = \left[\frac{(2.58)(\$1{,}000)}{\$100}\right]^2 = (25.8)^2 = 666$$

When determining the sample size for a proportion, we use the following formula:

$$n = p(1-p)\left(\frac{Z}{E}\right)^2 \qquad \textbf{9-9}$$

The value of p can be estimated from a pilot study. If no estimate is available, the value of 0.50 is used. Why? Because the term $p(1 - p)$ can never be larger than when $p = 0.50$. For example, if $p = 0.30$, then $p(1 - p) = 0.30(0.70) = 0.21$, but when $p = 0.50$, $p(1 - p) = 0.50(0.50) = 0.25$.

Problem We are planning a survey to find the proportion of cities that have private garbage collectors. We want a maximum error (E) of 0.10 and a 90% confidence level, and tentatively estimate the proportion at 0.5. What is the required sample size?

Solution Using formula 9–9:

$$n = (.50)(.50)\left[\frac{1.65}{.10}\right]^2$$

We need to contact a minimum of 69 cities.

Self-Review 9-4

We are conducting a study designed to estimate the mean number of hours worked per week by suburban housewives. A pilot study revealed that the population standard deviation is 2.7 hours. How large a sample should be selected to be 95% confident that the sample mean differs from the population mean by at most 0.2 hours?

Exercises

21. A company wishes to estimate the mean starting salaries for security personnel in their manufacturing operations. From a previous study, they estimate that the standard deviation is $2.50. How large a sample should they select to be 95% confident that the sample mean differs from the population mean by at most $0.50?

22. A meat packer is investigating the marked weight shown on links of summer sausage. A pilot study showed a mean weight of 11.8 pounds per link and a standard deviation of 0.7 pounds. How many links should the packer sample to be 95% confident that the sample mean differs from the population mean by at most 0.2 pounds?

23. A motel is studying the number of customers who stay more than one night. The manager would like to estimate the proportion within 0.05, with a 95% level of confidence. He estimates the proportion to be 0.30. How large a sample is required?

24. Refer to Exercise 23, but assume that no estimate is available regarding the population proportion. How large a sample is required?

SUMMARY

In this chapter, we emphasized that

1. if the population from which the sample is drawn is normal, the sampling distribution of the means will also be normally distributed;
2. if the population from which the sample is drawn is not normally distributed, the distribution of the sample means will become more and more normal as the number of members in the sample becomes larger and larger. This is usually referred to as the central limit theorem. In essence, it states that when the size of the sample is large enough, the distribution of the sample means is normal regardless of the shape of the population from which the sample is selected. A sample of 30 or more is usually considered "large enough."

Finally, based on the central limit theorem, a confidence interval can be set up for a sample mean and a sample proportion. It specifies an interval that has a high probability of containing the population mean μ. For a typical distribution,

1. 95% of the sample means or sample proportions selected from that population will lie within 1.96 standard deviations of the population mean;
2. 99% of the sample means or sample proportions will lie within 2.58 standard deviations of μ.

Intervals computed in this manner are called confidence intervals. The intervals are computed in the following manner:

For Means	For Proportions
$\overline{X} \pm z\frac{s}{\sqrt{n}}$	$p \pm z\sqrt{\frac{p(1-p)}{n}}$

where s is the standard deviation of the sample, z is the standard normal value corresponding to the desired level of confidence, p is the sample proportion, $\overline{X}$ is the sample mean, and n the sample size.

When the sample constitutes more than 5% of the population and samples are selected without replacement, a finite population correction factor is used. The formula for a confidence interval is then changed to:

For Means	For Proportions
$\overline{X} \pm z\frac{s}{\sqrt{n}}\left(\sqrt{\frac{N-n}{N-1}}\right)$	$p \pm z\left(\sqrt{\frac{p(1-p)}{n}}\right)\left(\sqrt{\frac{N-n}{N-1}}\right)$

In many research problems, the size of the sample needs to be determined. The required sample size is based on the allowable error, the confidence level, and an estimate of the population standard deviation:

For Means	For Proportions
$n = \left(\frac{z \cdot s}{E}\right)^2$	$n = p(1-p)\left(\frac{z}{E}\right)^2$

Exercises

25. A study dealing with divorced couples gathered data on the length of time from marriage to separation. A random sample of 100 divorced couples had an average length of marriage of 5.9 years, with a sample standard deviation of 2.0 years. Construct a 99% confidence interval for the mean length of time from marriage to separation for the population of all divorced couples.

26. Refer to Exercise 25. Suppose that this study dealt with only the couples who were divorced in a particular city. Assume that the population of divorced couples was 500. Compute the confidence interval.

27. A random sample of 80 terms of sentence for rape (first offense) showed a mean equal to 3.9 years, with a standard deviation of 1.8 years. Construct a 95% confidence interval for the mean term of imprisonment for all rape sentences.

28. Refer to Exercise 27. Assume that the study referred to only one prison in which there were 250 rape cases (first offense). Compute the confidence interval.

29. Sick-leave records obtained from a random sample of 200 social workers showed a mean number of days sick leave equal to 25.6 last year. If the sample standard deviation is 5.1, construct a 90% confidence interval for the mean number of days' sick leave for all social workers last year.

30. The 1990 costs per pupil for seven city school systems in Ohio are:

City	Cost per Pupil (in thousands of dollars)
Akron	1.4
Canton	1.3
Cincinnati	1.5
Cleveland	1.7
Columbus	1.5
Dayton	1.7
Toledo	1.4

a. Using the random number table in Appendix H, select two school systems at random.

b. Calculate the mean cost for your sample of two school systems.

c. Compare your sample mean with the population mean.

31. On a fair die, the numbers 1 through 6 each have a probability of 1/6 of occurring. Suppose two fair dice are thrown and the sample mean number of points showing is calculated. What is the probability distribution of this sample mean? Find its mean and standard deviation.

32. The California Department of Highways is analyzing traffic patterns on a busy section of Interstate 605, near Long Beach. They want to estimate the mean number of cars that pass this section each day. The requirements are that the estimate be within 20 cars per day of the population mean and that the analysts be 95% confident of the results. A similar study showed the standard deviation to be 75 cars per day. How large a sample is required?

33. A consultant to a large Idaho ski resort wants to estimate the mean daily amount of money spent by its guests. How large a sample should the consultant select in order to estimate the mean daily amount spent to within $2.00 with a 99% level of confidence? A reasonable estimate of the standard deviation is $5.00.

34. A political candidate wants to estimate his chances of winning the upcoming election. He wants to estimate—within 3 percentage points, with a 95% level of confidence—the proportion of voters who will vote for him. He has no estimate of the population proportion. How many voters should be contacted?

35. The International Council of Shopping Centers reported that a sample of 30 people spent an average of 69 minutes each visit to a mall or shopping center. The sample standard deviation was 49 minutes. Construct a 95% confidence interval for the mean shopping time.

36. During the month of May, the Department of Transportation monitored 80 flights on a regional airline. Seventy-four of those flights arrived on time. Use this sample data to construct the 99% confidence interval for the proportion of all flights which arrive on time.

37. Hrivnyak Literacy Action Group is planning a survey to determine the percent of adults in the United States who are illiterate. They wish a maximum error of .02. They plan to use a 95% confidence level. Hrivnyak tentatively estimates the proportion of illiterate adults at .15. What sample size would be required to achieve their goal?

38. A survey of 50 new-car dealers shows a sample mean of 49 reports of defects per month, with a sample standard deviation of 7 reports per month. Construct a 99% confidence interval for the population mean number of reports of defects per month.

Data Exercises

39. Refer to Real Estate data, which reports information on homes sold in Northwest Ohio during 1992.

a. Develop a 95% confidence interval for the mean selling price of the homes.

b. Develop a 95% confidence interval for the mean distance from the center of the city.

c. Develop a 95% confidence interval for the proportion of homes with a fireplace.

d. Develop a 95% confidence interval for the proportion of homes with an attached garage.

40. Refer to Salary data, which refers to a sample of middle managers in Sarasota, Florida.

a. Develop a 95% confidence interval for the mean salary of the middle managers.

b. Develop a 95% confidence interval for the mean number of employees supervised.

c. Develop a 95% confidence interval for the proportion of female managers.

WAS THE SAMPLE SIZE TOO LARGE?

Nancy Lapp, a graduate psychology student, has gathered data on the frequency of drug counseling visits experienced by an urban methadone clinic during a particular year. She is able to reject a null hypothesis that the mean number of visits is less than or equal to 24 with a test statistic (z-value) equal to 12. Her mentor suggests that a value of 12 is very unusual! Assuming no arithmetic errors in her work, is it possible she spent too much time recording data? All other things equal, could a smaller sample size achieve the same goal? If so, how much smaller could the sample be without affecting the conclusion?

CHAPTER ACHIEVEMENT TEST

Answer all the questions. The answers are at the back of the book.

MULTIPLE-CHOICE QUESTIONS. Select the response that best answers each of the questions.

1. Which of the following statements is *not* true?
   a. A point estimate is more accurate than an interval estimate.
   b. A point estimate is subject to sampling error.
   c. A point estimate says nothing about the error in estimation.
   d. An interval estimate includes the corresponding point estimate.
2. Which of the following is *not* needed to compute a confidence interval for the mean?
   a. The shape of the sampling distribution
   b. The sample size and population size
   c. The population standard deviation
   d. All of the above are needed.
3. If we have *a* sample of 50 items from a very large population, which of the following is *not* needed to compute a confidence interval for the proportion?
   a. The shape of the sampling distribution
   b. The sample size and population size
   c. The degree of confidence
   d. All of the above are needed.
4. The sample standard deviation based on 15 pieces of data is
   a. always larger than the standard error of the mean.
   b. always smaller than the standard error of the mean.
   c. another name for the standard error of the mean.
   d. None of the above
5. You should use the finite population correction factor whenever
   a. the sample is more than 5% of the population.

b. you are sampling without replacement.
c. the standard error of the mean is too large.
d. the population is finite.

6. For samples of one item, the sampling distribution of the sample mean is normally distributed if the
a. population is symmetric.
b. standard deviation is known.
c. population is normally distributed.
d. standard deviation is less than the mean.

7. The standard error of a proportion becomes larger as
a. π approaches zero.
b. n grows large.
c. π nears one-half.
d. π is close to one.

8. A 95% confidence interval indicates that 95 out of 100 similarly constructed intervals will include
a. a sampling error.
b. an estimation of the parameter.
c. Both a and b
d. None of the above

9. If the confidence level is lowered from 95% to 90% and everything else remains the same, the required sample size will
a. increase.
b. decrease.
c. not change.
d. There is too little information provided to determine the effect.

10. To find a sample size, the use of the finite population correction factor will
a. affect sampling with replacement.
b. decrease the required sample size.
c. increase the required sample size.
d. have more impact on large populations.

COMPUTATION PROBLEMS

11. To estimate the medical charges for an appendectomy, Blue Star Insurance has data from a random sample of 70 patients. The sample mean cost is $510, with a sample standard deviation of $70.
a. Find the standard error of the sample mean.
b. Construct a 95% confidence interval for the population mean cost.

12. A real estate agent records the ages of 50 randomly selected home buyers in her sales area. The mean age is 38 years, with a sample standard deviation of 10 years.
a. What is the standard error of the estimate?
b. Find a 99% confidence interval for the population mean age.

13. Larry Clark is a superstar for the Warren High School basketball team. In the game against Dubois, Larry took 20 shots. We want to estimate his long-term shooting percentage.

a. If he normally makes 70% of his shots, what is the standard error of your estimate?

b. If he makes 13 baskets, what is a 99% confidence interval for his long-term shooting percentage?

14. The police department reports that 375 of 500 randomly selected burglar alarms received at the station were false alarms.

a. What is the standard error of an estimate based on this data?

b. What is a 90% confidence interval for the population proportion of false alarms?

15. A study of the income of farming households in a particular state is undertaken. How large a sample is required to estimate the mean income within $200 with a 95% level of confidence? The standard deviation is estimated to be $3,000.

ANSWERS TO SELF-REVIEW PROBLEMS

9-1 **1.** $\sigma_{\overline{X}} = \dfrac{12}{\sqrt{900}} = \dfrac{12}{30} = 0.4$

Then $42 \pm 1.96(0.4) = 42 \pm 0.784 =$ 41.216 years and 42.784 years

2. $\sigma_{\overline{X}} = \dfrac{120}{\sqrt{312}} = \dfrac{120}{17.66352} = 6.79$

Then $560 \pm 2.58(6.79) = 560 \pm 17.52 =$ 542.48 and 577.52. These would probably be rounded to 542 and 578.

9-2 $p = \dfrac{10}{50} = 0.20$

$$0.20 \pm 1.96\left(\sqrt{\frac{.2(.8)}{50}}\right)$$

0.20 ± 0.11

9-3 $p = \dfrac{12}{30} = 0.40$

$$0.40 \pm 1.96\sqrt{\frac{(0.40)(0.60)}{30}}\left(\sqrt{\frac{200-30}{200-1}}\right)$$

0.40 ± 0.162

9-4 $n = \left[\dfrac{(1.96)(2.7)}{0.2}\right]^2$

$n = 701$

CHAPTER 10

Hypothesis Tests: Large-Sample Methods

OBJECTIVES

When you have completed this chapter, you will be able to

- describe the five-step hypothesis-testing procedure;
- distinguish between a one-tailed and a two-tailed statistical test;
- identify and describe possible errors in hypothesis testing;
- conduct a hypothesis test about a population mean;
- conduct a hypothesis test between two population means;
- test a hypothesis about a population proportion;
- test a hypothesis about two population proportions.

CHAPTER PROBLEM The Luck of the Irish

ARE PROFESSIONAL MONEY MANAGERS WORTH WHAT THEY'RE PAID? ONE NEW YORK ANALYST CONFESSED THAT HE BASED HIS STOCK PURCHASES ON THE TOENAIL MARKS HIS IRISH SETTER LEFT ON THE FINANCIAL PAGES! JOHN DORFMAN OF *THE WALL STREET JOURNAL* DECIDED TO STAGE A CONTEST. HE CHALLENGED FOUR INVESTMENT PROFESSIONALS TO CHOOSE PORTFOLIOS OVER A SIX-MONTH PERIOD WHILE HE "SELECTED" STOCKS BY THROWING DARTS AT AN "INVESTMENT" DARTBOARD. OUT OF 27 TRIALS, THE PROS DID BETTER THAN THE DARTS ONLY 15 TIMES. ARE THE SO-CALLED "FINANCIAL WIZARDS" JUST GUESSING, OR DOES A 56% SUCCESS RATE PROVE THAT THEY KNOW SOMETHING?

INTRODUCTION

In Chapter 6, we developed a discrete probability distribution to describe the possible outcomes of an experiment. Chapter 7 dealt with the normal probability distribution—a continuous distribution. We noted that it is symmetrical and bell-shaped and the tails taper off into infinity.

In Chapters 8 and 9, we used both the normal distribution and the notion of a probability distribution to develop a sampling distribution. It was also pointed out that it is often too expensive, or impossible, to study an entire population. Instead, a part of the population of interest, called a sample, is examined. The purpose of sampling is to learn something about the population. For example, we might want to approximate the mean income or the mean age of the population. We accomplished this by constructing confidence intervals within which the population mean might fall.

This chapter continues our study of the use of the normal distribution in sampling. Instead of building an interval in which a population parameter (such as the mean) is expected to fall, we test the validity of a statement about a population parameter. This is called **hypothesis testing.**

THE GENERAL IDEA

To illustrate the concept, let us first examine a nonstatistical case. It concerns private eye Ms. Sharpe as she tries to unmask a mysterious murderer. Upon her arrival at the scene of the crime, Ms. Sharpe observed that the victim had been struck from above by a left-handed person whose size 7 shoes had been covered with mud.

What course of action will our detective pursue? Naturally, she will first suspect that the butler performed the foul deed. Then, she will examine each piece of evidence in succession to see if it is consistent with the presumption that the butler might be the murderer. If the butler is a short, right-handed man who wears size 11 shoes, it is highly unlikely that he committed the crime, and he will be dismissed as a suspect.

As she considers the butler's innocence or guilt, what goes through the detective's mind? She realizes that she must arrest or release the prime suspect. Either way, the butler may in fact be either guilty or innocent. Thus, there are four possibilities that might occur when Ms. Sharpe finally reaches her decision:

1. She can arrest the butler when the butler actually committed the crime—a correct decision.
2. She can release the butler when the butler is innocent—again, a correct decision.
3. She can arrest the butler when the butler is actually innocent—an incorrect decision.

STAT BYTE

During 1992, 42.9 AIDS cases were diagnosed per 100,000 population in the state of New York. In contrast, North Dakota had only 0.2 cases diagnosed per 100,000 population. Is this a random difference?

4. She can release the butler when the butler is actually guilty—another incorrect decision.

These four possibilities facing our investigator can be diagrammed as follows:

	Arrest the butler.	Release the butler.
The butler did it.	Correct	Error!
The butler is innocent.	Error!	Correct

If Ms. Sharpe's problem were one involving statistics, the first thing she would do is to set up a **null hypothesis.**

Null Hypothesis A claim about the value of a population parameter.

The null hypothesis, designated H_0, in the murder case could be:

H_0: The butler is innocent.

The null hypothesis is a claim that is established for the purpose of testing. This claim is either rejected or not rejected. If the evidence is sufficient to reject the null hypothesis, then the **alternate hypothesis** is accepted.

Alternate Hypothesis A claim about the population parameter that is accepted if the null hypothesis is rejected.

In Ms. Sharpe's murder investigation, the alternate hypothesis, designated H_a, is:

H_a: The butler is not innocent

Note that the procedure is to test the null hypothesis. The alternate hypothesis is accepted if, and only if, the null hypothesis is rejected. The strategy is to make a decision first with respect to the null hypothesis.

After examining all the evidence, the detective might fail to reject the null hypothesis (the butler is innocent) and release him. Or she might reject the null hypothesis and have him arrested.

As noted before, the investigator could make two kinds of mistakes during her investigation. They are called **Type I** and **Type II** errors, respectively.

Type I Error An error that occurs when a true null hypothesis is rejected.

Type II Error An error that occurs when a false null hypothesis is not rejected.

If, for example, the butler had been arrested as a suspect when he was actually innocent of the murder, a Type I error would occur. But if a null hypothesis is not rejected when it is actually not true, a Type II error occurs. That is, if the butler had been released when he actually did commit the murder, a Type II error would occur. To summarize:

	ACTION	
	Fail to reject H_0	Reject H_0
H_0 is true.		Type I error
H_0 is false.	Type II error	

Of course, a Type I error could be avoided if we never reject H_0, the null hypothesis. In the crime problem, if Ms. Sharpe never rejected the H_0 that a suspect was innocent, she would never make the mistake of arresting an innocent man. Clearly, this is a rather extreme way for a crime fighter to avoid a Type I error! It also increases the chances that a Type II error will occur.

A Type II error could be avoided if we always reject the null hypothesis. If our private eye were always to reject the null hypothesis and accept the alternate hypothesis, she would never release a murderer. Either extreme position is unrealistic. As will be noted in the following sections, statistical theory deals with "decision rules" based on probability, which attempt to balance these two kinds of potential errors.

We will now leave Ms. Sharpe to her ongoing murder investigation. Many different types of statistical hypothesis-testing problems will now be considered in this and in subsequent chapters. As will become evident, however, the thinking and the procedure developed in the nonstatistical murder investigation are very similar to research problems involving statistics. A systematic five-step approach will be applied in solving each problem. A brief discussion of these five steps follows. Later, as each of the tests is presented, the five steps will be explained in greater detail.

STEP 1. State the **null hypothesis.** Along with the null hypothesis, make a second statement called the **alternate hypothesis.** Accept the alternate hypothesis if you reject the null hypothesis.

STEP 2. Choose a **level of significance.** Usually, select either the 0.05 or the 0.01 level. The level of significance refers to the probability of making a Type I error.

STEP 3. Choose a **test statistic.** A test statistic is a quantity calculated from the sample information. Its value will be used in Steps 4 and 5 to arrive at a decision regarding the null hypothesis.

STEP 4. Set up a **decision rule** based on the level of significance chosen in Step 2 and the sampling distribution of the test statistic from Step 3.

STEP 5. Select **one or more samples.** Then, using the sample results, *compute the value of the test statistic.* In this chapter, the standard normal distribution z will be the only test statistic employed. Finally, use the decision rule in Step 4 to **make a decision**—either to reject the null hypothesis or not to reject the null hypothesis.

As you can see, in statistics we employ a method of "proof by contradiction." Generally, we hope to prove that something is true (the alternate hypothesis) by rejecting the claim in the null hypothesis. If the sample evidence contradicts the null hypothesis, we can reject it and "believe" the alternate hypothesis. On the other hand, if the sample information does not contradict the null hypothesis, we still have not proved that the null hypothesis is true. We have merely found that the null hypothesis cannot be rejected.

A parallel with hypothesis testing is the legal practice in which a jury finds a defendant either "guilty" or "not guilty," but never declares a defendant innocent. "Not guilty" is the legal equivalent of a null hypothesis. A person is innocent until proven guilty. A prosecutor must present evidence to the jury so that they think that the probability the defendant is innocent is so small as to be unbelievable. Then the jury "rejects the null hypothesis" and says the defendant is "guilty."

STAT BYTE

According to the *Statistical Abstract of the United States,* in 1988 black males between 15 and 19 years of age were shot to death at the rate of 67.9 per 100,000. For young white males, the rate was 6.9 per 100,000. Is the difference significant?

A TEST INVOLVING THE POPULATION MEAN (LARGE SAMPLES)

One type of hypothesis-testing problem involves checking whether a reported mean is reasonable. To perform the test, we take a random sample from the population. The sample must be "large," of size 30 or greater. Note, however, that if we know the population is normally distributed, we can use any size sample.

Problem According to the Bureau of the Census, the mean annual income of government employees during a recent year was \$24,632. There was some doubt that this mean was representative of incomes of government employees living in the San Francisco Bay area during the same period. Is there sufficient evidence to conclude that the mean annual income of government employees living in the Bay area was different than the national average during that period?

Solution Even before we start to collect data, the first step is to state the hypothesis to be tested. Recall from the murder investigation that we call this statement the null hypothesis, usually expressed symbolically as H_0. The null hypothesis, in this case, is that the population mean will not be affected—or, stated another way, that the mean in the Bay area is equal to \$24,632. The mean is designated by the Greek lowercase letter μ.

Symbolically, then, the null hypothesis is:

$$H_0: \mu = \$24,632$$

The alternate hypothesis—that the mean annual income in the San Francisco Bay area is *not* equal to \$24,632—is written:

$$H_a: \mu \neq \$24,632$$

Note that the equality condition appears in the null hypothesis. This will always be the case.

The second step in testing a hypothesis is to select the **level of significance,** which is the probability of a Type I error.

Level of Significance The probability of rejecting the null hypothesis when it is true.

The crucial question in this income problem is to determine when to reject the null hypothesis. The significance level will define more precisely when the sample mean is too far removed from the hypothesized value of \$24,632 for the null hypothesis to be plausible. How do you decide what level of significance to select? You should consider the "cost" of being wrong. As an example of the "cost" of being wrong, it is usually a more serious error for meteorologists to predict sunshine and then have it rain than it is for them to predict rain and then have the sun shine. The most common levels of significance used in applied work are 0.05 and 0.01, although any value between 0 and 1 is possible. For this illustration, a 0.05 significance level has been selected. If rejecting a true hypothesis is relatively serious, then set the significance level quite low. If accepting a false null hypothesis is relatively more serious, then pick a high level of significance.

Suppose, for example, that a test had been devised to detect cancer. The null hypothesis might be that a patient is free of the disease (H_0: no disease). A Type I error would reject the null hypothesis when it is true—that is, telling a patient that he has cancer when in fact he does not. While such an error would cause much pain and suffering, in a case like this one it is not as serious or costly as a Type II error. Remember, a Type II error means that a false null hypothesis is not rejected. In this example, it means you tell a patient he is healthy when in fact he has cancer. Most physicians would agree that the Type II error here is much more serious than the Type I error; consequently, a high level of significance should be selected for this test—say, 0.10 instead of 0.01.

The third step is to select the appropriate **test statistic.**

Test Statistic A quantity, calculated from the sample information, used as a basis for deciding whether or not to reject the null hypothesis.

Recall from Chapter 9 that, according to the central limit theorem, the sampling distribution of the sample mean is approximately normal, and the standard deviation of the sampling distribution of means ($\sigma_{\overline{X}}$) is $\sigma/\sqrt{n}$. Hence, the following test statistic is appropriate:

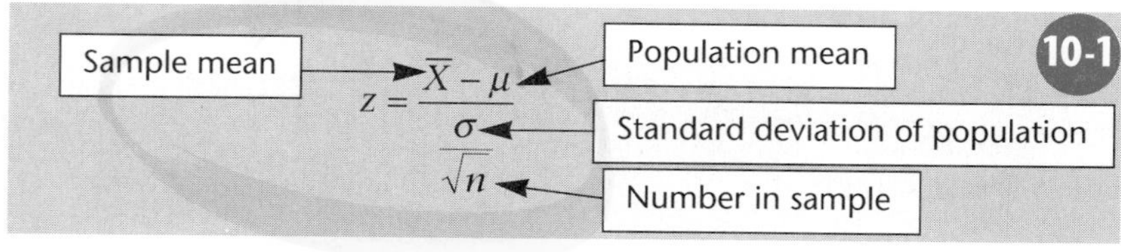

$$z = \frac{\overline{X} - \mu}{\frac{\sigma}{\sqrt{n}}} \qquad \textbf{[10-1]}$$

where

$\overline{X}$	is the sample mean.
μ	is the population mean.
$\frac{\sigma}{\sqrt{n}}$	is the standard error of the mean.
n	is the number in the sample.

Recall that σ is the standard deviation of the population from which the sample was drawn. For this test, we assume that either σ

is already known, based on prior studies, or that a good estimate of its value can be obtained from the sample data.

The fourth step is to formulate a **decision rule.**

> **Decision Rule** A statement of the condition or conditions under which the null hypothesis is rejected.

In the study of Bay area wages, the decision rule is an objective statement that will allow us to test the null hypothesis.

The question we are exploring is, does the mean annual income of government employees in the Bay area differ significantly from the national average? The mean in the Bay area could be larger or smaller than the national average for all government employees. The decision rule is designed to accommodate both these possibilities. Thus, the test is called a **two-tailed test.**

As described in Chapter 9, the distribution of the means of all possible samples of the same size selected from a population is normally distributed. When the null hypothesis is true, the test statistic z will also be normally distributed. Recall from Chapter 7 that 95% of the area in the distribution is between –1.96 and 1.96. Thus, if the significance level is 0.05, the region of rejection falls to the left of –1.96 and to the right of 1.96—that is, less than –1.96 and greater than 1.96. These two values are called the **critical values.**

> **Critical Value** A value or values that separate the region of rejection from the remaining values.

So the decision rule is this: Do not reject the null hypothesis if the computed value of z is in the region from – 1.96 to 1.96. When that is not the case, the null hypothesis is rejected and the alternate hypothesis is accepted. The decision rule can be shown in the form of a diagram:

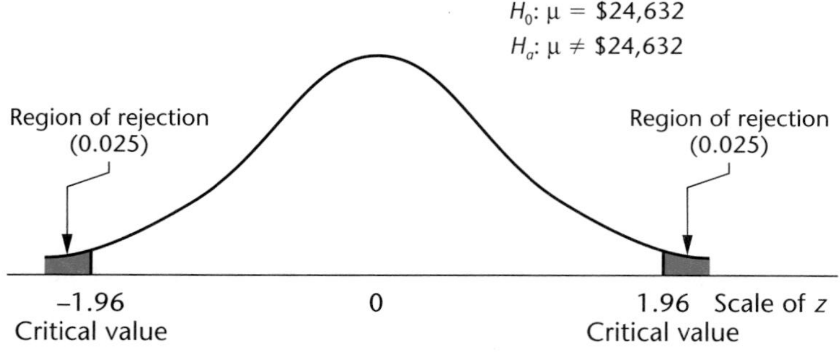

The final step is to take a sample of government employees in the Bay area, compute the sample mean $\overline{X}$, and determine the value of the test statistic z. Based on the computed value of z, the null hypothesis is either accepted or rejected. Let us say that the sample consisted of 49 employees, the mean of the sample was \$25,415, and the standard deviation of the sample was \$1,827. Because the sample is reasonably large, the sample standard deviation (s) is a good estimate of the population standard deviation (σ):

$$\begin{aligned} z &= \frac{\overline{X} - \mu}{\frac{\sigma}{\sqrt{n}}} \\ &= \frac{\$25{,}415 - \$24{,}632}{\frac{\$1{,}827}{\sqrt{49}}} = \frac{\$783}{\$261} \\ &= 3.0 \end{aligned}$$

Because 3.0 is in the rejection region beyond 1.96, the null hypothesis is *rejected* at the 0.05 level of significance. The alternate hypothesis, which states that the mean annual income of government employees in the San Francisco Bay area is *not* equal to \$24,632, is accepted. Such a large difference between the sample mean and the population mean cannot reasonably be attributed to chance.

In general, the correct course of action is either to reject or to fail to reject the null hypothesis. By failing to reject, the investigator takes the position that the evidence is not sufficient to rule out—that is, to reject—the null hypothesis. However, in practice we often think in terms of "accepting" the null hypothesis rather than "failing to reject" it. While this language is technically incorrect, its use is common.

Probability Values (*p*-values) in Hypothesis Testing

The five-step hypothesis-testing procedure compares the test statistic to a critical value. A decision is then made whether to reject the null hypothesis. For example, if the critical value is 1.96 and the test statistic is 1.97, the decision is to reject H_0. Likewise, if the computed value of z is 3.76, H_0 is still rejected. No additional information is reported if the value of the test statistic is barely into the rejection region or well into it.

In recent years, the availability of more sophisticated computer software has provided information regarding the "strength" of the rejection. Test results are often reported along with ***p*-values** or "observed" significance levels.

***p*-value** The probability of getting a value as extreme as that found in the sample when the null hypothesis is true.

STAT BYTE

In 1990, the average American consumed 111 pounds of fresh vegetables, including 27.8 pounds of iceberg lettuce, 18.6 pounds of onions, and 15.4 pounds of tomatoes. How do you compare?

This procedure compares the p-value or "observed" significance level with the significance level chosen by the researcher for the test. If the p-value is smaller than the significance level, the null hypothesis is rejected. This procedure not only allows a decision regarding the null hypothesis, but also gives additional insight into the strength of the rejection. A very small p-value—say, .0001—indicates a more significant result than a larger p-value of, say, .03. In general, for a one-sided test, p-values are calculated as the probability of observing a value of the test statistic greater than the absolute value of the observed value. For a two-sided test, the p-value is twice the probability for the corresponding one-sided test.

In the statistical test involving the mean annual income of government employees in the Bay area, the null hypothesis is rejected. The value of the test statistic z was computed to be 3.00. The probability of obtaining a z-value of 3.00 or greater is 0.0013, found by 0.5000 – 0.4987. To compute the p-value, we are concerned with values less than –3.00 as well as those greater than 3.00 because it was a two-tailed test. Hence, the p-value is 2(0.0013) = 0.0026. This describes the likelihood of observing a value in either tail of the test statistic that is more extreme than the observed test statistic.

The p-value approach not only provides a measure of the significance of the observed statistic, but it also lets each reader of the study result pick a significance level he or she is comfortable with in the particular situation.

Self-Review 10-1

Answers to the Self-Review problems are at the end of the chapter.

In his team's first report, Alfred Kinsey indicated that the average frequency of sexual relations was 9.5 times per month. The standard deviation was 3.9 times per month. Twenty years later, *Redbook* conducted a survey to establish whether there had been a change in the frequency of sexual relations since publication of the Kinsey Report. The 0.01 level of significance was used.

a. State the null hypothesis and the alternate hypothesis.
b. What is the level of significance?
c. What is the appropriate test statistic? Give its formula.
d. Show the decision rule in the form of a diagram.
e. A sample of 18,000 responded to the *Redbook* questionnaire. *Redbook* reported that Americans had sexual relations an average of 9.2 times per month. Compute z, assuming the standard deviation is 3.9. At the 0.01 level of significance, does this evidence indicate that there had been a change in the frequency of sexual relations in the 20 years since the Kinsey Report?
f. What is the p-value?

Exercises

Answers to the even-numbered exercises are at the back of the book.

1. The Public Health Service publishes the *Annual Data Tabulations, Continuous Air Monitoring Projects,* which recently indicated that a large midwestern city had an annual mean level of sulfur dioxide of 0.12 (concentration in parts per million). To change this concentration, many steel mills and other manufacturers installed antipollution equipment. Plans are to make about 36 random checks during the year to determine if there has been a change in the sulfur dioxide level. The 0.05 level is to be used.

a. State both the null hypothesis and the alternate hypothesis.
b. What is the level of significance?
c. What is the appropriate test statistic? Give its formula.
d. Show the decision rule in the form of a diagram.
e. Thirty-six random checks were made throughout the year. It was found that the sample mean was 0.10 and the sample standard deviation 0.03. At the 0.05 level, does this evidence indicate that there has been a change in the sulfur dioxide level in the city?
f. Compute the p-value.

2. During the past several years, frequent checks were made of the spending patterns of citizens returning from a vacation of 21 days or less to countries in Europe. Results indicated that travelers spent an average of $1,010 on items such as souvenirs, meals, film, and gifts. A new survey is to be conducted to determine if there has been a change in the average amount spent. The 0.01 level is to be used.

a. State the null hypothesis and the alternate hypothesis in the forms H_0 and H_a.
b. State the decision rule.
c. A survey of 50 travelers has a sample mean of $1,090. The standard deviation of the sample was $300. At the 0.01 level, is there evidence that there has been a change in the mean amount spent abroad, or is the increase of $80 probably due to chance?
d. Compute the p-value.

3. A West Coast study of a very large population showed that the mean time devoted to volunteer political and civic activities was 32 minutes per month, with a population standard deviation of 4.5 minutes. A smaller group of 49 similar individuals who lived in the South reported a mean of 34 minutes per month. Is this difference possibly due to sampling error, or does it represent a significant difference at the 0.05 level?

4. A local manufacturer of doors for use in home construction is attempting to determine if the average height of adult males in its market area has changed from 70 inches. A sample of 121 men was randomly selected. The average height was found to be 72 inches, with a standard deviation of 2 inches. At the 0.05 significance level, can we conclude that the mean height has changed?

5. A reading test for eight-year-old children is standardized on a large nationwide basis so that the mean is 50, with a standard deviation of 10. School authorities in California choose a statewide random sample of 300 eight-year-olds to compare reading skills in California with those in the rest of the nation.
   a. Formulate the null and alternate hypotheses.
   b. For a 0.05 level of significance, what is the decision rule?
   c. These children average 51.3 on the test. Compute the test statistic.
   d. Is the null hypothesis rejected?
   e. Explain the meaning of your results.

6. It is claimed that the distribution of monthly parking fees in a particular city is normal, with a mean of \$125 and a standard deviation of \$5. You plan to sample 400 people to test the truth of this statement.
   a. State your null and alternate hypotheses.
   b. What is the decision rule for a test with a 0.01 level of significance?
   c. The sample mean is \$124. Is the null hypothesis accepted or rejected? What is the *p*-value? Interpret.

One-Tailed Tests

The earlier example that dealt with the annual income of government employees in the San Francisco Bay area required a two-tailed test. Before the sample of 49 employees was taken, there was no knowledge that the average income would be above or below the national average. For the 0.05 level, there were two rejection regions, one to the right of 1.96, the other to the left of –1.96. (Consult the diagram on p. 338)

Another problem might require the application of a **one-tailed test.** Suppose we suspected that federal employees in the heavily populated Bay area earn *more* than typical government employees. In this case, the only concern is whether the Bay area employees earn significantly *more than* typical government employees. To test our suspicions, we would state in our null hypothesis that the population mean is *equal to or less than* \$24,632 or, stated symbolically:

$$H_0 : \mu \leq \$24{,}632$$

The alternate hypothesis is that the mean income for Bay area government employees is *greater than* \$24,632. It is written:

$$H_a : \mu > \$24{,}632$$

Rejection of the null hypothesis and acceptance of the alternate hypothesis will allow us to conclude that Bay area government employees earn higher salaries than the national average.

For the 0.05 level of significance, the critical value is 1.65. We find it by referring to Appendix C again and searching the body of the table for 0.4500, or the value closest to it. The critical z-value is in the margin. Note that although the null hypothesis contains an inequality ($\mu <$ \$24,632), it is the equality condition ($\mu =$\$24,632) that is tested. The decision rule states that the null hypothesis will not be rejected if the computed value of z is equal to or less than the critical value of 1.65. If the computed z is greater than 1.65, the null hypothesis will be rejected and the alternate hypothesis accepted. Shown in diagram:

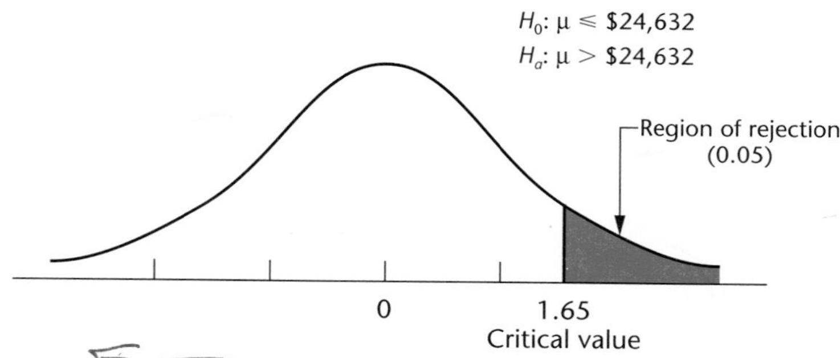

Problem Hyperactive children are often disruptive in the typical classroom setting because they find it difficult to remain seated for extended periods of time. Baseline data from a very large study show that the typical frequency of "out-of-seat behaviors" was 12.38 per 30-minute period, with a standard deviation of 3.52. A treatment known as covert positive reinforcement was applied to a group of 30 hyperactive children. The mean number of "out-of-seat behaviors" was reduced to 11.59 per 30-minute observation period. Using the 0.01 significance level, can we conclude that this decline in "out-of-seat behaviors" is significant?

Solution Use the five-step hypothesis-testing procedure.

STEP 1. The null hypothesis: The mean is equal to or greater than 12.38. That is:

$$H_o : \mu \geq 12.38$$

The alternate hypothesis: The mean is less than 12.38. That is:

$$H_a : \mu < 12.38$$

STEP 2. State the level of significance. It is 0.01.

STEP 3. Give the appropriate test statistic. It is formula 10–1:

$$z = \frac{\overline{X} - \mu}{\frac{\sigma}{\sqrt{n}}}$$

STEP 4. Since there is interest in demonstrating that the covert treatment *lowers* the mean value, a one-tailed test is appropriate. The solution to this problem requires a one-tailed test in the negative direction, so the critical value of z is on the left side of the curve. The critical value for the 0.01 significance level is –2.33 (refer to Appendix C and 0.4900):

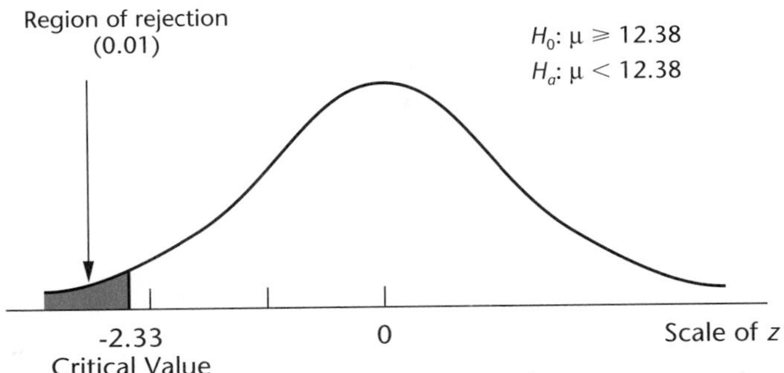

The decision rule: Reject the null hypothesis if the computed value of z is to the left of –2.33. Otherwise, accept the null hypothesis.

Note that the inequality sign in the alternate hypothesis (Step 1) "points" in the negative direction. Thus, the critical region will be in the left tail (the direction in which the inequality is pointing) and the critical value will have a negative sign. If a one-tailed test is employed, and the inequality sign in the alternate hypothesis points in the positive direction, the area of rejection will always appear in the positive tail and the sign of the critical value will be positive.

STEP 5. Compute z using formula 10–1 and arrive at a decision:

$$z = \frac{\overline{X} - \mu}{\frac{\sigma}{\sqrt{n}}} = \frac{11.59 - 12.38}{\frac{3.52}{\sqrt{30}}} = -1.23$$

Because the computed z-value of –1.23 is not in the rejection region, we fail to reject the null hypothesis at the 0.01 level. The difference between 11.59 and 12.38 "out-of-seat behaviors" can be attributed to sampling error. In this problem, the p-value is 0.1093, found by 0.5000 – 0.3907. In other words, there is about an 11% chance of obtaining a z-value this large or larger when H_0 is true. From a practical standpoint, it cannot be concluded that the covert positive reinforcement treatment reduced the "out-of-seat behaviors" of the hyperactive children.

Self-Review 10-2

Experience over a long period of time had shown that, on the average, a mother stayed 2.0 days in the Findlay Childrens Hospital after childbirth. The standard deviation was 0.5 days. Hospital administrators, doctors, and other groups decided to make a joint effort to reduce the average time a new mother spends in the hospital. Following their campaign, a sample of the files of 100 mothers revealed that the new mean length of stay was 1.8 days. Are mothers staying in the hospital less time, or could the difference between 2.0 and 1.8 be due to sampling error?

a. What are H_0 and H_a?
b. Using the 0.05 level, state the decision rule.
c. Arrive at a decision.

Exercises

7. A machine is set to fire 30.00 decigrams of chocolate pellets into a box of cake mix as it moves along the production line. Of course, there is some variation in the weight of the pellets. A sample of 36 boxes of mix revealed that the average weight of the chocolate pellets was 30.08 decigrams, with a sample standard deviation of 0.50 decigrams. Is the increase in the weight of the pellets significant at the 0.05 level? Apply the usual five steps to be followed in hypothesis testing. (*Hint:* This involves a one-tailed test, because there is interest in finding out only whether there has been an *increase* in weight.)

8. The board of education of a suburban school district wants to consider a new academic program funded by the Department of Education. For the school district to be eligible for the federal grant, the arithmetic mean income per household must not exceed \$16,000. The board has hired a research firm to gather the required data. In its report, the firm has indicated that the arithmetic-mean income in the district is \$17,000. The survey included 75 households, and the standard deviation of the sample was \$3,000. Use a one-tailed test and the 0.01 level of significance to decide if the board can argue that the larger household income (\$17,000) is due to chance.

9. A test was constructed to measure the degree of people's alienation. The mean score was 78 and the standard deviation of the population of scores was 16. The test was administered to a sample of 37 Vietnam veterans. Their mean score was 84. Test the hypothesis that Vietnam veterans are more alienated than the general population. Use the 0.05 significance level. Compute the p-value. Interpret.

10. According to informed Pentagon sources, they deploy an average of 90 warheads at each missile site. An international peacekeeping force plans to make 49 inspections at various sites around the country to determine if there has been an increase in the density of warheads. They will use the 0.10 level of significance.

a. State the null and alternate hypotheses.
b. Give the formula for the appropriate test statistic.
c. What is the decision rule?
d. The checks found a sample mean of 92 warheads, with a sample standard deviation of 9. What does this evidence indicate?
e. Compute the p-value. Interpret.

A TEST FOR TWO POPULATION MEANS (LARGE SAMPLES)

The previous section dealt with one large sample (30 or more). This section is concerned with *two population means*. The five-step hypothesis-testing procedure is the same one used for one-sample tests; however, the formula for the test statistic z is slightly different. The difference between two normal distributions is also a normal distribution.

STAT BYTE

Genes may play a large part in divorce, according to a recent survey by two psychologists at the University of Minnesota. They studied 722 pairs of identical twins and 794 pairs of fraternal twins to find out if the correspondence in divorce rates for a pair of identical twins (who are genetically more alike than fraternal twins) is higher than that for a pair of fraternal twins. The results: 45% for identical twins and 30% for fraternal twins.

Problem There have been complaints that resident physicians and nurses at the Las Palmas Hospital central desk respond slowly to emergency calls from senior citizens who are medical or surgical patients. It is claimed that other patients receive faster service. The 0.01 level of significance is to be used to test the hypothesis that the response times to emergency calls from senior citizens and from other patients are the same. The alternate hypothesis is that the response times for the senior citizens are *greater than* those for other medical or surgical patients.

Solution Unknown to the resident physicians and nurses, lengths of time it took them to respond to the calls of both senior citizens and other patients were recorded. The sample results are summarized as follows:

Patients	**Sample Mean**	**Sample Standard Deviation**	**Number in Sample**
Senior citizens	5.5 minutes	0.4 minutes	50
Other patients	5.3 minutes	0.3 minutes	100

The important question to be tested is whether the mean time to answer senior citizens' calls (5.5 minutes) really differs from that of the mean time for other patients (5.3 minutes). As in all problems involving sampling, we know that there is a distinct possibility that the difference between 5.5 minutes and 5.3 minutes is due to chance.

The null and alternate hypotheses are

$$H_0 : \mu_1 \leq \mu_2$$
$$H_a : \mu_1 > \mu_2$$

where μ_1 is the population-mean response time for the senior citizens and μ_2 is the population-mean response time for other patients. The way the alternate hypothesis is stated (the mean time for senior citizens is *greater than* that for other patients) reveals to us that this is to be a one-tailed test.

The test statistic z is:

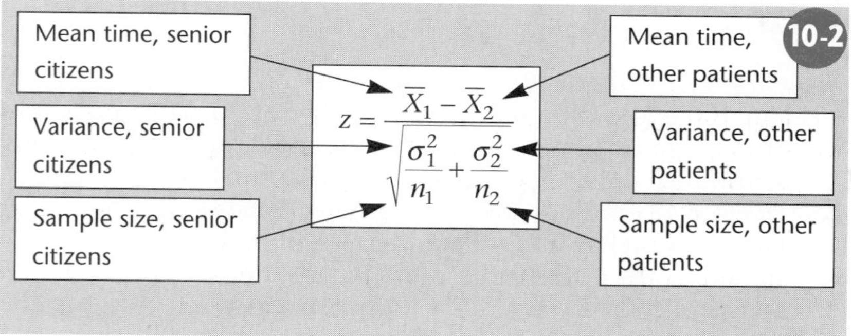

The decision rule for the 0.01 level: Do not reject the null hypothesis if the computed value of z is equal to or less than 2.33. Otherwise, reject the null hypothesis and accept the alternate hypothesis. The following diagram illustrates the decision rule:

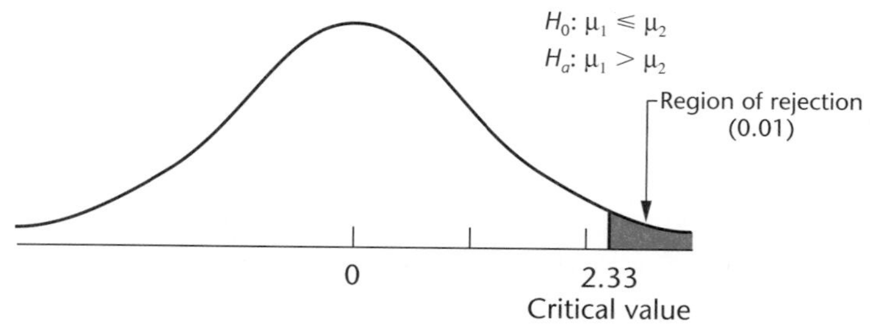

Computing z gives:

$$z = \frac{\overline{X}_1 - \overline{X}_2}{\sqrt{\frac{\sigma_1^2}{n_1} + \frac{\sigma_2^2}{n_2}}}$$

$$= \frac{5.5 - 5.3}{\sqrt{\frac{(0.4)^2}{50} + \frac{(0.3)^2}{100}}}$$

$$= \frac{0.2}{0.064}$$

$$= 3.13$$

The computed value of 3.13 is in the tail to the right of 2.33. Therefore, the null hypothesis is rejected at the 0.01 level of significance. It takes physicians and nurses longer to respond to calls from senior citizens as compared to calls from other patients. The probability that the difference of 0.2 minutes between the two means (5.5 – 5.3) is due to chance (sampling error) can be dismissed. In this case, the p-value is less than 0.0013.

It should be noted that to use the testing procedure for two means presented in the preceding paragraphs, two conditions must be met.

1. n_1 and n_2 must be 30 or more. In the preceding problem, 50 and 100 exceed the minimum number of 30. This restriction can be ignored, however, if the two populations are normally distributed and the population variances are known.
2. The samples must be *independent*. This means that the samples must be unrelated. For example, if senior citizen Smith were chosen in the sample of senior citizens, her selection must in no way affect the selection of any other patient, either in the senior citizen group or the other patient group.

Self-Review 10-3

The admissions officer at a university wants to investigate whether there is any difference between the scores on the mathematics placement test of students who attend classes primarily during the day and those of students who work during the day and attend evening classes.

a. Is this a one-tailed or a two-tailed test? Symbolically, what are the null hypothesis and the alternate hypothesis?

b. The 0.05 level of significance is to be used. State the decision rule.

c. The records of both day and evening students revealed the following:

Student	Mean Score	Standard Deviation	Number in Sample
Day	90	12	40
Evening	94	15	50

Compute z and arrive at a decision.

Exercises

11. Two machines fill bottles with a cough syrup. Workers check the performance of each periodically by testing bottles removed at random from the production line. The sample statistics for machine 1 are as follows: The mean weight of the contents of 40 bottles was 202.6 milligrams, and the standard deviation of the sample was 3.3 milligrams. For machine 2, the mean weight of 50 bottles was 200.0 milligrams, and the standard deviation was 2.0 milligrams. Using a two-tailed test and the 0.01 level of significance, test whether there is any difference in the performance of the two machines.

a. State both the null hypothesis and the alternate hypothesis.
b. State the decision rule.
c. Compute z and arrive at a decision.

12. You are conducting a study of the annual incomes of probation officers in metropolitan areas of fewer than 100,000 population and in metropolitan areas having greater than 500,000 population. Your sample data are:

	Population Less Than 100,000	Population Greater Than 500,000
Sample size	30	60
Sample mean	\$34,290	\$34,330
Sample standard deviation	\$135	\$142

Test the hypothesis that the annual incomes of probation officers in areas having greater than 500,000 population are significantly greater than those paid in areas of fewer than 100,000 population. Use the 0.05 level of significance. (*Hint:* Note that the problem includes the words "greater than.") Use a systematic approach by stating the null hypothesis, the alternate hypothesis, and so on.

13. A sociologist asserts that the mean length of courtship is longer before a second marriage than before a first. She bases this claim on the observation of (1) 80 first marriages in which the average courtship is 265 days, with a sample standard deviation of 60 days, and (2) 60 second marriages in which the average is 268 days, with a sample standard deviation of 50 days. Test her assertion by the five-step hypothesis-testing procedure. Use a 0.01 significance level. What is the p-value?

14. Planned Parenthood is investigating the differences between families in the Midwest and those on the Atlantic Coast. A particular study concerns the age of mothers at the birth of their last child. A random sample of 36 women from the Midwest had a sample mean of 32.9 years, with a sample standard deviation of 5.7 years. A second random sample of 49 women from the Atlantic Coast had a mean of 29.6 years, with a sample standard deviation of 5.5 years. Does this represent a significant difference between the two geographic regions? Use the 0.05 level of significance. Compute the p-value.

15. In an experiment to see if light from ultraviolet sun lamps affects muscle size, 50 pairs of laboratory animals were segregated into two groups. One group received ultraviolet light treatments daily for a month. The other did not. A particular muscle on each animal was then weighed.

a. State the null and alternate hypotheses.

b. Determine the decision rule for a 0.01 level of significance.

c. If the mean for the ultraviolet group is 89 milligrams, with a standard deviation of 9, and the mean for the control group is 57 milligrams, with a standard deviation of 7, what is the value of the test statistic?

d. Make a decision about the null hypothesis.

16. The burn rates for two different types of missile fuel are to be compared. Thirty observations on each will be obtained. You wish to design a test that will have a 0.01 significance level.

a. What are your hypotheses?

b. What is your decision rule?

c. The first sample mean is 20.44, with a standard deviation of 2.96. The second sample mean is 19.45, with a standard deviation of 2.13. What do you conclude?

A TEST DEALING WITH POPULATION PROPORTIONS

We introduced the concept of a proportion in Chapter 8. Now we will explore such questions as these: Will 80% of first offenders placed on probation commit a second crime? Do 20% of inner-city families live on a subsistence income? Is there a difference between the **proportion** of rural voters and urban voters who plan to vote for the incumbent governor in the forthcoming election?

Claims are often made that a proportion, or fraction, of a population possesses a certain characteristic. One claim might be that

25% of nonworking adults watched a daytime soap opera yesterday. Another claim might be that 30% of children under age 10 do not brush their teeth daily. A test to verify such claims follows. This test is appropriate when *both* $n\pi$ and $n(1-\pi)$ are *5 or more,* where π refers to the population proportion. If the data meet this requirement, we can use the normal approximation to the binomial, discussed in Chapter 7. The test statistic, which follows the z-distribution, is

$$z = \frac{\frac{X}{n} - \pi}{\sqrt{\frac{\pi(1-\pi)}{n}}} \qquad \textbf{10-3}$$

where

- X is the observed number in the sample possessing the trait.
- n is the size of the sample.
- π is the population proportion possessing the characteristic.

If we are to employ this test statistic, the items in the sample must be *selected independently* and each must have the *same chance of being selected.*

Problem We are investigating the following question: Is the proportion of inner-city families living on a subsistence income 20%, or 0.20?

Solution Use the five-step hypothesis-testing procedure.

STEP 1. State the null hypothesis and alternate hypothesis. In this case, the null hypothesis is that the proportion of all inner-city families living on a subsistence income is 0.20, which is a population parameter. The symbol π (Greek letter *pi*) is used as the symbol for a population proportion. Symbolically:

$$H_0 : \pi = 0.20$$

The alternate hypothesis is that the proportion of inner-city families living on a subsistence income is not 0.20. Symbolically:

$$H_a : \pi \neq 0.20$$

Because the alternate hypothesis does not state a direction, the test is two-tailed.

STEP 2. Select a level of significance. The significance level determines when the sample measurement is too far away from the hypothesized proportion to be believable. The 0.05 level has been chosen for the inner-city problem.

STEP 3. Choose a test statistic. For this problem, a sample of 200 inner-city families will be studied. $n = 200$ and $\pi = 0.20$, so $n\pi$ is $200(0.20) = 40$, and $n(1 - \pi)$ is $200(1 - 0.20) = 160$. Both $n\pi$ and $n(1 - \pi)$ are greater than 5. Therefore, the appropriate test statistic is z, given previously in formula 10–3:

$$z = \frac{\dfrac{X}{n} - \pi}{\sqrt{\dfrac{\pi(1-\pi)}{n}}}$$

STEP 4. Formulate a decision rule. According to the standard normal distribution in Appendix C, the critical value for the 0.05 level of significance is 1.96. We find it by locating 0.4750 in the body of the table and then moving to the margin to read the critical z-value of 1.96. The decision rule states that the null hypothesis will not be rejected if the computed value of z is in the interval between −1.96 and 1.96. Otherwise, it will be rejected and the alternate hypothesis accepted. Shown schematically:

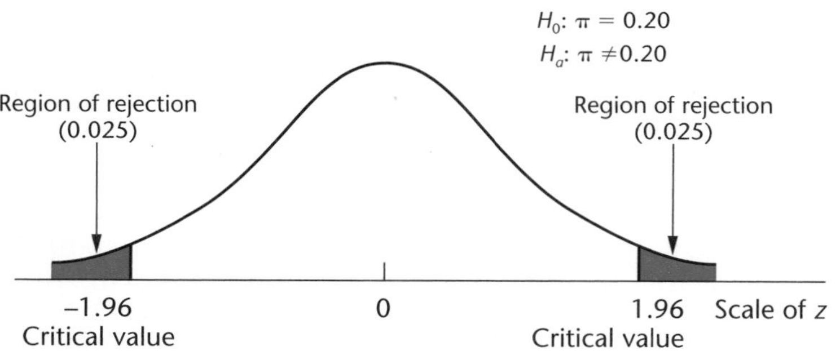

STEP 5. Select a sample from the population of all inner-city families, compute z, and then decide whether to reject H_0. A sample of 200 inner-city families revealed that 38 had incomes at the subsistence level. Computing z using formula 10–3, we get:

$$z = \frac{\dfrac{X}{n} - \pi}{\sqrt{\dfrac{\pi(1-\pi)}{n}}}$$

$$= \frac{\dfrac{38}{200} - 0.20}{\sqrt{\dfrac{0.20(1-0.20)}{200}}}$$

(continued)

$$= \frac{-0.01}{\sqrt{0.0008}}$$
$$= -0.35$$

The decision is not to reject the null hypothesis (that $\pi = 0.20$) because –0.35 is between –1.96 and 1.96. Although there is a difference between the sample proportion of 0.19 (found by 38/200) and 0.20, the apparent difference can be attributed to chance (sampling error). A helpful measure of the likelihood of this outcome is the *p*-value, as described earlier in this chapter. In this case, the *p*-value is 0.7264, found by 2(0.5000 – 0.1368). Note that the *p*-value is greater than the significance level, indicating that H_0 should not be rejected.

Problem Recall from the Chapter Problem the stock selection contest pitting dart throwing against professional investors. In that competition, the pros did better than the darts 15 times out of 27. Formulate and test an appropriate statistical test using this contest data.

Solution Use the five-step hypothesis-testing procedure.

STEP 1. In this case, the null hypothesis is that the proportion of contests won by the pros is 0.5. If the pros were just guessing, we would expect them to win 13.5 times and lose 13.5 times. The alternate hypothesis is that the proportion is greater than 0.5. This reflects the hope that the pros do have some advantage and means that the test is one-tailed.

STEP 2. We chose a significance level of 0.10 for this illustration.

STEP 3. Since n is 27 and π is 0.5, the normal approximation to the binomial would be appropriate. The test statistic is given in formula 10–3.

STEP 4. Consulting the standard normal table in Appendix C, we see that the critical value for a one-sided test with a 0.10 level of significance is 1.28. The decision rule becomes: Reject the null hypothesis if the test statistic exceeds 1.28.

STEP 5. Using formula 10–3, with $X = 15$ we get:

$$z = \frac{\frac{15}{27} - 0.500}{\sqrt{\frac{0.5(1-0.5)}{27}}} = \frac{0.556 - 0.500}{0.096} = \frac{0.056}{0.096} = 0.58$$

We fail to reject the null hypothesis because 0.58 is *not* greater than 1.28. Although the professionals won more than half of the contests, the difference between the sample proportion (0.556) and random guessing (0.500) could easily result from sampling error or "dumb luck." The p-value in this case is more than 25%, indicating that the test statistic could easily have happened by chance. There is thus no statistical evidence that the pros will top the dart throwing!

Self-Review 10-4

National figures reveal that the probability that a youthful offender on probation will commit another crime is 0.80. A special rehabilitation program was conducted for youthful offenders. A sample of 100 enrolled in the program showed that 75 subsequently committed another crime. The null hypothesis to be tested was $\pi \geq 0.80$. The alternate hypothesis was $\pi < 0.80$.

a. Based on the way the alternate hypothesis is stated, is this a one-tailed test or a two-tailed test?
b. Using the 0.01 significance level, show the decision rule graphically.
c. Arrive at a decision.

Exercises

For Exercises 17–20, use the five-step hypothesis-testing procedure.

17. You believe that 90% of the population would continue to work even if they inherited sufficient wealth to live comfortably without working. This assertion is questioned by a social research team. A list of persons who recently inherited more than $400,000 is obtained, and 350 persons are selected from the list. Out of the 350 selected, 308 indicate that they are still working. At the 0.05 level, is this sufficient evidence for the researchers to doubt your claim?

18. The local newspaper conducted a study of the welfare claims in the area. It reported that "at least 20% of the present welfare recipients are ineligible and should not be receiving monthly payments." You are hired by the welfare department to investigate the newspaper's claim. You randomly select 400 recipients from the files and carefully investigate each one. You find that 60 should not be receiving payments because they filed false claims, failed to report that they work, or committed other fraudulent acts. At the 0.01 level of significance, should the newspaper's claim be rejected? Back your decision with statistical evidence.

19. The Joint Monitoring Committee, a federal advisory panel, reports that 28% of adults in the United States are overweight. The Indian Health Commission of Arizona sampled 356 members of a tribe and found that 90 of them are overweight. At the 0.10 level of significance, can you conclude that a smaller proportion of the tribe members are overweight compared with all adults in the United States?

20. Out of 600 addicts admitted in fiscal 1990, 70 patients completed treatment in the California Alcohol Abuse Services Administration programs. By comparison, completion rates at federally funded clinics averaged 16%. Is the difference between the two proportions statistically significant at the 0.05 level? Compute the p-value. Interpret.

A TEST DEALING WITH TWO POPULATION PROPORTIONS

Some problems involve testing whether two population proportions are equal. To conduct such a test, first we select a random sample from each population. Then, if *both* samples meet the requirement that $n\pi$ and $n(1-\pi)$ are *5 or more,* we use the two-sample test that follows. The test statistic z follows the standard normal distribution and is computed by

$$z = \frac{\dfrac{X_1}{n_1} - \dfrac{X_2}{n_2}}{\sqrt{\bar{p}(1-\bar{p})\left(\dfrac{1}{n_1} + \dfrac{1}{n_2}\right)}} \qquad \textbf{10-4}$$

where

X_1 is the number possessing the trait in the first sample.
X_2 is the number possessing the trait in the second sample.
n_1 is the number in the first sample.
n_2 is the number in the second sample.
$\bar{p}$ is the proportion possessing the trait in the combined samples. It is referred to as the *pooled estimate* of the proportion:

$$\bar{p} = \frac{\text{Total number of successes}}{\text{Total number of samples}} = \frac{X_1 + X_2}{n_1 + n_2} \qquad \textbf{10-5}$$

Problem You know that a large group of homosexuals lives in a certain district of a large city. An in-depth study is to be made of this group. One of your objectives is to determine if there is a difference in the respective proportions of white homosexuals to the total white population and of black homosexuals to the total black population living in that district.

Solution Use the five-step hypothesis-testing procedure.

STEP 1. State the null and alternate hypotheses. The null hypothesis is that there is no difference between the two population proportions—that is, the two proportions are equal. Symbolically:

$$H_0 : \pi_1 = \pi_2$$

The alternate hypothesis is that the two population proportions are not equal. Symbolically:

$$H_a : \pi_1 \neq \pi_2$$

The problem does not imply a direction; therefore, a two-tailed test is used.

STEP 2. Give the level of significance. You decided on 0.05.

STEP 3. The test statistic for problems involving two proportions is z. The formula is:

$$z = \frac{\frac{X_1}{n_1} - \frac{X_2}{n_2}}{\sqrt{\bar{p}(1-\bar{p})\left(\frac{1}{n_1} + \frac{1}{n_2}\right)}}$$

STEP 4. The decision rule: Do not reject the null hypothesis if the value of the computed z is between – 1.96 and 1.96. Otherwise, reject the null hypothesis and accept the alternate hypothesis. A graphic presentation of the decision rule is:

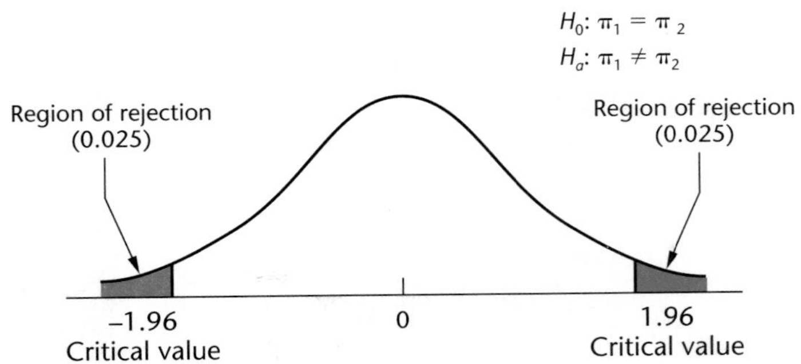

STEP 5. Take a sample from the white population living in the district and determine the number of male homosexuals in the sample. Then take a sample from the black population living in the area and determine the number of male homosexuals in the sample. The results are:

	Number in Sample	Number of Male Homosexuals	Proportion of Male Homosexuals in Population
White	804	575	0.715, found by 575/804
Black	175	111	0.634, found by 111/175

Repeating formula 10–4 for z gives:

$$z = \frac{\frac{X_1}{n_1} - \frac{X_2}{n_2}}{\sqrt{\bar{p}(1-\bar{p})\left(\frac{1}{n_1} + \frac{1}{n_2}\right)}}$$

The pooled estimate of the proportion designated by $\bar{p}$ is 0.70, found by using formula 10–5:

$$\bar{p} = \frac{X_1 + X_2}{n_1 + n_2} = \frac{575+111}{804+175} = \frac{686}{979}$$
$$= 0.70$$

Computing z, we obtain:

$$z = \frac{\frac{575}{804} - \frac{111}{175}}{\sqrt{0.70(1-0.70)\left(\frac{1}{804} + \frac{1}{175}\right)}}$$
$$= \frac{0.715 - 0.634}{\sqrt{0.001461}}$$
$$= \frac{0.081}{0.038} = 2.13$$

Because 2.13 is greater than the upper critical value of 1.96, the decision is to reject the null hypothesis. The conclusion is that there is a significant difference between the two population proportions. It is unlikely that the difference of 8.1% between the two sample proportions is due to chance (sampling error). The p-value is 0.0332, found by 2(0.5000 – 0.4834).

Self-Review 10-5

The effectiveness of a newly developed allergy-relief capsule is to be compared with that of one that has been on the market for a number of years. A sample of 250 persons using the new capsule revealed that 150 received satisfactory relief. Out of a group of 400 using the older capsule, 232 received satisfactory relief. Using the 0.02 level of significance, test the null hypothesis that the proportion receiving relief from the new capsule is equal to the proportion receiving relief from the old capsule. Use a two-tailed test.

a. State the null and alternate hypotheses, using the letters H_0 and H_a.
b. Show the decision rule graphically.
c. Compute z and arrive at a decision.

Exercises

21. Subsalicylate bismuth is the active ingredient in several non-prescription drugs for the relief of stomach upsets. Only 14 out of 62 persons sampled who took the bismuth during their three weeks' vacation away from home suffered from the gastric and intestinal discomfort that often afflicts travelers. Of 66 persons who did not take bismuth, 40 suffered from stomach upsets. Is the difference in the two proportions significant at the 0.05 level? Use a two-tailed test.

22. A major automobile insurance company claims that, compared with other drivers, a larger proportion of younger drivers take high-risk chances when driving. To investigate this claim, you give a driving-simulator test to a sample of young drivers and a sample of other drivers. Out of 100 young drivers tested, 30 took high-risk chances. Out of 200 other drivers tested, 55 took high-risk chances. Do you think there is sufficient evidence to support the insurance company's contention? Use the 0.05 level of significance.

23. Research into the use of physical punishment (spanking) in childrearing yielded the following information: Out of 100 children aged 3 to 9 years, 82 had been spanked by their parents during the previous month. Similarly, 40 out of 60 children aged 10 to 14 years had been spanked. Do parents use spankings significantly less often on older children? Test at the 0.01 level of significance.

24. A noted medical researcher has suggested that a heart attack is less likely to occur among men who actively participate in athletics. A random sample of 300 men is obtained. Of that total, 100 are found to be athletically active. Within this group, 10 had suffered heart attacks; among the 200 athletically inactive men, 25 had suffered heart attacks. Test the hypothesis that the proportion of men who are active and suffered heart attacks is equal to the proportion of men who are not active and suffered heart attacks. Use the 0.05 significance level.

SUMMARY

This chapter has presented the general idea of hypothesis testing. A five-step testing procedure was developed. The steps to follow are these:

1. State a null hypothesis and an alternate hypothesis.
2. Choose a level of significance. The most common levels are 0.05 and 0.01.
3. Select an appropriate test statistic. The test statistic in all problems in this chapter is z.
4. Set up a decision rule. The objective statement found in the decision rule gives a basis on which to make a decision. The test can be one-tailed or two-tailed, depending on how the alternate hypothesis is phrased.
5. Take one or more samples, depending on the problem, and then compute z. Based on the computed value of z and the critical values stated in the decision rule, decide whether to reject the null hypothesis.

Two types of errors are possible when we employ sampling: Type I errors and Type II errors. If we reject a true null hypothesis, then we have made a Type I error. We have made a Type II error if we accept a false null hypothesis.

Four statistical hypothesis tests were presented. All are based on "large samples" and use z as the test statistic. The first, a one-sample test, is used to determine if a reported mean is reasonable. The test statistic is:

$$z = \frac{\overline{X} - \mu}{\frac{\sigma}{\sqrt{n}}}$$

The second test compares two samples to find out if the two population means are equal. The test statistic is:

$$z = \frac{\overline{X}_1 - \overline{X}_2}{\sqrt{\frac{\sigma_1^2}{n_1} + \frac{\sigma_2^2}{n_2}}}$$

The third test compares a sample proportion and a hypothesized population proportion. The test statistic is:

$$z = \frac{\frac{X}{n} - \pi}{\sqrt{\frac{\pi(1-\pi)}{n}}}$$

If we are to apply this test, $n\pi$ and $n(1 - \pi)$ must be at least 5.

The other test of hypothesis compares two population proportions. The test statistic is:

$$z = \frac{\frac{X_1}{n_1} - \frac{X_2}{n_2}}{\sqrt{\bar{p}(1-\bar{p})\left(\frac{1}{n_1} + \frac{1}{n_2}\right)}}$$

Exercises

25. The Department of Health and Human Services reported that the mean number of years of school completed by adults in the United States was 11.5. The Tennessee Department of Employment Security wants to determine if the level of education of their employees is less than the national average.

a. State the null and alternate hypotheses both in words and using the letters H_0 and H_a.

b. Give the test statistic z.

c. State the decision rule in words, for a 0.05 significance level. Show it in a diagram.

d. A sample of 100 employees revealed that the mean is 13 years, with a standard deviation of 1.1 years. Compute z and arrive at a decision regarding the null hypothesis.

26. A psychologist wants to evaluate whether windowless schools lead to anxiety in the psychological development of children. A sample of children who are in windowless classrooms is selected and given an anxiety test. Higher scores indicate more anxiety. The same procedure is followed for children in classrooms with windows. The results:

	Windowless Schools	**Schools with Windows**
Mean	94	90
Standard deviation	8	10
Sample size	100	80

Using the five-step hypothesis-testing procedure, the 0.01 level, and a one-tailed test, arrive at a conclusion regarding the anxiety levels of the two groups.

27. A medical researcher contends that lung capacity varies significantly between smokers and nonsmokers. The mean capacity of a sample of 30 nonsmokers is 5.0 liters, with a sample standard deviation of 0.3 liters. Of the 40 smokers in the sample, the mean lung capacity is 4.5 liters, with a standard deviation of 0.4 liters. At the 0.01 significance level, is there sufficient evidence to conclude that lung capacity is larger among nonsmokers?

28. Patients entering a hospital have complained that it takes 30 minutes to fill out the forms required for admittance. As a result of their complaints, the forms and procedure have been revised. A recent analysis of a random sample of 40 incoming patients reveals that the mean time to fill out the forms now is 28.5 minutes, with a standard deviation of 5 minutes. Is there sufficient evidence at the 0.02 level of significance to show that the new system is an improvement?

29. A random sample of 100 freshmen entering Ivy Tech in the fall of 1985 had an average combined ACT score of 23, with a sample standard deviation of 0.80. A sample of 100 similar freshmen in the fall of 1992 showed a mean score of 24, with a sample standard deviation of 0.90. Has there been a significant increase in the test scores over the seven-year period? Use the 0.05 significance level.

30. A group of environmentalists claims that warm water from the Virginia Power nuclear plant at Lake Anna is limiting the growth of striped bass at the lake. Use the 0.01 level of significance.

- **a.** What are the appropriate null and alternate hypotheses?
- **b.** Show the test statistic.
- **c.** State the decision rule.
- **d.** A study of 50 striped bass at the lake revealed a mean weight of 20 pounds, with a sample standard deviation of 4 pounds. In a similar southeastern reservoir, a sample of 40 striped bass was found to average 40 pounds, with a standard deviation of 8 pounds. Do these data support the environmentalists? What do you conclude?

31. The Municipal Environmental Protection Agency claims there has been a change in the density of the ozone layer over the city. Data from previous monitoring programs indicate that the mean is 384 ppm (parts per million), with a standard deviation of 20 ppm. You want to perform an independent test of the density.

- **a.** State appropriate null and alternate hypotheses.
- **b.** If you want a 0.01 significance level, what is your decision rule?

c. You have collected a random sample of 100 observations, and their mean is 390 ppm. What do you conclude?

32. A soft drink manufacturing facility turns out "16-ounce" bottles of beverage on a production line that, based on many observations, has a mean filling rate of 16.01 ounces, with a standard deviation of 0.005 ounces. As a newly hired quality control inspector, you decide to draw a sample of 40 bottles for testing.

a. What are the null and alternate hypotheses?

b. What is the decision rule for a 0.10 level of significance?

c. Your sample has a mean of 15.97 ounces. Is that statistically significant?

33. A tradition in a particular fishing village indicates that a certain strain of fish will be found at a depth of 70 meters. You wish to test that lore by using modern equipment to locate the fish and recording the depth in the vicinity of the school.

a. State your hypotheses.

b. For a 0.05 level of significance, what is the decision rule?

c. If you get a mean of 73 meters, with a standard deviation of 2 meters from a sample of 35 measurements, what do you conclude?

34. A study dealing with comparative nutrition will record the heights of children at age 16 from two different countries. One hundred youths will be sampled in the first country and 150 in the second country.

a. What are the null and alternate hypotheses?

b. What is the decision rule for a 0.10 level of significance?

c. The mean from the first country is 52.7 inches, with a standard deviation of 2.5 inches. The mean from the second country is 51.8 inches, with a standard deviation of 2.6 inches. Complete the test of the hypothesis.

35. An investigation of two kinds of photocopying equipment showed that 60 failures of the first kind of equipment took, on the average, 82.4 minutes to repair, with a standard deviation of 19.4 minutes, while 60 failures of the second kind of equipment took, on the average, 91.6 minutes to repair, with a standard deviation of 18.8 minutes. Test at the 0.01 level of significance whether the difference between these two sample means is significant.

36. An EEOC complaint against the Betts Manufacturing Company alleges that men are paid more than women in a particular job category.

a. State the appropriate null and alternate hypotheses.

b. For a test at the 0.02 level of significance, what is the decision rule?

c. Sample data show that 80 men earned, on the average, $353 with a standard deviation of $18, while 80 women earned, on the average, \$315, with a standard deviation of \$21. State the appropriate conclusion to the complaint.

37. The League of Savings Institutions reported that the mean age of home buyers in 1985 was 35.8 years old, with a standard deviation of 14.24 years. You intend to sample 100 home buyers in 1990 and record their mean age.
   a. State appropriate null and alternate hypotheses.
   b. For a 0.02 level of significance, what is your decision rule?
   c. The mean age of your sample is 37.3. Is that statistically significant?

38. An urban planner claims that, nationally, 20% of all families renting condominiums move during the year. A random sample of 200 families renting condominiums in a large development revealed that 56 had moved during the past year. Does the evidence demonstrate that this development has a greater proportion of movers than the national average?

39. A drug manufacturer has developed a drug that is said to cure postnatal depression in 85% of cases. A random sample of 150 women who gave birth at the Colorado General Hospital and who used the drug in a two-year period revealed that 120 of them found it effective. Does this result contradict the manufacturer's claim? Use a 0.02 level of significance.

40. The American Council on Health Policy reports that 65% of citizens in the United States are covered by private health insurance. The health planning agency of Utah has data from a random sample of 200 residents of Utah. They wish to determine if the Utah proportion is consistent with the nationally reported percentage.
   a. Formulate appropriate null and alternate hypotheses.
   b. For the 0.10 level of significance, what is the decision rule?
   c. In the sample, 120 were covered by private health insurance. Compute the test statistic.
   d. What do you conclude?

41. Thirteen percent of the manuscripts submitted to the *New England Journal of Medicine* are accepted for publication. A prominent medical school has submitted 50 manuscripts and claims that their acceptance rate is better than average.
   a. What are the null and alternate hypotheses?
   b. State the decision rule for a test at the 0.05 level of significance.
   c. Ten of the papers are accepted. Compute the test statistic.
   d. Complete the statistical test.

42. The *Washington Post* conducted a poll after the 1986 death of a basketball star. A total of 729 of the 1,656 respondents said that they believed that the use of illegal drugs by athletes is a widespread problem. One year earlier, a similar poll showed only

301 of 1,432 respondents held the same view. Has there been a significant shift in the public's view of drug use? Use the 0.01 level of significance and the formal steps of hypothesis testing to address this question.

43. Citizens for Tax Justice reported that 42 of 250 companies surveyed paid no income tax this year. Last year, they said that 130 of the 250 firms legally avoided federal income taxes. Has the proportion of firms that avoid income tax declined or can the difference be attributed to sampling error?

44. During a baseball game, one team gets eight hits in 37 at-bats, while another gets seven hits in 38 at-bats. Is there a statistically significant difference in the proportion of hits for the two teams?

45. Past elections indicate that 40% of the central city votes are needed to pass any amendment that will increase taxes in the entire county. A preelection poll revealed that a tax increase is favored by 300 out of 1,000 central city voters. Is it likely that the difference between the sample proportion (0.30) and the proportion needed (0.40) is due to chance (sampling error)? Use the 0.01 level.

46. Suppose that a random sample of 1,000 American-born citizens revealed that 200 favor resumption of full diplomatic relations with Cuba. Similarly, 110 out of a sample of 500 foreign-born citizens favor it. Test at the 0.05 level the H_0 that there is no significant difference in the two proportions.

47. Past experience at Penta State University indicates that 50% of the students change their major area of study after their first year of college work. At the end of the past year, a sampling of 100 students revealed that 48 of them had changed their major at the end of their first year. Has there been a significant decrease in the proportion of students who change their major, or can the difference between the expected proportion (50%) and the sample proportion (48%) be attributed to chance (sampling error)? Use the 0.05 level of significance.

48. A survey dealing with car pooling in metropolitan Philadelphia found that during a particular October, 83 out of 420 vehicles had more than one occupant. Five years later, a similar sample of 423 cars showed that 146 had more than one occupant. Is this increase significant at the 0.01 level?

49. Preliminary research suggests that the proportion of young couples who use a particular method of birth control is 0.30. A detailed study conducted in Napanee, Indiana, showed a total of 46 users in a random sample of 200 young women. Are these results consistent with the preliminary research? Use the 0.10 level of significance.

50. A sportswriter comments that 50% of college football players suffer damage to their knees. To test this claim, you sample 300 players. If 143 of this group have knee injuries, is this statistically significant at the 0.05 level?

51. The Citizens Action Council claims that over 60% of the citizens of Fowlerville oppose a new highway construction plan. If 140 of 190 randomly selected residents are opposed to the plan, test the claim at the 0.01 level of significance.

52. To evaluate a new method for treating depression, researchers subjected 70 patients at a clinic to the new method; 60 others were treated by the traditional method. If 46 of the patients in the first group and 28 in the second group showed signs of improvement, what can we conclude at the 0.01 level of significance?

53. In a random sample of 200 Ohio residents, 25 reported allergic reactions during changing weather. In a similar sample of 150 Arizona residents, 10 reported experiencing this type of reaction. Use the 0.02 level of significance. Do the data support the idea that the proportion of persons having allergic reactions is not different in Arizona?

Data Exercises

54. Refer to Real Estate data, which reports information on homes sold in Northwest Ohio during 1992.

a. The *Detroit News* reported that "the typical Northwest Ohio home is now selling for less than $190,000." Does this claim appear valid based on Data Set 1? Write a brief summary of your findings. Use the .05 significance level. Be sure to comment on the *p*-value.

b. The Northwest Ohio Area Chamber of Commerce noted in one of its publications that "because of the good expressway system the typical homeowner lives more than 15 miles from the central city." Is this claim valid based on the information in Data Set 1? Write a brief summary of your findings. Use the 0.05 significance level. Be sure to comment on the *p*-value.

c. Do half of the homes sold have a fireplace, or is the actual proportion less? Use the 0.05 significance level to conduct the test.

d. An important factor in the decision to sell a home is what fraction the selling price is of the list price. At the .05 significance level, can we conclude what the mean fraction of the selling price is if the list price is different from 0.975?

55. Refer to Salary data, which refers to a sample of middle managers in Sarasota, Florida.

a. Can we conclude that the typical manager makes more than $60,000? Use the .01 significance level.

b. Do middle managers with less than 15 years of experience earn, on average, less than those with 15 or more years of experience? Create the necessary variables and follow the five-step hypothesis-testing procedure. Use the .05 significance level.

c. An obvious question of interest is whether there is a difference in the mean salaries of men and women. Explain why it is not appropriate statistically to perform the five-step test on the information in Data Set 2.

CASE: SHOULD ANNE FORRESTER BE DISMISSED?

Joseph Steop, the president of National Savings and Loan, is concerned about the default rate of the loans the bank has made to small businesses. He feels that Anne Forrester is responsible for an inordinate number of the defaults. Ms. Forrester claims that she is a victim of sex discrimination and that her default rate is no greater than those of her male counterparts. The board of directors has asked the human resources manager to investigate the dispute. He selects a random sample of small business loans made by male loan officers since 1988 and records whether they are in default. A similar sample of Ms. Forrester's loans is selected. The results are as follows ("d" stands for default, "nd" for no default):

	Male Loan Officers		Anne Forrester
nd	Rob's Beauty Shop	nd	AA Machine Shop
nd	Grimes Window Cleaning	nd	Jackson Printing
d	Gordon Metal Works	nd	Art and Ben's Concrete
nd	Computer Systems	nd	Arc Welding
nd	Glencoe Paving	d	Reproduction Furniture
nd	Christmas Shoppe	d	D and B Ceramics
d	The Art Shop	nd	Kramer's Antiques
nd	Korbin's Body Shop	nd	Kyle's Marina
d	Superior Refinishing	d	The Kandy Korner
nd	The Frame Shop	nd	Mall Shoe Repair
nd	Just Pens	d	Electronics and Stuff
d	Silk Flowers	nd	Baker's Pool Supply
nd	Package Delivery	d	Mannelli's Pizza
nd	Fergie's Men's Boutique	d	Garner's Car Tops
d	Beauty Supplies	nd	Phil's Plumbing Supplies
nd	Candolini's Restaurant	d	Ace Glass Co.
nd	Furniture Stripping	nd	Aldon Lamp Shades
d	All-American Cookies	d	Rugs, Rugs, Rugs
nd	Landings Travel	nd	Exquisite Gifts

nd	Clip, Clip Hair Stylists	d	Grow Fast Nursery
		nd	Shoes, Shoes
		d	The Snack Shop
		nd	Dela's Massage
		nd	The Tanning Salon
		nd	Windshields
		nd	Military Surplus
		nd	Tod's RV Repair
		nd	Corky's Dental Supplies
		nd	Wigs
		nd	Jamie's Book Store

Does this evidence help resolve the controversy? If you were the human resources manager, how would you interpret these data to the board?

CHAPTER ACHIEVEMENT TEST

Answer all the questions. The answers are at the back of the book.

MULTIPLE-CHOICE QUESTIONS. Select the response that best answers each of the questions.

1. A statement about the value of a population parameter is called a
   a. null hypothesis.
   b. sample statistic.
   c. test statistic.
   d. decision rule.

2. The probability of making a Type I error is called the
   a. power of a test.
   b. level of significance.
   c. alternate hypothesis.
   d. critical value.

3. The value computed from sample data is called the
   a. decision rule.
   b. critical value.
   c. test statistic.
   d. alternate hypothesis.

4. A Type II error occurs if you
   a. reject a false alternate hypothesis.
   b. accept a false alternate hypothesis.
   c. reject a false null hypothesis.
   d. accept a false null hypothesis.

5. A one-sample test for a population mean has a critical value of 1.96 and a test statistic of 1.71. The correct statistical conclusion is to
  a. reject the null hypothesis.
  b. reject the alternate hypothesis.
  c. not reject the null hypothesis.
  d. not reject the alternate hypothesis.

6. Tests for proportions require that $n\pi$ and $n(1 - \pi)$ be
  a. equal.
  b. 30 or more.
  c. 5 or more.
  d. 50 or more.

7. The "pooled" estimate of a proportion that results from combining samples will
  a. equal one of the sample proportions.
  b. be greater than either sample proportion.
  c. be between the two sample proportions.
  d. be less than the smaller of the two.

8. To determine if more than 20% of consumers will purchase a new product, a researcher should use a
  a. two-tailed test.
  b. sample of at least 25 consumers.
  c. one-tailed test.
  d. 20% level of significance.

9. Tests for proportions are based on which of the following probability distributions?
  a. Binomial approximation of the normal
  b. Poisson distribution
  c. Student's *t*
  d. Normal approximation of the binomial

10. A sample (as opposed to a population) proportion is found from
  a. the null hypothesis.
  b. sample data.
  c. the alternate hypothesis.
  d. the decision rule.

COMPUTATION PROBLEMS

11. Doctors in Sweden claim that people with a high waist-to-hip measurement ratio have more heart attacks. You have information from medical records at Delta Hospital covering a sample of 316 people with low waist-to-hip ratios and a sample of 214 with high waist-to-hip ratios. To evaluate the doctors' claim, you are to design a statistical test.
  a. State your null and alternate hypotheses in statistical symbols.
  b. For a significance level of 0.01, what is your decision rule?

c. Heart attacks were actually suffered by nine people with low waist-to-hip ratios and 11 people with high waist-to-hip ratios. Show the test statistic and essential calculations.

d. Can the null hypothesis be rejected? State your conclusion.

12. Kaiser Medical compares the hospital length of stay for a random sample of 40 patients in New York with a similar sample of 30 patients in Detroit. Do the data indicate a difference in the mean length of hospital stay in the two cities?

a. State in symbolic form the appropriate null and alternate hypotheses.

b. At the 7% level of significance, what is the decision rule?

c. The mean length of stay is 5.21 days in New York, with a standard deviation of 2.3 days. The Detroit average is 4.65, with a sample standard deviation of 1.9 days. Show the test statistic and essential calculations.

d. What do you conclude? State your findings in a sentence.

13. Congresswoman Linda Jones says that 38% of the voters in her district can name the leaders of Brazil, Argentina, and Chile. To test her claim, a random sample of 25 voters is selected.

a. In symbols, state appropriate null and alternate hypotheses.

b. What is the decision rule at a significance level of 0.02?

c. Only 10 out of 25 in the sample could identify the leaders. Give the test statistic and show the essential calculations.

d. Should the null hypothesis be rejected? State your conclusion in a sentence.

ANSWERS TO SELF-REVIEW PROBLEMS

10-1 **a.** $H_0 : \mu = 9.5$

$H_a : \mu \neq 9.5$

b. 0.01

c. $$z = \frac{\overline{X} - \mu}{\frac{\sigma}{\sqrt{n}}}$$

d.

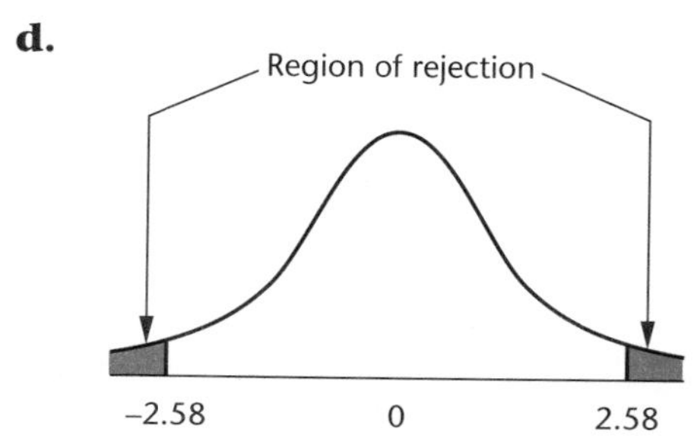

e. $$z = \frac{9.2 - 9.5}{\frac{3.9}{\sqrt{18,000}}} = \frac{-0.3}{\frac{3.9}{134.16}} = \frac{-0.3}{0.029} = -10.3$$

Reject H_0, accept H_a. There was a change.

f. The *p*-value is virtually zero. The probability of a *z*-value less than −10.3 or greater than 10.3 is less than 0.0001.

10-2 a. $H_0 : \mu \geq 2.0$

$H_a : \mu < 2.0$

b. Reject H_0 if computed z is to the left of -1.65. Otherwise, do not reject H_0 at the 0.05 level.

c.
$$z = \frac{1.8 - 2.0}{\frac{0.5}{\sqrt{100}}} = -4.0$$

Reject H_0. Mothers are staying less time.

10-3 a. Two-tailed

$H_0 : \mu_1 = \mu_2$

$H_a : \mu_1 \neq \mu_2$

b. Accept H_0 if computed z is between -1.96 and $+1.96$.

c.
$$z = \frac{90 - 94}{\sqrt{\frac{(12)^2}{40} + \frac{(15)^2}{50}}} = \frac{-4.000}{2.846} = -1.41$$

Do not reject H_0. There is no difference between the two groups.

10-4 a. One-tailed

b.

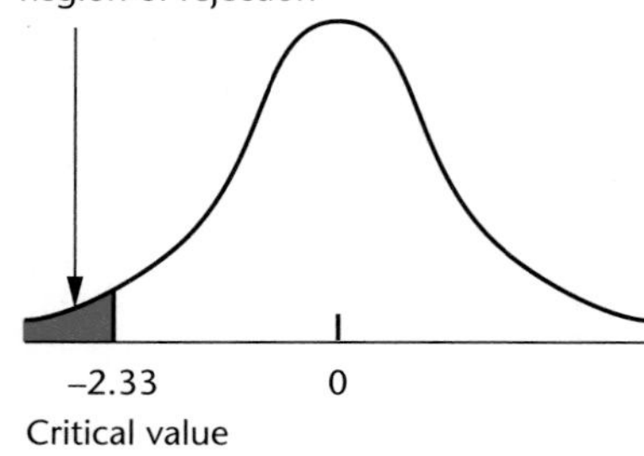

c.
$$z = \frac{\frac{75}{100} - 0.80}{\sqrt{\frac{0.80(1-0.80)}{100}}} = -1.25$$

Do not reject the null hypothesis. The rehabilitation program is not effective.

10-5 a. $H_0 : \pi_1 = \pi_2$

$H_a : \pi_1 \neq \pi_2$

b.

Region of rejection (0.01) Region of rejection (0.01)

–2.33 Critical value 0 2.33 Critical value

c.
$$z = \frac{\frac{150}{250} - \frac{232}{400}}{\sqrt{0.59(1-0.59)\left(\frac{1}{250} + \frac{1}{400}\right)}} = \frac{0.60 - 0.58}{\sqrt{0.00157}} = \frac{0.02}{0.0396} = 0.505$$

Fail to reject the null hypothesis.

UNIT REVIEW

The last three chapters dealt with the fundamental concepts of statistical inference. They described (1) how we can use sampling to develop a range of values within which a population parameter is likely to occur, and (2) the procedures for testing hypotheses about population parameters.

Key Concepts

1. **Sampling** is important because complete knowledge regarding the population is seldom available. The basic purpose of sampling is to provide an estimate about a population parameter based on sample evidence.
2. Sampling is necessary because
   a. it may be impossible to check the entire population;
   b. the cost to study the entire population may be prohibitive;
   c. to contact the entire population would be too time-consuming;
   d. the tests may destroy the product.
3. A **confidence interval** is a range of values within which the population parameter is expected to fall for a preselected level of confidence.
4. The **sampling distribution of the mean** is a probability distribution that describes all possible sample means of a given size selected from a population.
5. The **central limit theorem** states that the distribution of the sample means is approximately normal regardless of the shape of the population.
6. The **standard error of the sample mean** is the standard deviation of the distribution of sample means.
7. The **standard error of the sample proportion** is the standard deviation of the sampling distribution of sample proportions.
8. A **point estimate** is the value computed from a sample, which is used to estimate a population parameter. An **interval estimate** is a range of values within which we have some assurance that the population parameter lies.
9. A **proportion** is a fraction, percent, or ratio that indicates what part of a sample or population has a certain trait.
10. When the sample constitutes more than 5% of the population, the **finite population correction factor** is used. Its purpose is to reduce the standard error because the sample constitutes a large portion of the population. The term $\sqrt{(N-n)/(N-1)}$ is multiplied by the standard error of the mean or proportion.

11. **Hypothesis testing** is an extension of the concept of interval estimation and offers a strategy for choosing among alternative courses of action. For the purpose of testing, two claims are made about a population parameter. One is called the **null hypothesis** and the other is called the **alternate hypothesis.**
12. Two types of errors may be committed during a test of the null hypothesis. If the null hypothesis is rejected when it is true, a **Type I error** is committed. If the null hypothesis is accepted when it is false, a **Type II error** is committed.
13. The **level of significance** is the probability of rejecting the null hypothesis when it is true. The **test statistic** is a quantity calculated from sample information and used as a basis for deciding whether or not to reject the null hypothesis. The **decision rule** is a statement of the conditions under which the null hypothesis is rejected. A **critical value** or **critical values** separate the rejection region from the remaining values.
14. The ***p*-value** is the probability (assuming the null hypothesis is true) of getting a value of the test statistic at least as extreme as that found in the sample.

Key Terms

Probability sampling
Sample frame
Sample design
Sample survey
Census
Control group
Experimental group
Random sample
Table of random digits
Systematic random sample
Stratified random sample
Sampling error
Sampling distribution of the sample mean
Sampling distribution of the sample proportion
Central limit theorem
Confidence interval
Standard error of the sample mean
Decision rule
Two-tailed test
Critical value
Point estimate
Interval estimate
Standard error of the sample proportion
Finite population correction factor
Null hypothesis
Alternate hypothesis
Type I error
Type II error
Level of significance
Test statistic
p-value
One-tailed test
Proportion

Key Symbols

$\sigma_{\overline{X}}$ The standard error of the mean.
σ_p The standard error of the sample proportion.
E Maximum allowable error.
H_0 The null hypothesis.

H_a The alternate hypothesis.

π The proportion of objects or things in the population that possess a particular trait.

p The sample proportion (X/n).

$\bar{p}$ The pooled estimate of the population proportion based on the two samples.

Problems

1. A special 10-question examination was given to a population of five students. The number of correct responses by each student was as follows:

Student	Number Count
Taylor	5
Tigranian	6
Sims	7
Sobsak	9
Hilix	8

   a. How many different samples of two are possible from this population?
   b. List the possible samples of two and compute their respective means.
   c. Organize the sample means into a probability distribution.
   d. Compare the mean of the sampling distribution to the mean of the population.
   e. Compare the spread of the sampling distribution to that of the population.

2. The Department of Mathematics consists of seven faculty members. Their salaries are as follows:

Professor	Salary (in thousands of dollars)
Hause	$23.0
Delgado	26.0
Gallaghar	29.0
Jackson	21.0
Ramsdall	20.0
Tang	19.0
Elsass	23.0

   a. How many different samples of two are possible?
   b. List the possible samples of two and compute their respective means.

c. Organize the sample means into a probability distribution.
d. Compare the mean of the sampling distribution to the mean of the population.
e. Compare the spread of the sampling distribution to that of the population.

3. The daily demand for coffee at the Koffee Korner is to be estimated. A random sample of 40 days from the first year of operation showed mean sales of 240 cups per day, with a standard deviation of 20 cups. Construct a 95% confidence interval for the daily demand.
4. A manufacturer of men's slacks is considering placing TV ads during the upcoming Super Bowl. The manufacturer would like an estimate of the proportion of men who will watch the game. A sample of 500 men revealed that 300 planned to watch the game. Develop a 95% confidence interval for the proportion of men who plan to watch the game.
5. A study has been conducted on the movie-going habits of young adults. A random sample of 50 reveals the mean number of movie-viewing hours per month to be 9.0 hours. The standard deviation is 2.8 hours. Develop a 95% confidence interval for the mean number of hours per month of movies viewed.
6. The Levision Brothers Department Store recently made an analysis of its delinquent accounts. A random sample of 40 of its 300 delinquent accounts showed a mean amount of $78.65, with a standard deviation of $36.51. Construct a 90% confidence interval for the population mean.
7. The nurses at Memorial Hospital are concerned with working hours, wages, and seniority issues. Before approaching management, the nurses conduct an opinion survey among the nursing staff. One of the questions is, "Would you join a union?" There are 250 nurses on the staff. A sample of 60 revealed that 45 would join a union. Develop a 90% confidence interval for the proportion of nurses who would join a union.
8. The Security Trust Company owns a fleet of 200 cars that are used by bank executives. The bank wants to study the yearly cost of maintaining these cars. A random sample of 40 of the fleet showed that the mean maintenance cost per year per car was $1,240, with a standard deviation of $50. Construct a 98% confidence interval for the mean repair cost per year per car.
9. A large health insurer considers raising its rates. However, the decision is based on the mean annual family expenditure for medical care. The insurer wants to be 99% confident that the sample is within $70 of the correct mean amount spent for medical care. The standard deviation is estimated to be $250. How large a sample is required?
10. The manager of Clem's Supermarket wants to estimate the mean amount spent in his store per customer. He estimates the population standard deviation to be $25.00, and he would like

to be within $4 of the population mean. Assume a 99% level of confidence. How large a sample is required?

11. Coffee cans are filled to a net weight of 32 ounces, but there is some variability. A random sample of 36 cans revealed a mean weight of 31.8 ounces, with a standard deviation of 0.4 ounces. Test the hypothesis, at the 1% level of significance, that the net weight of the coffee is actually 32 ounces.

12. Records maintained by Orange Cross Insurance Company indicate that the mean length of hospital confinement for a routine appendectomy is 4.0 days. St. Mark's Memorial Hospital selected a sample of 40 patients who had a routine appendectomy and found that the mean length of confinement was 3.5 days, with a standard deviation of 1.2 days. At the 0.05 significance level, can it be concluded that St. Mark's patients are released significantly earlier than insurance records suggest?

13. The president of Wattsburg Community College reported in a meeting with students: "When I came here four years ago, the typical student traveled only 7 miles to class." He further stated that "a recent survey by the Admissions Office staff showed that in a sample of 35 students, the mean distance traveled to class was 9.2 miles, and the standard deviation was 6.0 miles." Does this recent study show that students are now driving significantly farther to attend class? Use a 0.01 level of significance.

14. The student newspaper interviewed a random sample of 200 male students and found that 150 were in favor of having a homecoming dance. A random sample of 400 females revealed that 312 were in favor of the dance. Does the sample evidence show (at the 0.10 significance level) that there is a difference between the proportion of males and of females favoring a dance?

15. Shoplifting has become a national problem. One report indicates that 10% of the customers who enter a large discount store will steal something. The store manager of the Jamesway Discount Store randomly selected 150 customers and observed them from behind a one-way window. Twenty of these shoppers attempted to steal something. Do these data suggest that shoplifting is more of a problem for Jamesway than for the typical large discount store? Use a 0.05 significance level.

16. A random sample of 40 male students and 42 female students revealed the following information about the number of bottles of cola (12-ounce size) they consumed last week:

	Male Students	**Female Students**
Mean	8.75	7.25
Standard deviation	2.30	1.95
Sample size	40	42

At the 0.05 significance level, can it be shown that the male students drink more cola?

17. A machine is set to produce ball bearings that have a radius of 1.5 cm. A sample of 36 produced by the machine had a radius of 1.504 cm, with a sample standard deviation of 0.009 cm. Is there reason to believe that the machine is producing ball bearings with a mean radius greater than 1.5 cm? Use the 0.05 significance level.

18. Mrs. Knitt, a producer of candy assortments, packages a holiday mix of nut candies, chewy candies, and dark chocolates. The mix is advertised to contain 50% nut candies, 35% chewy pieces, and 15% dark chocolates. A random sample from several boxes shows 125 nut candies, 85 pieces of chewy candy, and 30 dark chocolates. Test at the 0.05 level to determine if the actual mix contains the advertised proportion of nut candies.

19. You study the stock market and believe that the market behaves differently on Fridays than it does on other days of the week. You carefully record whether the Dow Jones Industrial Average advances or declines over 18 Fridays and 72 other weekdays. The market actually advances on 10 Fridays and 30 other weekdays among your sample. Test your conjecture at the 1% level of significance.

20. Accountants at the Spartan Soap Company claim that their mean salary is significantly below the national mean of $19,500 for all accountants. You randomly select the personnel files of 50 Spartan accountants.

a. State the null and alternate hypotheses in statistical symbols.

b. Using a 10% significance level, what is the decision rule?

c. The sample mean salary is $18,750, with a sample standard deviation of $1,581. Show the test statistic and essential calculations.

d. What do you conclude? State your findings in a sentence.

Using Statistics

Situation A labor negotiator in a particular industry estimates that 80% of all major union contract disputes are settled without a strike. A management representative claims, on the other hand, that the proportion is only 40%. During a particular, randomly selected year in which 10 major contracts came up for renewal, settlements were achieved before a strike on three occasions.

Which estimate is more likely to be correct? What range of values would you suggest as plausible for the population parameter?

Discussion If we temporarily accept the hypothesized population proportion as 0.40, we can find the sampling distribution for a particular sample of 10 by examining a binomial probability distribution. See the section with $n = 10$ and $\pi = 0.4$ in Appendix A. There you find the following potential outcomes and their associated probabilities of occurrence:

0	0.006
1	0.040
2	0.120
⋮	
6	0.111
7	0.042
8	0.011
9	0.002

For a significance level of 0.05 and a two-tailed test, we put a probability of 0.025 in each of the two extremes, or tails, of the distribution. Because the possible outcomes are discrete, we can only approximate the 0.05 value. We would like to have a probability near 0.025 in each tail. The graph below shows that on the low end of the scale, the probability of 0 is only 0.006, a very unlikely occurrence. However, the probability of 1 is 0.04, which exceeds 0.025. So, the lower rejection region would contain only the outcome 0. At the high end of the scale, the outcome 10 has virtually 0 probability. The outcome 9 has a probability of only 0.002, and 8 has probability 0.011. These three have a probability of occurrence of 0.013. The outcome 7, on the other hand, has a probability of 0.042, which is too large to be included in the rejection region. Thus, the upper tail of the rejection would be composed of outcomes 8, 9, and 10. The decision rule for this test would be: If the number of occurrences in the sample data is not between 1 and 7, inclusive, then we should reject the null hypothesis. Finally, applying this rule to our data in which three strikes were avoided, we see that we *cannot* reject the null hypothesis. It is reasonable to claim that 40% of the contracts will be settled without a strike.

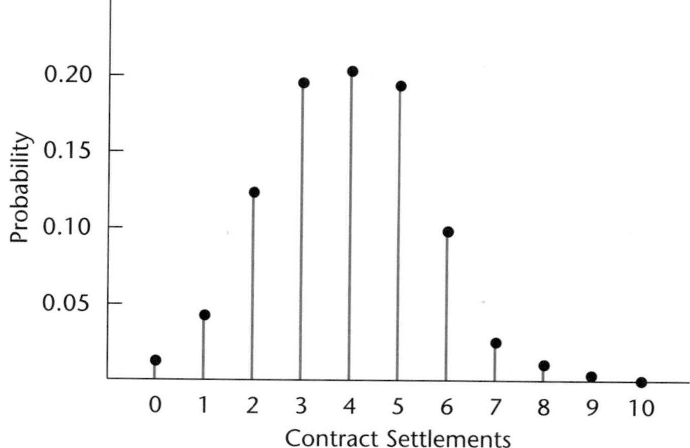

If the null hypothesis is changed to claim that 80% of the contracts are settled without a strike, the sampling distribution for our test statistic is also changed. In this case, consult the binomial table with $n = 10$ and $\pi = 0.8$. You find, in part:

3	0.001
4	0.006
5	0.026
⋮	
10	0.107

In this situation, the cumulative probabilities for 0, 1, 2, 3, and 4 "successes" is 0.007, a very small value. All of these values are included in the rejection region. At the other end of the distribution, the largest possible value is 10. Its probability is 0.107. This is too large to be included in the rejection region. So, there is no rejection area on the high end of the scale! For this hypothesis test, the decision rule is that if 4 or fewer occurrences are seen, we should reject the null hypothesis.

We observed that only 3 contract negotiations were settled without a strike. Thus, we are led to *reject* the hypothesis that the population proportion is 0.80. This claim is not substantiated by the data.

To create a confidence interval of approximately 95% for the population proportion, examine a binomial table with the parameter $n = 10$ for the whole range of possible values of π. Those values of π that have probabilities for the event (3 or more occurrences) of less than 0.05, or those that have probabilities for the event (3 or fewer occurrences) of less than 0.05, are considered unbelievable values of π.

Look at the section of the table with $n = 10$ and $\pi = 0.05$. What is the probability of 3 or more "successes" in this case? It is 0.011. This is too small to be credible. Therefore, we would not include this value in our confidence interval.

Next, examine the section with $n = 10$ and $\pi = 0.1$. Here the probability of 3 more "successes" is 0.069. This number is large enough so that we cannot ignore it. From this, it is possible that the population proportion is 0.1. This value would be included in our confidence interval.

Looking at larger values of π, we see that for $\pi = 0.6$ the probability of 3 or fewer "successes" is 0.055, a number that is too large to ignore. On the other hand, for $\pi = 0.7$ the corresponding probability is 0.010, which we consider insignificant. As a result, we would include 0.6 in our confidence interval, but exclude 0.7.

Based on our data of 3 "successes" in 10 trials, a confidence interval of approximately 95% for the proportion of contracts settled without a strike would range from 0.1 up to 0.6.

CHAPTER 11

Hypothesis Tests: Small-Sample Methods

OBJECTIVES

When you have completed this chapter, you will be able to

- describe the t distribution;
- conduct a hypothesis test for a population mean when the sample size is less than 30 and the population standard deviation is unknown;
- conduct a hypothesis test for the difference between two population means when either sample is less than 30 and the population standard deviations are unknown;
- conduct a hypothesis test for the difference between paired observations when the sample size is less than 30.

CHAPTER PROBLEM "We'll mufflerize in 30 minutes or less!"

TIDY MUFFLER CLAIMS ON THEIR NATIONAL RADIO AND TV COMMERCIALS THAT THEY CAN "MUFFLERIZE" ANY CAR IN 30 MINUTES OR LESS. THE CONSUMER PROTECTION AGENCY WOULD LIKE TO INVESTIGATE THE VALIDITY OF THIS CLAIM. IS THE MEAN TIME IT TAKES TIDY TO INSTALL A NEW MUFFLER MORE THAN 30 MINUTES?

INTRODUCTION

In Chapter 10, hypothesis tests for normally distributed interval-level data were considered. The population standard deviation σ either was known or was estimated from a "large" sample—generally considered by most statisticians to be of size 30 or larger. In many situations, however, there is interest in testing a hypothesis about a normally distributed interval-level variable, but the population standard deviation is not known and the sample size is small—that is, under 30. In such cases, the *z*-statistic is not appropriate because σ is unknown or cannot be accurately estimated based on only a few observations. In its place, **Student's *t* distribution** is used. This distribution first appeared in the literature in 1908 when W. S. Gosset, an Irish brewery employee, published a paper about the distribution under the pseudonym "Student." In his paper, Gosset assumed that the samples were taken from normal populations. Approximate results are also obtained in practice when we sample from nonnormal populations. Let's look first at the major characteristics of the *t* distribution.

CHARACTERISTICS OF THE *t* DISTRIBUTION

The *t* distribution is similar to the *z* (standard normal) distribution in some respects, but is quite different in others. It has the following major characteristics:

1. It is a continuous distribution like the standard normal or *z* distribution used in Chapter 10.
2. It is bell-shaped and symmetrical, again similar to the standard normal distribution. The mean of both *z* and *t* is zero.
3. There is only one standard normal distribution *z*, but there is a "family" of *t* distributions. That is, each time the size of the sample changes, a new *t* distribution is created.
4. The *t* distribution is more spread out at the center (that is, "flatter") than the normal distribution.

Because the *t* distribution has a greater spread than the *z* distribution, the critical values of *t* for a particular significance level will be greater in magnitude than the corresponding critical values of *z*. As the size of the sample increases, the value of *t* approaches the value of *z* for a particular significance level. For example, the critical values of a one-tailed test with a significance level of 0.05 are shown in Figure 11–1. The sample sizes of 3, 15, and the *z*-value corresponding to a one-tailed test are shown. Note that the *t*-value for a sample size of 3 is 2.920. For a sample of 15, the *t*-value is 1.761. Because the first sample size is smaller than the second, its critical value is larger.

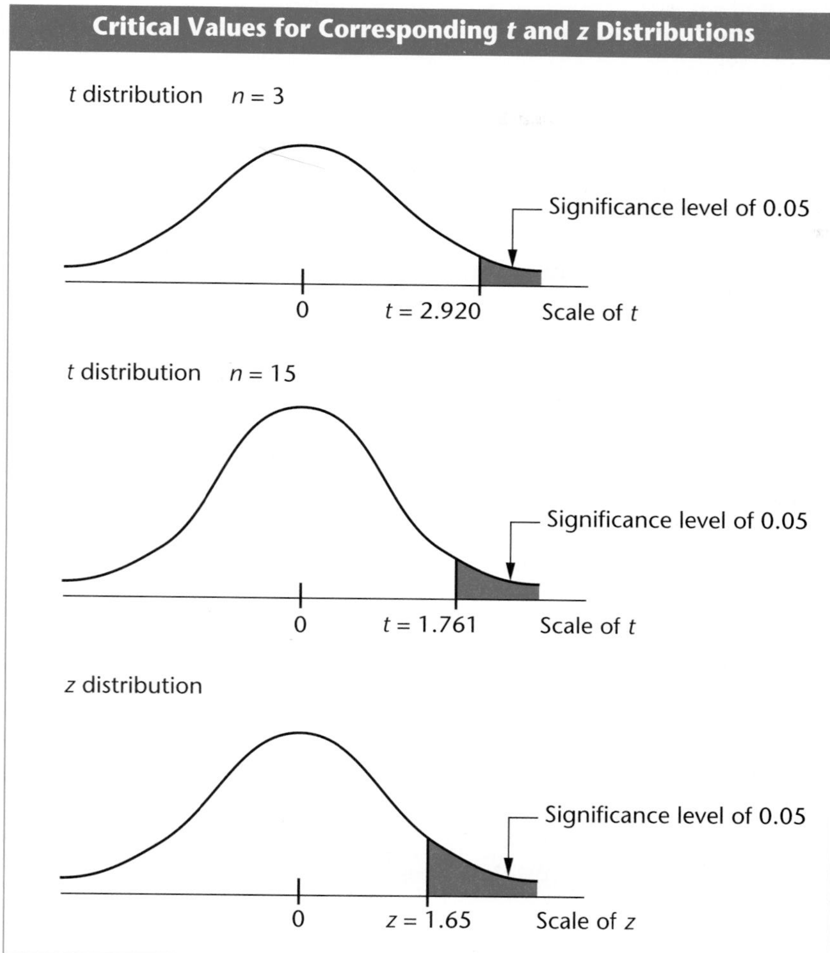

Figure 11-1

TESTING A HYPOTHESIS ABOUT A POPULATION MEAN

The t distribution is used to test a hypothesis about a population mean when the population standard deviation is not known and the sample size is small. The test statistic is

$$t = \frac{\overline{X} - \mu}{s / \sqrt{n}} \quad \text{(11-1)}$$

STAT BYTE

Do you live to work or work to live? A recent poll of 802 working Americans revealed that among those who considered their work a career, the mean number of hours worked per day was 8.7 and among those who considered work a job, the mean number of hours worked per day was 7.6.

where

$\overline{X}$ is the sample mean.
μ is the hypothesized population mean.
s is the sample standard deviation.
n is the number of items in the sample.

Note that in the formula for t the sample standard deviation (s) is used instead of the population standard deviation (σ).

Problem Recall from the Chapter Problem that Tidy Muffler Shops claim they can "mufflerize" any car in 30 minutes or less. The Consumer Protection Agency decides to investigate this claim. Ten of its fleet of cars, all unmarked and in need of a new muffler, are sent to various Tidy Muffler shops throughout a large city. The time it took, in minutes, to replace the muffler on each of the 10 vehicles was:

26 32 24 37 28 29 33 31 34 36

Does this evidence indicate that the mean time for Tidy to replace a muffler is more than 30 minutes, or are they meeting their advertised claim?

Solution Use the five-step hypothesis-testing procedure.

STEP 1. As usual, the null and the alternate hypotheses are formulated. The null hypothesis is that the mean time to "mufflerize" a car is 30 minutes or less. The alternate hypothesis is that the mean time is more than 30 minutes. Stated symbolically:

$$H_0 : \mu \le 30$$
$$H_a : \mu > 30$$

The way the alternate hypothesis is stated dictates a one-tailed test.

STEP 2. Select the level of significance. It is 0.05.

STEP 3. Choose the appropriate test statistic. The t distribution is used because the population standard deviation is not known and the sample size is small (10). The value of the test statistic is computed by using formula 11–1.

$$t = \frac{\overline{X} - \mu}{s / \sqrt{n}}$$

STEP 4. State the decision rule. A critical value separates the region where the null hypothesis is rejected from the region where it is not rejected. Appendix D gives the critical values for Student's t distribution. To use the table, we must determine the number of *de-*

grees of freedom (df). The number of degrees of freedom is equal to the sample size minus the number of samples. One sample of size 10 is used in this problem. Thus, the degrees of freedom is $n - 1 = 10 - 1 = 9$. Why is 1 subtracted from the sample size? Suppose a sample of four measurements is obtained and the values are 4, 6, 8, and 10. The mean of the sample is 7. If any three of the values are changed, then the fourth value is automatically fixed so that the mean is still 7. Suppose that the first three values were changed to 3, 5, and 9. The last number must be 11 for the four numbers to have a mean of 7—that is, $(3 + 5 + 9 + 11)/4 = 28/4 = 7$. Thus, only three out of the four numbers are free to vary, and so it is said that there are three "degrees of freedom." For this example, the degrees of freedom could be determined by $n - 1 = 4 - 1 = 3$ df.

The critical values for Student's t distribution are given in Appendix D. Table 11–1 excerpts a few of these critical values. Note that both two-tailed and one-tailed values are given. To locate the

Table 11-1

Portion of the *t* Distribution Table

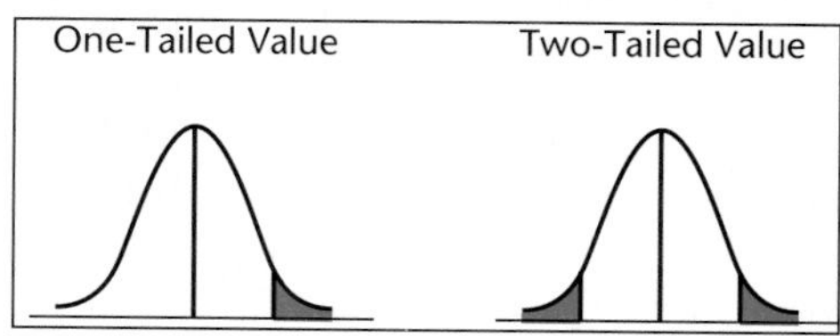

Appendix D
Student's *t* Distribution

DEGREES OF FREEDOM (df)	ONE-TAILED VALUE					
	0.25	0.10	0.05	0.025	0.01	0.005
	TWO-TAILED VALUE					
	0.50	0.20	0.10	0.05	0.02	0.01
1	1.000	3.078	6.314	12.706	31.821	63.657
2	0.816	1.886	2.920	4.303	6.965	9.925
3	0.765	1.638	2.353	3.182	4.541	5.841
4	0.741	1.533	2.132	2.776	3.747	4.604
5	0.727	1.476	2.015	2.571	3.365	4.032
6	0.718	1.440	1.943	2.447	3.143	3.707
7	0.711	1.415	1.895	2.365	2.998	3.499
8	0.706	1.397	1.860	2.306	2.896	3.355
9	0.703	1.383	1.833	2.262	2.821	3.250
10	0.700	1.372	1.812	2.228	2.764	3.169

critical value of t, go down the left margin in Table 11–1, labelled "degrees of freedom" (df), until you locate $n - 1$ degrees of freedom. There are 10 cars being "mufflerized" in the sample, so $10 - 1 = 9$ degrees of freedom. Then move to the right and read the value given under the "One-Tailed Value" column headed by 0.05. The critical value for 9 df and a one-tailed test is 1.833 for the 0.05 level of significance. Therefore, the decision rule is: Reject H_0 if the computed value of t is greater than 1.833. Shown schematically:

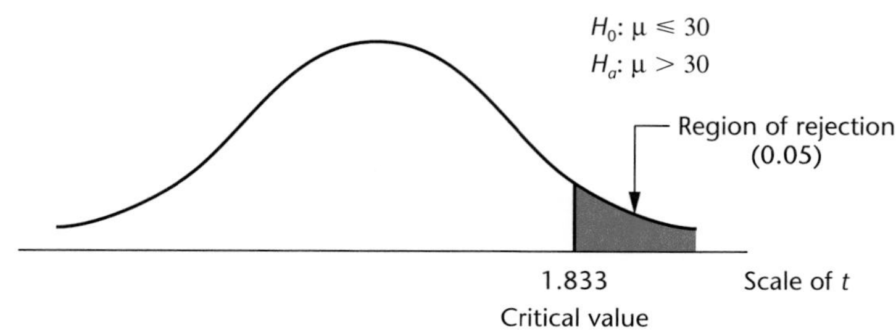

STEP 5. Make the statistical decision. The mean length of time it took Tidy Muffler to replace the mufflers on the Consumer Protection Agency cars is 31 minutes, found by 310/10. The standard deviation of the sample of 10 is 4.24 minutes, found by applying formula 4–8:

$$s = \sqrt{\frac{\Sigma X^2 - \frac{(\Sigma X)^2}{n}}{n-1}}$$

The essential calculations are shown in Table 11–2.

The computed value of t is 0.746, found by using formula 11–1:

$$t = \frac{\overline{X} - \mu}{s/\sqrt{n}} = \frac{31 - 30}{4.24/\sqrt{10}} = 0.746$$

Comparing the computed t-value of 0.746 with the critical value of 1.833, we fail to reject the null hypothesis. Although there is a difference between the hypothesized mean (30) and the sample mean (31), it is attributed to sampling error. The Consumer Protection Agency does not have sufficient statistical evidence to disprove Tidy Muffler's claim that they are able to "mufflerize" a car in 30 minutes on the average.

STAT BYTE

The typical American worker drinks an average of 2.25 cups of coffee and spends an average of 21 minutes getting to work each day. Wouldn't it be nice if journalists who report this kind of information could tell us something about the variability of the data—like maybe the standard deviation!

Table 11-2

Essential Calculations for the Sample Standard Deviation

Time (in minutes) X	X^2
26	676
32	1,024
24	576
37	1,369
28	784
29	841
33	1,089
31	961
34	1,156
36	1,296
310	9,772

$$s = \sqrt{\frac{\Sigma X^2 - \frac{(\Sigma X)^2}{n}}{n-1}} = \sqrt{\frac{9772 - \frac{(310)^2}{10}}{10-1}} = 4.24$$

In the above Problem, we used the five-step hypothesis-testing procedure to illustrate the t distribution. We can also use the probability, or p-value, approach to do this test. Recall that the p-value is the probability of a value of the test statistic as large or larger than the computed value of t, when the null hypothesis is true. Computer programs such as MINITAB can determine exact p-values. However, the t distribution in Appendix D and Table 11–1 includes only selected critical values, so we usually cannot find the exact p-value. However, we can find limits on p-values by scanning the appropriate row of the table based on the degrees of freedom and comparing the computed value of the test statistic with the critical values found in the table.

In the case where $t = 0.746$, we used a one-tailed test with 9 degrees of freedom. Look across the row with 9 degrees of freedom. The value of the test statistic is between 0.703 and 1.383, which corresponds to one-tailed significance levels of 0.25 and 0.10, respectively. So the p-value is more than 0.10 but less than 0.25. The exact p-value is 0.24, calculated by MINITAB (see p. 387). Thus, by using Appendix D we can get a reasonable estimate of the exact p-value.

Self-Review 11-1

Answers to the Self-Review problems are at the end of the chapter.

The American Association of Retired Persons reported that typical (average) senior citizens living at high altitudes claim their systolic blood pressure is lower than the arithmetic mean of 160. To test this claim, 16 senior citizens living at high altitudes were selected at random and their blood pressures checked. Their mean systolic pressure was 151; the standard deviation of the sample was 12. Do senior citizens living at high altitudes have significantly *lower* systolic blood pressure? Use the 0.05 level.

a. State the null hypothesis and the alternate hypothesis symbolically.
b. Give the formula for the test statistic.
c. How many degrees of freedom are there?
d. Give the decision rule.
e. Compute *t*.
f. Arrive at a conclusion.
g. What is the *p*-value? Interpret it.

Exercises

Answers to the even-numbered exercises are at the back of the book.

1. Find the critical value of *t* for the following hypothesis-testing situation when the sample size is 20. Use the 0.05 significance level.

$$H_0 : \mu \leq 20$$
$$H_a : \mu > 20$$

2. Find the critical value of *t* for the following hypothesis-testing situation when the sample size is 15. Use the 0.01 significance level.

$$H_0 : \mu = 50$$
$$H_a : \mu \neq 50$$

3. A car manufacturer asserts that with the new collapsible bumper system, the mean body repair cost for the damage sustained in a collision impact of 15 miles per hour does not exceed $400. To test the validity of this claim, six cars are crashed into a barrier at 15 miles per hour and their repair costs recorded. The mean and the standard deviation are found to be $458 and $48, respectively. At the 0.05 level of significance, do the test data contradict the manufacturer's claim that the repair cost does not exceed $400? Find the *p*-value. Interpret it.

4. It is claimed that a new treatment for prolonging the lives of cancer patients is more effective than the standard one. Records of earlier research show the mean survival period to have been 4.3 years with the standard treatment. The new treatment is administered to a sample of 20 patients and the duration of their survival is recorded. The sample mean is 4.6 years, and the standard deviation is 1.2 years. Is the claimed effectiveness of the new method supported at the 1% level of significance?

5. A manufacturer of strapping tape claims that the tape has a mean breaking strength of 500 pounds per square inch (psi). A random sample of 16 specimens is drawn from a large shipment of tape, and a mean of 480 psi is computed with the sample standard deviation of 50 psi. Can we conclude from these data that the mean breaking strength for this shipment is less than that claimed by the manufacturer? Use the 0.05 significance level.
6. An anthropologist is studying the heights of the adult members of an ancient population. The conventional theory is that the mean height of adult men from the population is 56 inches. A sample of 12 men showed a mean of 53.5 inches, with a standard deviation of 2.0 inches. Can the anthropologist conclude that the mean height is less than 56 inches? Use the 0.05 significance level. What is the *p*-value? Interpret it.

MINITAB can be used to perform the calculations in a hypothesis test. The first step is to enter the data. In this example, the data are entered in column C10, but of course any column can be used. The procedure TTEST is used with MU = 30, which indicates the population value being tested. A ";" at the end of the line means a subcommand is to be employed, in this case, "ALTERNATE = 1." The ALTERNATE allows for one-tailed tests. A "1" is used for a one-tailed test in the positive direction and a " –1" for a one-tailed test in the negative direction. The period must appear at the end of the subcommand.

```
MTB  >   set c10
DATA >   26,32,24,37,28,29,33,31,34,36
DATA >   end
MTB  >   ttest mu=30 c10;
SUBC >   alternate=1.

         TEST OF MU = 30.000 VS MU G.T. 30.000

          N      MEAN     STDEV    SE MEAN      T    P VALUE
C10      10    31.000     4.243      1.342   0.75       0.24
```

The t statistic of 0.75 is the same as calculated earlier. Included on the output is the p-value. The value of 0.24 indicates the probability of t greater than 0.75, with 9 degrees of freedom. The p-value is larger than the significance level of 0.05, so the null hypothesis is not rejected.

Self-Review 11-2

A new toy has been developed, and its manufacturer hopes to market it for the coming Christmas season. Before going into full production, the company handcrafts a large number of toys and sends them to five test market areas. The manufacturer plans to start full production if the monthly sales in the test markets average more than \$20,000 during the one-month trial period. The results, in thousands of dollars, were \$20, \$16, \$25, \$19, and \$24. For a 0.05 level of significance, can it be shown that the mean monthly sales volume is greater than \$20,000?

a. State the null and alternate hypotheses.
b. Is this a one-tailed or a two-tailed test?
c. Why is Student's t being used?
d. Calculate the value of the test statistic.
e. What is your decision regarding the null hypothesis?
f. Interpret your results for the toy manufacturer.
g. What is the p-value? Interpret it.

Exercises

7. A city health department wishes to determine if the mean bacteria count per unit-volume of water at Siesta Lake Beach is below the safety level of 200. Researchers have collected 10 water samples and have found the bacteria counts per unit-volume to be 175, 180, 215, 188, 194, 207, 211, 195, 198, and 190. Do the data warrant cause for concern? Use the 0.10 level of significance.

8. Sorenson Pharmaceutical has been conducting restricted studies on small groups of people to determine the effectiveness of a measles vaccine. The following measurements are readings on the antibody strength for five individuals injected with the vaccine: 1.2, 2.5, 1.9, 3.0, and 2.4. Use the sample data to test at the 0.01 level of significance the hypothesis that the mean antibody strength for individuals vaccinated with the new drug is more than 1.6.

9. In five test runs, a truck operated 8, 10, 7, 9, and 10 miles with one gallon of gasoline. At the 0.01 significance level, is this sufficient evidence to show that the truck is operating at a mean rate less than 11.5 miles per gallon? Find the p-value.

10. The credit manager of a discount store chain believes the mean age of the stores' customers is less than 40 years. A random sample of eight customers revealed the following ages: 38, 46, 30, 35, 29, 40, 40, and 46. At the 0.05 significance level, can we conclude that the mean age of customers is less than 40 years?

COMPARING TWO POPULATION MEANS

A t-test can be used to compare two population means to determine if the samples were obtained from the same or equal populations. We assume that

1. each population is normally distributed;
2. their population standard deviations are equal but unknown;
3. the two samples are unrelated (independent);

The t statistic for the two-sample case is similar to that employed for the two-sample z statistic in Chapter 10. An additional calculation is required. Two computed sample standard deviations are pooled to form a single estimate of the population standard deviation. This test is normally employed if either sample has fewer than 30 observations. The formula for t is

STAT BYTE

Supplying soldiers with enough water was a major problem during Operation *Desert Storm.* The typical American consumes an average of eight 8-oz. glasses of water per day. The typical soldier in *Desert Storm* consumed an average of 96 8-oz. glasses of water. Is the increase in water consumption significant?

$$t = \frac{\overline{X}_1 - \overline{X}_2}{s_p\sqrt{\dfrac{1}{n_1} + \dfrac{1}{n_2}}} \qquad \textbf{11-2}$$

where

$\overline{X}_1$ is the mean of the first sample.
$\overline{X}_2$ is the mean of the second sample.
n_1 is the number in the first sample.
n_2 is the number in the second sample.
s_p is a pooled estimate of the population standard deviation. Its formula is

$$s_p = \sqrt{\frac{(n_1 - 1)s_1^2 + (n_2 - 1)s_2^2}{n_1 + n_2 - 2}} \qquad \textbf{11-3}$$

where

s_1 is the standard deviation of the first sample.
s_2 is the standard deviation of the second sample.

Recall that the number of degrees of freedom is equal to the total number of items sampled minus the number of samples. The sample sizes for this problem are n_1 and n_2, and there are two samples. Hence, there are $n_1 + n_2 - 2$ degrees of freedom appearing in the denominator of the pooled standard deviation.

Problem We are studying the third-grade class of 11 children at the Toth Elementary School to determine if two different teaching methods make a difference in student comprehension. We randomly assign the 11 children to two groups. Then they are taught the basic concepts of multiplication by the same teacher, but she uses two different teaching methods. Finally, after the first week of instruction, we administer the same 10-question examination to both groups. The numbers of correct answers out of 10 are listed in Table 11–3. Is there a significant difference in the performance under the two teaching methods? Use the 0.01 significance level.

Solution The null hypothesis is that there is no difference in the mean scores of the two groups. The alternate hypothesis is that there is a difference in the mean scores. Stated symbolically:

$$H_0 : \mu_1 = \mu_2$$
$$H_a : \mu_1 \neq \mu_2$$

The alternate hypothesis indicates that this is a two-tailed test.

Again, the decision rule depends on the number of degrees of freedom. In this case, that number is equal to the combined number of observations in the two samples minus the number of samples. This is expressed as $n_1 + n_2 - 2$. For this problem, $n_1 + n_2 - 2 = 5 + 6 - 2 = 9$ degrees of freedom.

The critical value for the 0.01-level, two-tailed test is 3.250 (from Appendix D). The decision rule is that we reject H_0 if $t < -3.250$ or $t > 3.250$. Shown schematically:

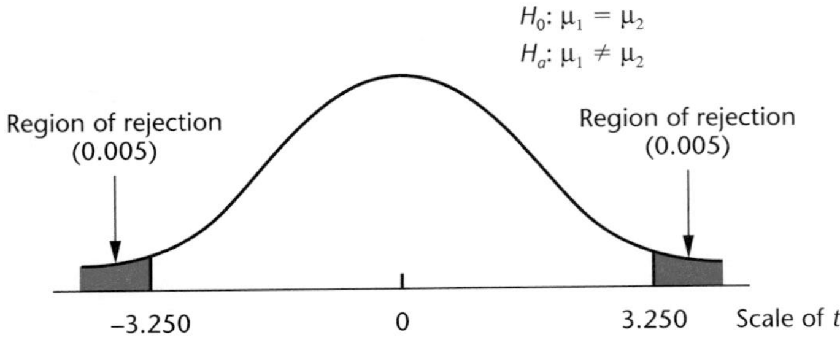

The calculation of Student's t can be accomplished in three stages. First, calculate the standard deviation of each sample. Second, "pool" these standard deviations into a single estimate of the

Table 11-3

Number Correct on a 10-Question Examination			
Teaching Method I	**Score**	**Teaching Method II**	**Score**
Sally	4	Olga	9
James	3	Orville	6
Abdul	5	Peter	8
Jackson	7	Rachel	4
Noreen	6	Susan	7
		Andrew	5

population standard deviation using formula 11–3. Third, calculate t by using formula 11–2.

STAGE 1. Calculate the sample standard deviations using these sums and squares.

Method 1		**Method 2**	
X_1	X_1^2	X_2	X_2^2
4	16	9	81
3	9	6	36
5	25	8	64
7	49	4	16
6	36	7	49
		5	25
25	135	39	271

$$s_1 = \sqrt{\frac{\Sigma X_1^2 - \frac{(\Sigma X_1)^2}{n_1}}{n_1 - 1}} = \sqrt{\frac{135 - \frac{(25)^2}{5}}{5-1}} = 1.58 \qquad s_2 = \sqrt{\frac{\Sigma X_2^2 - \frac{(\Sigma X_2)^2}{n_2}}{n_2 - 1}} = \sqrt{\frac{271 - \frac{(39)^2}{6}}{6-1}} = 1.87$$

STAGE 2. Pool the standard deviations by using formula 11–3:

$$s_p = \sqrt{\frac{(n_1 - 1)s_1^2 + (n_2 - 1)s_2^2}{n_1 + n_2 - 2}}$$
$$= \sqrt{\frac{(5-1)(1.58)^2 + (6-1)(1.87)^2}{5+6-2}}$$
$$= \sqrt{3.05}$$
$$= 1.746$$

STAGE 3. Calculate t using formula 11–2:

$$t = \frac{\overline{X}_1 - \overline{X}_2}{s_p\sqrt{\frac{1}{n_1} + \frac{1}{n_2}}} = \frac{5.0 - 6.5}{1.746\sqrt{\frac{1}{5} + \frac{1}{6}}} = -1.42$$

Because –1.42 falls between –3.250 and 3.250, the null hypothesis cannot be rejected. We conclude that there is no significant difference in the mean scores of the children taught under the two methods. The difference in comprehension evidenced in the two samples could be attributed to sampling error.

To find the p-value for this problem, refer to the row with 9 degrees of freedom in Appendix D or Table 11–1. The test statistic 1.42 is greater than 1.383, but smaller than 1.833. For a two-sided test, those values have significance levels of 0.20 and 0.10, so the p-value is between 0.10 and 0.20.

A Computer Example The MINITAB output for the Toth Elementary School example software is shown below. The procedure is called TWOSAMPLE, and a subcommand POOLED is required.

```
MTB  >  set c1
DATA >  4,3,5,7,6
DATA >  end
MTB  >  set c2
DATA >  9,6,8,4,7,5
DATA >  end
MTB  >  twosample c1 c2;
SUBC >  pooled.

TWOSAMPLE T FOR C1 VS C2
        N      MEAN     STDEV     SE MEAN
C1      5      5.00      1.58       0.71
C2      6      6.50      1.87       0.76
```

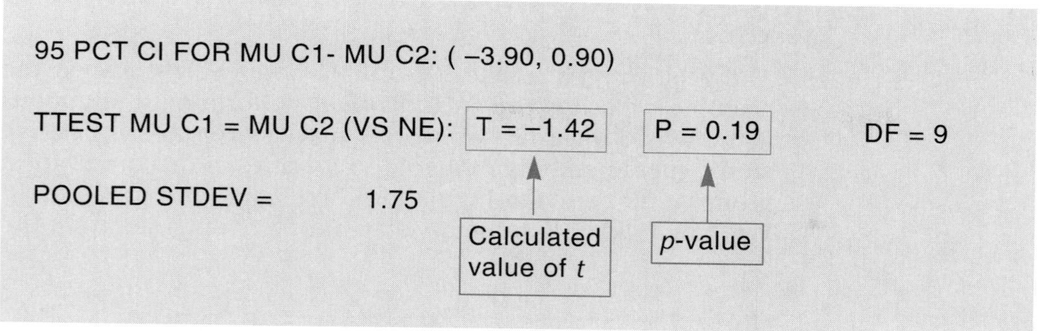

Self-Review 11-3

An orthodontist wants to investigate the effectiveness of the interceptive treatment she prescribes for some of her patients. (Interceptive treatment is dental work performed on relatively young patients in hopes of forestalling more extensive treatment later.) She compares the length of time a sample of interceptive patients must wear braces with a random sample of noninterceptive patients. The results are (time is in months):

Time Wearing Braces

Interceptive		Noninterceptive	
Joseph	12	Sally	16
Karen	13	George	22
Pam	11	Enrico	14
Peter	12	Aldine	18
Nickie	14	Jenny	19
Rosa	9		
Kurt	11		

Is there sufficient evidence to indicate that interceptive patients spend less time in braces? Use the 0.05 significance level.

a. State the null and alternate hypotheses. Use the subscript $_1$ to refer to the interceptive group.
b. How many degrees of freedom are there in this problem?
c. Calculate the two sample standard deviations.
d. Compute the pooled estimate of the standard deviation.
e. What is the decision rule?
f. Calculate the value of the test statistic.
g. What is your decision regarding the null hypothesis?
h. Interpret the results.

Exercises

11. The peak oxygen intake per unit of body weight, called the "aerobic capacity," of an individual performing a strenuous activity is a measure of work capacity. For a comparative study, measurements of aerobic capacities are recorded for a group of 20 Peruvian highland natives and for a group of 10 U.S. The following summary statistics were obtained from the data:

	Peruvians	Americans
Mean	46.3	38.5
Standard deviation	5.0	5.8

At the 0.05 level of significance, does the sample data indicate a significant difference in the mean aerobic capacity?

12. To compare the effectiveness of isometric and isotonic exercise methods of abdomen reduction, 20 overweight business executives are included in an experiment. Ten use each type of exercise, and after 10 weeks the reductions in abdomen measurements are recorded in centimeters:

	Isometric	Isotonic
Mean	2.5	3.1
Standard deviation	0.8	1.0

At the 0.01 level of significance, do these data support the claim that the isotonic method is more effective? Find the p-value.

13. Do cars traveling on the right lane of I-75 travel slower than those in the left lane? The following sample information was obtained. Data are reported in miles per hour.

Right Lane	Left Lane
65	69
70	72
60	68
62	73
68	64
	68

Use the 0.05 significance level.

14. The budget director of a large firm compared the luncheon expense vouchers for executives in the sales department with those for executives in the production department. The following sample information was gathered (rounded to the nearest dollar):

Sales:	\$17, 23, 26, 30
Production:	\$19, 15, 14, 17, 20

At the 0.05 significance level, can we conclude that luncheon expenses are less for executives in the production department?

A TEST OF PAIRED OBSERVATIONS

In some experiments, the investigator is concerned with the *difference* in a pair of related observations. For example, persons enrolled in a physical fitness class are weighed both before the course starts and after it is completed. The purpose of the experiment is to examine the effectiveness of the fitness program. Therefore, the t test focuses on the weight loss of each person and not on the means of the two populations. In such cases, the test is based on the difference in each pair of observations, instead of on the value of the individual observations. The distribution of this population of differences is assumed to be normal, with an unknown standard deviation. The mean of this population of differences is designated μ_d. As pointed out earlier, it is often impossible to study the entire population of differences. Therefore, a sample is selected. The symbol d is used to designate a particular observed difference and $\bar{d}$ the mean difference. The formula for t is

$$t = \frac{\bar{d}}{s_d / \sqrt{n}} \qquad \textbf{11-4}$$

where

$\bar{d}$ is the mean difference between the paired observations.

s_d is the standard deviation of the difference between the paired observations.

n is the number of paired observations.

Problem ETM Chemical Company is considering the purchase of a new Colton word processor. Colton claims that their new-model word processor increases typing efficiency. Ten typists randomly selected from the typing pool are assigned to test the new model. The results:

Typist	Proposed Model	Current Model
Woodstock	61	55
LaRoche	60	54
Hayes	56	47
Kootz	63	59
Kinney	56	51
Markas	63	61
Chung	59	57
Jaynes	56	54
Shue	62	63
Keller	61	58

Using a 0.01 level of significance, what would you conclude?

Solution Before making the change to the new model, the company would like to know that the typists can type *faster* on the proposed model. The difference d is $X - Y$, where X is the speed on the proposed model and Y the speed on the current model. The null hypothesis is that the typists will perform at least as well on the currently used model. Colton hopes the null hypothesis will be rejected. Thus, a one-tailed test is necessary. The null and alternate hypotheses are:

$$H_0 : \mu_d \leq 0$$
$$H_a : \mu_d > 0$$

To reject H_0 and accept H_a will indicate that the difference is not zero. The sample size n is equal to the number of paired observations (10). In this example, there are 9 (10 – 1) degrees of freedom. The decision rule is to reject H_0 if the computed value of t is to the right of 2.821 (see Appendix D).

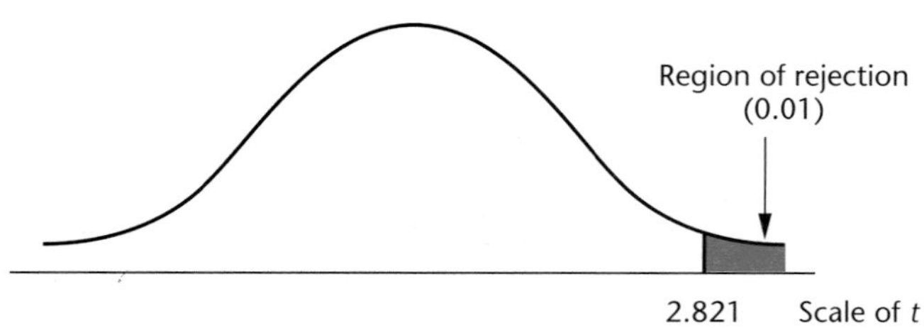

The sample data and the calculations are given in Table 11–4, followed by the computations. The mean difference is computed by formula 3–1, substituting d for X:

Table 11-4

Comparative Results of Efficiency

Typist	Proposed Model X	Current Model Y	$X - Y = d$	d^2
Woodstock	61	55	6	36
LaRoche	60	54	6	36
Hayes	56	47	9	81
Kootz	63	59	4	16
Kinney	56	51	5	25
Markas	63	61	2	4
Chung	59	57	2	4
Jaynes	56	54	2	4
Shue	62	63	–1	1
Keller	61	58	3	9
			38	216

$$\bar{d} = \frac{\Sigma d}{n} = \frac{38}{10} = 3.8$$

The $\bar{d}$ of 3.8 indicates that the average typist increased his or her speed by 3.8 words per minute. The sample standard deviation is found by using formula 4–8, repeated below. Inserting the sums from Table 11–4, the standard deviation of the differences is 2.82 minutes:

$$s_d = \sqrt{\frac{\Sigma d^2 - \frac{(\Sigma d)^2}{n}}{n-1}} = \sqrt{\frac{216 - \frac{(38)^2}{10}}{10-1}} = 2.82$$

The value of the test statistic t is 4.26, found by using formula 11–4.

$$t = \frac{\bar{d}}{s_d / \sqrt{n}} = \frac{3.8}{2.82/\sqrt{10}} = 4.26$$

Because 4.26 is to the right of 2.821, the null hypothesis is rejected. Since 4.26 > 3.25, the p-value is less than 0.005! The alternate hypothesis is accepted ($\mu_d > 0$). Therefore, the conclusion is that the typists perform more effectively on the proposed model.

Note that in Table 11–4, X represents the proposed model and Y the current model. The difference d is $X - Y$, where X is expected to be larger than Y. This leads to the positive direction of the alternate hypothesis. (The entire test could have been reversed if d were found by $Y - X$. In this case, the d values would have been negative and the direction of the alternate hypothesis would have been negative.)

Self-Review 11-4

Advertisements for the Sylph Physical Fitness Center claim that completion of their course will result in a loss of weight. A random sample of recent students revealed the following body weights before and after completion of the course. At the 0.01 level, can it be concluded that the course will result in a significant weight loss?

Name	Before	After
Wellman	155	154
Gersten	228	207
Tamayo	141	147
Miller	162	157
Ringman	211	196
Garbe	185	180
Monk	164	150
Heilbrunn	172	165

a. State the null and alternate hypotheses.
b. What is the critical value of t?
c. Calculate the value of the test statistic.
d. What is the decision regarding the null hypothesis?
e. Interpret the results.

Exercises

15. Two methods of memorizing difficult material are being tested. Nine pairs of students are matched according to IQ and background and then assigned to one of the two methods at random. A test is finally given to all the students, with the following results:

	Pair								
	1	2	3	4	5	6	7	8	9
Method A	90	86	72	65	44	52	46	38	43
Method B	85	87	70	62	44	53	42	35	46

Using the 0.05 level of significance, test to determine if there is a difference in the effectiveness of the two methods. What is the p-value?

16. Measurements of the left-handed and right-handed gripping strengths of 10 left-handed persons are recorded:

	Person									
	1	2	3	4	5	6	7	8	9	10
Left hand	140	90	125	130	95	121	85	97	131	110
Right hand	138	87	110	132	96	120	86	90	129	100

Do these data provide evidence, at the 0.01 level of significance, that those tested have greater gripping strength in their dominant hand?

17. A manufacturer of shock absorbers is comparing the durability of his model with those of his competitors. To conduct the test, the manufacturer installed one of his and one of the competitor's shock absorbers on each of the nine cars selected at random. Each of the cars was driven 20,000 miles. Then, each shock was measured for strength. The results are shown below.

Car	Manufacturer's Shock	Competitor's Shock
1	10.0	9.6
2	11.7	11.2
3	13.7	13.1
4	9.9	9.4
5	9.8	9.0
6	14.4	14.0
7	15.1	14.6
8	10.6	10.3
9	9.8	9.4

At the 0.05 significance level, do the data provide evidence that the manufacturer's shocks last longer?

18. The Austin-Hall Department Store plans to renovate one of its floors, increasing either the men's or women's clothing department. The final decision will be based on mean sales, and the department that has the greater sales will be enlarged. The last 12 months of sales data (in thousands of dollars) are shown below.

Month	Women's Clothing	Men's Clothing
Jan.	$20.0	$22.5
Feb.	14.0	13.9
Mar.	29.0	24.8
Apr.	30.0	27.5
May	16.0	24.9

Month	Women's Clothing	Men's Clothing
June	17.0	27.4
July	20.1	21.4
Aug.	20.5	20.7
Sept.	21.3	20.6
Oct.	22.0	22.8
Nov.	25.7	23.5
Dec.	42.2	44.2

At the 0.05 significance level, can the department store conclude that there is a difference in the sales? Find the p-value.

SUMMARY

The hypothesis-testing methods in this chapter assume that the populations are normally distributed, that the population standard deviation is unknown, and that the sample size is small (under 30). Under these conditions, the appropriate test statistic is Student's t distribution. The t distribution is based on the number of degrees of freedom. (The number of degrees of freedom is equal to the sample size minus the number of samples.) The critical value of t can be obtained from Appendix D.

Three hypothesis-testing situations were considered:

1. comparing a single sample against the population mean;
2. comparing two sample means, where the samples are independent and have the same population standard deviation;
3. comparing the difference between two related pairs of observations.

In each situation, the usual five-step hypothesis-testing procedure is used.

Exercises

Use the five-step hypothesis-testing procedure in exercises 19–34.

19. The university library is interested in determining whether the mean number of books checked out per visit has increased. In the past, the mean was 3.0 books per student visit. A random sample of 10 students revealed a mean of 4.1 books, with a standard deviation of 2.0 books. At the 0.05 level of significance, does this information provide sufficient evidence to show that students are now checking out more books per visit?

20. A recent newspaper article claims that the typical American is 20 pounds overweight. To test this claim, 15 randomly selected persons were weighed. They were an average of 18 pounds overweight, with a standard deviation of 5 pounds. At the 0.05 significance level, is there sufficient evidence to reject the newspaper's claim?

21. An association of college textbook publishers recently reported the mean retail cost of its members' books to be $35.00. A group of students lobbying for increased state support to students because of higher education costs has challenged this claim. A random sample of 20 books is selected. Calculations indicate that the mean cost is $35.80, with a standard deviation of $3.80. At the 0.05 level of significance, is there sufficient evidence to reject the claim of the publishers' association? Can the students assert that the mean cost is higher? Find the p-value.

22. Your new car has an EPA rating of 26.0 miles per gallon. The mileage figures actually obtained on six trips were 24.3, 25.2, 24.9, 24.8, 25.6, and 25.4. Is there sufficient evidence, at the 0.01 level of significance, to conclude that the car performs below the EPA specifications?

23. A recent newspaper headline indicated that the typical residential home in a certain community is now selling for less than $100,000. A sample of 10 recent transactions revealed the following selling prices (in thousands of dollars): $96.0, $92.0, $95.0, $97.0, $103.0, $105.0, $94.0, $97.0, $96.0, and $95.0. At the 0.01 significance level, can we conclude that the mean selling price of these homes is less than $100,000?

24. The National Weather Service reports that the high temperature on July 1 is as likely to be above 25°C as below. Test this claim at the 0.10 significance level, given the following July 1 temperature readings over the past 16 years: 22, 26, 28, 24, 27, 20, 29, 32, 28, 21, 25, 27, 26, 28, 30, 22. What is the p-value?

25. A toothpaste manufacturer claims that children who brush their teeth with Bianca will have fewer cavities than those who brush with Sparkle. In a carefully supervised study, the number of cavities that occurred with each of the two brands is compared. Is there sufficient evidence, at the 0.01 level of significance, to support the manufacturer's claim?

	Cavities
Bianca	1, 2, 3, 4, 2, 0, 2
Sparkle	4, 5, 4, 2, 1, 2, 4

26. The following data are the weight gains, measured in pounds, of babies from birth to age six months. All babies in the sample weighed between seven and eight pounds at birth. One group of babies was breast-fed and the other was fed a specific formula. Is there evidence that the weight gains are different between the two groups? Use a 0.01 significance level.

Breast-Fed	Formula-Fed
7	9
8	10
6	8
10	6
9	7
8	8
9	

27. Do managers in the retail industry earn less than those in the auto industry? A sample of 15 retail managers shows a mean salary of $47,250, with a standard deviation of $5,500. A sample of 10 auto managers shows the mean salary to be $56,720, with a standard deviation of $6,200. Use the 0.05 significance level. What is the *p*-value?

28. A manufacturer of wrist supports for bowlers maintains that use of this support will improve a bowler's average. A sample of 12 bowlers who have a league-sanctioned average of over 150 roll two complete games, one with the wrist support and one without it. Does the evidence support the manufacturer's claim? Assume a significance level of 0.01.

Bowler	With Wrist Band	Without Wrist Band
Clarke	230	217
Redenback	225	198
Simmons	223	208
Pelton	216	222
Griffin	229	223
Farthy	201	214
Hawkins	205	187
Nugent	193	187
Bryan	177	178
Hucklebury	201	195
Baker	178	169
Berry	207	194

29. Six junior executives were sent to a class to improve their verbal skills. To test the quality of the program, the executives were tested before and after taking the class, with the following results:

	Before	After
Levin	18	30
Baker	38	70
Craft	8	20
Denfrey	10	4
Longhi	12	10
Foster	12	20

Do these records indicate a significant improvement in verbal skills at the 0.10 significance level?

30. To measure the effectiveness of his sales training program, a car dealer selects at random eight sales representatives to take the course. The weekly sales volume of each sales representative is shown below. At the 0.05 significance level, can it be concluded that the new program is effective in increasing sales? Find the p-value.

Sales Rep	Gross Sales after the Course (in thousands of dollars)	Gross Sales before the Course (in thousands of dollars)
1	$14.0	$13.5
2	10.7	11.4
3	12.4	10.7
4	11.1	11.1
5	10.9	9.8
6	10.5	9.6
7	10.8	10.7
8	13.0	11.7

31. Dhondt Doors manufactures garage doors. Mr. Dhondt is considering two sources for the springs used in the doors. The research department of Dhondt Doors conducted strength tests of the two springs, with the following results:

Sevits Manufacturing	Molna Industries
$n = 20$	$n = 32$
$\overline{X} = 66.0$ kg	$\overline{X} = 68.3$ kg
$s = 2.1$	$s = 1.4$

At the 0.05 significance level, is there a difference in the strength of the two springs?

32. Last year a mean of 6.00 new homes per month were started in the Perrysburg area. Because of good weather, increased demand for housing, and lower mortgage rates, the Perrysburg Builders Association claims that the number of housing starts has increased this year. The mean number of new homes started for the first six months of this year is 7.2, with a standard deviation of 1.05. Assuming the data from the first six months to be a sample, can we conclude that the mean number of housing starts has increased? Use the 0.01 significance level.

33. Kazimer's Supermarkets is considering locating a new store in either Whitehouse or Waterville. Before making a final decision, the company has hired a marketing research firm to determine the mean amount spent per family unit on groceries in the two communities.

Whitehouse	Waterville
$n = 12$	$n = 15$
$\overline{X} = \$132$	$\overline{X} = \$140$
$s = \$13$	$s = \$16$

At the 0.05 significance level, can Kazimer's conclude that there is a difference in the mean amount spent on groceries in the two communities? Find the p-value.

34. Do sellers generally receive the asking price when selling their home? The following prices (in thousands of dollars) were obtained from a sample of 12 homes recently sold in the Pittsburgh, Pennsylvania, area.

Single-Family Home	Asking Price	Selling Price
1	$103.0	$ 99.3
2	127.0	124.3
3	114.9	110.8
4	102.0	102.0
5	84.6	80.0
6	160.5	158.8
7	99.6	95.4
8	173.0	167.3
9	212.5	210.1
10	89.2	86.3
11	99.9	96.8
12	138.0	132.6

At the 0.01 significance level, can we conclude that sellers normally take less than their asking price?

Data Exercises

35. Refer to Real Estate data, which reports information on homes sold in Northwest Ohio during 1992.

a. There are 21 homes with a fireplace. At the .05 significance level, can we conclude that the mean selling price of these homes is different from $185,000?

b. Consider the homes with an attached garage. Is there a difference in the mean selling price of those homes with a fireplace and those without a fireplace? Use the .05 significance level.

36. Refer to Salary data, which refers to a sample of middle managers in Sarasota, Florida.

a. Is the mean salary of the men different from $68,000? Use the .05 significance level.

b. Is there a difference between the mean salary of the men and the mean salary of the women? Use the .05 significance level. Compare these results to those you would get if you used the "large" sample methods. Give at least two reasons why the results are approximately the same.

c. Create a table where sex is one variable, and group the managers who are 49 years or younger (called "young") and those who are 50 or older (called "old") as the other variable. Can we conclude that the mean salary of the "young" women is significantly less than that of the "old" men?

ARE NONSMOKING ROOMS LESS POLLUTED?

Dr. Debruil, an environmental health professional, has been hired by Hotels International to investigate whether the air in rooms designated as nonsmoking contains less airborne pollution. He has developed a sensing device that measures the volume of airborne cigarette pollutants in parts per million. This device is installed in a sample of 20 rooms. Ten are located on the west side of a hall and the remainder on the east side. Those on the west side are reserved for nonsmoking patrons. The readings are as follows:

Smoking Room Number	Pollutants (ppm)	Nonsmoking Room Number	Pollutants (ppm)
312	6.36	313	4.72
314	7.43	315	5.52

Smoking Room Number	Pollutants (ppm)	Nonsmoking Room Number	Pollutants (ppm)
316	8.14	317	6.57
318	8.40	319	6.60
320	10.11	321	8.17
322	9.62	323	8.30
324	10.34	325	8.72
326	14.09	327	12.89
328	12.19	329	10.62
330	12.97	331	10.96

Compare the two types of rooms. Does this sample information indicate that nonsmoking rooms are less polluted?

CHAPTER ACHIEVEMENT TEST

Answer all the questions. The answers are at the back of the book.

MULTIPLE-CHOICE QUESTIONS. Select the response that best answers each of the questions.

1. The t test for the difference between the means of two independent samples assumes that
   a. the samples were obtained from normal populations.
   b. the population standard deviations are equal.
   c. the samples were obtained from independent populations.
   d. All of the above
2. If we are testing for the *difference* between the means of two related samples with $n = 15$, the number of degrees of freedom is equal to
   a. 28.
   b. 30.
   c. 15.
   d. 14.
   e. None of the above
3. The t distribution approaches which distribution as the sample size increases?
   a. Binomial
   b. Normal
   c. Poisson
   d. None of the above
4. A random sample of 10 is selected from a normal population. The population standard deviation is unknown. If a two-tailed test of significance is to be used at the 0.01 significance level, the null hypothesis is not rejected if

a. z is between –2.58 and 2.58.
b. t is between –3.250 and 3.250.
c. t is between –3.169 and 3.169.
d. t is less than –2.764.
e. None of the above

5. The two-sample t test and the t test for paired observations will
a. always yield the same results.
b. always have the same degrees of freedom.
c. always have the same sample sizes.
d. None of the above

For questions 6–11, use the following information.

A U.S. congressman claims that the mean enrollment in two-year public institutions of higher learning is fewer than 4,500 students. To test this claim, a random sample of six schools is selected. The enrollments at the selected schools are shown below.

School	Number of Students (in thousands)
Central CC	3.2
Edison CC	1.7
Lacy State	5.6
Northside Tech	1.4
Shawnee State	1.7
Washington CC	0.7

6. The appropriate null and alternate hypotheses are
a. H_0: $\overline{X}_1 \geq \overline{X}_2$
H_a: $\overline{X}_1 < \overline{X}_2$
b. H_0: $\mu \geq 4.5$
H_a: $\mu < 4.5$
c. H_0: $\mu = 4.5$
H_a: $\mu \neq 4.5$
d. H_0: $\overline{X} \geq 4.5$
H_a: $\overline{X} < 4.5$

7. Assuming that the population approximates a normal distribution, the test statistic is t because
a. the population standard deviation is not known.
b. the sample size is less than 30.
c. the enrollments are about equal.
d. Both a and b
e. None of the above

8. The population standard deviation must be estimated from the sample information. The best estimate is
a. 1.77.
b. 1.62.

c. 15.75.
d. 3.15.
e. None of the above

9. If the congressman assumed a 0.05 significance level, the null hypothesis would be rejected if
a. t is to the left of –1.476.
b. t is to the left of –2.015.
c. t is outside the interval from –2.015 to 2.015.
d. z is to the left of –1.65.
e. None of the above

10. The computed value of the test statistic is
a. $t = -2.93$.
b. $z = -1.25$.
c. $t = 2.93$.
d. None of the above

11. At the 0.05 significance level, which of the following would be a correct conclusion?
a. Enrollment is not fewer than 4,500.
b. Enrollment is at least 4,500.
c. Enrollment is fewer than 4,500.
d. You cannot determine.
e. None of the above

COMPUTATION PROBLEMS. Use the five-step hypothesis-testing procedure in problems 12–14.

12. The Food and Drug Administration is conducting tests on a certain drug to determine if it has the undesirable side effect of reducing the body's temperature. It is known that the mean human temperature is 98.6°F. The new drug is administered to 25 patients and the patients' mean temperature drops to 98.3°F, with a standard deviation of 0.64°. At the 0.05 significance level, is there sufficient reason to conclude that the drug reduces body temperature?

13. A home builder claims that the addition of a heat pump will reduce electric bills in all-electric homes. To support his claim, he tests the electric bill for the month of January for two consecutive years, one before the heat pump was installed and one after. Is there sufficient evidence to show that heating bills were reduced at the 0.01 level?

Customer	Before	After
Garcia	$180	$160
Huffman	156	164
Johnson	188	172
Palmer	132	130
Kerby	208	200
Beard	196	190
Sauve	190	184

14. The fire chief for Slocum County is evaluating two plans for the location of emergency medical equipment. One plan calls for supplies to be kept near the engines most often used by paramedical personnel. The second one calls for storage near the crew's sleeping quarters. To decide objectively if one location is better than the other, the chief tries each of the two locations, clocking the time it takes paramedics to collect their equipment under emergency conditions. The results, shown in seconds, are listed below. Use the 0.05 significance level to determine if there is a significant difference in the time needed for the paramedics to collect their equipment. What is the *p*-value?

Plan 1	Plan 2
10	15
55	9
30	47
30	3
53	34
	41
	30
	29

ANSWERS TO SELF-REVIEW PROBLEMS

11-1 a. $H_0 : \mu \geq 160$
$H_a : \mu < 160$

b. $t = \dfrac{\overline{X} - \mu}{s / \sqrt{n}}$

c. 15, found by 16 – 1.

d. Reject H_0 if computed t falls to the left of –1.753.

e. $t = \dfrac{151 - 160}{12 / \sqrt{16}} = -3$

f. Reject H_0. Senior citizens living in high altitudes do have a lower systolic blood pressure.

g. The p-value is less than 0.005.

11-2 a. $H_0 : \mu \leq 20$
$H_a : \mu > 20$

b. One-tailed

c. σ is unknown, the sample is small.

d. $$s = \sqrt{\frac{2218 - \frac{(104)^2}{5}}{5 - 1}} = 3.70$$

$$t = \frac{20.8 - 20.0}{3.70 / \sqrt{5}} = 0.48$$

e. Fail to reject H_0, because 0.48 is less than the critical value of 2.132.

f. Although the sample mean of 20.8 is greater than 20, the difference could be due to sampling error. The product should not be marketed.

g. The p-value is greater than 0.25.

11-3 a. $H_0 : \mu_1 \geq \mu_2$

$H_a : \mu_1 < \mu_2$

b. df = 7 + 5 – 2 = 10

c. $$s_1 = \sqrt{\frac{976 - \frac{(82)^2}{7}}{6}} = 1.60$$

$$s_2 = \sqrt{\frac{1621 - \frac{(89)^2}{5}}{4}} = 3.03$$

d. $$s_p = \sqrt{\frac{(6)(1.60)^2 + (4)(3.03)^2}{7+5-2}}$$

$s_p = 2.28$

e. Reject H_0 if t is less than (to the left of) –1.812.

f. $$t = \frac{11.714 - 17.80}{2.28\sqrt{\frac{1}{7} + \frac{1}{5}}} = -4.56$$

g. Reject H_0.

h. Interceptive treatment is effective in reducing the time in braces.

11-4 a. $H_0 : \mu_d \leq 0$

$H_a : \mu_d > 0$

b. $t = 2.998$

c. $$t = \frac{7.75}{8.60/\sqrt{8}} = 2.55$$

d. H_0 is not rejected.

e. The program cannot be shown to result in a significant weight loss.

CHAPTER 12

An Analysis of Variance

OBJECTIVES

When you have completed this chapter, you will be able to

- describe the F distribution;
- conduct a test of hypothesis to determine if two sample variances are equal;
- construct an ANOVA table;
- test for a difference among two or more population means;
- construct a confidence interval for treatment means.

CHAPTER PROBLEM **Leaving the Nest**

DR. BARBARA WAITE IS A CLINICAL PSYCHOLOGIST STUDYING THE AGE AT WHICH PEOPLE BECOME PSYCHOLOGICALLY INDEPENDENT OF THEIR FAMILIES. SHE WANTS TO IDENTIFY FACTORS THAT AFFECT THE AGE OF INDEPENDENCE, SUCH AS THE NUMBER AND EDUCATIONAL LEVEL OF PARENTS LIVING IN THE HOUSEHOLD, RELIGIOUS AFFILIATION, THE NUMBER OF SIBLINGS, AND SO ON. SHE BELIEVES THAT THERE ARE NO DIFFERENCES IN THE MEAN AGES OF THE DIFFERENT RELIGIOUS GROUPS. IS SHE CORRECT?

INTRODUCTION

This chapter continues the discussion of hypothesis testing introduced in Chapter 10. Recall that in Chapter 10 we developed the general theory of hypothesis testing and applied it to the means of two normally distributed populations using the *z* distribution. We also tested the difference between two proportions, again employing the *z* statistic. In Chapter 11, we analyzed differences between means of two normally distributed populations, but we used Student's *t* distribution instead of the *z* statistic.

In this chapter, we will describe the ***F* distribution.** This probability distribution is used as the test statistic under two conditions. First, we use it when testing two population variances to determine if they are equal. Second, we use the distribution to simultaneously compare two or more population means. This simultaneous comparison of several population means is called **analysis of variance (ANOVA).** In both situations, the data must be interval scale and the populations must be normal.

THE *F* DISTRIBUTION

In Chapter 11, we described Student's *t* distribution. The major characteristics of the *t* distribution were these: It is continuous and symmetrical, both tails of the *t* distribution approach the *X*-axis but never reach it, and it is specified by the degrees of freedom, so that there are many *t* distributions.

This chapter uses another probability distribution as its test statistic—the *F* distribution. What are the major characteristics of the *F* distribution?

1. The value of *F* is not negative.
2. The *F* distribution is continuous. Its values may range from zero with no upper limit. It is positively skewed.
3. There is a "family" of *F* distributions. A particular member of the family is determined by two parameters, namely, a pair of degrees of freedom. The following graph, in which three *F* distributions are shown schematically, illustrates the point. There is one *F* distribution for the combination of 32 degrees of freedom in the numerator and 30 degrees of freedom in the denominator. There is a second, different *F* distribution for the pair of 20 degrees of freedom in the numerator and 8 degrees of freedom in the denominator. The third *F* distribution has 8 *df* in the denominator and 8 *df* in the numerator.

Two tests of hypothesis utilizing the *F* distribution are presented in this chapter. Both tests involve the ratio of two sample variances. The first case involves testing whether two variances are equal. The second test compares population *means* by utilizing estimates of the population variance.

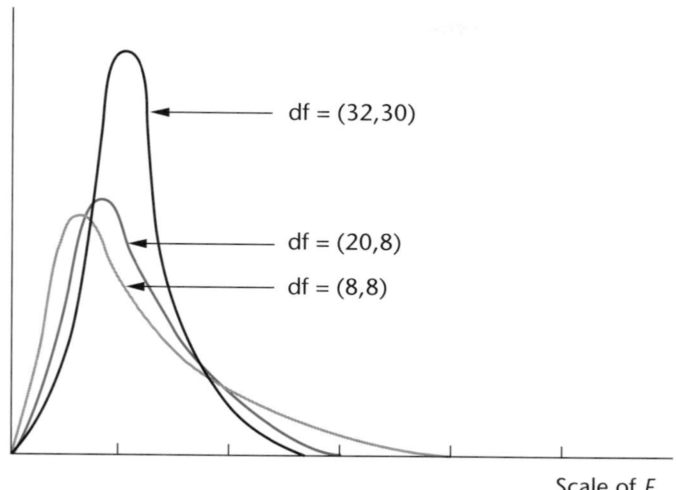

COMPARING TWO POPULATION VARIANCES

The *F* distribution can be used to test the hypothesis that the variance of one normal population equals the variance of another normal population. This test determines whether one population is more variable than another. For example, a nursing supervisor is interested in whether there is more variation in the temperatures taken with one type of thermometer than with another.

This test also determines whether the assumptions underlying certain statistical tests are valid. Recall that in the independent *t* test to determine whether two population means are the same, it was necessary to assume that the variances of the two populations are equal. This assumption can be validated by the following test.

In the null hypothesis, the variance of one normal population, σ_1^2, equals the variance of the other normal population, σ_2^2. To conduct this hypothesis test, we obtain a random sample of n_1 observations from one population and a sample of n_2 observations from the second population.

The test statistic is s_1^2/s_2^2, where s_1^2 and s_2^2 are the respective sample variances. If the null hypothesis (H_0: $\sigma_1^2 = \sigma_2^2$) is true, the test statistic is the *F* distribution with $n_1 - 1$ and $n_2 - 1$ degrees of freedom. For a two-tailed test, the larger sample variance is placed in the numerator. We find the critical value of *F* by dividing the level of significance by two, written $\alpha/2$, and then referring to the appropriate number of degrees of freedom in Appendix E.

STAT BYTE

Have you ever waited in line for a telephone, and it seemed like the person using the phone talked on and on? Apparently, people tend to spend more time on a public telephone when someone is waiting than when they are alone. In a recent study, researchers measured the length of time 56 shoppers in a mall spent on the phone when (1) they were alone, (2) when a person was using an adjacent telephone, and (3) when a person was using an adjacent phone *and* someone was waiting to use a phone. The study revealed that shoppers talked *less* when alone than when a person was using an adjacent phone or someone was waiting!

Problem A study involves the number of absences per year among union and nonunion workers. Of concern is whether the two populations have equal variances. A sample of 16 union workers has a sample standard deviation of 3.0 days. A sample of 10 nonunion workers has a standard deviation of 2.5 days. At the 0.02 significance level, can we conclude that there is a difference in variation between the two groups?

Solution Use the five-step hypothesis-testing procedure.

STEP 1. State the null and alternate hypotheses. This test is two-tailed because we are looking for a difference in the variances. We are not testing whether one variance is less than the other, which would be a one-tailed test.

$$H_0 : \sigma_1^2 = \sigma_2^2$$
$$H_a : \sigma_1^2 \neq \sigma_2^2$$

STEP 2. The significance level is 0.02.

STEP 3. The appropriate test statistic is F.

STEP 4. Obtain the decision rule from Appendix E, a portion of which is shown in Table 12–1. Because we are using a two-tailed test, the significance level is 0.01, found by $\alpha/2 = 0.02/2$. There are $n_1 - 1 = 16 - 1 = 15$ degrees of freedom in the numerator, and $n_2 - 1 = 10 - 1 = 9$ degrees of freedom in the denominator. To obtain the critical value, move horizontally across the top portion of the F table, given in Table 12–1, to 15 degrees of freedom in the numerator. Then move down that column to the critical value opposite 9 degrees of freedom in the denominator. The critical value of F is 4.96. If the ratio of the sample variances exceeds 4.96, H_0 is rejected.

STEP 5. The computed value of the test statistic is 1.44, found by $(3.0)^2/(2.5)^2$. The null hypothesis that the population variances are equal cannot be rejected. The data do not indicate a difference in the variation of days absent for union and nonunion workers.

For a one-tailed test, the numerator is determined from the statement of the alternate hypothesis. For example, if H_a: $\sigma_2^2 > \sigma_1^2$, the appropriate test statistic is $F = s_2^2/s_1^2$. The critical value of F is obtained for α (not $\alpha/2$) and the given degrees of freedom. The decision rule: Reject H_0 if the computed statistic F exceeds the critical value.

If the problem involved determining whether union members were absent more irregularly than nonunion members, the test would be one-tailed. Appendix E gives only the 0.05 and 0.01 critical values; therefore, if we use the text only these significance levels can be selected. For other significance levels, use a computer program like MINITAB.

Table 12-1

Portion of *F* Table

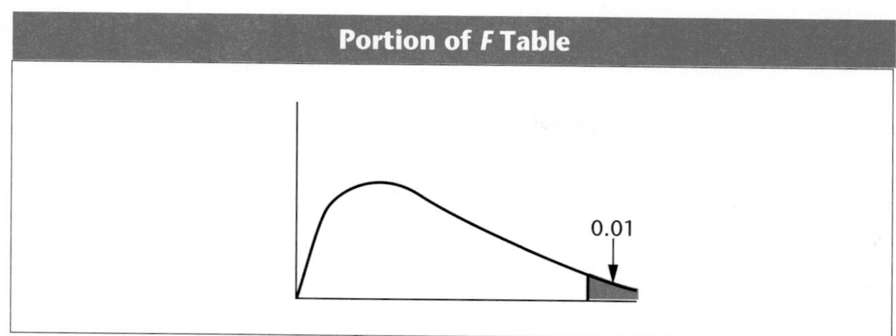

Degrees of Freedom in Denominator	Degrees of Freedom in Numerator: 10	12	15	20	24
1	6,056	6,106	6,157	6,209	6,235
2	99.4	99.4	99.4	99.4	99.5
3	27.2	27.1	26.9	26.7	26.6
4	14.5	14.4	14.2	14.0	13.9
5	10.1	9.89	9.72	9.55	9.47
6	7.87	7.72	7.56	7.40	7.31
7	6.62	6.47	6.31	6.16	6.07
8	5.81	5.67	5.52	5.36	5.28
9	5.26	5.11	4.96	4.81	4.73
10	4.85	4.71	4.56	4.41	4.33
11	4.54	4.40	4.25	4.10	4.02
12	4.30	4.16	4.01	3.86	3.78
13	4.10	3.96	3.82	3.66	3.59
14	3.94	3.80	3.66	3.51	3.43
15	3.80	3.67	3.52	3.37	3.29

Self-Review 12-1

Answers to the Self-Review problems are at the end of the chapter.

The Ace Construction Company has two front-end loaders and employs two workers to operate them. The dirt is scooped from the construction site and deposited in a waiting truck. There seems to be a difference in the volume of dirt scooped up by the two operators. A random sample of nine loads from Operator A showed a standard deviation of 0.90 tons, while a sample of six loads from Operator B showed a standard deviation of 0.40 tons. Can we conclude, at the 0.05 significance level, that there is more variation in Operator A's work?

Exercises

Answers to the even-numbered exercises are at the back of the book.

1. Find the critical value of F (using Appendix E) under the following conditions:

a. H_0: $\sigma_1^2 = \sigma_2^2$; H_a: $\sigma_1^2 \neq \sigma_2^2$, where $n_1 = 6$ and $n_2 = 10$. Use the 0.02 significance level.

b. H_0: $\sigma_1^2 \leq \sigma_2^2$, H_a: $\sigma_1^2 > \sigma_2^2$, where $n_1 = 6$ and $n_2 = 10$. Use the 0.05 significance level.

2. Find the critical value of F (using Appendix E) under the following conditions:

a. H_0: $\sigma_1^2 = \sigma_2^2$; H_a: $\sigma_1^2 \neq \sigma_2^2$, where $n_1 = 8$ and $n_2 = 7$. Use the 0.10 significance level.

b. H_0: $\sigma_1^2 \leq \sigma_2^2$; H_a: $\sigma_1^2 > \sigma_2^2$, where $n_1 = 4$ and $n_2 = 10$. Use the 0.01 significance level.

3. A study is made concerning hours spent reading the newspaper. A sample of 10 men showed a standard deviation of 4 hours per week. A sample of eight women revealed a standard deviation of 5.7 hours per week. Can we conclude that there is more variation among the women? Use the 0.05 significance level.

4. The raw materials supplied by two vendors are compared. The vendors seem to provide materials that are normally distributed with the same mean, but the variability is a matter of concern. A sample of 16 lots from Vendor A yields a variance of 150, and a sample of 21 lots from Vendor B reveals a variance of 225. Are the population variances equal? Use the 0.10 significance level.

UNDERLYING ASSUMPTIONS FOR ANOVA

The second use of the F distribution is in the **analysis-of-variance (ANOVA)** technique. To use ANOVA, we assume the following conditions:

1. The populations being studied are normally distributed.
2. The populations have equal standard deviations (σ).
3. The samples selected from those populations are independent and random.

When those conditions are met, the F statistic is used (instead of z or t) to test if the *means* of the populations are equal. Whenever the assumptions about the normality of the population distributions and equal standard deviations cannot be met, an analysis-of-variance technique developed by Kruskal and Wallis, to be discussed in Chapter 17, is used.

Because analysis of variance had its beginnings in agriculture, the term **treatment** is generally used to identify the different populations being examined.

Treatment A specific source or cause of variation in a set of data.

Two illustrations will help to clarify the term *treatment* and to demonstrate the application of the ANOVA technique.

1. WHEAT YIELDS. A farmer wants to use the brand of fertilizer that will produce the maximum yield per acre of wheat. Assume that three different commercial brands—Prothro, Scotts, and Anderson—are to be applied. As an experiment, the farmer divides his field into 15 plots of equal size. The 15 areas are planted in the same manner and at the same time, but he randomly assigns Prothro to five plots, Scotts to five plots, and Anderson to five plots. At the end of the growing season, the number of bushels of wheat produced by each plot is recorded. The results are:

Prothro	Anderson	Scotts
40	72	51
45	71	55
47	68	60
50	75	57
47	66	54

Do the treatments differ? In this case, "treatment" refers to the different fertilizers that are applied. Are the mean yields of wheat different among the three populations? Figure 12–1 is an illustration of how the means would appear if the yields were *different*.

Suppose that the means were, in fact, identical. From a practical standpoint, this would indicate that the fertilizers all produced the same yield of wheat. Figure 12–2 illustrates three yields with equal population means ($\mu_1 = \mu_2 = \mu_3$). Note that the distributions are approximately normal and that the dispersion of each is about the same.

2. INSTRUCTIONAL MODES. The instructor of a nursing arts course wants to know if the extent of learning in the course differs with the type of instruction. Four methods (treatments) are proposed: (1) lecture, (2) movie, (3) lecture and experience, and (4) movie and experience. (The experience is to be obtained by having the students follow the particular nursing procedure with the

Figure 12-1

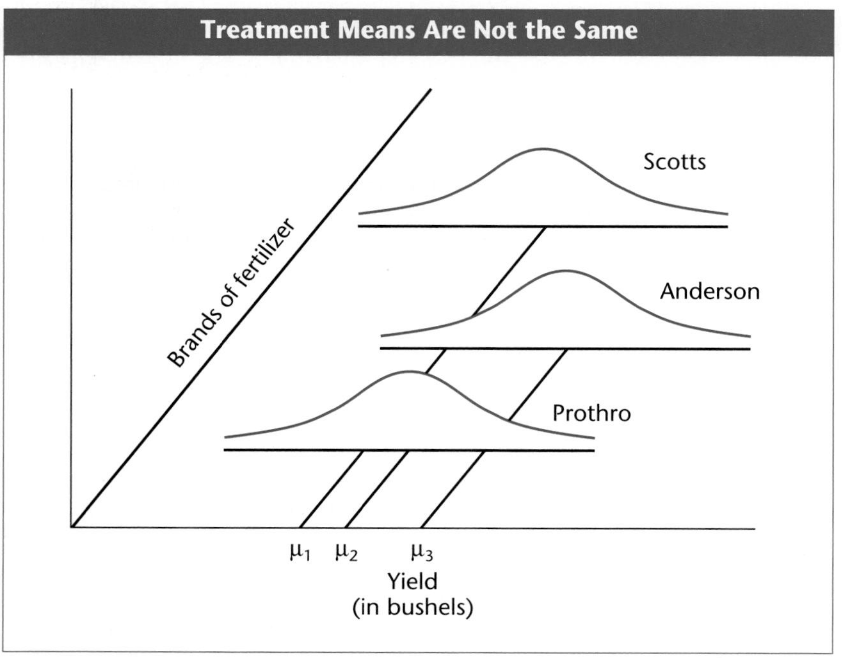

Figure 12-2

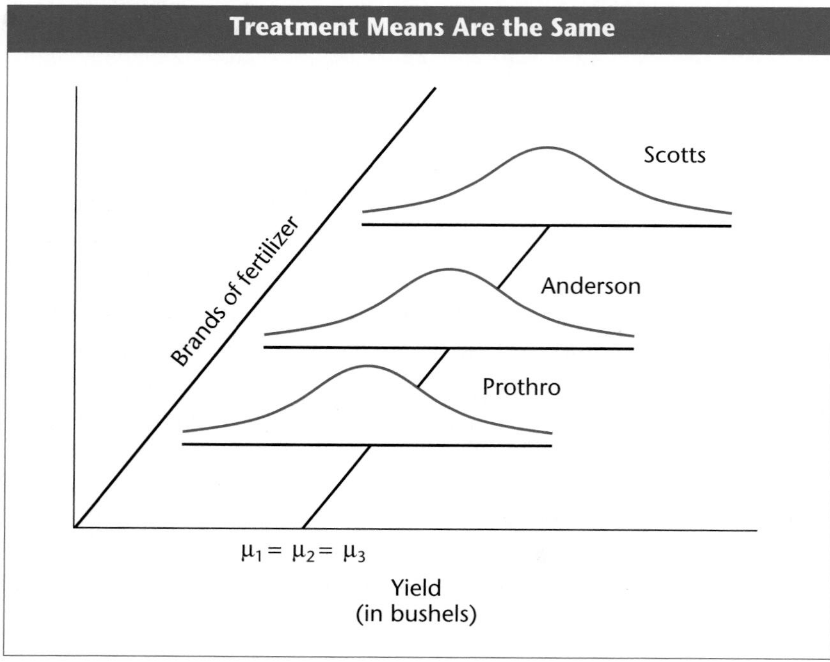

instructor's guidance.) As an experiment, the 19 students in the current class were assigned randomly to the four groups. After the course, the same final examination was administered to all students. The scores were:

Lecture	Movie	Lecture and Experience	Movie and Experience
80	59	85	81
72	65	84	84
69	68	77	76
75	61	69	71
	70	73	72

Do these data demonstrate to the instructor that there is a difference in the extent of learning achieved with different types of instruction?

THE ANOVA TEST

The purpose of this section is to explore some of the reasoning behind ANOVA. First, ANOVA breaks down the total variation into two parts. One part measures the variation between sample means (treatments). The other part measures the variation of the observations from their treatment mean; that is, it measures the variability within each of the sampled populations. Next, we compare this relationship between the two sources of variation by forming the F ratio in the following manner:

STAT BYTE

A recent study by *The Wall Street Journal* revealed that advertisements placed at the end of TV shows (27 minutes after the hour) are seen by more viewers than those placed at the beginning (eight minutes after the hour).

$$F = \frac{\text{Estimate of the population variance based on differences between sample means}}{\text{Estimate of the population variance based on variation within samples}}$$

The following problem provides additional insight into the analysis-of-variance technique.

Problem Recall from the Chapter Problem that Dr. Waite, a clinical psychologist, is studying the age at which people become psychologically independent of their families. One factor that may affect this variable is religious orientation. To test this possibility, she selects a sample of 39 young people and classifies each by religion. The possible categories are: Protestant, Catholic, Jewish, and Other. A tally of the religious affiliations of those sampled and the age at which they became independent of their families is given in Table 12–2.

Table 12-2 **Comparison of Religious Affiliation and Age at Which Those Sampled Became Independent**

Religious Affiliation			
Jewish	Catholic	Protestant	Other
22	27	20	18
19	25	18	16
13	22	21	24
19	27	21	19
23	19	16	22
15	23	17	22
16	21	20	24
18	28	18	
20	23	17	
20	25	19	
	27	18	

Dr. Waite's hypothesis is: There is no difference in the mean age at which independence was achieved for the four religious groups.

Solution Use the five-step hypothesis-testing procedure.

STEP 1. The null hypothesis is that the mean age is the same for each faith:

$$H_0 : \mu_1 = \mu_2 = \mu_3 = \mu_4$$

The alternate hypothesis is that the means are not all equal:

$$H_a : \text{ Not all means are equal.}$$

STEP 2. The significance level is 0.01.

STEP 3. The appropriate test is based on the F statistic.

STEP 4. Formulate the decision rule. Remember that in order to arrive at a decision rule we need to identify the *critical value.* The critical values for the F statistic can be found in Appendix E, a portion of which follows in Table 12–3. You will find the critical values for the 0.05 significance level on the first page of Appendix E and the values for the 0.01 significance level on the second. To use the table, you need to know two numbers: the degrees of freedom in the numerator and the degrees of freedom in the denominator. The de-

Table 12-3

Portion of F Table

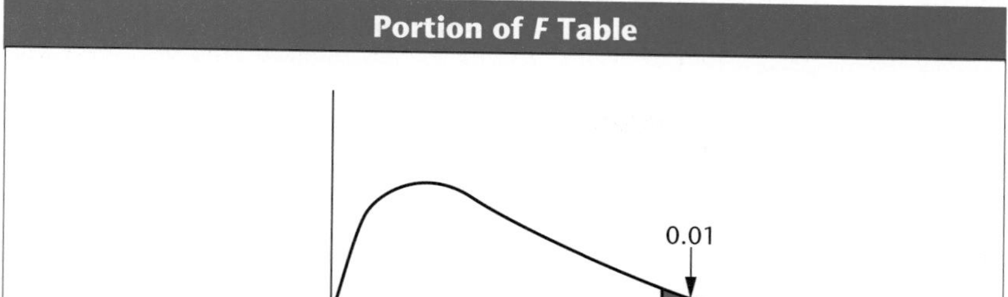

		Degrees of Freedom in Numerator				
		1	2	3	4	5
Degrees of Freedom in Denominator	21	8.02	5.78	4.87	4.37	4.04
	22	7.95	5.72	4.82	4.31	3.99
	23	7.88	5.66	4.76	4.26	3.94
	24	7.82	5.61	4.72	4.22	3.90
	25	7.77	5.57	4.68	4.18	3.86
	30	7.56	5.39	4.51	4.02	3.70
	40	7.31	5.18	4.31	3.83	3.51
	60	7.08	4.98	4.13	3.65	3.34
	120	6.85	4.79	3.95	3.48	3.17
	∞	6.63	4.61	3.78	3.32	3.02

grees of freedom in the numerator refer to the number of treatments, designated k, minus 1. That is, the degrees of freedom are found by $k - 1$. The degrees of freedom in the denominator refer to the total number of observations, designated N, minus the number of treatments, that is, $N - k$. For this problem, there are four treatments and a total of 39 observations. Thus:

Degrees of freedom in numerator = $k - 1 = 4 - 1 = 3$
Degrees of freedom in denominator = $N - k = 39 - 4 = 35$

Refer to Appendix E or Table 12–3 and the 0.01 level of significance. Move horizontally at the top of the table to 3 degrees of freedom in the numerator. Then move down that column to the critical value opposite 40 degrees of freedom (the closest to 35) for the denominator. The critical value is approximately 4.31. The decision rule, therefore, is to reject the null hypothesis if the computed value of F is greater than 4.31. Recall that the F distribution is positively skewed. Shown diagrammatically, the decision rule is:

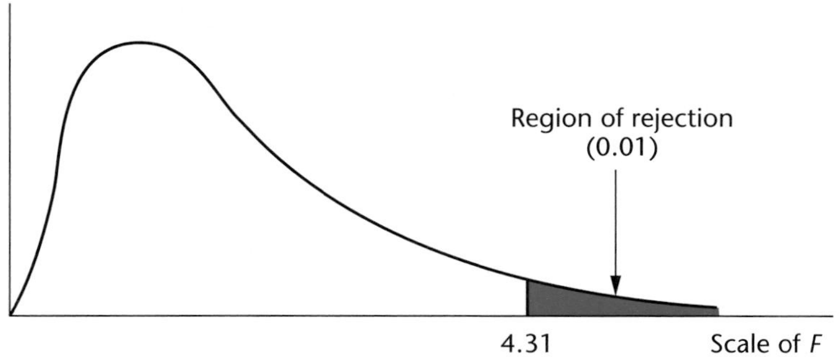

STEP 5. Compute F and make a decision. It is often convenient to record the computations for the F statistic in an ANOVA table. The general form is:

ANOVA Table

Source of Variation (1)	Sum of Squares (2)	Degrees of Freedom (3)	Mean Square (4)	F (5)
Treatments	SST	$k - 1$	$\frac{SST}{k-1}$	$\frac{SST/(k-1)}{SSE/(N-k)}$
Within	SSE	$N - k$	$\frac{SSE}{N-k}$	
Total	SS total			

Referring to the general format for the analysis of variance, note that three totals, called **sums of squares,** are needed to compute F—namely, *SST, SSE,* and *SS* total.

1. *SST* is the abbreviation for the sum of squares due to the treatment effect and is found by

$$SST = \sum\left(\frac{T_c^2}{n_c}\right) - \frac{(\Sigma X)^2}{N} \qquad \textbf{12-1}$$

where

T_c is the column total for all observations in the treatment.
n_c is the number of observations (sample size) for each respective treatment.
ΣX is the sum of all observations.

k is the number of treatments.
N is the total number of observations.

2. *SSE* is the abbreviation for the sum of squares within (error). *SSE* is computed by:

$$SSE = \Sigma\left(X^2\right) - \sum\left(\frac{T_c^2}{n_c}\right) \tag{12-2}$$

3. *SS total* is the total variation. That is, it is the sum of *SST* and *SSE*.

$$SS \text{ total} = SST + SSE$$

As a check, *SS* total is calculated by:

$$SS \text{ total} = \Sigma\left(X^2\right) - \frac{(\Sigma X)^2}{N} \tag{12-3}$$

To compute *F:*

$$F = \frac{\dfrac{SST}{k-1}}{\dfrac{SSE}{N-k}} \tag{12-4}$$

The calculations for *F* are shown in Table 12–4.
The entries for the ANOVA table are computed:

$$\begin{aligned} SST &= \sum\left(\frac{T_c^2}{n_c}\right) - \frac{(\Sigma X)^2}{N} \\ &= \frac{(185)^2}{10} + \frac{(267)^2}{11} + \frac{(205)^2}{11} + \frac{(145)^2}{7} - \frac{(802)^2}{39} \\ &= 234.93 \end{aligned}$$

Table 12-4 **Calculations Needed for Computed F**

	Religious Affiliation				
	Jewish	Catholic	Protestant	Other	**Grand Totals**
	Ages (X)				
	22	27	20	18	
	19	25	18	16	
	13	22	21	24	
	19	27	21	19	
	23	19	16	22	
	15	23	17	22	
	16	21	20	24	
	18	28	18		
	20	23	17		
	20	25	19		
		27	18		
Column totals (T_c)	185	267	205	145	802 ← ΣX
Sample size (n_c)	10	11	11	7	39 ← N
Squared totals (ΣX^2)	3,509	6,565	3,849	3,061	16,984 ← $\Sigma(X^2)$

$$
\begin{aligned}
SSE &= \Sigma\left(X^2\right) - \sum\left(\frac{T_c^2}{n_c}\right) \\
&= \left(22^2 + 19^2 + \cdots + 24^2\right) \\
&\quad - \left[\frac{(185)^2}{10} + \frac{(267)^2}{11} + \frac{(205)^2}{11} + \frac{(145)^2}{7}\right] \\
&= 256.66 \\
SS \text{ total} &= SST + SSE \\
&= 234.93 + 256.66 \\
&= 491.59
\end{aligned}
$$

As a check, use formula 12–3:

$$
\begin{aligned}
SS \text{ total} &= \Sigma(X^2) - \frac{(\Sigma X)^2}{N} \\
&= 16{,}984 - \frac{(802)^2}{39} \\
&= 491.59
\end{aligned}
$$

Insert these values into an ANOVA table:

Source of Variation	Sum of Squares	Degrees of Freedom	Mean Square
Treatments	234.93	3	78.31
Within	256.66	35	7.33
Total	491.59		

Now compute F by applying formula 12–4:

$$F = \frac{\frac{SST}{k-1}}{\frac{SSE}{N-k}} = \frac{78.31}{7.33} = 10.68$$

Because the computed F value of 10.68 is greater than the critical F value of 4.31 (determined in Step 4), the null hypothesis is rejected at the 0.01 level. Therefore, the p-value is less than 0.01. This indicates that it is quite unlikely that the differences in the four means could have occurred by chance. Thus, we conclude that the population means are not all equal. Apparently, religious affiliation affects the age at which people become independent of their families. To determine which pairs of means differ, additional analysis is required. How to recognize which pairs are significantly different is illustrated in the following example.

A Computer Example

MINITAB has several procedures for computing a one-way analysis of variance. The output that follows was obtained with the procedure AOVONEWAY. The four columns of data are entered in c1, c2, c3, and c4.

```
MTB >     set c1
DATA >     22,19, . . . , 20
DATA >     end
MTB >     set c2
DATA >     27,25, . . . , 27
DATA >     end
MTB >     set c3
DATA >     20,18, . . . , 18
DATA >     end
MTB >     set c4
DATA >     18, 16, . . . , 24
```

```
DATA >   end
MTB >    name c1 'Jewish' c2 'Catholic' c3 'Prot.' c4 'Other'
MTB >    aovoneway c1 c2 c3 c4

ANALYSIS OF VARIANCE
SOURCE    DF       SS        MS         F        P
FACTOR     3   234.93     78.31     10.68    0.000
ERROR     35   256.66      7.33
TOTAL     38   491.59
                                  INDIVIDUAL 95 PCT CI'S FOR MEAN
                                  BASED ON POOLED STDEV

LEVEL      N     MEAN    STDEV  ---+---------+---------+---------+---
Jewish    10   18.500    3.100  (-----*-----)
Catholic  11   24.273    2.901                        (-----*-----)
Prot.     11   18.636    1.690   (-----*-----)
Other      7   20.714    3.094        (-----*-----)
                                ---+---------+---------+---------+---

POOLED STDEV = 2.708              17.5      20.0      22.5      25.0
```

Note that the computer output gives the computed value of F, namely 10.68. It also reports a p-value of 0.0000. MINITAB supplies some additional information, such as the sample size, the mean, and the standard deviation for each of the four religious categories. Further, a confidence interval for the mean of each category is given. The endpoints for each interval are designated by "("and ")". If H_0 is rejected, we can use these confidence intervals to determine which of the particular pairs of means differ. Note that the intervals for Jewish and Catholic do not overlap. That is, the upper limit for Jewish is about 21 years and the lower limit for Catholic is about 22.5 years. Hence, we can conclude that there is a difference in these groups. Continuing, we conclude that Catholic and Protestant differ, but that Jewish and Protestant, Jewish and other, Protestant and other, and Catholic and other do not differ.

Problem The head nurse at University Hospital has the responsibility of assigning personnel to the emergency room. Present policy calls for the same number of registered nurses to be assigned to all three shifts. The head nurse, however, thinks that the number of emergencies handled may not be the same for each shift. It is decided to use the ANOVA technique to investigate whether the same number of emergencies are handled on each shift.

A random sample of five days from each shift is selected. The results are shown. The "treatment" in this problem is the shift. Note that the sample sizes are equal. ANOVA follows the five-step hypothesis-testing procedure outlined in Chapter 10.

Number of Emergency Cases Reported per Shift

	Day	Afternoon	Night
	44	33	39
	53	42	24
	56	15	30
	49	30	27
	38	45	30
Mean	48	33	30

Solution Repeating this table with all necessary calculations, we get

	Day		**Afternoon**		**Night**			
	X	X^2	X	X^2	X	X^2	**Grand Totals**	
	44	1,936	33	1,089	39	1,521		
	53	2,809	42	1,764	24	576		
	56	3,136	15	225	30	900		
	49	2,401	30	900	27	729		
	38	1,444	45	2,025	30	900		
Column totals (T_c)	240		165		150		555	← ΣX
Sample size (n_c)	5		5		5		15	← N
Sum of squares (ΣX^2)		11,726		6,003		4,626	22,355	← $\Sigma(X^2)$

STEP 1. The null hypothesis is that the mean number of emergencies is the same for each shift. That is, $\mu_1 = \mu_2 = \mu_3$. The alternate hypothesis is that the means are not all equal.

STEP 2. The level of significance is 0.05.

STEP 3. The appropriate statistical test is F.

STEP 4. Formulate the decision rule. There are 2 degrees of freedom in the numerator and 12 in the denominator. So, we reject H_0 if F exceeds 3.89.

STEP 5. The F statistic is determined by formula 12–4:

$$F = \frac{\frac{SST}{k-1}}{\frac{SSE}{N-k}}$$

where k is the number of treatments and N the total number of items sampled.

Calculate SST (the abbreviation for the sum of squares due to the treatment effect):

$$\begin{aligned} SST &= \sum\left(\frac{T_c^2}{n_c}\right) - \frac{(\Sigma X)^2}{N} \\ &= \frac{(240)^2}{5} + \frac{(165)^2}{5} + \frac{(150)^2}{5} - \frac{(555)^2}{15} \\ &= 930 \end{aligned}$$

Next, calculate SSE, the abbreviation for sum of squares within (error):

$$\begin{aligned} SSE &= \Sigma(X^2) - \sum\left(\frac{T_c^2}{n_c}\right) \\ &= 22{,}355 - \left(\frac{(240)^2}{5} + \frac{(165)^2}{5} + \frac{(150)^2}{5}\right) \\ &= 22{,}355 - 21{,}465 = 890 \end{aligned}$$

We can now determine the F statistic:

$$F = \frac{\frac{930}{3-1}}{\frac{890}{15-3}} = 6.27$$

The decision rule, formulated in Step 4, stated that H_0 is to be rejected if the computed F value exceeds 3.89. Because 6.27 is greater than 3.89, we reject the null hypothesis at the 0.05 level of significance. To put it another way, the differences in mean number of emergencies handled per shift (48, 33, and 30) cannot be attributed to chance. From a practical standpoint, it may be concluded that the number of emergency cases handled on the three shifts is not the same. If the 0.01 significance level were used, the critical value would be 6.93. Since F is between 3.89 and 6.93, the p-value is between 0.01 and 0.05.

Self-Review 12-2

Energy shortages have caused many schools to turn down the heat in the classroom. The principal at Penn Street Elementary School is concerned that this may have an effect on achievement. To investigate this question further, students in the fifth-grade mathematics class were randomly assigned to one of three groups. The three groups were then separated and placed in rooms having different temperatures. Each group received televised instruction in long division. At the end of the lesson, the same 10-question examination was given to all three groups. The results, scoring the number of correct answers out of 10, were:

Temperature		
65°F	72°F	78°F
3	7	4
5	6	6
4	8	5
3	9	7
4	6	6
	8	5
	8	4
		3

Is there any significant difference in the mean scores? Use the 0.05 significance level.

a. Compute *SST.*
b. Compute *SSE.*
c. Compute *SS* total.
d. Arrange the values in an ANOVA table and compute *F.*
e. Arrive at a decision.
f. What is the *p*-value?

Exercises

5. The quantity of oxygen dissolved in water is a measure of water pollution. Samples are taken at three locations in a lake and the quantity of dissolved oxygen is recorded as follows (lower readings indicate greater pollution):

Location	Quantity
Bayview	6.5, 6.4, 6.9
Council Bluffs	6.7, 7.1, 6.9, 7.3
Swan Creek	7.4, 6.9, 7.2

Do these data indicate a significant difference in the mean amount of dissolved oxygen at the three locations? Use the 0.05 significance level in determining your answer.

6. There are three banks in Warren, Pennsylvania. Customers are randomly selected from each bank and their waiting times before being served are recorded:

Bank	Waiting Time (in minutes)
Pittsburgh Bank	3.8, 4.5, 5.3
Mountain Trust	6.9, 9.6, 7.3, 8.2
Western National	4.3, 5.1

Do these data indicate a significant difference among the mean waiting times at the three banks? Use the 0.05 significance level. Find the *p*-value.

7. Deals-on-Wheels, a manufacturer of mobile homes, is interested in the ages of buyers of each of five available floor designs. The president suspects that certain designs tend to appeal to younger buyers more than others. A random sample of 18 records is selected from last year's sales files. The respective ages of the principal buyers from the files are:

Floor Design				
A	B	C	D	E
30	48	54	52	44
32	52	60	50	48
31	45	56	43	50
	38	50	42	

a. State the null and alternate hypotheses.
b. How many degrees of freedom are there in the numerator and the denominator?
c. For the 0.05 level of significance, what is the critical value of *F*?
d. Compute *F*.
e. State your decision.
f. Interpret the results.

8. A social researcher wants to evaluate the ethical behavior of attorneys in four major regions of the United States. Samples are selected and an index of ethical behavior computed for each attorney in the study. Refer to the data on the following page. At the 0.01 level of significance, is there a statistically significant difference in ethical behavior among the four regions? Use the five-step hypothesis-testing procedure.

West	South	North Central	East
12	19	34	19
16	20	29	21
12	18	31	17
14	9	19	24
26	22	26	
	19		

A final note: In an analysis of variance, the null hypothesis is that *the means are all equal.* The alternate hypothesis is that *at least one of the means is different.* Rejection of the null hypothesis and acceptance of the alternate hypothesis leads to the conclusion that there is a significant difference between at least one pair of means. Rejecting the null hypothesis does not pinpoint which pairs—or how many pairs—of means differ significantly. It only indicates that there is a significant difference between at least one pair of means. Multiple-comparison tests, discussed in more advanced texts, may be used to identify the treatment that differs significantly.

SUMMARY

The major characteristics of the F distribution are these: It is nonnegative; it is continuous and positively skewed; there is a family of F distributions; and F is based on two sets of degrees of freedom, one in the numerator and one in the denominator.

Two tests of hypothesis were presented in this chapter. Both use the F distribution and the ratio of two variances. The first test compared the ratio of two variances to determine if they are equal. The second test is the analysis of variance (ANOVA). The ANOVA technique is used to test simultaneously whether two or more population *means* are equal. It assumes that the populations are normally distributed with equal standard deviations. It also assumes that the samples are independent.

F is computed by

$$\frac{\dfrac{SST}{k-1}}{\dfrac{SSE}{N-k}}$$

where

$$SST = \sum\left[\frac{T_c^2}{n_c}\right] - \frac{(\Sigma X)^2}{N}$$

and

$$SSE = \Sigma\left(X^2\right) - \sum\left[\frac{T_c^2}{n_c}\right]$$

In an ANOVA test, the critical values of F are located in the right-hand tail of the distribution.

Exercises

9. Two brands of cigarettes have the same mean nicotine content. However, there seems to be a difference in the amount of variation in the two brands. A random sample of 13 cigarettes of Brand A showed a standard deviation of 3.5 grams. A random sample of eight cigarettes of Brand B showed a standard deviation of 2.0 grams. At the 0.05 significance level, can we conclude that there is more variation in Brand A? Find the p-value.

10. Last year's graduates in Accounting and Sales had the same mean starting salaries. However, the standard deviation was $2,000 (sample of eight) among the accountants and $4,200 (sample of seven) among the sales majors. At the 0.05 significance level, can we conclude there is more variation among the sales graduates?

11. A physician randomly selects 18 patients among those she is treating for high blood pressure. These patients are randomly assigned to three groups and treated with three different drugs, all designed to reduce blood pressure. The amount of reduction, in millimeters of mercury, is shown. At the 0.01 level of significance, is there sufficient evidence to show that the drugs act differently?

Drug A	Drug B	Drug C
10	13	9
10	14	8
9	11	6
10	10	10
7	9	10
6	10	7

12. There are four different instructors for the history course, Introduction to Western Civilization, taught at Scandia Tech. Following is the number of pages of reading assigned by each instructor every week for the first five weeks of the course. At the 0.05 level of significance, is there sufficient evidence to show a difference in the average length of the readings assigned by the four instructors? State your conclusion, and interpret the results. What is the p-value?

Mr. Barr	Dr. Sedwick	Dr. Reading	Dr. Faust
25	35	30	28
29	20	27	32
30	20	18	33
42	17	19	35
35	30	26	24

13. The merchandising manager for Food Mart grocery stores is analyzing the effect of various placements of candy displays within the company's stores. He decides to conduct an experiment by locating the display in different areas in each of four Food Mart outlets. The amount in pounds sold in the various locations each week is recorded. Is there sufficient evidence to indicate that there is a difference in sales at the various locations? Use a 5% significance level. State your conclusion, and interpret the results.

Back of Store, Near Bread	Top Shelf, Near Cookies	End of Aisle, Near Meat	Third Shelf, Near Soda Pop
76	73	89	96
75	70	82	92
83	81	85	104
87	78	79	89
81	76	80	94

14. A psychologist wants to investigate the effect of social background on the time (in minutes) it takes freshmen to solve a puzzle. A random sample of students from different backgrounds is selected, resulting in the following data. Use the 0.05 level of significance to test the hypothesis that social background has no effect on the time required to solve the puzzle. State your conclusion, and interpret the results.

Inner City	Urban	Suburban	Rural
16.5	10.9	18.6	14.2
5.2	5.2	8.1	24.5
12.1	10.8	6.4	14.8
14.3	8.9		24.9
	16.1		5.1

15. A wholesaler is interested in comparing the weights in ounces of grapefruit from Florida, Texas, and California:

Florida	Texas	California
12.6	12.8	16.0
13.8	13.2	15.1
14.0	12.4	13.9
	13.2	14.3
		15.0

a. What are the null and alternate hypotheses?
b. Fill in an ANOVA table.
c. What is the critical value of *F*, assuming a 0.01 level of significance?
d. What decision should the wholesaler make? Describe your conclusion. What is the *p*-value?

16. Theft in student rooms is a major problem at a large university. In an effort to reduce the problem, the university conducted an experiment in which 15 dormitories were randomly divided into three groups. In the first group of four dorms, all students were involved in structured discussion groups about the theft problem. Informal "peer" group meetings were held for the students in five other dorms. The six dorms in the third group were not subjected to any changes. The number of reported thefts in each dorm after one semester is shown by group:

Cluster A Structured	Cluster B Informal	Cluster C Control
35	18	14
20	10	3
32	21	16
27	14	10
	13	11
		12

Do these data present sufficient evidence to indicate a difference among the three methods? Use a 1% level of significance. Describe your conclusion.

17. An oncologist—a physician who specializes in the treatment of tumors—has 24 patients with advanced lung cancer. He is aware of three treatments, reported in medical journals, that may gain remission for his patients. To assess the effect of the treatments, the doctor randomly assigns patients to each treatment, then keeps careful records on the number of days the patients live after treatment starts. At the 0.05 level, can it be concluded that there is any difference in the effect of the treatments?

Laetrile	Chemotherapy	Radiation
75	80	64
88	82	90
62	64	58
97	45	64
62	67	82
81	84	71
93	55	59
	39	66
	60	

18. Applicants for a position as a school bus driver are given a psychological test to determine their ability to stay calm under difficult conditions. The following table gives the scores of the 13 applicants by age:

Age of Applicant		
25–34	35–44	45–54
48	61	78
67	75	82
58	82	81
56	59	80
	69	

At the 0.05 significance level, do the means of the three age groups differ? Find the p-value.

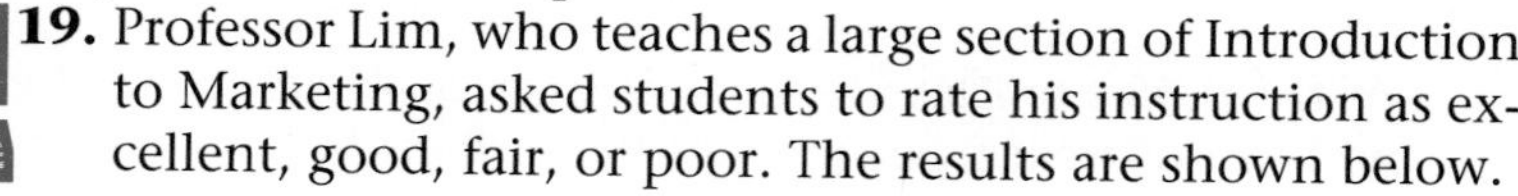

19. Professor Lim, who teaches a large section of Introduction to Marketing, asked students to rate his instruction as excellent, good, fair, or poor. The results are shown below.

Excellent	Good		Fair	Poor
85	80	88	73	71
77	70	75	71	75
74	78	73	70	76
77	72	80	79	81
90	74	82	73	79
94	77	83	76	70
89	79	74	76	79
	78	76	68	82
	82	91	80	
		78	78	

A graduate student collected the ratings and assured the students that Professor Lim would not receive them until after final grades were filed in the Registrar's Office. After the grades were filed, the rating given by each student was matched with the student's score on the final examination. At the 0.05 significance level, is there a difference in the mean scores of the four rating groups?

20. An independent testing agency has been hired by a large manufacturer of tires. Specifically, the tire manufacturer would like to know if there is a difference in the tread wear of a tire on a variety of road surfaces. To assure uniformity, the testing agency measured the wear on only the right front tire, drove the cars in all types of traffic, and attempted to ensure that the weight was the same in all test vehicles. The data, in millimeters of wear, are shown below.

Type of Road Surface			
Concrete	Composite	Brick	Gravel
8.8	10.1	11.9	13.4
9.6	10.1	11.1	13.0
8.3	10.3	11.0	11.9
9.3	9.8	12.1	12.6
9.1	9.9	12.6	12.7
8.3	10.6	10.9	13.0
8.4	10.8	11.8	
	10.3	12.9	
		12.3	

At the 0.05 significance level, is there a difference in the mean amount of wear from the various surfaces?

Data Exercises

21. Refer to the Real Estate Data, which reports information on homes sold in Northwest Ohio during 1992.

a. Is there a difference in the mean selling prices of the homes with different numbers of bathrooms? Use the number of bathrooms as a treatment variable and test the hypothesis to determine if there is no difference in treatment means. Use the .05 significance level.

b. Is there a difference in the mean selling prices of the homes with different numbers of bedrooms? Use the number of bedrooms as a treatment variable and test the hypothesis to determine if there is a difference in treatment means. Use the .05 significance level.

22. Refer to the Salary Data, which refers to a sample of middle managers in Sarasota, Florida.

a. The managers were sampled from five different industries. Is there a difference in the mean salaries of the managers by industry? Use the .05 significance level.

b. The samples were obtained from four regions of the Sarasota area. Is there a difference in the mean salaries of the managers by region? Use the .05 significance level.

c. In the Data Exercises in Chapter 11, Problem 36b, we compared the mean salary of the men and women managers. With sex as the treatment variable, use the ANOVA technique to conduct the same test. Compare the p-values of the two tests. Are the tests the same? (*Hint:* Check the values for MSE and s_p^2.) Comment on your findings.

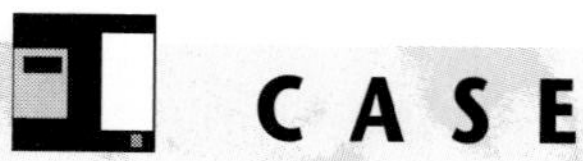

WHAT SHOULD DANFORTH ELECTRONICS DO?

Danforth Electronics manufactures subassemblies for large firms such as IBM, Panasonic, and NCR. Danforth has a rather unique assembly organization, inaugurated many years ago by its president, Charles M. Danforth. Some assembly "gangs," as they are called, consist entirely of men, others entirely of women. Upon the suggestion of a young manager, Danforth was convinced to try mixed gangs. To investigate the issue, random samples from all-men gangs, all-women gangs, and mixed gangs were selected. Several statistics summarizing the number of units produced per shift of the three types of gangs are shown below. Do they indicate any differences among the three types of gangs?

	Units Produced			
Gang	Mean	Median	Standard Deviation	Number
All men	111.25	112	9.51	8
All women	125.40	120	11.37	10
Mixed	114.25	113	10.45	12

CHAPTER ACHIEVEMENT TEST

Answer all the questions. The answers are at the back of the book.

MULTIPLE-CHOICE QUESTIONS. Select the response that best answers each of the questions.

Questions 1–4 refer to the following problem. Mr. Tourtellotte can drive to work along one of three different routes. The following data show the number of minutes it takes to make the trip on five different occasions for each route. Is there sufficient evidence to indicate that there is a difference in the mean time it takes to drive the three routes?

Via Expressway	Via Downtown	Past the University
33	22	14
35	26	21
34	17	24
32	18	25
38	20	15

1. The number of treatments is
   a. 3.
   b. 2.
   c. 12.
   d. 15.
   e. 14.

2. The number of degrees of freedom for the denominator of the F distribution is
   a. 3.
   b. 2.
   c. 12.
   d. 15.
   e. 14.

3. If a 0.05 significance level is used, the critical value of F is
   a. 19.4
   b. 3.49.
   c. 3.89.
   d. 6.93.
   e. 1.96.

4. The value of F was computed to be 23.07. The correct conclusion is to
   a. not reject H_0.
   b. reject H_a.
   c. reject H_0 and accept H_a.
   d. find the samples too small to be appropriate.

5. The ANOVA technique was used to compare three population means based on samples of size nine drawn from each population. It was found that $SST = 90$ and $SSE = 200$. At the 0.05 level of significance, the correct conclusion is:
   a. $F = 0.50$, do not reject H_0.
   b. $F = 2.22$, do not reject H_0.
   c. $F = 5.40$, reject H_0.

d. $F = 12$, reject H_0.
e. The value of F cannot be determined from the information given.

6. In a particular ANOVA test, the calculated value of F is between zero and the value of the F table. The correct conclusion is:
a. Do not reject H_0 and conclude that the treatment means being tested are not significantly different.
b. Do not reject H_0 and conclude that treatment means being tested differ significantly.
c. Reject H_0, accept H_a, and conclude that the treatment means are not significantly different.
d. Reject H_0, accept H_a, and conclude that the treatment means are significantly different.

7. If the sample means for each of the treatment groups were identical, the value of the F statistic would be
a. equal to 1.00.
b. 0.
c. infinite.
d. a negative number.
e. a number between 0 and 1.00.

8. The F distribution
a. cannot be negative.
b. is positively skewed for small samples.
c. is determined by two parameters.
d. All of the above

9. Suppose the within-samples (error) estimate of the variance was found to be negative. This means that
a. the null hypothesis is not rejected.
b. the null hypothesis is rejected.
c. we should have used the t distribution.
d. a mistake was made in arithmetic, because the variance cannot be negative.

10. Rejecting H_0 in an ANOVA test indicates that there is a significant difference
a. between at least one pair of means.
b. between all means.
c. between the F values.
d. between *SST* and *SS* total.

COMPUTATION PROBLEM

11. A manufacturer of automobiles is testing a new design of brakes. The director of engineering reports that test data are available for both the existing design and two proposed designs. Analysts obtained test results by measuring the stopping distance, in feet, of the cars traveling at a speed of 15 miles per hour:

Existing Design	First Proposal	Second Proposal
5	5	8
7	5	4
6	8	5
	7	9
	6	

a. What are the null and alternate hypotheses?
b. Complete an ANOVA table.
c. What is the critical value of F, assuming a 0.05 level of significance?
d. Can we conclude that there is a difference in the mean stopping distances?
e. What is the p-value?

ANSWERS TO SELF-REVIEW PROBLEMS

12-1 H_0: $\sigma_a^2 \leq \sigma_b^2$
H_a: $\sigma_a^2 > \sigma_b^2$
H_0 is rejected if $F > 4.82$.
$F = (0.90)^2/(0.40)^2 = 5.0625$
H_0 is rejected. There is more variability in Operator A's work.

12-2 a.

3	7	4	
5	6	6	
4	8	5	
3	9	7	
4	6	6	
	8	5	
	8	4	
		3	
19	52	40	$\Sigma X = 111$

Then:

$$SST = \frac{(19)^2}{5} + \frac{(52)^2}{7} + \frac{(40)^2}{8} - \frac{(111)^2}{20}$$
$$= 42.44$$

b.
$$SSE = 3^2 + 5^2 + 4^2 + \cdots + 3^2 - \left[\frac{(19)^2}{5} + \frac{(52)^2}{7} + \frac{(40)^2}{8}\right]$$
$$= 681.000 - 658.49 = 22.51$$

c. $42.44 + 22.51 = 64.95$

d.

Source	Sum of Squares	df	Mean Square
Between	42.44	2	21.22
Within	22.51	17	1.32
Total	64.95		

$$F = \frac{21.22}{1.32} = 16.08$$

e. The critical value of F is 3.59, found by 0.05 level, 2 df in numerator, 17 df in denominator. Since $16.08 > 3.59$, H_0 is rejected. The means are not all equal.

f. Since 16.08 is also greater than 6.11, the p-value is less than 0.01.

UNIT REVIEW

In the last two chapters, we examined small-sample methods of hypothesis testing for both means and variances. First, we used Student's *t* distribution as the test statistic when we compared a single sample mean to a population mean. Next, we compared two sample means, again using the *t* distribution, to determine if they came from populations with the same mean. We used the *F* distribution to test the hypothesis that two populations had the same variance. We also used the *F* distribution when comparing more than two populations to determine if they have the same mean.

Key Concepts

1. **Student's *t* distribution** is used to test hypotheses about the mean of a single population, the means of two populations, or the difference in related observations. In each of these instances, the population is normal with an unknown standard deviation. Usually, the sample size is smaller than 30. The *t* distribution has the following characteristics:
   - **a.** It is a continuous distribution.
   - **b.** It is somewhat bell-shaped and symmetrical.
   - **c.** There is a "family" of *t* distributions. That is, each time the sample size changes, the *t* distribution changes.
   - **d.** As the sample size increases, the *t* distribution approaches the normal distribution. Often, when $n > 30$ the standard normal distribution is used instead of *t*, because the values of these two statistics are close.
   - **e.** The *t* distribution is more spread out than the normal distribution.
2. When testing the means of more than two populations simultaneously, we use the **analysis of variance (ANOVA).** The test statistic follows the ***F*** distribution, which has the following characteristics:
   - **a.** It is either zero or positive (it cannot be negative).
   - **b.** It is a continuous distribution.
   - **c.** It is positively skewed.
   - **d.** The *F* distribution is based on the ratio of two variances. There is a different *F* distribution each time either of the sample sizes is changed.
3. For the *F* distribution to be employed to test the difference in more than two populations, the following conditions must be met:
   - **a.** The data must be at least of interval scale.

b. The populations should be approximately normally distributed.
c. The population variances should be equal.
d. The samples must be randomly selected.

4. The F distribution is also used to test the hypothesis that two populations have equal variances.

Key Terms

Student's t distribution
Degrees of freedom
Pooled variance estimate
Independent samples
Dependent samples
F distribution
Treatment
Sum of squares total
Sum of squares error
Sum of squares treatment
Analysis of variance (ANOVA)

Key Symbols

t	Depending on its usage, t refers either to the t test itself, the computed value of t, or the critical value of t.
df	Degrees of freedom.
s_p	Pooled estimate of the population standard deviation.
s_d	Standard deviation of the paired differences.
F	The F probability distribution.
k	The number of treatments.
SST	Sum of squares treatment (between columns).
SSE	Sum of squares error (within columns).
SS total	Sum of squares total.

Problems

1. To service her customers more effectively, Ms. Dodd, owner of Damschroders, an exclusive women's boutique, wants to know their mean age. A random sample of 15 customers revealed the mean to be 46.3, years with a standard deviation of 10.6 years. When Ms. Dodd opened her boutique, the manager of the mall where the shop is located had reported the mean age of shoppers to be 40. Can Ms. Dodd conclude that her clients are older? Use the 0.05 significance level.
2. A government testing agency routinely tests various foods to ensure that they meet label requirements. A random sample of 10 one-liter bottles of a soft drink actually contained the following amounts when tested: 0.93, 0.97, 0.96, 1.02, 1.05, 1.01, 1.02, 0.97, 0.98, and 0.97. At the 0.05 significance level, can the testing agency show that the soft-drink manufacturer is underfilling the product?
3. The personnel manager of a large corporation believes that people are now retiring at a later age. A sample of 20 employees who retired in 1970 revealed that their mean age at retirement was 63.7 years, with a standard deviation of 3.2 years. A sample of 15 employees who retired last year revealed the

mean to be 66.5 years and the standard deviation 4.3 years. At the 0.01 significance level, can the manager conclude that the mean age at retirement has increased? Find the p-value.

4. The dean of Northern University believes that student grade point averages have increased in recent years. She obtains a sample of seven student grade point averages for 1985 and eight for this year. Based on these data, can the dean conclude that grades have increased? Use the 0.01 significance level.

Student Grade Point Averages

1985	This Year
2.90	2.70
2.85	3.35
2.67	3.60
1.98	2.75
3.20	2.35
2.65	2.90
2.35	2.98
	3.01

5. Eight upper-middle-class families were surveyed to determine their annual medical expenses. The mean amount was $960 for last year, with a sample standard deviation of $135. Can we conclude that the mean of the population's medical expenses is greater than $900 per year? Use the 0.05 significance level.

6. A study was designed to determine if drinking affects a driver's reaction time. A random sample of 12 people was given a driving-simulator test. Each person was then asked to drink two ounces of whiskey and to repeat the test. The "errors" made in each test were tabulated. At the 0.05 significance level, can it be concluded that people make more errors after having drunk two ounces of whiskey? Find the p-value.

Subject	Before Drinking	After Drinking
A	8	9
B	9	12
C	10	13
D	8	14
E	11	15
F	6	11
G	12	12
H	15	14
I	10	13
J	7	12
K	8	13
L	10	19

7. A large university is concerned about salary discrimination on the basis of sex. To investigate, a sample of 11 female instructors has been obtained and their salaries are determined. For each female selected, a male faculty member with similar tenure status, academic rank, discipline, and so on is obtained and his salary is paired with that of the female instructor. The data (in thousands of dollars) are shown in the table that follows. Can it be concluded that female instructors earn significantly less? Use the 0.05 significance level.

Faculty Pair	Male	Female
1	$46.0	$45.9
2	42.9	43.5
3	43.9	42.7
4	47.8	41.7
5	45.5	44.7
6	44.3	43.0
7	40.7	40.8
8	45.4	44.8
9	46.4	49.5
10	51.0	47.2
11	38.3	38.1

8. All Chicken and Eggs Restaurants are rated for cleanliness and food quality by a corporate quality control department. For a sample of eight restaurants, the ratings by two inspectors are obtained. At the 0.05 significance level, can we conclude that Inspector Powell consistently rates the restaurants higher than does Inspector Saner?

	Inspector	
Restaurant	Powell	Saner
Gallion	79	75
Maumee	70	61
Oak Harbor	46	32
Sylvania	55	59
Rossford	65	58
Russell	72	70
Bradford	63	60
Erie	65	65

9. Suppose that the United States has two sources for short-range rockets. The rockets built by the first source have a mean target error of 68 feet, with a standard deviation of 18 feet, for a sam-

ple of 8 rockets. Those built by the second have a mean target error of 48 feet, with a standard deviation of 12 feet, for a sample of 6 rockets. At the 0.05 significance level, can we conclude that the population variances are the same? If yes, can we conclude that there is more mean error in the first source's rockets at the 0.05 significance level? Find the *p*-value.

10. An investor is trying to decide between oil stocks and utility stocks. He is concerned that there is more variability in the oil stocks. For a sample of 10 oil stocks, the standard deviation of the dividends paid was \$12.51. For a sample of 8 utility stocks, the standard deviation of the dividends paid was \$4.56. At the 0.05 significance level, can we conclude there is more variability in the oil stocks?

11. The Plumber's Union has gathered data on the hourly wages of random samples of plumbers in four southwestern cities. At the 0.05 significance level, can it be shown that the mean hourly wage differs in the four cities?

City			
A	B	C	D
\$14.20	\$14.40	\$16.20	\$15.40
15.40	15.20	15.40	15.80
14.80	15.80	16.00	16.60
13.80	16.20	16.00	14.80
	16.00	15.80	
	15.60		

12. A health spa has two one-week programs for grossly overweight persons. A client may select either of the two. To assess the effectiveness of one program over the other, a sample of 300-pound persons was selected and their weight losses (in pounds) recorded. Seven were in Program A and nine in Program B.

Weight Losses	
Program A	Program B
40	39
38	21
41	29
52	42
27	43
32	28
40	29
	28
	46

Use the 0.05 level of significance to test whether there is a statistically significant difference between the two mean weight-loss programs. What is the *p*-value? Explain what it indicates.

13. The health spa also has three programs designed to lower stress. An incoming group of patients was randomly assigned to the three programs. The reduction in stress (in percent) after the week-long programs is shown below.

Stress Reduction

Program A	Program B	Program C
27	17	22
21	26	31
18	33	16
32	26	18
26		27
		32

Test at the 0.01 level the hypothesis that there is no difference in the effectiveness of the three stress-reducing programs.

14. A fertilizer-mixing machine is set to give 10 pounds of nitrate for every 100 pounds of fertilizer. Eight 100-pound bags were examined and the pounds of nitrate were 8, 10, 9, 11, 7, 10, 9, and 10. Is there reason to conclude that the mean is not equal to 10 pounds? Use a 0.01 significance level.

15. A random sample of 12 women not employed outside the home was asked to estimate the selling prices of two 25-inch color TV sets. Their estimates are:

Homemaker	Model A	Model B
1	$715	$810
2	830	650
3	815	620
4	770	760
5	650	830
6	680	720
7	770	800
8	760	830
9	990	830
10	550	900
11	670	620
12	760	630

Is there sufficient evidence, at the 0.05 level of significance, to claim that homemakers perceive Model B to be more expensive than Model A? Find the p-value.

16. The manager of Sally's Hamburger World suspects that the mean number of customers between 4 p.m. and 7 p.m. differs by day of the week. The data show the number of customers during that period for randomly selected days of the week. At the 0.01 significance level, can it be concluded that the number of customers differs by day of the week?

Monday	Tuesday	Wednesday	Thursday	Friday
86	77	69	78	84
96	102	91	77	88
78	54	86	90	94
66	98	74	84	102
100		82	72	96
		78	74	
		84		

Using Statistics

Situation. Bumper laws vary from state to state. Many states prohibit more than three inches of variation in the heights of bumpers. Some states require only that a vehicle have a bumper in place. Still others have no laws at all. Some municipal executives claim that states with stricter laws generally have stricter law enforcement as well. Statistics on the number of traffic citations issued per thousand registered vehicles is one indicator of the enforcement level. The 50 states have been classified by level of enforcement. Statistics within each group were then computed, with the following results:

Type of Enforcement	Mean Number of Citations	Standard Deviation	Number of States
Very strict	260	30	20
Moderate	280	40	18
Lax	250	29	12

Is there a significant difference among the three enforcement categories in terms of the mean number of citations issued, or could these apparent differences be due to chance?

Discussion Because there are more than two sample means, the ANOVA technique is used. The null hypothesis is that the mean number of citations is the same for the three levels of enforcement. The alternate hypothesis is that the mean number of citations is not the same for the three levels of enforcement. Stated symbolically:

H_0: $\mu_1 = \mu_2 = \mu_3$

H_a: The population means are not all the same

There are three populations, so k is 3 and the number of degrees of freedom in the numerator of the F statistic is 3 – 1 = 2. There is a total of 50 observations (states), so the number of degrees of freedom in the denominator is 50 – 3 = 47. Using the 0.05 significance level and Appendix E, the critical value is estimated to be 3.20.

At this point, we would look for the familiar ANOVA formulas:

$$SS \text{ total} = \Sigma X^2 - \frac{(\Sigma X)^2}{N}$$

$$SST = \sum\left(\frac{T_c^2}{n_c}\right) - \frac{(\Sigma X)^2}{N}$$

$$SSE = \Sigma X^2 - \sum\frac{T_c^2}{n_c}$$

We do not know some of the values, such as ΣX^2 and ΣX, but we can determine them through an understanding of $\overline{X}$ and s^2.

Recall that in formula 3–1, $\overline{X} = \Sigma X/n$. With a little algebra, $n(\overline{X}) = \Sigma X$. Therefore, for the very strict enforcement states $n = 20$ and $\overline{X} = 260$, so $\Sigma X = 20(260) = 5200$. Similarly, the other values are 18(280) = 5040 for the moderate enforcement group and 12(250) = 3000 for the lax enforcement group. These three values are the column totals (T_c). When the three values are added, the result is 13,240, which is the sum of all Xs, or ΣX.

Now we can use formula 12–1 to determine the sum of squares due to the treatments:

$$SST = \sum\left(\frac{T_c^2}{n_c}\right) - \frac{(\Sigma X)^2}{N}$$

$$= \left[\frac{(5{,}200)^2}{20} + \frac{(5{,}040)^2}{18} + \frac{(3{,}000)^2}{12}\right] - \frac{(13{,}240)^2}{50} = 7{,}248$$

From formula 4–8, the sample variance is $(\Sigma X^2 - (\Sigma X)^2/n)/(n - 1)$. We could rearrange this equation to find $\Sigma X^2 - (\Sigma X)^2/n$. It is equal to $(n - 1)s^2$. For example, for the very strict group, $(n - 1)s^2 = (20 - 1)(30)^2 = 17{,}100$. Applying this to each of the three populations, we find SSE:

$$SSE = \Sigma X^2 - \sum \frac{T_c^2}{n_c} = (20-1)(30)^2 + (18-1)(40)^2 + (12-1)(29)^2 = 53,551$$

This information can be placed in the familiar ANOVA table and the details of the ANOVA test carried out:

Source	Sum of Squares	df	Mean Square	F
Treatment	7,248	2	3624	3.18
Error	53,551	47	1139	
Total	60,799	49		

Because the computed value of F (3.18) is less than the critical value of 3.20, the null hypothesis is not rejected. The data show no significant differences among the categories of enforcement. Note, however, that the computed value is very close to the critical value. A different decision would be made at the 0.10 significance level. In other words, the p-value is between 0.05 and 0.10.

CHAPTER 13

Correlation Analysis

OBJECTIVES

When you have completed this chapter, you will be able to

- describe the relationship between two variables;
- compute Pearson's coefficient of correlation;
- compute the coefficients of determination and nondetermination;
- test the statistical significance of the coefficient of correlation;
- compute the Spearman coefficient of rank correlation.

CHAPTER PROBLEM **Our Books are Priceless!**

HANNAH SIMPSON IS THE PRESIDENT OF STUDENT GOVERNMENT AT NORTH CENTRAL STATE UNIVERSITY. FOR SOME TIME, SHE AND HER COLLEAGUES HAVE BEEN CONCERNED WITH THE INCREASING COST OF TEXTBOOKS. AT A RECENT MEETING, MS. SIMPSON DECIDED TO FORM A COMMITTEE TO EXAMINE THE MATTER. WHAT FACTORS AFFECT THE COST OF A TEXTBOOK? THE NUMBER OF PAGES? THE NUMBER OF PICTURES? THE NUMBER OF FORMULAS? SHOULD THE STUDENT COMMITTEE LOOK AT BOOKS IN TECHNICAL AREAS, SUCH AS MATHEMATICS AND CHEMISTRY, TO FIND IF THEY COST MORE THAN, SAY, SOCIOLOGY AND PSYCHOLOGY TEXTBOOKS?

INTRODUCTION

In Chapters 10–12, we dealt with hypothesis tests involving means and proportions. The techniques concentrated on only a *single* feature of the sampled item, such as income. This chapter begins a study of the relationship between *two or more* variables. We may want to determine if there is any relationship between the number of years of education completed by federal government employees and their incomes. Or, we may want to explore one of these questions: Does the crime rate in inner cities vary with the unemployment rate in those cities? Is there a relationship between the amount of money spent advertising a product such as a toothpaste and its sales? Is there any relationship between the number of hours studied and a student's grade on an examination? Note that in each case there are two separate characteristics—years of education and income, for example.

CORRELATION ANALYSIS

The study of the relationship between two variables is called **correlation analysis.** The objective of correlation analysis is to determine the degree of correlation (relationship) between variables, from zero (no correlation) to perfect (complete) correlation, designated as 1.0. Our attention will focus first on the correlation between two interval-scaled variables, then on the relationship between two ordinal-scaled variables.

> **Correlation Analysis** The statistical techniques used to determine the strength of the relationship between two variables.

The Scatter Diagram

One tool that is very useful for visualizing the relationship between two variables is the **scatter diagram.**

> **Scatter Diagram** A graphic tool that portrays the relationship between two variables.

Problem Recall from the Chapter Problem that a student committee was formed to look into the factors affecting the cost of textbooks. The committee selected a sample of textbooks currently in use at North Central State. After considerable discussion and consultation with outside experts, it decided to consider the relationship between cost and length first.

Solution A random sample of six textbooks was chosen from those on sale in the student book store. As Table 13–1 indicates, the selling price of the textbooks is somehow determined by, or dependent upon, the number of pages.

We call the selling price the dependent variable and the number of pages the independent variable. The **dependent variable** is thus the variable being estimated and the **independent variable** is the variable used as the estimator. It is traditional to put the dependent variable on the vertical axis (Y) and the independent variable on the horizontal axis (X) of the scatter diagram. We plot the paired data for Title A ($X = 400$, $Y = 44$) by moving to 400 on the X-axis, then moving vertically to a position opposite 44 on the Y-axis, and placing a dot at that intersection (see the scatter diagram that follows). This process is continued for all the books in the sample. The completed scatter diagram is shown as Figure 13–1.

Table 13-1

Number of Pages and Selling Price

Title	Number of Pages X	Selling Price Y
A	400	$44
B	600	47
C	500	48
D	600	48
E	400	43
F	500	46

Figure 13-1

Scatter Diagram of Number of Pages and Selling Price

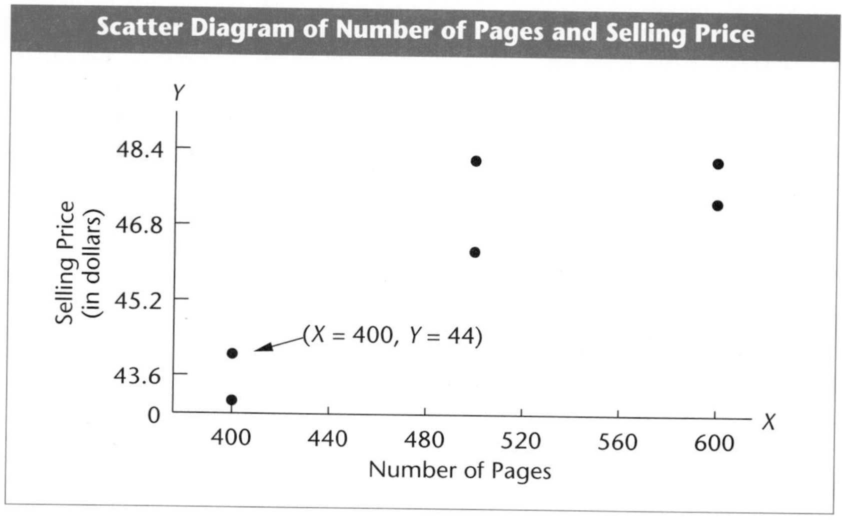

As will be explained shortly, there is indeed a very strong relationship (correlation) between the number of pages in each book and its selling price. Books like Title D, which is relatively long (600 pages), sell for a higher-than-average price ($48).

Self-Review 13-1

Answers to the Self-Review problems are at the end of the chapter.

An astute college recruiter noted that the enrollment figures at the local campus of the state university seemed to fluctuate with the unemployment rate for the region. To establish whether or not his suspicions were justified, the recruiter collected relevant unemployment and enrollment figures for the same time periods. His findings were:

Percent Unemployed X	State University Campus Enrollment Y
3.4	13,500
3.3	14,500
4.6	15,200
5.1	14,900
4.5	14,700
3.5	14,300
5.7	15,700
7.6	17,100

a. Construct a scatter diagram for the paired data.
b. As unemployment increases, does enrollment appear to increase, decrease, or remain the same?

STAT BYTE

The town of Ocean City, Maryland, relies heavily on summer-tourist revenue. Officials need an accurate estimate of the number of visitors each weekend to judge the success of promotional campaigns. Using regression analysis, they found a correlation between the number of visitors each day and the volume of water used. Now, instead of counting the cars coming over the bridge into the city, they simply read the municipal water meter. Can you imagine what it might look like during halftime at the Super Bowl?

Exercises

Answers to the even-numbered exercises are at the back of the book.

1. A Peace Corps agronomist is studying the relationship between the mean temperature and the yield in bushels per acre for a crop. He collected the following information for several regions:

Region	Temperature (in °C) X	Yield (in bushels per acre) Y
1	4	1
2	8	9
3	10	7

Region	Temperature (in °C) X	Yield (in bushels per acre) Y
4	9	11
5	11	13
6	6	7

a. Construct a scatter diagram for the paired data.

b. As temperature increases, does yield appear to increase, decrease, or remain the same?

2. A human resources trainee believes that there is a relationship between the number of years of employment with the company and the number of days a year an employee is absent from work. She obtains the following information from the company records of six employees picked at random:

Employee	Length of Employment (in years) X	Absences Last Year (in days) Y
Phuong	1	8
Fadale	5	1
Sasser	2	7
Dilone	4	3
Thelin	4	2
Tortelli	3	4

a. Construct a scatter diagram for the paired data.

b. As the years of employment increase, do absences appear to increase, decrease, or remain the same?

3. A retailer of men's apparel wants to determine the correlation between the number of suits sold during a day and the number of salespeople. A random sample of the company's files resulted in the following paired information:

Salespeople on Duty X	Suits Sold Y
3	7
1	5
2	6
3	6
4	9
5	10
1	4

Plot a scatter diagram with the number of salespeople on the X-axis and the number of suits sold on the Y-axis. Does it appear that more suits are sold when more salespeople are on duty?

4. A pair of dice are rolled five times and the number of spots appearing face up is noted each time. The results are:

First Die X	Second Die Y
6	4
4	4
5	1
6	2
1	3

Develop a scatter diagram, scaling the first die on the X-axis and the second die on the Y-axis. What comments can you make? Would you expect to find a relationship?

THE COEFFICIENT OF CORRELATION

About 1900, Karl Pearson, who made significant contributions to the science of statistics, developed a measure that describes the relationship between two sets of *interval-scaled variables.* It is called the coefficient of correlation and is designated by the letter r.

Coefficient of Correlation A measure of the strength of the association between two variables.

Pearson's r is also known as the product-moment correlation coefficient, partly to distinguish it from other correlation coefficients. It is a valid measure of correlation if the relationship between the variables is *linear.* As illustrated by the scatter diagram in Figure 13–2, if all data points lie on a straight line, the correlation coefficient is either +1.00 or –1.00, depending on the direction of the slope of the line. Coefficients of +1.00 or –1.00 describe *perfect correlation.*

If there is no relationship between X and Y, r will be zero, and the dots on the corresponding scatter diagram will be randomly scattered. This situation and several others are portrayed in Figure 13–3.

Figure 13-2

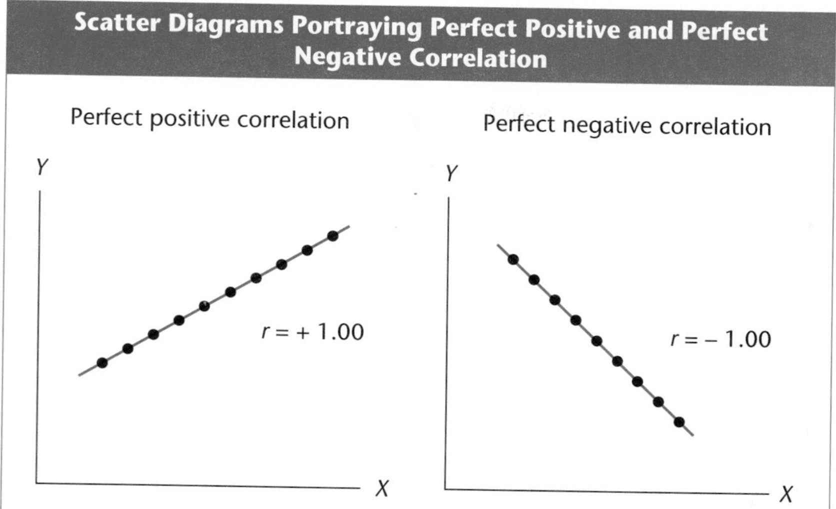

Figure 13-3

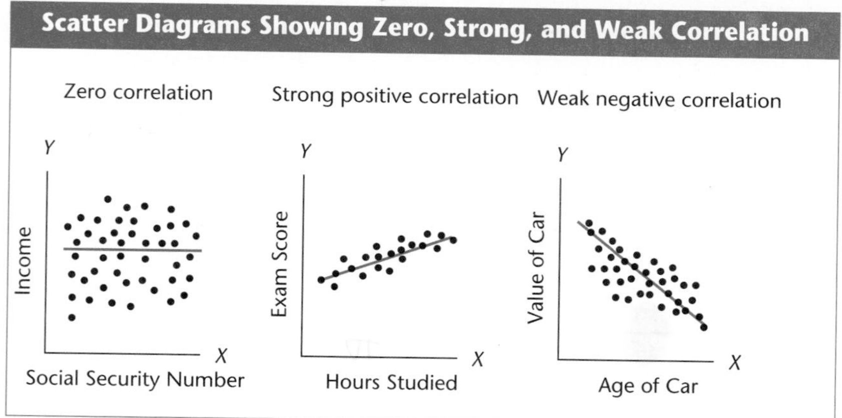

The strength and direction of the coefficient of correlation are summarized in the diagram on the following page. Negative numerical values such as –0.92 or –0.48 signify inverse correlation, whereas positive numerical values such as +0.83 and +0.46 indicate direct correlation. The closer Pearson's r is to 1.00 in either direction, the greater the strength of the correlation. Note, however, that *the strength of the correlation is not dependent on the direction.* Therefore, –0.12 and +0.12 are equal in strength (both weak). Coefficients of +0.94 and –0.94 are also equal in strength (both very strong).

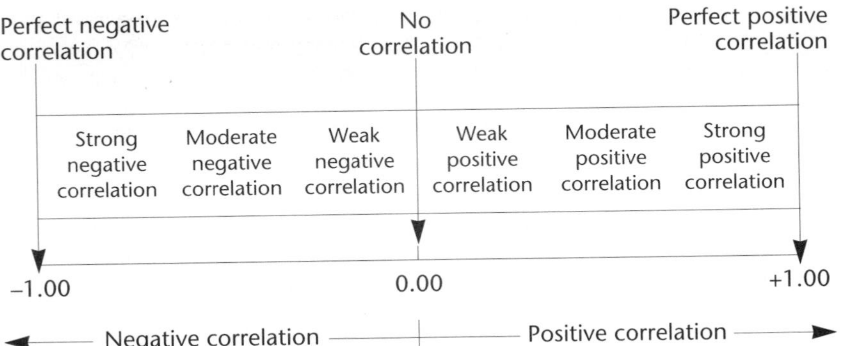

How is the value of the coefficient of correlation determined? Suppose we draw a vertical line through the mean X-value and a horizontal line through the mean Y-value. For example, look at Figure 13–4, where the data on the number of pages and the selling

4 Quadrants made

Figure 13-4

Computation of the Coefficient of Correlation

Y (Cost): 0, 42.0, 44.0, 46.0, 48.0

X (Pages): 400, 440, 480, 520, 560, 600

$\bar{X}$, $\bar{Y}$

$X - \bar{X} = 600 - 500 = 100$

$Y - \bar{Y} = 47 - 46 = 1$

$Y - \bar{Y} = 44 - 46 = -2$

$X - \bar{X} = 400 - 500 = -100$

I, II, III, IV

above the mean

Below the mean

the mean cost of the 6 books

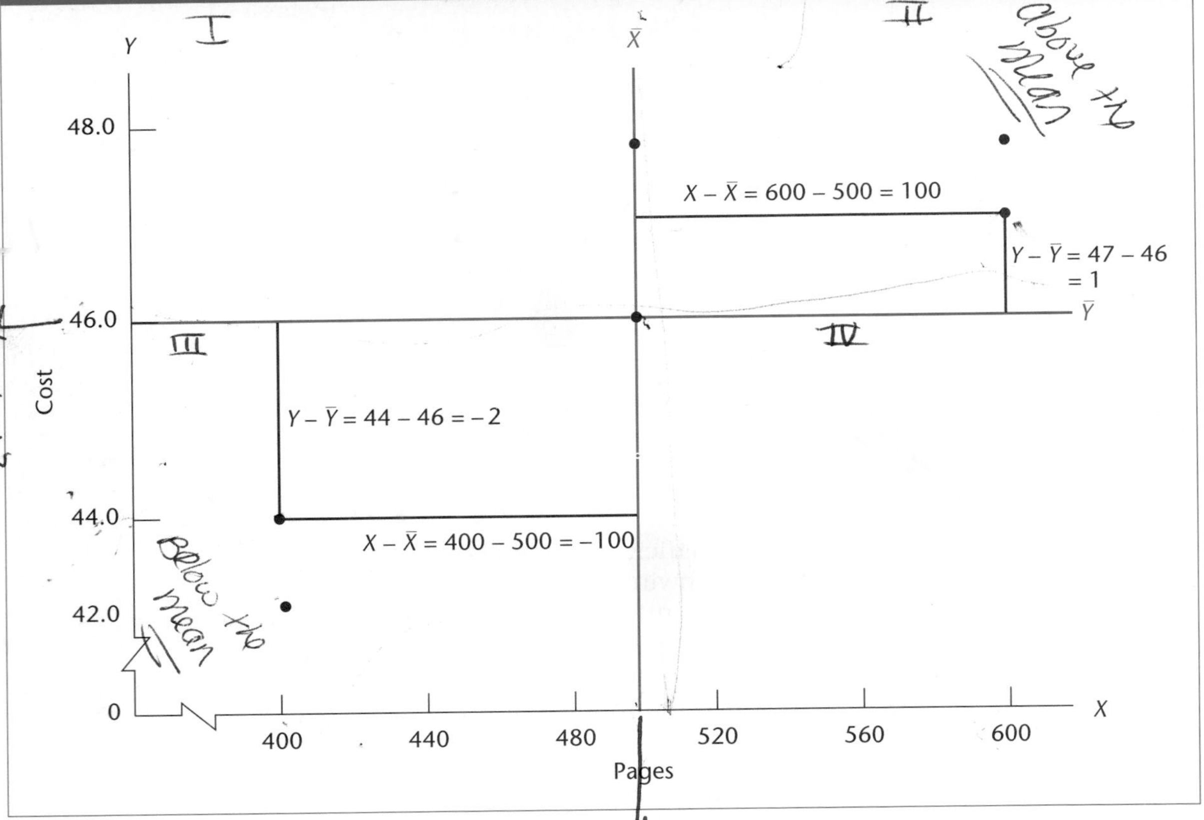

500

the mean number of text books pages for the 6 books.

price of the sampled textbooks are presented. A vertical line is added through 500 pages, the mean number of pages for the six textbooks in the sample. A horizontal line is drawn through $46, the mean cost of the six sampled texts. These lines pass through the "center" of the data and divide the scatter diagram into four quadrants.

If the two variables are directly related, when the selling price is above the mean, the number of pages is as well. These points appear in the upper-right quadrant. Similarly, when the selling price is below the mean, so is the number of pages; these points fall in the lower-left quadrant. Textbook A, for example, is 100 pages shorter than the mean length and sells for $2 less than the mean selling price. It is therefore located in the lower-left quadrant. Textbook B is 100 pages longer than the mean length and sells for $1 more than the mean selling price. It is therefore located in the upper-right quadrant. The deviations from the mean number of pages and from the mean selling price are summarized in Table 13–2 for the entire sample of the six textbooks.

In both the upper-right and lower-left quadrants, the product of $(X - \bar{X})(Y - \bar{Y})$ is positive because both of the factors have the same sign. This happens for all the points in the textbook example. The resulting correlation coefficient is a positive value.

If the two variables are inversely related, one variable will be above the mean and the other below the mean. Most of the points in this case occur in the upper-left and lower-right quadrants. Now, $(X - \bar{X})$ and $(Y - \bar{Y})$ will have opposite signs, so their product is negative. The resulting correlation coefficient is negative.

What happens if there is no relationship between the two variables? The points on the scatter diagram will appear in all four quadrants. The positive products of $(X - \bar{X})(Y - \bar{Y})$ will be offset by the negative products, and their sum will be near zero. This leads to a correlation coefficient near zero.

Pearson also wanted r to be unaffected by the particular scales being used. For example, if we had used cents instead of dollars in

Table 13-2

Deviations from the Mean and Their Products

Textbook	Pages	Cost	$X - \bar{X}$	$Y - \bar{Y}$	$(X - \bar{X})(Y - \bar{Y})$
A	400	44	–100	–2	200
B	600	47	100	1	100
C	500	48	0	2	0
D	600	48	100	2	200
E	400	43	–100	–3	300
F	500	46	0	0	0
					800

our textbook example, we would want the value of the correlation coefficient to be the same value. The correlation coefficient is independent of the scale used if we divide the term $\Sigma(X - \overline{X})(Y - \overline{Y})$ by the sample standard deviations. It is also made independent of the sample size and bounded by the values +1.00 and –1.00, if we divide by $(n - 1)$.

This reasoning leads to the following formula:

$$r = \frac{\Sigma(X - \overline{X})(Y - \overline{Y})}{(n-1)(s_y)(s_x)} \qquad \text{13-1}$$

To compute the coefficient of correlation, we use the standard deviations of the sample of six page lengths and six selling prices. The standard deviation of the length is 89.44 pages, and the standard deviation of the selling price is \$2.10. First, formula 4–8 is used to compute the sample variance; then we take the square root to get the sample standard deviation:

$$s_x^2 = \frac{\Sigma X^2 - \frac{(\Sigma X)^2}{n}}{n-1} = \frac{1{,}540{,}000 - \frac{(3000)^2}{6}}{6-1} = 8000$$

$$s_x = \sqrt{8000} = 89.44$$

$$s_y^2 = \frac{\Sigma Y^2 - \frac{(\Sigma Y)^2}{n}}{n-1} = \frac{12{,}718 - \frac{(276)^2}{6}}{6-1} = 4.40$$

$$s_y = \sqrt{4.40} = 2.10$$

We now insert these values into formula 13–1 to determine the coefficient of correlation:

$$r = \frac{\Sigma(X - \overline{X})(Y - \overline{Y})}{(n-1)(s_y)(s_x)} = \frac{800}{(6-1)(89.44)(2.10)} = 0.85$$

The coefficient of correlation r can also be computed by

$$r = \frac{n(\Sigma XY) - (\Sigma X)(\Sigma Y)}{\sqrt{\left[n(\Sigma X^2) - (\Sigma X)^2\right]\left[n(\Sigma Y^2) - (\Sigma Y)^2\right]}} \qquad \text{13-2}$$

where

n is the number of paired observations.
ΣX is the sum of the X-variable.
ΣY is the sum of the Y-variable.
ΣX^2 is the X-variable squared and the squares summed.
$(\Sigma X)^2$ is the X-variable summed and the sum squared.
ΣY^2 is the Y-variable squared and the squares summed.
$(\Sigma Y)^2$ is the Y-variable summed and the sum squared.

Problem The data on selling prices and number of pages in the textbooks are repeated below.

Title	Number of Pages X	Selling Price Y
A	400	\$44
B	600	47
C	500	48
D	600	48
E	400	43
F	500	46

Determine the coefficient of correlation.

Solution The totals and the sums of squares needed are in Table 13–3 at the top of the following page. Computing gives:

$$r = \frac{n(\Sigma XY) - (\Sigma X)(\Sigma Y)}{\sqrt{\left[n(\Sigma X^2) - (\Sigma X)^2\right]\left[n(\Sigma Y^2) - (\Sigma Y)^2\right]}}$$

$$= \frac{6(138{,}800) - (3{,}000)(276)}{\sqrt{\left[6(1{,}540{,}000) - (3{,}000)^2\right]\left[6(12{,}718) - (276)^2\right]}}$$

$$= \frac{4800}{\sqrt{(240{,}000)(132)}} = 0.85$$

Because 0.85 is close to the perfect correlation value of 1.00, the student committee at North Central State can conclude that there is a strong relationship between the number of pages in a book and its selling price.

Table 13-3 **Calculations Needed to Determine the Coefficient of Correlation**

Title	X	Y	XY	X^2	Y^2
A	400	44	17,600	160,000	1,936
B	600	47	28,200	360,000	2,209
C	500	48	24,000	250,000	2,304
D	600	48	28,800	360,000	2,304
E	400	43	17,200	160,000	1,849
F	500	46	23,000	250,000	2,116
	3,000	276	138,800	1,540,000	12,718

Self-Review 13-2

Is there a relationship between the number of votes received by candidates for public office and the amount spent on their campaigns? The following sample information was gathered for a recent election:

n Candidate	Amount Spent on Campaign (in thousands of dollars) X	Votes Received (in thousands) Y
Weber	$3	14
Taite	4	7
Spencer	2	5
Lopez	5	12

a. Draw a scatter diagram.
b. Compute the Pearson coefficient of correlation.
c. Interpret the correlation coefficient.

Exercises

5. In Exercise 1, an agronomist was studying the relationship between the average temperature (in °C) and the yield in bushels per acre for a certain fall crop. The data are repeated below.

Region	Temperature (in °C) X	Yield (in bushels per acre) Y
1	4	1
2	8	9

Region	Temperature (in °C) *X*	Yield (in bushels per acre) *Y*
3	10	7
4	9	11
5	11	13
6	6	7

Compute the coefficient of correlation and interpret it.

6. In Exercise 2, a human resources trainee was studying the relationship between the number of years of employment with the company and the number of days absent from work last year. The following information was gathered from company records:

Employee	Length of Employment (in years) *X*	Absences Last Year (in days) *Y*
Phuong	1	8
Fadale	5	1
Sasser	2	7
Dilone	4	3
Thelin	4	2
Tortelli	3	4

Compute the coefficient of correlation and interpret it.

7. Is age related to the length of stay of surgical patients in a hospital? The following sample data were obtained to study the relationship:

Age	Days in Hospital
40	11
36	9
30	10
27	5
24	12
22	4
20	7

a. Draw a scatter diagram. X is age and Y is days in hospital.
b. Compute the coefficient of correlation.
c. Interpret the results.

8. Is there a relationship between the number of golf courses in the United States and the divorce rate? The following sample information was collected for several regions:

Number of Golf Courses *X*	Divorce Rate (per 1,000 Population) *Y*
28	2.2
38	2.5
47	3.5
54	4.1
62	4.8
66	5.0

a. Draw a scatter diagram.
b. Compute the coefficient of correlation.
c. Interpret the results.

THE COEFFICIENT OF DETERMINATION

The coefficient of correlation allowed us to make statements such as "The relationship between the two variables is very strong." But how can one measure "very strong"? A measure of correlation that does have a more precise meaning is the **coefficient of determination.**

Coefficient of Determination The proportion of the total variation in one variable that is explained by the other variable.

This technique results in a proportion, or percent, that makes it relatively easy to arrive at a precise interpretation. We compute it by squaring the coefficient of correlation. The coefficient of determination may vary from 0 to 1.00 or, converted to a percent, from 0 to 100%. It is usually represented by r^2.

To illustrate its computation and meaning, let us return to the Chapter Problem where the relationship between the number of pages and the selling price of a sample of six textbooks was considered. The coefficient of correlation r is computed to be 0.85. The coefficient of determination r^2 is 0.72, found by $(0.85)^2$. Thus, the student committee can conclude that the variation in page lengths of the textbooks explains, or accounts for, 72% of the variation in selling price.

The **coefficient of nondetermination** is the proportion of the total variation *not* explained. It is $1 - r^2$. For this problem, it is 0.28, found by $1 - (0.85)^2 = 1 - 0.72 = 0.28$. Interpreting, we conclude that 28% of the total variation in selling price is not explained (accounted for) by the number of pages in the textbook.

The magnitude of the coefficient of determination is smaller than that of the coefficient of correlation. It is a more conservative measure of the relationship between two variables, and for this reason many statisticians prefer it. To put it another way, the coefficient of correlation tends to overstate the association between two variables. A correlation coefficient of 0.70, for example, would suggest that there is a fairly strong relationship between two variables. Squaring r, however, gives a coefficient of determination of 0.49, which is a somewhat smaller value.

Self-Review 13-3

In Self-Review 13–2, you computed the coefficient of correlation between campaign expenses and number of votes received to be 0.43.

a. What is the coefficient of determination?
b. What is the coefficient of nondetermination?
c. Interpret the two measures.

Exercises

9. A study is made of the relationship between the number of passengers on an aircraft (X) and the total weight in pounds of luggage stored in the aircraft's baggage compartment (Y). The coefficient of correlation is computed to be 0.94.
a. What is the coefficient of determination?
b. What is the coefficient of nondetermination?
c. Interpret these two measures.

10. The coefficient of correlation between the number of people on the beach at 4 p.m. and the high temperature for that day at Sunner's Creek is computed to be 0.96.
a. What is the coefficient of determination?
b. Find the coefficient of nondetermination.
c. Interpret the two measures.

11. In a government study of the relationship between the tar content and nicotine content (in milligrams) of a cigarette, the coefficient of correlation is 0.43.
a. Find the coefficient of determination.
b. Compute the coefficient of nondetermination.
c. Explain the meaning of these two numbers.

12. At Middletown High, a coefficient of correlation of -0.68 was found between the number of high-school activities offered and the number of drug-related suspensions.

a. Determine the coefficient of determination.
b. What is the coefficient of nondetermination?
c. Interpret the two measures.

TESTING THE SIGNIFICANCE OF THE COEFFICIENT OF CORRELATION

The student committee at North Central State University selected only six textbooks for a study of the relationship between selling prices and number of pages. The coefficient of correlation was calculated to be 0.85, indicating a very strong, positive association between the two variables. The question arises, however, whether it is possible—due to the small size of the sample—that the correlation in the population is really zero and the apparent relationship is due to chance. The "population" in this case might be *all* textbooks sold at the university.

To decide formally whether the correlation in the population could be zero, we apply the five-step hypothesis-testing procedure used in Chapters 10–12. The statement that the coefficient of correlation in the population is, in fact, zero becomes our null hypothesis. The alternate hypothesis states that the coefficient of correlation in the population is not zero. The population correlation coefficient is usually represented by the Greek letter rho (ρ). Symbolically, then, the null hypothesis and the alternate hypothesis are:

$$H_0 : \rho = 0$$
$$H_a : \rho \neq 0$$

The appropriate test statistic follows the t distribution with $n - 2$ degrees of freedom and has the formula

STAT BYTE

In many books and movies, a dying person often holds onto life until some special event occurs. For example, a mother waits until her son returns from war. Is this merely a device in romantic fiction? There is strong statistical evidence that some people actually postpone their deaths until their birthdays!

$$t = \frac{r\sqrt{n-2}}{\sqrt{1-r^2}} \qquad \text{13-3}$$

where

r is the sample coefficient of correlation.
n is the number of items in the sample.

If the computed t-value lies in the rejection region, we reject the null hypothesis that there is no correlation between the variables. Therefore, a t-value in the rejection region indicates that there is a significant correlation in the population between the two variables. Shown in a diagram:

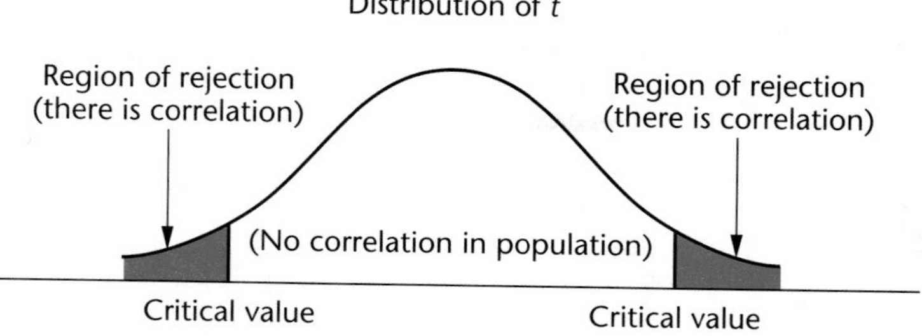

Problem Recall from the Chapter Problem that the coefficient of correlation for the sample of six textbooks at North Central State University was computed to be 0.85. Is it possible that, due to the small size of the sample, the coefficient of correlation in the population is zero? We will now test this hypothesis at the 0.05 level of significance.

Solution Use the five-step hypothesis-testing procedure.

STEP 1. State the null and alternate hypotheses. They are:

$$H_0 : \rho = 0$$
$$H_a : \rho \neq 0$$

STEP 2. Select the level of significance. The null hypothesis is to be tested at the 0.05 level. Remember that the level of significance is the same as the probability of a Type I error. A Type I error is rejecting a null hypothesis when it is actually true.

STEP 3. Select the appropriate test statistic. It is formula 13–3:

$$\boxed{t = \frac{r\sqrt{n-2}}{\sqrt{1-r^2}}} \quad \text{with } n-2 \text{ degrees of freedom}$$

As noted, $n - 2$ degrees of freedom are connected with this test statistic. There are six paired observations in the problem. Thus, n is 6 and $n - 2 = 6 - 2 = 4$ degrees of freedom.

STEP 4. Formulate a decision rule. The alternate hypothesis states that ρ (the coefficient of correlation in the population) is not zero. Because it does not specify a direction, we apply a two-tailed test. The critical values of t are given in Appendix D. Go down the left column to 4 degrees of freedom. Then, move horizontally to the critical value in the column for the 0.05 level of significance, two-tailed test. The value is 2.776. The decision rule, therefore, is: Do not reject the null hypothesis that ρ is zero if the computed value of t is between –2.776 and 2.776. Otherwise, reject the null hypothesis and accept the alternate hypothesis that ρ is not zero. Shown graphically:

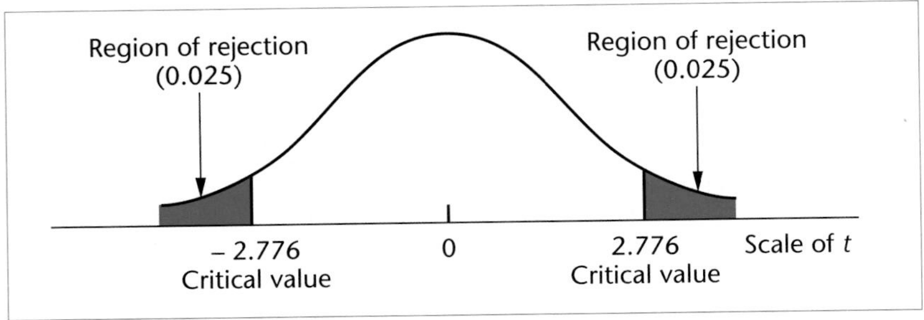

STEP 5. Make a decision. In testing the hypothesis that $\rho = 0$, the final step is to compute t and to reach a decision based on the decision rule in the previous step. Computing t gives:

$$t = \frac{r\sqrt{n-2}}{\sqrt{1-r^2}} = \frac{0.85\sqrt{6-2}}{\sqrt{1-(0.85)^2}} = 3.23$$

The computed t-value of 3.23 is in the rejection region. Therefore, H_0—that the population correlation coefficient is zero—is rejected at the 0.05 level of significance. H_a is accepted. The conclusion is that the correlation between selling price and number of pages is not zero. The p-value is between .05 and .02.

Self-Review 13-4

In Self-Review 13–3, we found that the correlation between number of votes received and campaign expenses was 0.43.

a. Using a 0.01 significance level, test the hypothesis that correlation in the population is zero. The alternate hypothesis that ρ is not zero.

b. Explain your results.

Exercises

13. A major airline selected a random sample of 25 flights and found that the correlation between the number of passengers and the total weight in pounds of luggage stored in the luggage compartment is 0.94. Using a 0.05 significance level, test the hypothesis that the correlation in the population is zero.

14. A sociologist claims that the success of students in college (measured by their GPA) is unrelated to their family's income. For a random sample of 20 students, the coefficient of correlation was computed as 0.40. Using the 0.01 significance level, test the hypothesis that the coefficient of correlation in the entire population of students is zero.

15. An Environmental Protection Agency study of 12 cars revealed a correlation of 0.47 between engine size and performance. Use the 0.01 significance level to test the hypothesis that the population correlation coefficient is zero.

16. A study of college soccer games revealed the correlation between shots attempted and goals scored to be 0.21 in a sample of 20 games. Use the 0.05 significance level to test the hypothesis that the population correlation coefficient is zero.

SPEARMAN'S RANK-ORDER CORRELATION

In the previous section, the scale of measurement was interval. However, measures of correlation are also available for ordinal (ranked) data. The most widely used one was introduced in 1904 by the British statistician Charles Edward Spearman (1863–1945). Spearman's rank-order correlation coefficient is similar to Pearson's r but is designated by the symbol r_s. It is determined by

$$r_s = 1 - \frac{6\Sigma d^2}{n\left(n^2 - 1\right)} \qquad \textbf{13-4}$$

where

d is the difference between the ranks for each pair of observations.

n is the number of paired observations.

Spearman's rank-order correlation coefficient ranges (as does Pearson's r) from –1.00 to +1.00, with a +1.00 indicating that the ranks are in perfect agreement. A –1.00 indicates that the variables are inversely related but in perfect agreement. An r_s of 0 reveals that there is no relationship between the ranks. The scatter diagrams in Figures 13–5 and 13–6, on the following page, portray perfect agreement and perfect disagreement, respectively.

Problem The director of nursing at a city hospital asked two staff nurses to rank 10 patients according to the difficulty of care required. A "difficult" assignment would include giving the patient a complete bath and monitoring the intravenous equipment. A less difficult assignment would be to give medication to a patient only at bedtime. Table 13–4 on page 471 summarizes the ranking of each patient by the two nurses.

Figure 13-5

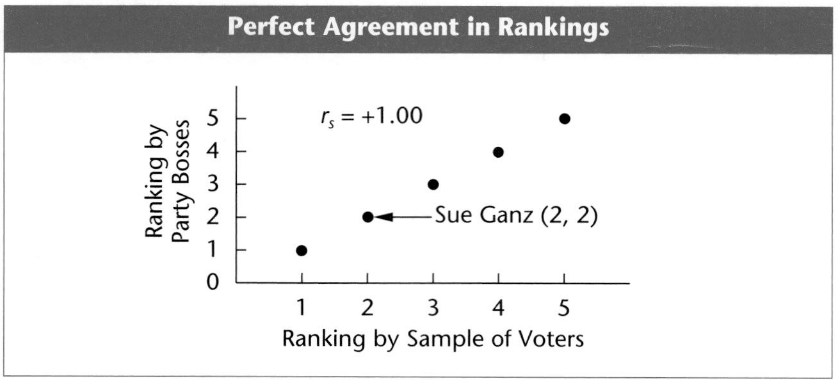

Figure 13-6

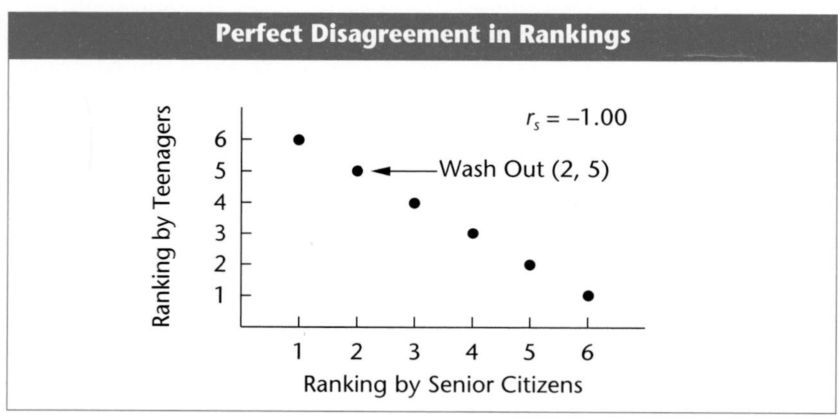

1. Compute Spearman's rank-order correlation coefficient.
2. At the 0.05 significance level, can we conclude that the ranks are positively correlated in the population?

Solution The computations needed to determine the rank-order correlation coefficient are in Table 13–5.

Determining Spearman's rank-order correlation coefficient, we get:

$$r_s = 1 - \frac{6\Sigma d^2}{n(n^2 - 1)} = 1 - \frac{6(28)}{10(99)} = 1 - 0.170 = 0.83$$

The coefficient of 0.83 indicates a very strong agreement between the two nurses with respect to the rankings given the patients.

Table 13-4

Two Nurses' Rankings of Difficulty of Care for 10 Patients

Patient	Nurse Scott's Ranking	Nurse Palmer's Ranking
Ms. Garcia	1	2
Ms. Hibner	2	1
Ms. Schaefer	3	5
Ms. Gerko	4	3
Ms. Owens	5	7
Mr. Bunt	6	6
Mr. Kolby	7	4
Mr. Cassow	8	10
Mr. Bianchi	9	9
Mr. Loftop	10	8

Table 13-5

Two Nurses' Rankings and Calculations Needed for the Rank-Order Correlation Coefficient

Patient	Nurse Scott's Ranking	Nurse Palmer's Ranking	Difference Between Rankings d	Difference Squared d^2
Ms. Garcia	1	2	−1	1
Ms. Hibner	2	1	1	1
Ms. Schaefer	3	5	−2	4
Ms. Gerko	4	3	1	1
Ms. Owens	5	7	−2	4
Mr. Bunt	6	6	0	0
Mr. Kolby	7	4	3	9
Mr. Cassow	8	10	−2	4
Mr. Bianchi	9	9	0	0
Mr. Loftop	10	8	2	4
			0	28

Test of Hypothesis about a Correlation Coefficient

Now we will determine if the ranks are positively correlated in the population. When the number of paired observations is 10 or more, we can use Student's t to determine the significance of Spearman's correlation coefficient. The sampling distribution follows the t distribution with $n - 2$ degrees of freedom. The computed value of t is found by:

$$t = r_s\sqrt{\frac{n-2}{1-r_s^2}}$$

In this example, the null and alternate hypotheses are:

H_0: There is no correlation among the ranks in the population.
H_a: There is positive rank correlation in the population.

The decision rule, determined from Appendix D, calls for rejecting the null hypothesis if t is greater than 1.860. This critical value is determined by using 8 *df*, and a one-tailed test at the .05 significance level.

The computed value of t is 4.21, found by:

$$t = r_s\sqrt{\frac{n-2}{1-r_s^2}} = 0.83\sqrt{\frac{10-2}{1-(0.83)^2}} = 4.21$$

The null hypothesis is rejected and the alternate is accepted because $4.21 > 1.86$. There is positive agreement between the rankings of the two nurses as to the more difficult patients in the population. The p-value is less than .005, for a one-tailed probability. Recall that a p-value reports the probability of a value of the test statistics this large or larger when the null hypothesis is true.

In the previous Problem, each nurse *ranked* a patient with respect to difficulty of care. In some problems, we may need to convert data to ranks before computing the rank-order correlation coefficient. For example, two executives rated the performance of six subordinate executives on a scale of 0 to 100, with 0 representing very undesirable performance and 100 representing outstanding performance (see the Ratings columns in Table 13–6). Because the data are not known to be of interval scale, Pearson's r is not appropriate. However, if we convert each of the executive ratings to rankings, the rank-order correlation may be used to describe the agreement between the two executives' ratings.

Referring to the first column of Table 13–6, note that Harn rated Kingman highest (81). Therefore, Kingman is ranked 1. Wong was rated next-highest (76) and is ranked 2, and so on. Kander's ratings are treated the same way—that is, O'Hara, with a 90 rating, is ranked 1, Wong with an 86 rating is ranked 2, and so on. (Note that the ratings for each executive were ranked from high to low. However, had each set of ratings been ranked from low to high, the rank-order correlation coefficient would be the same.)

Suppose now that Harn had rated Kingman's and Wong's performance as equal, assigning 83 points to each. How would their respective ranks be determined? Both would be ranked 1.5, which is the average of ranks 1 and 2. Had Kander rated O'Hara, Wong, and

Table 13-6

Ratings of Subordinates by Two Executives, with Conversion to Rankings

	Ratings		**Rankings**	
Subordinate	Harn's	Kander's	Harn's	Kander's
Glasser	42	55	4	3
Kingman	81	40	1	4
Cardellini	27	10	6	5
O'Hara	60	90	3	1
Compton	36	8	5	6
Wong	76	86	2	2

Glasser highest, with 91 points each, how would their ranks be determined? These three would tie for ranks 1, 2, and 3. The average of the three tied ranks would be 2, found by $(1 + 2 + 3)/3$. Therefore, each of the three would be given the rank 2.

Self-Review 13-5

A professional and a blue-collar worker were asked to rank 12 occupations according to the social status they attached to each. A ranking of 1 was assigned to the occupation with the highest status, a 2 to the occupation with the next-highest status, and so on. Their rankings are given below.

a. Compute the rank-order coefficient of correlation.
b. Interpret the coefficient.
c. Can we conclude that there is a positive rank correlation at the 0.05 significance level?

Occupation	Rank of Professional	Rank of Blue-Collar Worker
Physician	1	1
Dentist	4	2
Attorney	2	4
Pharmacist	6	5
Optometrist	12	9
School teacher	8	12
Veterinarian	10	6
College professor	3	3
Engineer	5	7
Accountant	7	8
Health care administrator	9	11
Government administrator	11	10

Exercises

17. There is an opening for police sergeant. Applicants for promotion to sergeant are rated separately by both the police board and the county commissioners. On a scale of 1 to 50, the ratings of six applicants are:

Applicant	Police Board's Rating	County Commissioners' Rating
Arbuckle	42	40
Cantor	36	40
Silverman	16	21
Trepinski	42	40
Lopez	49	47
Condon	8	9

a. Rank the ratings of the applicants. (Watch the ties.)
b. Compute Spearman's rank-order correlation coefficient.

18. A sales manager and a personnel manager were asked to rank graduates of the company's sales training program in terms of the likelihood of their success on the job. Determine the coefficient of rank correlation.

Trainee	Sales Manager	Personnel Manager
A	3	2
B	1	5
C	8.5	9
D	2	1
E	4	6
F	8.5	12
G	5	3
H	7	4
I	10	11
J	6	7
K	11	8
L	12	10

a. Compute Spearman's rank-order correlation coefficient.
b. Can it be shown statistically that there is positive rank correlation?

CAUTION: REASONING ABOUT CAUSE AND EFFECT. If a strong correlation, such as 0.85, is found between two variables, it is tempting to assume that a change in one variable *causes* the other variable to change. This may not be true. For example, it has been discovered that the birth rate goes up when beer consumption goes up. An increase in the consumption of beer, however, does not cause the birth rate to increase. This only indicates that there is a relationship between the two variables. Results of this nature are called nonsense, or spurious, correlations.

SUMMARY

Several techniques are available to describe the relationship between two variables. One of the simplest, the scatter diagram, allows us to visualize the potential association, or correlation, between two variables. If the data are interval-scaled, the Pearson's coefficient of correlation can be computed to measure the strength of the relationship. Perfect agreement between two variables will result in a coefficient of either –1.00 or +1.00. Coefficients of –0.92 or +0.92, for example, would reveal that the correlation is very strong, while –0.21 or +0.21 would show it to be very weak.

Other measures of the strength of the association between two interval-scaled variables are the coefficient of determination and the coefficient of nondetermination. They can assume any values from 0 to +1.00 or 0 to 100%. The coefficient of determination is found by r^2 and the coefficient of nondetermination by $1 - r^2$. The coefficient of determination is the proportion of the total variation in one variable explained, or accounted for, by the variation in the other variable. The coefficient of nondetermination is the proportion not accounted for. Instead of the correlation coefficient, many researchers prefer to use the coefficient of determination to measure the relationship because it is a smaller value and thus does not overstate the association between the two variables.

A small sample selected from the population often invites the question, "Based on the sample coefficient of correlation, is it possible that the correlation in the population is zero?" We can answer this question by applying the five-step hypothesis-testing procedure examined in earlier chapters. Let us repeat the five steps in brief. (1) The null hypothesis tested is ρ (the population coefficient of correlation) = 0. The alternate hypothesis is $\rho \neq 0$. (2) A level of significance is selected. Usually, it is 0.05 or 0.01. (3) An appropriate test statistic is selected. In this case, it is t. (4) A decision rule is formulated. (5) Based on sample data, the decision is made either to reject or to fail to reject the null hypothesis.

The strength of the relationship between two sets of ordinal-scaled data is measured by Spearman's rank-order correlation coefficient. This measure also ranges from –1.00 to +1.00. As implied by its name, the rank-order coefficient is obtained from data organized in ranked order.

Exercises

19. The following table shows the percentage of the vote actually received by the candidates for mayor in five large cities and the corresponding percentages predicted by a national polling organization:

Predicted	Actual Percentage
59	62
47	51
43	42
55	56
57	57

a. Plot the data on a scatter diagram.
b. Compute the coefficients of correlation, determination, and nondetermination.
c. Test the hypothesis that the coefficient of correlation in the population is 0. Use a 0.05 significance level.
d. Interpret your findings.

20. The following data relate family size to annual food expenditures for a sample of 10 families:

Family	Food Expenditures (in hundreds of dollars)	Family Size
1	$19	5
2	20	6
3	15	5
4	6	2
5	11	3
6	17	3
7	18	3
8	15	3
9	13	2
10	16	3

a. Compute the coefficient of correlation.
b. Can we conclude that there is a relationship between food expenditures and family size?

21. A researcher studied the relationship between the years of education beyond high school and monthly salary of a sample of 27 welfare employees. The coefficient of correlation was computed to be 0.25. Test the hypothesis that there is no correlation in the population, using the 0.01 significance level. Use a two-tailed test.

22. The crime rate in a large city has been increasing. A special task force studying the problem has suggested that more full-time law enforcement personnel be employed. The police commissioner, however, is not convinced that this is the best strategy; he doubts that there is a relationship between the number of full-time law enforcement personnel and the annual number of crimes. To investigate, 11 large cities are randomly selected and both the total number of reported crimes and the total number of full-time enforcement officers are obtained. The results are:

City	Number of Offenses (in thousands) *X*	Full-Time Law Enforcement Personnel (in thousands) *Y*
Atlanta	49.5	1.9
Buffalo	29.9	1.4
Chicago	214.1	14.8
Cleveland	53.1	2.3
Denver	52.9	1.7
Houston	106.3	3.0
Louisville	23.5	1.0
Milwaukee	37.0	2.3
Pittsburgh	32.0	1.4
Seattle	40.0	1.4
Tampa	27.7	0.8

a. Compute the coefficients of correlation, determination, and nondetermination.
b. Test the hypothesis that the coefficient of correlation is 0; use the 0.05 significance level. Use a two-tailed test.
c. Interpret the findings.

23. The design department of a large automobile manufacturer developed 10 different mock-up models of a mid-sized automobile. The design group is interested in determining the opinions of teenagers and senior citizens with respect to the models. A group of teenagers was asked to rank each model. Based on their rankings, a composite rank for each model was determined. For example, the teenagers ranked mock-up model E the most appealing, and model J the next most appealing. The same procedure was followed for the senior citizens. The rankings of the two groups are:

Model	Rankings by Teenagers	Rankings by Senior Citizens
A	6	8
B	4	1
C	8	10
D	3	3
E	1	2
F	10	7
G	7	9
H	5	5
I	9	6
J	2	4

a. Draw a scatter diagram. Place the rankings of the teenagers on the X-axis and the rankings of the senior citizens on the Y-axis.
b. Compute Spearman's rank-order correlation coefficient.
c. Interpret your findings.

24. Professor P. Dant believes that students who complete her examinations in the shortest time receive the highest grades and that those who take the longest to complete them receive the lowest grades. To verify her suspicion, she numbers the midterm examination paper as each student completes it.

Name	Order of Completion	Grade (50 possible)
Gorney	1	48
Gonzales	2	48
McDonald	3	43
Sadowski	4	49
Jackson	5	50
Smythe	6	47
Carlson	7	39
Archer	8	30
Namath	9	37
MacFearson	10	35

a. Rank the grades.
b. Draw a scatter diagram. Put order of completion on the X-axis and the rankings of the grades on the Y-axis.
c. Compute the rank-order correlation coefficient.
d. Interpret your findings.

25. Two panels, one composed of all men and the other of all women, were asked by a consumer-testing agency to rank eight colas according to taste. A rank of 1 was given to the best-tasting cola and a rank of 8 to the least desirable. The purpose of the test was to determine if there is any relationship between men's and women's tastes for cola.

Brand	Panel of Men	Panel of Women
Red Ribbon Cola	8	7
Bluebell Cola	6	5
Steel City Cola	2	4
Glatz Cola	1	2
Pearl Cola	3	1
Deer Run Cola	4	3
Boca Cola	5	6
Krolla Cola	7	8

Compute the rank-order correlation coefficient and interpret your findings.

26. There has recently been much debate about the cost of traveling on the Ohio Turnpike and whether or not the tolls should be reduced. As part of a citizens' investigation, the following information on other turnpikes and toll roads was obtained:

Toll Road	Length (in miles)	Cost per Mile (in cents)
Massachusetts Turnpike	123.0	4.15
Pennsylvania (east–west)	358.9	4.10
Pennsylvania (north–south)	111.1	3.74
New Jersey Turnpike	118.0	3.89
Florida Turnpike	265.0	3.75
New York Thruway (Mainline)	390.0	3.10
New York Thruway (Erie)	67.0	3.13
New York Thruway (Saratoga)	24.0	2.97
Indiana Toll Road	156.9	2.96
Maine Turnpike	106.0	2.92
Oklahoma (Turner)	86.0	2.33
Oklahoma (Will Rodgers)	88.5	2.26
Ohio Turnpike	241.2	2.03

a. Determine Pearson's coefficient of correlation between the length of the toll road and the cost per mile to travel on the toll road.

b. Determine the coefficient of determination.

c. At the .05 significance level, can we conclude that the correlation between the length of the typical toll road and the cost per mile on that toll road is different from zero?

27. A real estate agent has sample data on the selling price and distance from the center of the city for 75 homes. The mean distance from the center of the city is 14.9 miles, with a standard deviation of 4.9 miles. The mean selling price of the homes is \$167,000, with a standard deviation of \$36,000. The sum of the products of the deviations from the mean is –4821.

a. Determine the coefficient of correlation.

b. At the .05 significance level, can we conclude that the distance from the center of the city is inversely related to the selling price?

28. The following table reports the qualifying speed, starting position, and order of finish for each of the drivers in the 1991 Indianapolis 500.

Driver	Speed (in miles per hour)	Starting Position	Order of Finish
Mears, Rick	224.113	1	1
Foyt, A. J.	222.443	2	28
Andretti, Mario	221.818	3	7
Rahal, Bobby	221.401	4	19
Andretti, Michael	220.943	5	2
Unser, Al	219.823	6	4
Andretti, John	219.059	7	5
Crawford, Jim	218.947	8	26
Sullivan, Danny	218.343	9	10
Cheever, Eddie	218.122	10	31
Andretti, Jeff	217.632	11	15
Goodyear, Scott	216.751	12	27
Bettenhausen, Gary	224.468	13	22
Luyendyk, Arie	223.881	14	3
Fitipaldi, Emerson	223.064	15	11
Cogan, Kevin	222.844	16	29
Fox, Stan	219.501	17	8
Groff, Michael	219.015	18	24
Brayton, Scott	218.627	19	17
Bettenhausen, Tony	218.188	20	9
Jourdain, Bernard	216.683	21	18
Brabham, Geof	214.859	22	20
Lazier, Buddy	218.692	23	33

Driver	Speed (in miles per hour)	Starting Position	Order of Finish
Matsushita, Hiro	218.351	24	16
Paul, John Jr.	217.952	25	25
Palmroth, Tero	215.648	26	23
Pruett, Scott	214.814	27	12
Guerrero, Roberto	214.027	28	30
Ribbs, Willy T.	217.997	29	32
Dobson, Dominic	215.326	30	13
Lewis, Randy	215.043	31	14
Carter, Pancho	214.012	32	21
Johncock, Gordon	213.812	33	6

a. What is the coefficient of rank correlation between the starting position and the order of finish? At the .05 significance level, is the correlation significant?

b. Determine the coefficient of rank correlation between the order of finish and the rank of the qualifying speed. At the .05 significance level, is the correlation significant?

29. The following table reports the amount of tax for a family of four with a median income of $61,372, by state, for the year 1992. The sales tax for each of the 50 states and the District of Columbia are also included.

State	Amount of Tax	Sales Tax (percent)
Alabama	$2,450	4.000
Alaska	196	0.000
Arizona	2,400	5.000
Arkansas	2,923	4.000
California	2,577	6.000
Colorado	2,394	3.000
Connecticut	300	8.000
Delaware	2,777	0.000
District of Columbia	4,036	6.000
Florida	164	6.000
Georgia	3,034	4.000
Hawaii	4,463	4.000
Idaho	3,744	5.000
Illinois	1,715	6.250
Indiana	2,268	5.000
Iowa	2,721	4.000

State	Amount of Tax	Sales Tax (percent)
Kansas	2,453	4.250
Kentucky	2,479	5.000
Louisiana	1,887	4.000
Maine	3,498	5.000
Maryland	3,782	5.000
Massachusetts	2,944	5.000
Michigan	2,846	4.000
Minnesota	3,548	6.000
Mississippi	2,331	6.000
Missouri	2,598	4.225
Montana	2,993	0.000
Nebraska	2,215	4.000
Nevada	231	5.750
New Hampshire	132	4.000
New Jersey	1,614	6.000
New Mexico	2,774	4.750
New York	3,690	4.000
North Carolina	3,479	3.000
North Dakota	1,366	6.000
Ohio	1,899	5.000
Oklahoma	3,104	4.000
Oregon	4,095	0.000
Pennsylvania	1,480	6.000
Rhode Island	2,017	6.000
South Carolina	3,316	5.000
South Dakota	410	4.000
Tennessee	627	5.500
Texas	213	6.000
Utah	3,717	6.000
Vermont	2,043	4.000
Virginia	2,912	4.500
Washington	250	7.000
West Virginia	2,763	6.000
Wisconsin	3,605	5.000
Wyoming	316	3.000

a. Determine Pearson's coefficient of correlation between the annual tax amount and the percent of sales tax. Are you surprised that it is negative? Comment.

b. At the .05 significance level, can we conclude that the correlation is negative?

c. Determine the coefficient of determination. Are you surprised that it is rather low? Comment.

30. For several large cities, the typical size of downtown offices (in mean number of square feet per worker) and the average monthly rental rate (in dollars per square foot) are listed below.

City	Square Footage	Rental Rate
Austin	350	$14.52
Dallas	344	13.10
Houston	327	12.32
Billings	306	12.78
New Orleans	302	10.95
Seattle	300	13.73
San Jose	293	15.47
Atlanta	288	10.62
Portland, OR	277	13.91
Phoenix	275	11.76
Cleveland	272	12.00
Birmingham	268	14.34
Kansas City, MO	268	10.35
Pittsburgh	268	15.30
Jacksonville	266	17.75
Omaha	265	14.13
Denver	262	12.96
Chicago	257	18.07
St. Louis	254	12.61
Minneapolis	251	15.26
San Diego	249	12.31
Miami	246	16.97
Orlando	242	12.11
Philadelphia	229	16.43
Sacramento	223	18.87
Milwaukee	215	14.91
Charlotte	206	14.93
Columbus, OH	202	15.27
Honolulu	201	17.30
Baltimore	191	16.69
Detroit	187	12.78
Indianapolis	182	13.40

a. Develop a scatter diagram.
b. Compute the coefficient of correlation.
c. What is the coefficient of determination?
d. Interpret your findings.
e. Using a 0.05 significance level, test the hypothesis that the correlation in the population is zero.

Data Exercises

31. Refer to Real Estate Data, which reports information on homes sold in Northwest Ohio during 1992.
  a. Determine the coefficient of correlation between the selling price and the area of each home in square feet. At the .05 significance level, can we conclude that there is a positive association between the two variables?
  b. Determine the coefficient of correlation between the selling price and the distance from the center of the city of each home. At the .05 significance level, can we conclude that there is a negative association between the two variables?
  c. Determine the coefficient of correlation between the selling price and the number of bedrooms in each home. At the .05 significance level, can we conclude that there is a positive association between the two variables?

32. Refer to Salary Data, which refers to a sample of middle managers in Sarasota, Florida.
  a. Determine the coefficient of correlation between the salary and the age of the middle managers. At the .05 significance level, can we conclude that there is a positive correlation between the two variables?
  b. Determine the coefficient of correlation between the salary and the number of employees supervised by the middle managers. At the .05 significance level, can we conclude that there is a positive correlation between the two variables?
  c. Is the correlation between salary and region or the correlation between salary and type of industry a useful comparison? Explain why or why not.

IS ANYTHING WRONG HERE?

Harvey Maertin, a health department staff analyst for Arcanum, New York, gathered the following data on medical expenses last year for a sample of the 5,000 city residents. The 1990 appraised value of their respective homes is also listed. Based on this data, he found a significant inverse correlation between home value and medical expenses. He included this observation in a report to the executive director of the health department. When Harvey arrived at work this morning, he found that the study had been returned to his desk. On the cover, in red ink, was a note: "WHERE DID YOU LEARN STATISTICS?" Was Harvey correct? Explain.

Family Surname	Medical Expenses	Home Value
Miller	$5,032	$154,692
Hendrix	4,508	151,773
Sines	2,326	150,995
Kennedy	4,147	148,928
Abernathy	2,867	150,691
Lennox	3,352	150,310
Tusko	3,538	147,243
Lutz	2,784	151,651
Sell	15,000	145,038
Brigden	3,489	151,157
Hewitt	4,289	150,635

CHAPTER ACHIEVEMENT TEST

Answer all the questions. The answers are at the back of the book.

MULTIPLE-CHOICE QUESTIONS. Select the response that best answers each of the questions.

Questions 1–5 are based on the following graphs:

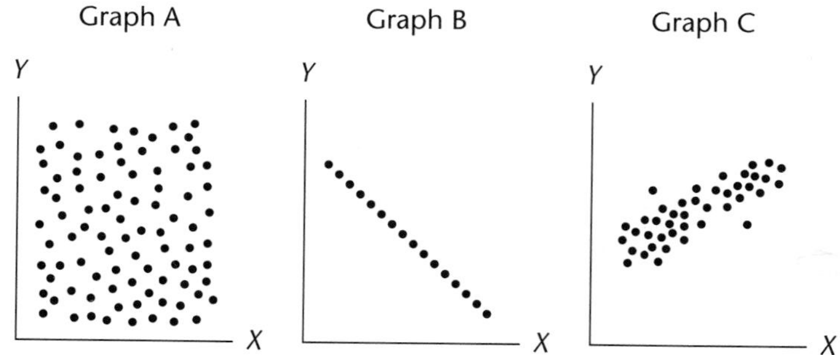

1. These graphs are called
a. tree diagrams.
b. shotgun blasts.

c. scatter diagrams.
d. frequency polygons.
e. None of the above

2. Graph A. The coefficient of correlation, if computed, would be approximately
a. 100.
b. –1.00.
c. +1.00.
d. 0.
e. None of the above

3. Graph B. The coefficient of nondetermination, if computed, would be approximately
a. –1.00.
b. 0.50.
c. 0.
d. +1.00.
e. None of the above

4. Graph B. The linear relationship between the two variables
a. is very weak.
b. is moderate.
c. is perfect.
d. cannot be determined from this graph.
e. None of the above

5. Graph C. The coefficient of determination, if computed, would be
a. about 60%.
b. negative.
c. about 0.
d. near infinity.

6. Pearson's product-moment coefficient of correlation for a set of paired observations was computed to be 0.70.
a. The coefficient of nondetermination is 0.30.
b. The coefficient of determination is 0.49.
c. The correlation in the population is 0.
d. The correlation in the population is 0.49.
e. None of the above

7. Pearson's r was computed to be –0.55. Which of the following values of r represents a stronger relationship than –0.55?
a. 0
b. +0.50
c. –0.70
d. 20.9
e. None of the above

8. Spearman's rank-order correlation coefficient
a. can range from –1.0 to 1.0.
b. is always positive.
c. is positively skewed.
d. All of the above

9. The term $\Sigma(X - \overline{X})(Y - \overline{Y})$ affects the
   a. sign of r.
   b. standard deviation of y.
   c. standard deviation of x.
   d. size of the sample.
10. Spearman's rank-order correlation coefficient was computed to be –0.76 for a sample of 15 observations. Which of the following are correct statements?
    a. Something is wrong with the calculation, because the rank correlation coefficient cannot be negative.
    b. As one ranking increases, so does the other.
    c. As one ranking increases, the other decreases.
    d. None of the above

COMPUTATION PROBLEMS

11. The ages and corresponding prices of five bottles of wine selected at random and sold at a wholesale auction are shown in the following list. The relationship between age and price is to be explored.

Age (in years) *X*	Price (in dollars) *Y*
2	$ 5
6	16
10	18
5	9
8	15

    a. Portray the paired data in a graph.
    b. Compute Pearson's coefficient of correlation.
    c. Compute both the coefficient of determination and the coefficient of nondetermination.
    d. Test the hypothesis that the coefficient of correlation in the population is 0. Use a 0.05 significance level.
    e. Interpret your findings.
12. Early in the basketball season, 12 teams appeared to be outstanding. A panel of sportswriters and a panel of college basketball coaches were asked to rank the 12 teams. Their composite rankings were as follows :

Team	Coaches	Sportswriters
Notre Dame	1	1
Indiana	2	5
Duke	3	4

Team	Coaches	Sportswriters
North Carolina	4	6
UCLA	5	2
Louisville	6	3
Ohio State	7	10
Syracuse	8	11
Georgetown	9	7
LSU	10	12
Clemson	11	8
St. John's	12	9

a. Compute the appropriate coefficient of correlation to evaluate the agreement between the sportswriters and coaches with respect to the rankings.
b. Interpret your findings.
c. Can we conclude that there is a significant correlation between the rankings of the coaches and the sportswriters at the 0.05 significance level?

ANSWERS TO SELF-REVIEW PROBLEMS

13-1 a.

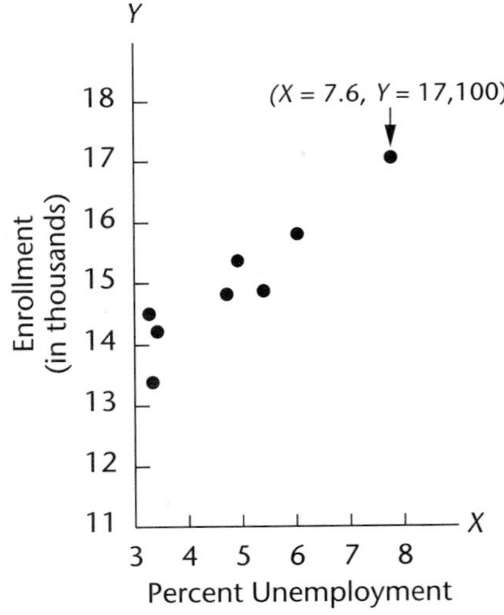

b. As unemployment in the region increases, enrollment at the state university campus also increases.

13-2 a.

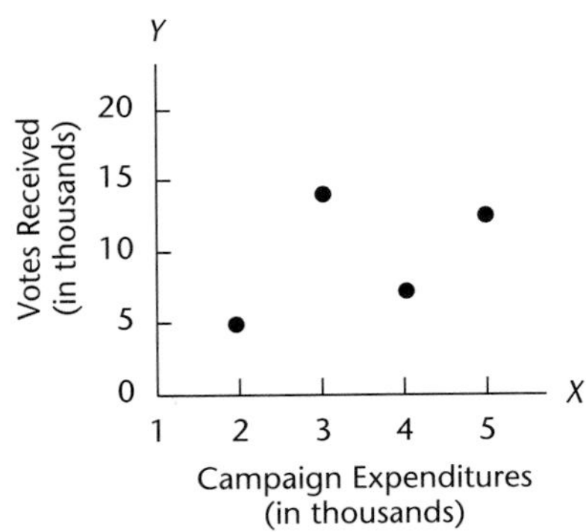

b.

$$r = \frac{4(140) - (14)(38)}{\sqrt{\left[4(54) - (14)^2\right]\left[4(414) - (38)^2\right]}}$$

$$= \frac{560 - 532}{\sqrt{[216 - 196][1{,}656 - 1{,}444]}}$$

$$= \frac{28}{\sqrt{[20][212]}} = \frac{28}{\sqrt{4,240}}$$

$$= \frac{28}{65.12} = 0.43$$

c. There is a moderate to weak relationship between the two variables.

13-3 a. 0.18, found by $(0.43)^2$

b. 0.82, found by $1 - (0.43)^2$

c. Only 18% of the variation in the number of votes is accounted for by the level of campaign expenses; 82% is not.

13-4 a.

$$t = \frac{0.43\sqrt{4-2}}{\sqrt{1-(0.43)^2}}$$

$$= \frac{0.6081}{0.9028}$$

$$= 0.67$$

With 2 degrees of freedom for the 0.01 significance level and a two-tailed test statistic, the critical values are –9.925 and +9.925. We do not reject the null hypothesis.

b. Correlation in the population could be zero.

13-5 a.

Rank	Rank	d	d^2
1	1	0	0
4	2	2	4
2	4	–2	4
6	5	1	1
12	9	3	9
8	12	–4	16
10	6	4	16
3	3	0	0
5	7	–2	4
7	8	–1	1
9	11	–2	4
11	10	1	1
		0	60

$$r_s = 1 - \frac{6(60)}{12\left[(12)^2 - 1\right]} = 1 - \frac{360}{1716} = 0.79$$

b. There is quite strong agreement between the two rankings.

c. H_0: No correlation in the ranks.
H_a: Positive correlation in the ranks.
H_0 is rejected if $t > 1.812$.

$$t = 0.79\sqrt{\frac{12-2}{1-(0.79)^2}} = 4.07$$

H_0 is rejected. There is positive correlation in the ranks.

CHAPTER 14

Regression Analysis

OBJECTIVES

When you have completed this chapter, you will be able to

- describe a linear relationship between two variables;
- determine the linear equation using the method of least squares;
- develop a measure of the error around the regression equation;
- establish confidence intervals for predictions.

CHAPTER PROBLEM Our Books are Priceless

RECALL FROM CHAPTER 13 THAT A COMMITTEE OF THE STUDENT GOVERNMENT AT NORTH CENTRAL STATE UNIVERSITY BEGAN A STUDY OF THE COST OF TEXTBOOKS. THEY FOUND A STRONG POSITIVE RELATIONSHIP BETWEEN THE NUMBER OF PAGES IN A TEXTBOOK AND ITS SELLING PRICE. THEY WOULD LIKE TO DEVELOP AN EQUATION FROM THE SAME SAMPLE DATA THAT CAN BE USED TO ESTIMATE THE SELLING PRICE OF A TEXTBOOK BASED ON THE NUMBER OF PAGES.

INTRODUCTION

Chapter 13 dealt with measures of correlation between two variables—Pearson's product-moment correlation coefficient and Spearman's rank-order correlation coefficient. In this chapter, we will continue the study of two related variables. We will develop a mathematical equation that allows us to estimate one variable based on another variable. For example, if a student studied three hours for an examination, it may be possible to estimate his or her score on the examination. Similarly, the sales of a product may be predicted by the level of advertising expenditures. The technique used to make these estimations is called *regression analysis.*

LINEAR REGRESSION

STAT BYTE

Sir Francis Galton compared the heights of parents and their children. He converted each parent and child height to a standard score (z-value) and noted that the children's z-scores deviated less from the mean than those of their parents. This regression to the mean formed the foundation of modern regression and correlation analysis.

The word *regression* was first used by Sir Francis Galton (1822–1911) in the late nineteenth century. A study by Galton revealed that the height of the children of tall parents was above average, but seemed to fall back, or regress, toward the mean height of the population. Thus, the general process of predicting one variable (such as a child's height) based on another variable (the parents' height) became known as *regression.*

The usual first step in studying the relationship between two variables is to plot the paired data in a scatter diagram, discussed in Chapter 13. To illustrate the concept of regression, we will examine the Chapter Problem relating the length of a book to its selling price. The data are repeated in Table 14-1 and plotted in the form of a scatter diagram in Figure 14-1. This time we are interested in *estimating* a selling price based on the number of pages in the text. Recall from Chapter 13 that the variable being predicted (selling price) is the *dependent* variable—designated Y—and the variable used to make the prediction (number of pages) is the *independent* variable—designated X.

Table 14-1 **Number of Pages and Selling Price of Selected Books**

Title	Pages	Price
A	400	$44
B	600	47
C	500	48
D	600	48
E	400	43
F	500	46

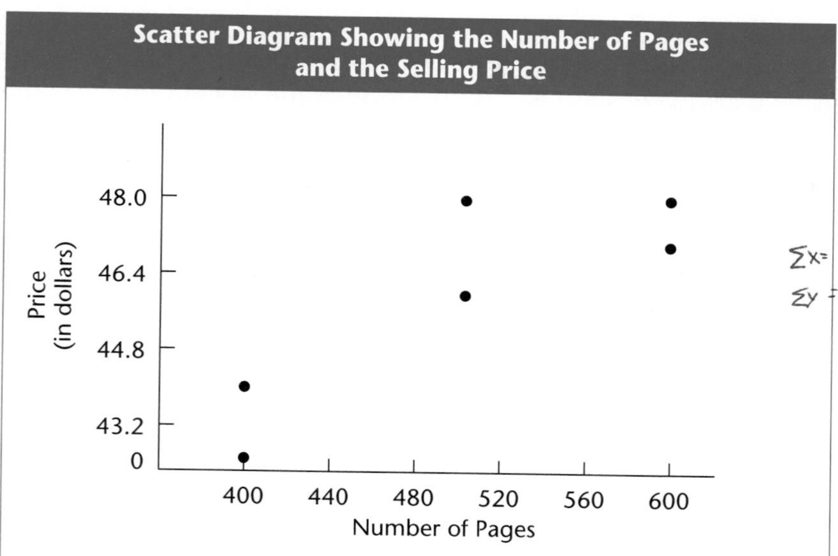

Figure 14-1

The objective is to find a line that—when translated into equation form—can describe the relationship between the two variables. An important use of the equation for the line is to approximate a value of the dependent variable based on a selected value of the independent variable. For example, in the Chapter Problem, where we are concerned with the selling price and the number of pages, we may want to approximate the selling price of a book that has 450 pages. The equation for that line is called the **regression equation.**

Regression Equation A mathematical equation that defines the relationship between two variables.

The linear equation that expresses the relationship between two variables is

$$Y' = a + bX \qquad \textbf{14-1}$$

where

Y' (read "Y prime") is the predicted value of Y for a selected X value. In the example, it is the selling price for a given number of pages.

a is a constant: the value at which the straight line intersects the *Y*-axis. It is also the value of *Y* when $X = 0$.

b is also a constant and is the **slope** of the straight line. It is the change in Y' for each change of one (either increase or decrease) in *X*. In this problem, *b* is the increase in the selling price for an increase of one page in the length.

X is any value of *X* that is selected. In this problem, *X* is the number of pages.

To further illustrate the meaning of *a* and *b*, suppose that the regression equation in the form of $Y' = a + bX$ for the data in Figure 14-2 had been computed to be $Y' = 20 + 10X$ (in thousands of dollars). When $X = 0$, the line intercepts the *Y*-axis at $20,000. This is *a*, which is sometimes called the ***Y*-intercept.**

Figure 14-2 also shows that weekly sales increase from $40,000 a week when two sales representatives call on clients, to $50,000 when three representatives call. Thus, with each increase of one sales representative, weekly sales increase by $10,000. This is *b*.

The regression equation $Y' = 20 + 10X$ (in thousands of dollars) can be used to estimate the average weekly sales if four sales representatives are employed, $X = 4$. Then $Y' = 20 + 10(4) = 60$, or $60,000.

Self-Review 14-1

Answers to the Self-Review problems are at the end of the chapter.

Some data dealing with the education and weekly income of residents of Precinct 13 have been collected. The purpose of the study is to predict incomes based on education.

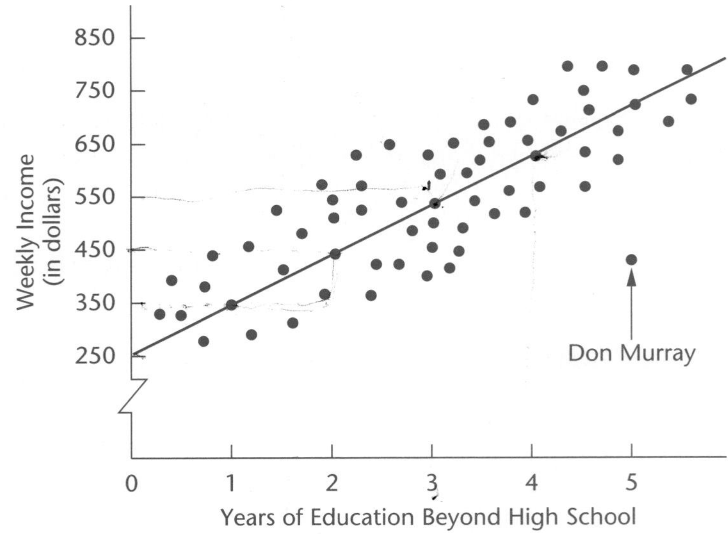

a. Interpret the meaning of the point for Don Murray.
b. What is the value of *a* in the regression equation $Y' = a + bX$?

(continued)

c. What is the value of b in the regression equation? What does it indicate?
d. What is the regression equation?
e. Based on the regression equation, estimate the average weekly income for a person with four years of education beyond high school.

Figure 14-2

Scatter Diagram and Line of Regression for Weekly Sales and Number of Sales Representatives

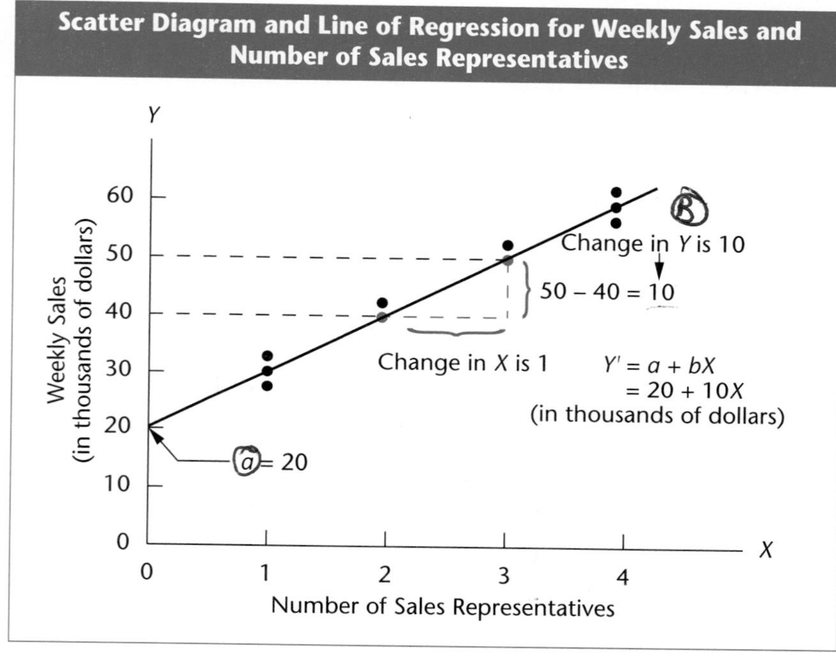

Exercises

Answers to the even-numbered exercises are at the back of the book.

1. The length of confinement in days (X) and the dollar cost (Y) for six patients at St. Matthew's Hospital were tabulated and then plotted on a scatter diagram:

Length of Confinement (in days) X	Dollar Cost Y
5	\$1,110
2	490
12	2,500
4	880
7	1,530
1	270

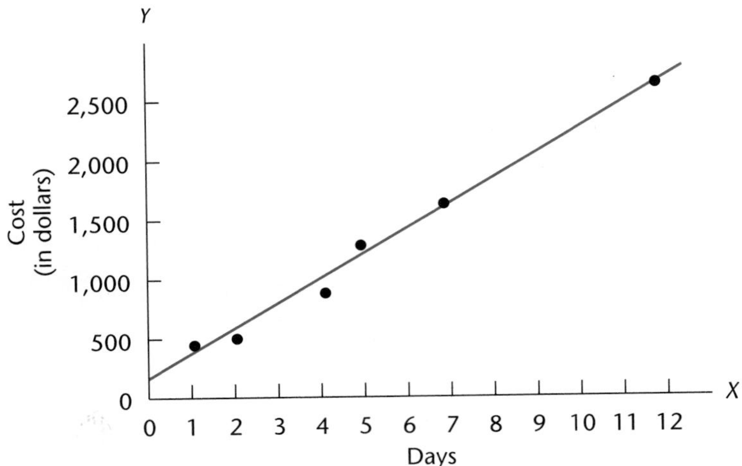

a. What is the general form of the regression equation?
b. Based on the scatter diagram, what is a?
c. What is b (the slope) of this regression line?
d. If a patient stayed six days, what would you estimate his or her bill to be?
e. Interpret in words the values of a and b.

2. A study attempting to relate the number of minority employees (Y) to the number of nonminority employees (X) in eight firms revealed the following:

Firm	Number of Nonminority Employees X	Number of Minority Employees Y
London Mfg.	4,628	182
ABC Chemicals	4,272	299
Cork Floors, Inc.	1,663	173
Sanson Industries	691	44
Martin Trucking	5,355	1,069
Cable Lead	5,618	777
Parson Farms	2,934	802
Tayo Electric	3,650	790

a. What is the general form of the regression equation?
b. Based on the following scatter diagram, what is the Y-intercept?
c. What is b, the slope of the regression line?
d. If a firm employs 3,000 nonminority persons, estimate the number of minority employees.
e. Interpret the values of a and b.

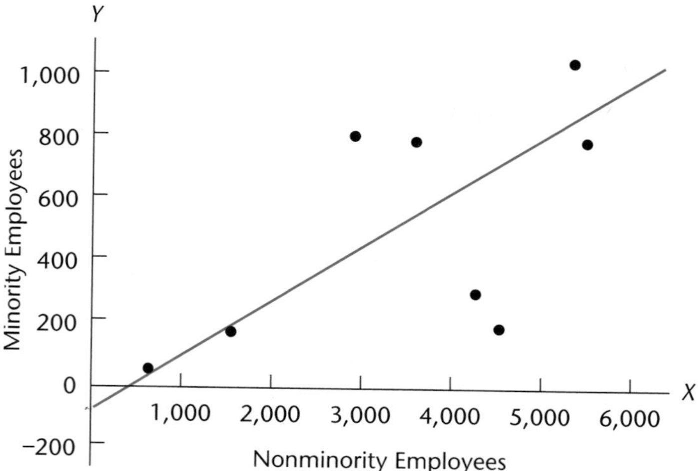

3. An equation that relates the annual cost to heat a single-family residence (Y) to its size in square feet of living area (X) is $Y' = -200 + 0.3X$.
   a. Estimate the cost of heating a 3,500-square-foot home.
   b. What is the slope of this line?
   c. What is the Y-intercept?
   d. Interpret in words the meaning of a and b.
4. The equation $Y' = 39 - 0.002X$ relates the annual birth rate per thousand population (Y) of a country to their median annual income (X).
   a. Estimate the birth rate for a country whose median annual income is $7,000.
   b. What is the Y-intercept of the line of regression?
   c. What is the slope of the line?
   d. Interpret the meaning of b.

FITTING A LINE TO DATA

How should the line through the plots on a scatter diagram be drawn? We could simply place a ruler on the scatter diagram and draw a line through the middle of the points. Yet if this job of drawing the best-fitting line "right through the middle of the points" were given to three persons, their lines would probably all be different. Consider the Chapter Problem concerned with predicting selling price on the basis of the number of pages in a book. The information for the six books comes from Table 14–1. Figure 14–3 shows how three different people might draw a straight line. Each could be different! What is needed is a precise method of arriving at the best-fitting straight line. Then the identical line will be obtained every time, regardless of who does the calculations.

Figure 14-3

Three Attempts at Fitting a Straight Line to the Data on Selling Price and Number of Pages

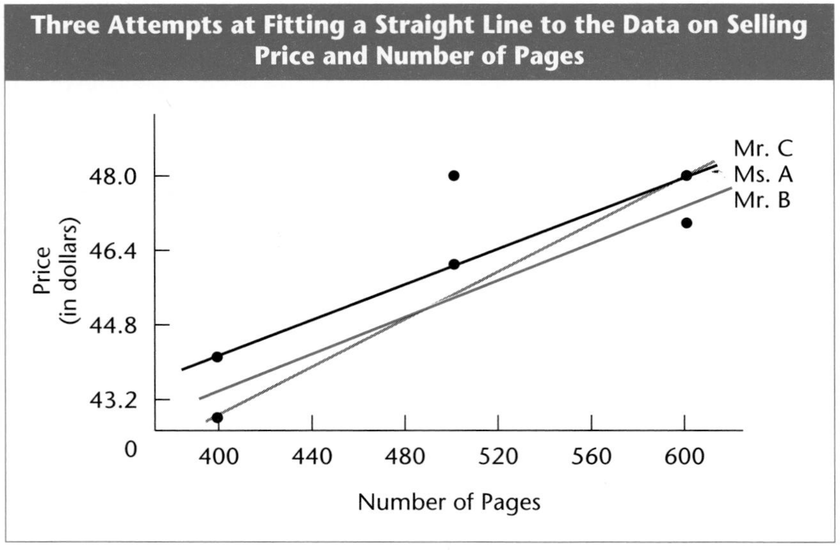

The Least-Squares Principle

We can avoid having to rely on personal judgment as a way to find the exact location of the line by employing a method called the **least-squares principle,** which will provide us with the "best-fitting" line. How does it work? Note that in Figure 14–2 (p. 495) the plotted points were not exactly on the straight line. The difference between an actual value (a plotted point) and a predicted value (a corresponding value on the line) can be thought of as the "error" we make when using the regression equation to predict a specific value of the dependent variable. Put another way, this difference, or error, can be considered to be the deviation of the actual value from the predicted value.

By using this principle we obtain a *unique* regression equation that is "best" in the sense that the squared errors around the regression line are smaller than around any other line. That is, the sum of the squares of the deviations between the actual values (Y) and the predicted values (Y') is minimized. The regression line is unique—meaning that there is only one least-squares line, and this line can be determined by anyone doing the calculations.

Least-Squares Principle A method used to determine the regression equation by minimizing the sum of the squares of the distances between the actual Y values and the predicted values of Y. $\Sigma(Y-Y')^2$

Figure 14–4 illustrates this concept. The line was determined by the method of least squares. The first plot ($X = 3$, $Y = 8$) deviates vertically by 2 units from the predicted value of 10 on the straight line.

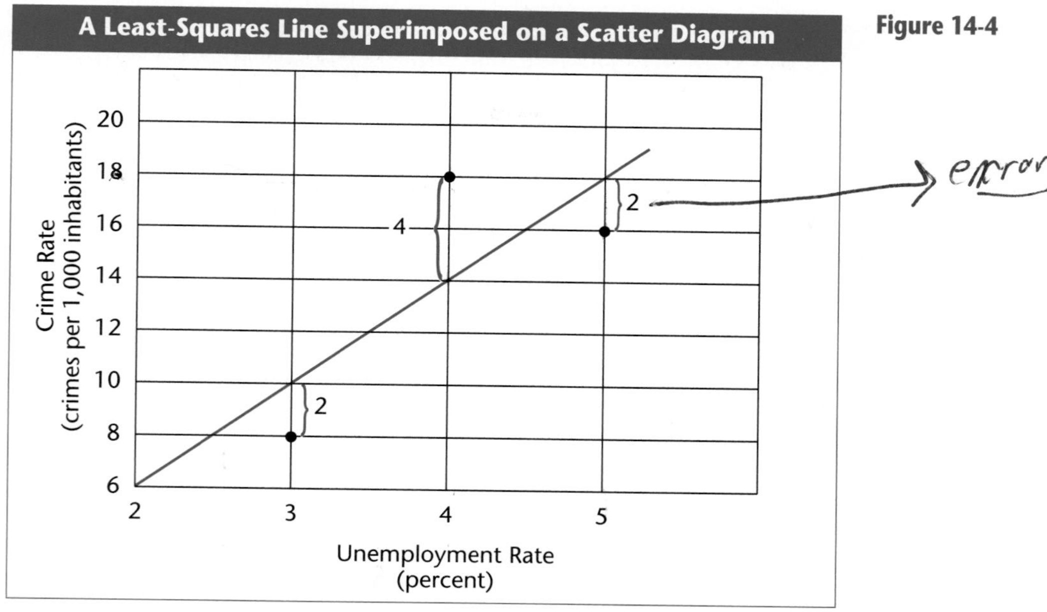

Figure 14-4

The squared deviation is 4, found by 2^2. The squared deviation for $X = 4$, $Y = 18$ is 16, found by $(18 - 14)^2$. The squared deviation for $X = 5$, $Y = 16$ is 4. The sum of the squared deviations is 24, found by $4 + 16 + 4$.

The same crime and unemployment data in Figure 14-4 also appear in Figure 14-5. The line through the data in Figure 14-5,

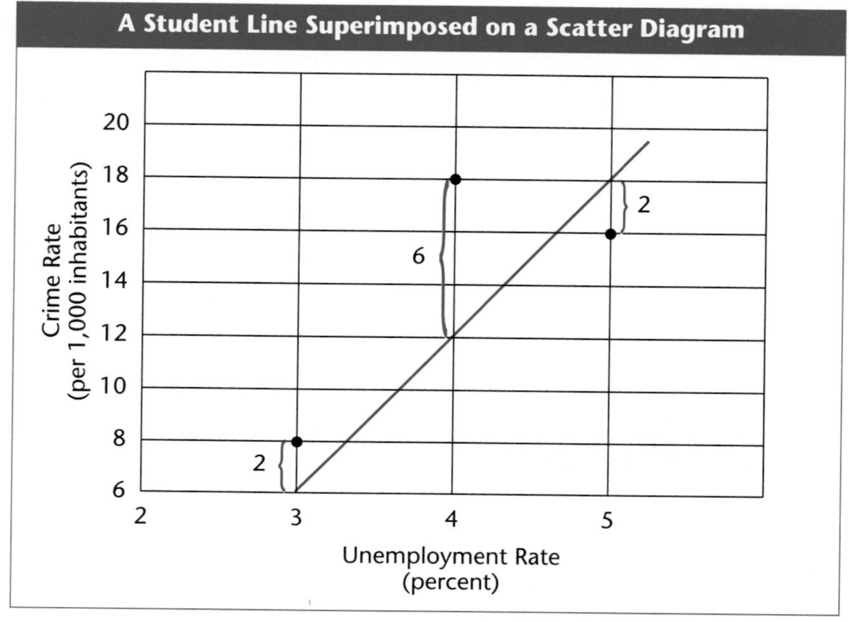

Figure 14-5

however, is *not* determined by the least-squares method. A beginning student thought the line should be positioned as shown. The sum of the vertical squared deviations from the freehand line is 44, found by $2^2 + 6^2 + 2^2$. The 44 is greater than 24, the sum of the squared deviations around the least-squares line.

Self-Review 14-2

The same crime and unemployment data used in Figures 14–4 and 14–5 appear again in the following scatter diagram. The line was carelessly drawn by a student.

a. Determine the sum of the squared vertical deviations from the line.
b. Is the sum greater or smaller than the sum from Figure 14-4? Explain.
c. What is the dependent variable in this diagram?

THE METHOD OF LEAST SQUARES

In the previous section, we stated that the "best" equation is the one that minimizes the sum of the squared deviations. Now we will show how this line is determined. The determination is made by the mathematical process of minimization. That is, among all possible lines, the one where $\Sigma(Y - Y')^2$ is the smallest is obtained, and it is unique. The following formulas, based on the least-squares principle, are used to compute the unique values of b and a:

$$b = \frac{n(\Sigma XY) - (\Sigma X)(\Sigma Y)}{n(\Sigma X^2) - (\Sigma X)^2} \qquad \text{14-2}$$

$$a = \frac{\Sigma Y - b(\Sigma X)}{n} \qquad \text{14-3}$$

where

X is a value of the independent variable.
Y is a value of the dependent variable.
n is the number of items in the sample.

Problem The data on the lengths of books and their selling prices from the Chapter Problem are repeated below.

Title	Number of Pages X	Selling Price Y
A	400	\$44
B	600	47
C	500	48
D	600	48
E	400	43
F	500	46

Using the method of least squares and formulas 14–2 and 14–3, determine the linear regression equation that could be used to estimate the selling prices for these books based on the number of pages in each. Then estimate the selling price for a book with 450 pages.

Solution The products, squares, and totals needed for the method of least squares are shown in Table 14–2. Solving for b and a using formulas 14–2 and 14–3, we obtain:

Table 14-2 **Calculations Needed for Least-Squares Regression Equation**

Title	X	Y	XY	X^2	Y^2
A	400	$44	17,600	160,000	1,936
B	600	47	28,200	360,000	2,209
C	500	48	24,000	250,000	2,304
D	600	48	28,800	360,000	2,304
E	400	43	17,200	160,000	1,849
F	500	46	23,000	250,000	2,116
	3,000	276	138,800	1,540,000	12,718

$$b = \frac{n(\Sigma XY) - (\Sigma X)(\Sigma Y)}{n(\Sigma X^2) - (\Sigma X)^2}$$

$$b = \frac{6(138{,}800) - (3{,}000)(276)}{6(1{,}540{,}000) - (3{,}000)^2} = \frac{4{,}800}{240{,}000} = 0.02$$

$$a = \frac{\Sigma Y - b(\Sigma X)}{n}$$

$$a = \frac{276 - 0.02(3{,}000)}{6} = \frac{216}{6} = 36$$

The regression equation is $Y' = 36 + 0.02X$. Thus, the estimated selling price for a book with 450 pages is $45, found by $Y' = a + bX = 36 + 0.02(450) = 36 + 9 = 45$.

DRAWING THE LINE OF REGRESSION

Now that the regression equation has been determined, we can plot the least-squares line of regression on the scatter diagram. The first step is to select an X value, say, 0. When $X = 0$, $Y' = 36$, found by $Y' = 36 + 0.02(0)$. The computations for some other points on the line are as follows:

When X is	Y' is	found by
200	40	$Y' = 36 + 0.02(200)$
400	44	$Y' = 36 + 0.02(400)$
600	48	$Y' = 36 + 0.02(600)$

Figure 14-6 shows the original scatter diagram with the line of regression superimposed on it.

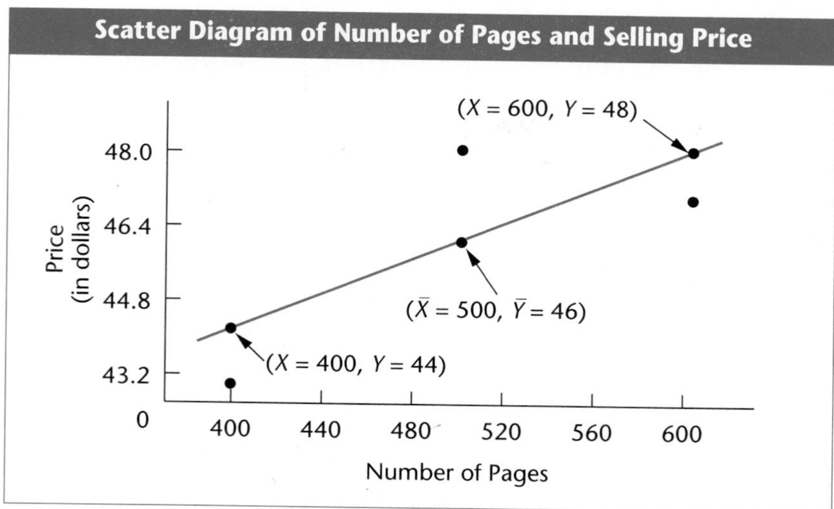

Figure 14-6

Self-Review 14-3

There is interest in determining if the number of votes that will be received in an election can be predicted, based on the amount of money spent during the campaign. A tabulation of available data follows.

Candidate	Amount Spent on Campaign (in thousands of dollars) X	Votes Received (in thousands) Y
Weber	\$3	14
Taite	4	12
Spence	2	5
Henry	5	20

a. Draw a scatter diagram.
b. Determine the regression equation using the method of least squares.
c. How many additional votes could a candidate expect for each additional \$1,000 spent?

(This problem will be continued; save your calculations.)

Exercises

5. The following table lists the populations of five counties (X) and the number of physicians (Y) who practice there:

County	Population (in thousands)	Number of Physicians
Lucas	38	21
Wood	76	35
Samborn	36	13
Atlee	62	24
Ohio	32	18

a. Draw a scatter diagram.
b. Find the least-squares regression equation.
c. Interpret the values of *a* and *b*.
d. Estimate the number of physicians for a county with a population of 45,000.

6. The number of adult movie theaters (X) and the annual number of sex-related crimes (Y) reported for five cities are:

Number of Adult Movie Theaters	Number of Sex-Related Crimes
15	175
20	220
10	120
12	152
16	181

a. Draw a scatter diagram.
b. Find the least-squares regression equation.
c. Interpret the values of *a* and *b*.
d. Estimate the number of sex-related crimes for a city with 18 adult movie theaters.

7. The mean nighttime low temperature (X) and an index of collective aggression (Y) are recorded for six urban areas:

Mean Nighttime Low (in °F)	Index of Collective Aggression
73	51
82	63
90	67
60	35
51	23
40	20

a. Draw a scatter diagram.
b. Find the least-squares regression equation.

c. Interpret the values of a and b.
d. Predict the value of the index when the mean nighttime low is 80.

8. Patients of various ages (X) undergoing the same medical treatment suffer asthma attacks of different durations (Y).

Age (in years)	Duration of Attack (in minutes)
30	15
25	28
65	30
50	22
40	24

a. Draw a scatter diagram.
b. Find the least-squares regression equation.
c. Interpret the values of a and b.
d. Predict the duration of an attack if the patient is 42 years old.

THE STANDARD ERROR OF ESTIMATE

If all the data points fall precisely on the line of regression, then our prediction is exact. In the previous chapter, this situation was described as perfect correlation. The coefficient of correlation would be either –1.00 or +1.00. The coefficient of correlation, then, is a possible measure of how well the line of regression fits the paired data. Figures 14–7 and 14–8 show the cases of perfect correlation and perfect prediction.

Perfect predictions of Y based solely on X, as illustrated in the two figures, are practically nonexistent in real-world situations. Instead, there is usually some scatter around the regression line. The **standard error of estimate** is another measure of the scatter of the observed Y values around the Y' values on the line of regression.

Standard Error of Estimate A measure of the variability of the observed values around the regression line.

The standard error, as it is usually called, is an indicator of how well the predicted Y' -values compare to the actual Y-values. It is designated by $s_{Y \cdot X}$ and has the formula

Figure 14-7 Scatter Diagram Showing Positive Correlation

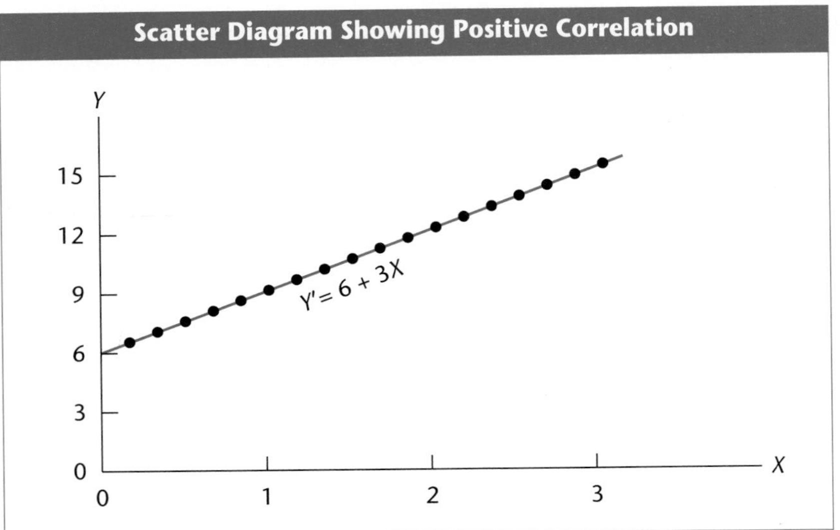

Figure 14-8 Scatter Diagram Showing Negative Correlation

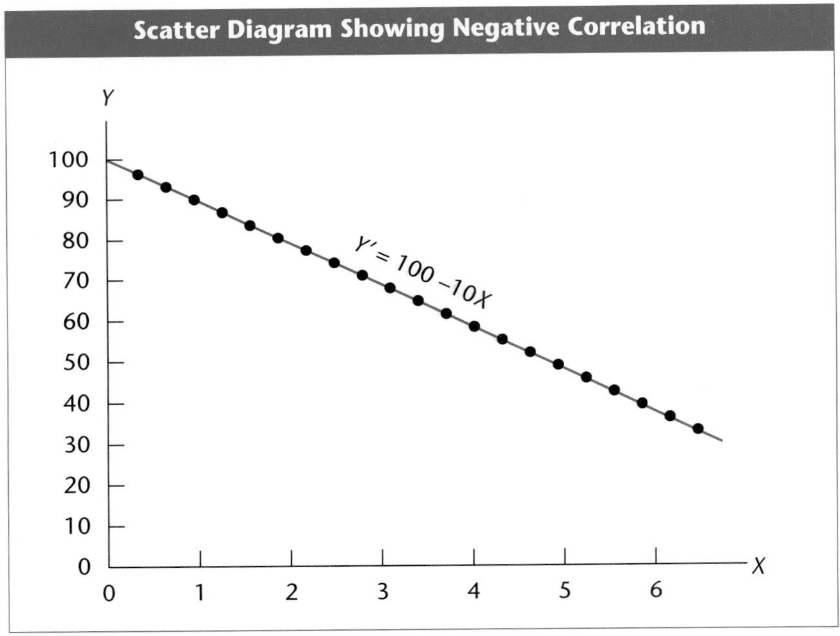

$$s_{Y \cdot X} = \sqrt{\frac{\Sigma(Y - Y')^2}{n-2}} \quad \text{14-4}$$

where the $\Sigma(Y - Y')^2$ component refers to the sum of the squared differences between the actual Y value and the value predicted for Y, called Y'. The subscript $Y \cdot X$ reminds us that we are predicting Y based on values of X, and n refers to the size of the sample.

The standard error measures variation in a manner similar to the standard deviation, except that, instead of summing the squared deviations from the mean as the standard deviation does $\Sigma(Y - \overline{Y})^2$, it employs the squared deviation between the actual values and the predicted values of Y'—that is, $\Sigma(Y - Y')^2$. In fact, it may be thought of as a conditional standard deviation of the dependent variable Y for a given value of the independent variable X.

The table below shows the calculation of the standard error based on the price and length of a textbook.

Title	Pages X	Price Y	Prediction Y'	Error $Y - Y'$	Squared Error $(Y - Y')^2$
A	400	\$44	\$44	0	0
B	600	47	48	–1	1
C	500	48	46	2	4
D	600	48	48	0	0
E	400	43	44	–1	1
F	500	46	46	0	0
					6

The standard error of estimate is \$1.22, found by using formula 14-4:

$$s_{Y \cdot X} = \sqrt{\frac{\Sigma(Y - Y')^2}{n - 2}}$$

$$= \sqrt{\frac{6}{6 - 2}} = 1.22$$

The previous formula for the standard error requires that the difference between each Y observation and its corresponding Y'-value be determined and then squared. These are very time-consuming and error-prone calculations, especially for large samples. An equivalent form is:

$$s_{Y \cdot X} = \sqrt{\frac{\Sigma Y^2 - a\Sigma Y - b\Sigma XY}{n - 2}} \qquad \textbf{14-5}$$

Problem Recall the Chapter Problem. The selling price and number of pages in a textbook, as well as the needed calculations for the standard error of estimate, are shown in Table 14-2 (p. 502).

Solution Earlier, the regression equation was found to be $Y' = a + bX = 36 + 0.02X$. The sums needed to determine a and b are in Table 14–2. They are $\Sigma X = 3{,}000$, $\Sigma Y = 276$, $\Sigma XY = 138{,}800$, $\Sigma X^2 = 1{,}540{,}000$ and $\Sigma Y^2 = 12{,}718$. Inserting the appropriate values in formula 14–4 for the standard error of estimate, we get:

$$s_{Y \cdot X} = \sqrt{\frac{\Sigma Y^2 - a\Sigma Y - b\Sigma XY}{n-2}}$$

$$= \sqrt{\frac{12{,}718 - 36(276) - 0.02(138{,}800)}{6-2}}$$

$$= \sqrt{1.5}$$

$$= \$1.22 \text{ (same as calculated before)}$$

Further interpretation of this value will follow shortly.

Self-Review 14-4

The problem begun in Self-Review 14-3 is continued here. The data are:

Candidate	Amount Spent on Campaign (in thousands of dollars) X	Votes Received (in thousands) Y	XY	X^2	Y^2
Weber	$ 3	14	42	9	196
Taite	4	12	48	16	144
Spence	2	5	10	4	25
Henry	5	20	100	25	400
	14	51	200	54	765

The regression equation is $Y' = -2.3 + 4.3X$. Compute the standard error of estimate.

Exercises

9. Refer to Exercise 5. The relationship between the number of physicians (Y) and the population (X) is

$$Y' = 3.259 + 0.388X,\ n = 5,\ \Sigma Y = 111,$$
$$\Sigma XY = 5{,}990, \text{ and } \Sigma Y^2 = 2{,}735$$

Compute the standard error of estimate.

10. Refer to Exercise 7 on the index of aggression and its relationship to the nighttime low temperature in July.

$$Y' = -25.7 + 1.04X,\ n = 6,\ \Sigma Y = 259,$$
$$\Sigma XY = 18{,}992, \text{ and } \Sigma Y^2 = 13{,}213$$

What is the standard error of estimate?

11. Sample data on the unemployment rate (X) and the crime rate (Y) are as follows:

X	Y
3	8
4	18
5	16

a. Calculate the regression equation.
b. Find the standard error of estimate.

12. The data from Exercise 1, showing length of confinement in days (X) and hospital cost (Y) at St. Matthew's Hospital are repeated here:

X	Y
5	$1,110
2	490
12	2,500
4	880
7	1,530
1	270

a. Calculate the regression equation.
b. Find the standard error of estimate.

ASSUMPTIONS ABOUT REGRESSION

The techniques of regression and correlation are based on several assumptions. For the techniques to be properly employed, the following assumptions should be met, or should be approximately true:

STAT BYTE

During World War II, statistics was often used to improve military operations. For example, one study counted every hole in the aircraft returning from bombing flights to determine where more protection was needed. Investigators were pleased with the progress of their study until one officer asked, "Shouldn't we be looking for the holes in the planes that *didn't* come back?"

1. For a given value of X, the Y observations are normally distributed around the line of regression.
2. The standard deviation of each of these normal distributions is the same. This common standard deviation is estimated by the standard error of estimate ($s_{Y \cdot X}$).
3. The deviations from the regression line are independent. There is no pattern in the size nor the direction of these deviations.

These assumptions are graphically summarized in Figure 14–9, wherein each of the distributions depicted is normal and has the same standard deviation, estimated by $s_{Y \cdot X}$.

Assuming that the observed Y values are normally distributed around the line of regression, we can say

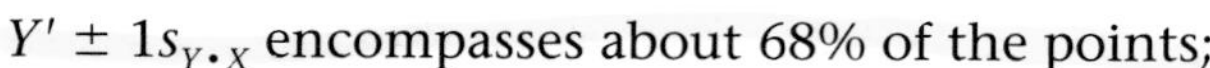

$Y' \pm 1s_{Y \cdot X}$ encompasses about 68% of the points;
$Y' \pm 2s_{Y \cdot X}$ encompasses about 95.5% of the points;
$Y' \pm 3s_{Y \cdot X}$ encompasses about 99.7% of the points;

What does the standard error of estimate in the Chapter Problem tell us? It is a measure of the accuracy of our prediction. Suppose we were to repeat this experiment with a large number of books and

Figure 14-9

An Illustration of the Regression Assumptions

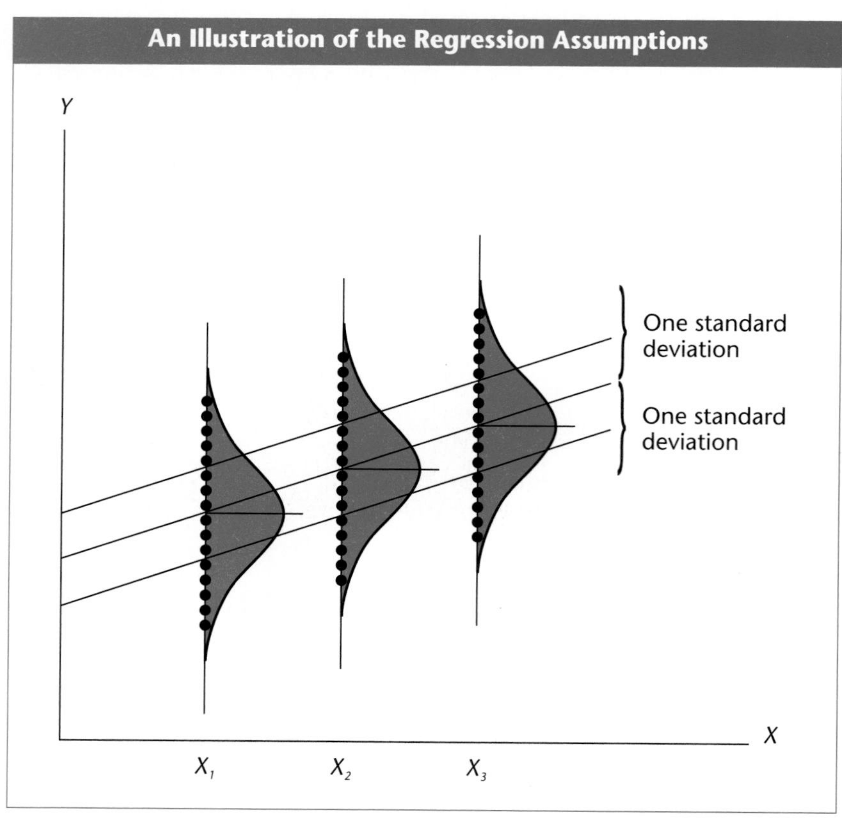

predict their selling prices. Then, the difference between our predicted selling price and the actual selling price would be less than one standard error of the estimate ($1.22) in 68% of the cases. Ninety-five percent of our predictions would be "off" by no more than 2($1.22) = $2.44. And virtually all of our predictions would be "off" by no more than 3($1.22) = $3.66.

CONFIDENCE AND PREDICTION INTERVALS

Suppose we wanted to calculate a confidence interval for our Chapter Problem. We would consider two factors. One is the size of the sample. The other is that errors of prediction are larger as we move away from the mean value of the independent variable (X).

We will consider two applications. In the first case, the *mean* value of Y is estimated for a given value of X and is referred to as a **confidence interval.** In the other case, an *individual* value of Y is estimated for a given value of X and is called a **prediction interval.**

First, for the confidence interval for the *mean value* of Y, the formula is

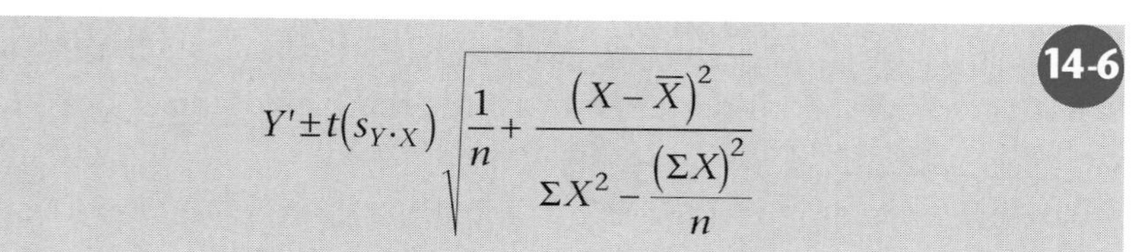

$$Y' \pm t(s_{Y \cdot X})\sqrt{\frac{1}{n} + \frac{(X - \overline{X})^2}{\Sigma X^2 - \frac{(\Sigma X)^2}{n}}} \qquad \textbf{14-6}$$

Second, the formula of the prediction interval for an *individual value* of Y is

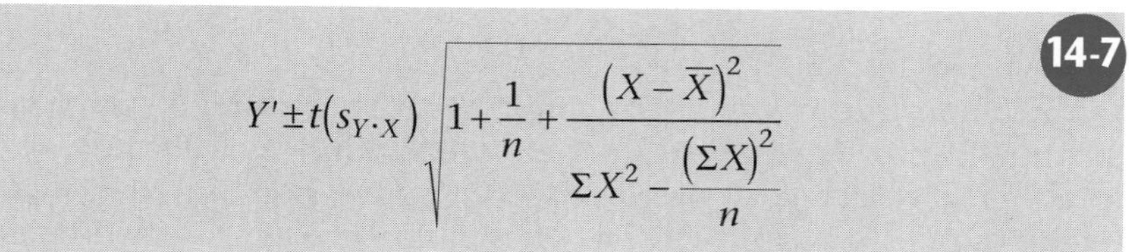

$$Y' \pm t(s_{Y \cdot X})\sqrt{1 + \frac{1}{n} + \frac{(X - \overline{X})^2}{\Sigma X^2 - \frac{(\Sigma X)^2}{n}}} \qquad \textbf{14-7}$$

where

t	is the value from Student's t distribution in Appendix D. For the two-variable linear case, two degrees of freedom are lost because two least-squares coefficients, a and b, are being estimated.
X	is any selected value of the independent variable.
$\overline{X}$	is the mean of the independent variable.

n is the number of paired observations.
Y' is the point estimate of the dependent variable.
$s_{Y \cdot X}$ is the standard error.

Note that there is only a slight difference in the two formulas. There is a 1 in the formula for finding the prediction interval. Logically, the inclusion of 1 under the radical widens the confidence interval for the individual. This is reasonable, because individuals are more unpredictable than group averages.

Problem Recall the Chapter Problem concerning the selling prices and lengths of textbooks at North Central State University.

Title	X	Y	XY	X^2	Y^2
A	400	$ 44	$ 17,600	160,000	1936
B	600	47	28,200	360,000	2209
C	500	48	24,000	250,000	2304
D	600	48	28,800	360,000	2304
E	400	43	17,200	160,000	1849
F	500	46	23,000	250,000	2116
	3,000	276	138,800	1,540,000	12,718

What is the 95% confidence interval for the *mean* selling price for all books with 450 pages?

Solution Using the regression equation computed previously, we find that the predicted selling price for a book with 450 pages is $45, found by:

$$\begin{aligned} Y' &= 36 + 0.02X \\ &= 36 + 0.02(450) \\ &= 45 \end{aligned}$$

We obtain the value of t by referring to Student's t distribution in Appendix D. Read down the left column until you locate $n - 2$, or $6 - 2 = 4$ degrees of freedom. Then move horizontally to the t-value under the 0.05 level of significance for a two-tailed test ($t = 2.776$).

Other values needed to set the confidence-interval estimate were computed previously: $s_{Y \cdot X} = \$1.22$ and, from Table 14–2 (p. 502), $n = 6$, $\Sigma X = 3000$, $\Sigma X^2 = 1{,}540{,}000$, and $\overline{X} = 3{,}000/6 = 500$. Inserting these values gives:

$$Y' \pm t(s_{Y \cdot X})\sqrt{\frac{1}{n} + \frac{(X - \overline{X})^2}{\Sigma X^2 - \frac{(\Sigma X)^2}{n}}}$$

$$= \$45 \pm (2.776)(\$1.22)\sqrt{\frac{1}{6} + \frac{(450-500)^2}{1{,}540{,}000 - \frac{(3{,}000)^2}{6}}}$$

$$= \$45 \pm 3.4\sqrt{.23}$$

$$= \$45 \pm 1.63$$

$$= \$43.37 \text{ and } \$46.63$$

Interpreting, we find that the mean selling price for a group of books, all of whose lengths are 450 pages, is \$45, and the 95% confidence interval is from \$43.37 to \$46.63.

Problem Estimate a 95% prediction interval for the selling price of a *specific* book that is 450 pages long. This differs from the previous illustration in that we are concerned about a specific book that is 450 pages long rather than the mean of all the books that are 450 pages long.

Solution The formula for a prediction interval for an individual value is:

$$Y' \pm t(s_{Y \cdot X})\sqrt{1 + \frac{1}{n} + \frac{(X-\overline{X})^2}{\Sigma X^2 - \frac{(\Sigma X)^2}{n}}}$$

For a specific book that is 450 pages long:

$$\$45 \pm (2.776)(\$1.22)\sqrt{1 + \frac{1}{6} + \frac{(450-500)^2}{1{,}540{,}000 - \frac{(3{,}000)^2}{6}}} = \$45 \pm 3.4\sqrt{1.23}$$

$$= \$45 \pm 3.77$$

$$= \$41.23 \text{ and } \$48.77$$

Interpreting, we conclude that for a specific book 450 pages long, the 95% confidence interval for the selling price is between \$41.23 and \$48.77.

Self-Review 14-5

In Self-Reviews 14-3 and 14-4, which dealt with campaign expenditures and the numbers of votes received, we computed these values:

$Y' = -2.3 + 4.3X$, $s_{Y \cdot X} = 3.3$, $n = 4$, $\Sigma X^2 = 54$, $\Sigma X = 14$, and $\overline{X} = 3.5$

a. Determine the 90% confidence interval for the *mean* number of votes gained by all candidates who spent \$3,800 on their campaigns.

b. Interpret your answer to part a.

c. Compute the 90% confidence interval for Sarah Norbett, whose campaign cost \$3,800.
d. Interpret your answer to part c.
e. Compare your responses to parts b and d.

Exercises

13. Refer to the data in the table on page 502.
  a. What is the 90% confidence interval for the mean selling price for all books with 600 pages?
  b. What is the 90% prediction interval for the selling price of a specific book that is 600 pages long?

14. Exercise 7 dealt with data relating an aggression index (Y) to the mean nighttime low temperature (X). The equation was $Y' = -25.7 + 1.04X$. The predicted value of the index was 57.5 when the temperature was 80°F. In Exercise 10, the standard error of estimate was found to be 3.6, $n = 6$, $\Sigma X = 396$, and $\Sigma X^2 = 27{,}954$.
  a. Construct a 95% confidence interval for the *mean* value of the aggression index when the temperature stands at 80° F.
  b. On a *particular* night the temperature is 80° F. Find the 95% prediction interval for the aggression index on this evening.

15. Exercise 5 presented some data relating the population of a county (X) to the number of physicians who practice in that county (Y). Exercise 9 asked you to compute the standard error of estimate.
  a. Based on these data, construct a 95% confidence interval for the mean number of physicians practicing in counties with a population of 45,000.
  b. Based on these data, construct a 95% prediction interval for the number of physicians practicing in a particular county with a population of 45,000.

16. Refer to the data from St. Matthew's Hospital (in Exercises 1 and 13).
  a. Construct a 90% confidence interval for the *mean* cost of a six-day stay.
  b. Construct a 90% prediction interval for the cost of an individual whose length of stay is six days.

A Comprehensive Example

In Chapters 12 and 13, we discussed methods that help us to predict one variable based on another and to measure the association between two variables. The following Problem reviews these topics. As you have observed, the calculation can be both lengthy and tedious. Therefore, we will use a statistical software package.

Problem Annually, *Forbes* magazine presents a summary of the richest Americans. Included is information on the net worth of the individuals as well as their ages. The objective of the following analysis is to determine the association between net worth and age. Does net worth tend to be greater among older people? In addition, we would like to estimate the net worth of these wealthy Americans, based on their age. The list of 36 in Table 14–3 was selected as an illustration. Treat it as a sample of all wealthy Americans.

a. Develop a scatter diagram. Based on this plot, does there appear to be a linear relationship between age and net worth?
b. Find the coefficient of correlation. Conduct a test of hypothesis using the null hypothesis of no correlation in the population. Can we conclude that the correlation is significantly greater than zero?
c. Compute the coefficients of determination and nondetermination. Interpret.
d. What is the standard error of estimate? What does it mean?
e. Determine the regression equation. Estimate the net worth for a 60-year-old wealthy person.
f. Develop a 95% confidence interval for a typical 60-year-old wealthy person.

Solution

a. MINITAB was used to develop the following scatter diagram. The independent variable (age) is plotted along the horizontal axis and the net worth along the vertical axis. There does appear to be a positive relationship: As age increases, net worth also increases.

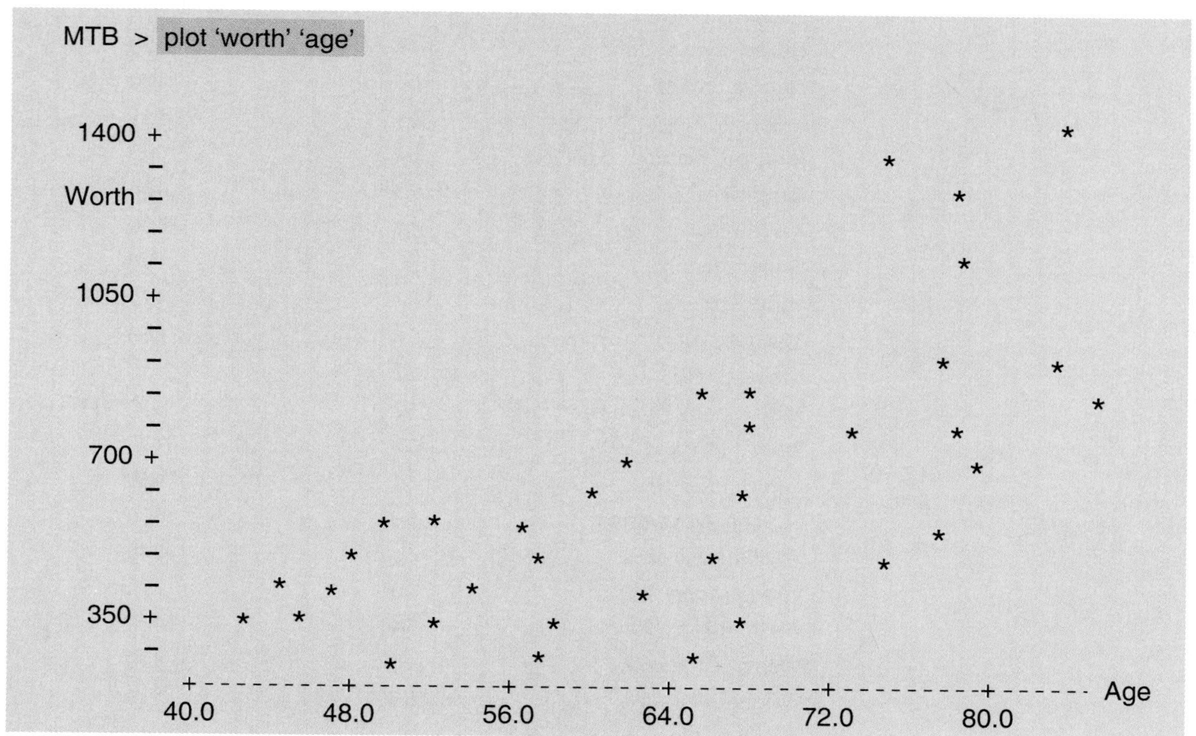

Table 14-3 **Age and Net Worth of 36 of the Richest Americans**

Name	Age	Net Worth (in millions of dollars)
Annenberg, Walter	81	$1,400
Carlson, Curtis	75	925
Cooke, Jack Kent	76	1,250
Copley, Helen	66	865
Coulter, Wallace	77	700
Cox, William	58	320
Disney, Roy	59	660
Field, Marshall	48	460
Ford, William Clay	64	860
Fribourg, Michel	76	800
Getty, John Paul, Jr.	57	300
Graham, Katharine	72	490
Griffin, Mervyn	64	300
Hearst, George Randolph, Jr.	62	438
Hearst, William Randolph, Jr.	81	876
Hess, Leon	75	560
Hewlett, William	76	1,100
Hollingsworth, John	71	750
Hunt, Caroline	66	800
Ingersoll, Ralph	43	345
Johnson, Samuel	61	690
Kauffman, Ewing	73	1,300
Kravis, Henry	45	400
Lauren, Ralph	50	550
Lurie, Robert	47	425
Marriott, John	57	466
Mellon, Paul	82	850
Monaghan, Thomas	52	530
Ryan, Patrick	52	330
Smith, Richard	64	460
Spanos, Alexander	66	600
Spelling, Aaron	66	345
Van Kampen, Robert	50	290
Washington, Dennis	54	400
Weber, Charlotte	46	320
Wrigley, William	56	530

b. Using statistical software, we find that the correlation between age and net worth is 0.724, indicating a rather strong association between the two. Could this association have occurred by chance? The following test of hypothesis is used:

$$H_0: \rho \leq 0$$
$$H_a: \rho > 0$$

At the 0.05 significance level, we reject the H_0 if the computed value of t is greater than 1.69. The computed value of t is 6.12, found by:

$$t = \frac{r\sqrt{n-2}}{\sqrt{1-r^2}} = \frac{0.724\sqrt{36-2}}{\sqrt{1-(0.724)^2}} = 6.12$$

Because 6.12 > 1.69, we reject the null hypothesis and conclude that there is a positive association between net worth and age.

c. The following output was generated:

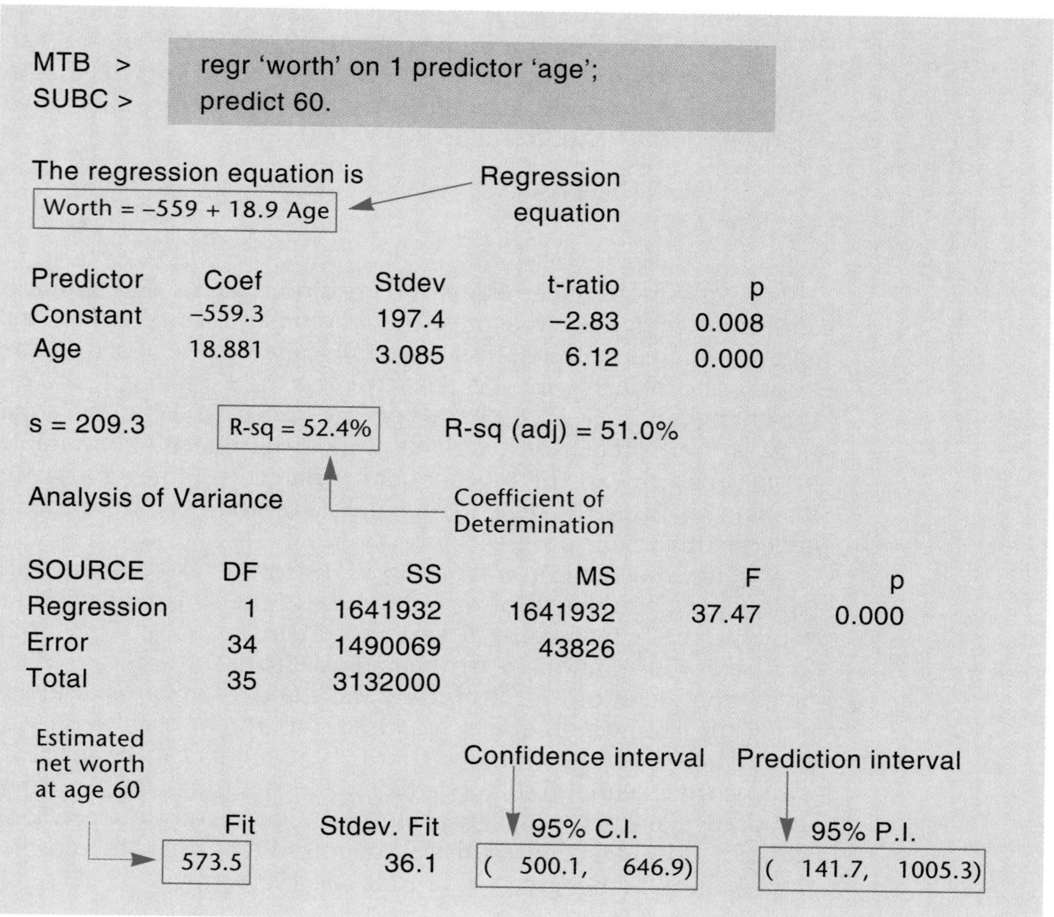

MTB > regr 'worth' on 1 predictor 'age';
SUBC > predict 60.

The regression equation is
Worth = −559 + 18.9 Age

Predictor	Coef	Stdev	t-ratio	p
Constant	−559.3	197.4	−2.83	0.008
Age	18.881	3.085	6.12	0.000

s = 209.3   R-sq = 52.4%   R-sq (adj) = 51.0%

Analysis of Variance

SOURCE	DF	SS	MS	F	p
Regression	1	1641932	1641932	37.47	0.000
Error	34	1490069	43826		
Total	35	3132000			

Fit	Stdev. Fit	95% C.I.	95% P.I.
573.5	36.1	(500.1, 646.9)	(141.7, 1005.3)

The coefficient of determination is 52.4%, which means that 52.4% of the variation in net worth is explained by the variation in age. The coefficient of nondetermination is 0.476, found by 1 – 0.524, which means that 47.6% of the variation in net worth is *not* explained by age and must be due to other causes or to random variation.

d. The standard error of estimate is 209.3. This is in millions of dollars, so about 95% of the values should be within \$418.6 million of the regression line—that is, 2(209.3).

e. The regression equation, found in part c, is $Y' = -559.3 + 18.88X$. In order to obtain the estimated net worth of a 60-year-old wealthy person, we substitute 60 for X in the regression equation:

$$Y' = 559.3 + 18.88(60) = 573.5$$

Thus, a wealthy 60-year-old's net worth is estimated to be \$573,500,000.

f. The formula for determining the confidence interval for all 60-year-old wealthy people is found on page 511, reported in the computer printout under the column "95% C.I." The values are 500.1 and 646.9. In other words, the odds are 95 to 5 that, for all wealthy 60-year-olds, the mean net worth is in this interval between \$500,100,000 and \$646,900,000.

SUMMARY

This chapter was concerned with regression analysis. The intent of regression analysis is to express the relationship between two variables through a mathematical equation, called the regression equation, of the form $Y' = a + bX$. The letter a is the Y-intercept, and b is the change in Y for each change of one unit in X. With the equation, we can predict one variable, called the dependent variable, based on a value of the independent variable. In plotting a scatter diagram, we scale the dependent variable on the Y-axis and the independent variable on the X-axis.

A regression equation was developed using the least-squares principle. This technique minimizes the sum of the squared deviations between the actual point Y and the predicted point Y', written $\Sigma(Y - Y')^2$ = minimum. To arrive at a measure of the accuracy of the prediction equation, we determine the standard error of estimate, under the assumption discussed, which can also be used to provide confidence-interval estimates. That is, statements can be made regarding the likelihood that a predicted (dependent) variable will be in a particular interval, given a value of the predictor (independent) variable. Confidence intervals can be found for either the mean of the Y-values for any given X, or for a specific X-value.

Exercises

17. A tanning booth survey has produced the following information on length of time in a booth compared to results. Each person surveyed was asked how long his or her average stay lasted. Then the employees scored each tan on a scale of 1 to 10. The data:

Name	Tanning Time (in minutes)	Score
Jill	30	7
Jodi	20	6
Mark	40	9
Ted	15	5
Tami	25	7

a. Find the least-squares regression equation.
b. Interpret the values of a and b.
c. Predict the rating that a customer who spends 50 minutes in the tanning booth will receive.

18. Data show the following relationship between a man's height and shoe size:

Height (in inches)	Shoe Size
76	12
72	11
65	8
78	13
82	16
67	9

a. Calculate the regression equation. Shoe size is the dependent variable.
b. Find the standard error of estimate.

19. The number of requests for information (X) and actual enrollments (Y) for the past six years at Nordam University are:

Requests for Information (in thousands)	Actual Enrollments (in thousands)
3.0	3.3
3.5	4.1
4.2	5.6
4.8	5.2
5.0	5.9
5.1	5.5

a. Draw a scatter diagram.
b. Determine the least-squares regression equation.
c. Calculate the standard error of estimate.
d. Compute a 90% confidence interval for the mean enrollment when the requests for information number 4,500.

20. A student of criminology, interested in predicting the age at incarceration (Y) using the age at first police contact (X), collected the following data:

Age at First Contact X	Age at Incarceration Y
11	21
17	20
13	20
12	19
15	18
10	23
12	20

a. Draw a scatter diagram.
b. Find the regression equation by the least-squares method.
c. Compute the standard error of estimate.
d. Estimate the age at incarceration for an individual who at age 14 had his first contact with police. Develop a 95% prediction interval for your estimate.

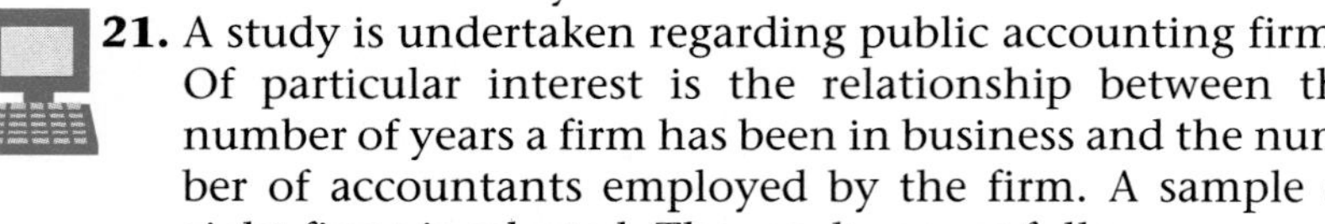

21. A study is undertaken regarding public accounting firms. Of particular interest is the relationship between the number of years a firm has been in business and the number of accountants employed by the firm. A sample of eight firms is selected. The results are as follows:

Firm	Years	Accountants
A & G Accounting	12	36
Kramer Associates	18	72
Whitte and Whitte	27	160
Englekay and Jerdonic	5	13
Huffman and Clark	7	18
Grasser and Lewis	22	68
M & M Accountants	15	40
Bookkeeping, Inc.	10	33

a. Draw a scatter diagram. Let age be the independent variable. Does there appear to be a relationship between years and number of accountants?

b. Determine the regression equation.
c. Compute the standard error of estimate.
d. Estimate the number of accountants in a 10-year-old firm.
e. Develop a 95% confidence interval for all 10-year-old firms.

22. The length of time children are exposed to a common virus (X) and the percentage of the group (Y) who contract the disease are:

Weeks X	Percentage Y
2	0.1
3	0.3
3	0.5
4	0.8
4	1.2
6	1.8
7	2.5
8	3.4

a. Draw a scatter diagram of the data.
b. Compute the least-squares line of regression.
c. Find the standard error of estimate.
d. Make a prediction about the percentage of a group who will contract the virus after a five-week exposure.
e. Determine a 99% prediction interval for the percent of a group exposed five weeks that contracts a virus.

23. The length of various magazine articles (X) and an individual's respective reading time (Y) were recorded:

Length of Article (in pages)	Reading Time (in minutes)
5	13
6	15
6	15
7	18
4	10
3	8
7	18
12	25

a. Draw a scatter diagram.
b. Calculate the regression line.
c. What is the standard error of estimate?

d. Predict the average reading time for a particular 10-page article, and construct a 95% prediction interval about your estimate.

24. A statistics instructor wishes to study the relationship between the number of cuts (absences) students take and their final course grades. The data obtained are:

Number of Absences	Grade
1	98
2	90
4	83
3	88
5	71
2	85
4	76
3	81
6	71

a. Draw a scatter diagram and determine the regression equation.
b. Compute the standard error of estimate.
c. Develop a 90% confidence interval for all students with three absences.
d. If a particular student, Kerry O'Keene, has three absences, what is the 90% prediction interval for her score?

25. The marriage-license bureau records the ages of men and women when they apply for a marriage license. You suspect that the age of the prospective wife can be predicted from her prospective husband's age. You gather the following data to verify your hypothesis:

Husband's Surname	Husband's Age	Wife's Age
Cain	25	25
Behner	22	21
Freeman	26	23
Rops	37	24
Sarantou	30	25
Crosby	38	36
Gasser	24	20
Labash	46	35
Bailey	19	18
Saelzler	31	34

Husband's Surname	Husband's Age	Wife's Age
Smith	41	22
Baden	21	17
Snyder	21	20
Quilter	23	23
Cole	24	21
Schwamberger	20	18
Mominee	34	28
Sweney	21	20

a. Draw a scatter diagram and determine the regression equation.
b. Compute the standard error of estimate.
c. Predict the prospective wife's age for the average 40-year-old man and construct 90% confidence limits for your estimate.

26. A study was made of the relationship between the height and weight of a sample of 14 college men:

Height (in inches) X	Weight (in pounds) Y
64.7	165
67.2	116
71.6	158
65.0	153
72.0	149
71.8	181
73.0	173
64.5	120
68.3	125
72.4	163
66.3	125
72.5	173
68.0	146
67.0	139

a. Determine the regression equation.
b. Compute the standard error of the mean.
c. Determine a 95% confidence interval for the weight of all men who are six feet tall.
d. John Kuk is six feet tall. What is the 95% prediction interval for his weight?

27. The Department of Transportation is investigating the relationship between the number of bidders on a highway construction project and the winning bidder's price. Does the number of bidders decrease the price of the project? A sample of 20 recent construction projects revealed the following information:

Project	Number of Bidders	Winning Bid (in millions of dollars)
1	5	8.7
2	4	9.5
3	3	8.8
4	1	10.9
5	6	7.1
6	7	7.4
7	3	9.3
8	5	7.4
9	6	6.8
10	5	8.7
11	3	8.9
12	6	6.2
13	9	6.3
14	4	9.0
15	6	6.5
16	3	9.3
17	2	9.7
18	5	6.3
19	4	9.2
20	4	8.0

a. Determine the regression equation.
b. Note that the sign of b is negative. Does this surprise you? Comment.
c. How much does the cost of a project decline for each additional bidder?
d. Determine a 95% confidence interval for all highway projects with five bidders.
e. Determine a 95% prediction interval for the project of building a bridge across the Conewango River. There are five bidders for the project.

28. The Placement Director at Arkansas Southern College is studying the relationship between the grade point average (GPA) of students in the College of Social Science and the starting salary of their first full-time job after graduation. A sample of 15 students who graduated last year is selected. Their GPAs and starting salaries were as follows:

Student	GPA	Salary (in thousands of dollars)
1	3.18	$30.0
2	2.63	28.0
3	2.70	25.7
4	3.09	31.0
5	2.87	28.1
6	3.21	19.4
7	2.89	26.9
8	2.79	24.8
9	2.60	24.4
10	2.42	20.5
11	3.40	31.0
12	2.30	22.4
13	3.78	34.0
14	2.53	24.8
15	2.92	24.4

a. Determine the regression equation.
b. How much does the starting salary change for an increase of 0.1 in the GPA?
c. Determine the 95% confidence interval for the starting salary of all students with a GPA of 3.00.
d. David Romaker has just graduated from Arkansas Southern. His GPA was 3.00. Determine a 95% prediction interval for his starting salary.

29. Does the baseball team with the highest team batting average win the most games? Can we estimate the number of games a team will win by knowing its team batting average? To answer these questions, the number of games won and team batting average are shown below for each of the 26 major league clubs for the 1992 season. (The table is continued on 526.)

Team	Wins	Team Batting Average
Toronto	96	.263
Milwaukee	92	.268
Baltimore	89	.259
Cleveland	76	.266
New York Yankees	76	.261

(continued)

Team	Wins	Team Batting Average
Detroit	75	.256
Boston	73	.246
Oakland	96	.258
Minnesota	90	.277
Chicago White Sox	86	.261
Texas	77	.250
California	72	.243
Kansas City	72	.256
Seattle	64	.263
Pittsburgh	96	.255
Montreal	87	.252
St. Louis	83	.262
Chicago Cubs	78	.254
New York Mets	72	.235
Philadelphia	70	.253
Atlanta	98	.254
Cincinnati	90	.260
San Diego	82	.255
Houston	81	.246
San Francisco	72	.244
Los Angeles	63	.248

a. Draw a scatter diagram, with the number of games won on the Y-axis and the team batting average on the X-axis. Do any of the teams seem to be vastly different compared with the others? Which teams?

b. Determine the regression equation. How many wins would you estimate for a team with an average of .260?

c. Determine the 95% confidence interval for the number of games won for all teams with an average of .260.

30. The Yuppie Inn is an exclusive restaurant that accepts reservations for Saturday night only. The owner would like to estimate the number of meals to be served on the basis of the number of reservations. The following is a sample of recent experience:

Reservations	Meals Served
48	79
56	69
65	98
67	82
65	89
63	91
76	95

Reservations	Meals Served
46	94
92	91
84	115
52	70
90	114
46	86
32	70
100	110
80	91
60	96
90	110
81	95
64	79
47	69
45	83
69	81
85	87
50	75

a. Draw a scatter diagram. Does there appear to be a relationship between the two variables? Is it direct or inverse?

b. Determine the regression equation. Estimate the number of meals to be served if there are 100 reservations.

c. There are 100 reservations for this Saturday night. Determine a 95% confidence interval for the number of meals to be served.

d. Using the material from Chapter 13, determine the coefficient of correlation. At the .05 significance level, can we conclude that there is positive association between the number of reservations and the number of meals served?

31. A real estate salesperson in a metropolitan area is studying the relationship between the size of a home (in square feet) and the selling price of the property. A random sample of 20 homes is selected:

Selling Price (in thousands of dollars)	Area (in square feet)
$ 88	1,400
102	1,600
109	1,450
65	1,500
67	1,400
93	1,800
107	1,560

(continued)

Selling Price (in thousands of dollars)	Area (in square feet)
71	1,540
86	1,490
105	1,700
58	1,330
91	1,810
98	1,700
120	1,750
68	1,360
106	1,650
75	1,505
80	1,475
100	1,550
90	1,490

a. Draw a scatter diagram and determine the regression equation.
b. Compute the standard error of estimate.
c. What selling price would you estimate for all homes with 1,500 square feet?
d. Develop the 95% confidence interval for your prediction.

Data Exercises

32. Refer to the Real Estate Data Set, which reports information on homes sold in Northwest Ohio during 1992.

a. Determine the regression equation with the selling price as the dependent variable and the area of the home as the independent variable. Write a brief interpretation of the regression equation.

(1) Estimate the selling price of a home with an area of 2,500 square feet.
(2) Develop a 95% confidence interval for the mean selling price of all homes with 2,500 square feet.
(3) Develop a 95% prediction interval for the mean selling price of a particular home with an area of 2,500 square feet.

b. Determine the regression equation with the selling price as the dependent variable and the distance from the center of the city as the independent variable. Write a brief interpretation of the regression equation.

(1) Estimate the selling price of a home located 20 miles from the center of the city.
(2) Develop a 95% confidence interval for the mean selling price of all homes located 20 miles from the center of the city.

(3) Develop a 95% prediction interval for the mean selling price of a particular home located 20 miles from the center of the city.

33. Refer to the Salary Data Set, which refers to a sample of middle managers in Sarasota, Florida.

a. Determine the regression equation with salary as the dependent variable and the age of the manager as the independent variable. Write a brief interpretation of the regression equation.

(1) Estimate the salary of a 50-year-old manager.

(2) Develop a 95% confidence interval for the mean salary of all 50-year-old managers.

(3) Develop a 95% prediction interval for the mean salary of a particular 50-year-old manager.

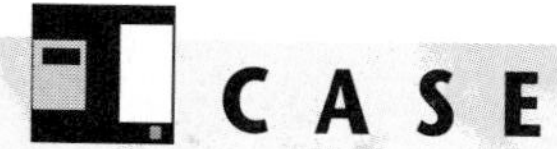

DID THE PROFESSORS SUCCEED?

Many employers administer a standard test called the CEA to people applying for positions with their firm. A person's score is used to predict his or her performance. Professors Garth and Tudor have designed a simple 18-item test, modestly named after them as the GATU test. To validate the GATU test, a random sample of 20 recently hired employees was given both the CEA and the GATU. The scores and the performance ratings of these people after one year of employment are presented below.

Rating	CEA	GATU
31	75	10
27	71	11
30	78	9
30	76	10
26	66	13
38	95	4
21	58	16
30	76	10
21	77	12
29	83	8
36	87	7
30	76	10
30	78	8
33	82	8
30	75	10
25	65	14
33	89	6
30	76	10
30	82	8
26	70	12

Which test, if either, will better predict performance? Can you develop a simple formula that will relate performance to the score on the test you prefer?

CHAPTER ACHIEVEMENT TEST

Answer all the questions. The answers are at the back of the book.

MULTIPLE-CHOICE QUESTIONS. Select the response that best answers each of the questions.

1. A variable about which predictions or estimates are made is called the
   a. dependent variable.
   b. discrete variable.
   c. independent variable.
   d. correlation variable.
2. In the regression equation, the value that gives the amount by which Y changes for every unit in X is called the
   a. coefficient of correlation.
   b. coefficient of determination.
   c. slope.
   d. Y-intercept.
3. In the equation $Y' = a + bX$, the letter a stands for the
   a. coefficient of correlation.
   b. coefficient of determination.
   c. slope of the regression line.
   d. Y-intercept of the regression line.
4. What kind of relationship exists if Y decreases as X increases?
   a. Inverse
   b. Direct
   c. Significant
   d. No relationship
   e. None of the above

Questions 5–7 are based on the following situation. Boys of varying ages were asked to perform chin-ups. The data was used to develop the regression equation $Y' = -1.0 + 1.5X$.

Boy	Age (X)	Chin-ups (Y)
Adam	16	22
Bob	8	12
Chuck	14	22
Denny	10	12
Felix	12	17

5. Which of the five data points comes closest to the regression line?
   a. Adam's **b.** Bob's **c.** Chuck's **d.** Denny's **e.** Felix's
6. The maximum difference between the actual and predicted numbers of chin-ups is
   a. 0. **b.** 1. **c.** 2. **d.** 3. **e.** 4.
7. What is the sum of the squared errors?
   a. 0 **b.** 3.16 **c.** 6 **d.** 10 **e.** 100
8. The standard error of estimate is a measure of
   a. the variation around the mean of X.
   b. the variation around the regression line.
   c. the explained variation.
   d. All of the above
9. Which of the following does not affect the width of a confidence interval?
   a. The standard error of estimate
   b. The size of the sample
   c. The value of X for which the estimate is being made
   d. The Y-intercept

COMPUTATION PROBLEM

10. A commercial grower wants to study the relationship between the mean height of a new variety of dahlia and the number of days since emergence above ground of a sample of seven plants:

Days Above Ground X	Height (in centimeters) Y
6	10
22	19
34	31
42	39
45	47
48	58
47	66

   a. Draw a scatter diagram.
   b. Determine the least-squares regression equation.
   c. Determine the standard error of estimate.
   d. Compare a 90% confidence interval for the height of all dahlias that have been above ground 25 days.

ANSWERS TO SELF-REVIEW PROBLEMS

14-1 a. Don has five years of education beyond high school; his weekly income is \$450.

b. $a = 250$

c. $b = 100$. One way to compute: When $X = 3$, $Y = 550$ and when $X = 4$, $Y = 650$; then, $650 - 550 = 100$. It indicates that for each one additional year beyond high school, weekly income increases \$100.

d. $Y' = 250 + 100X$ (in dollars)

e. \$650, found by $Y' = 250 + 100(4)$

14-2 a. 132, found by:

X	Deviation	Squared Deviation
3	8	64
4	2	4
5	8	64
		132

b. 132 is greater than 24. The sum of squared deviations from the least-squares line is always smaller.

c. The crime rate (it is on the Y-axis)

14-3 a.

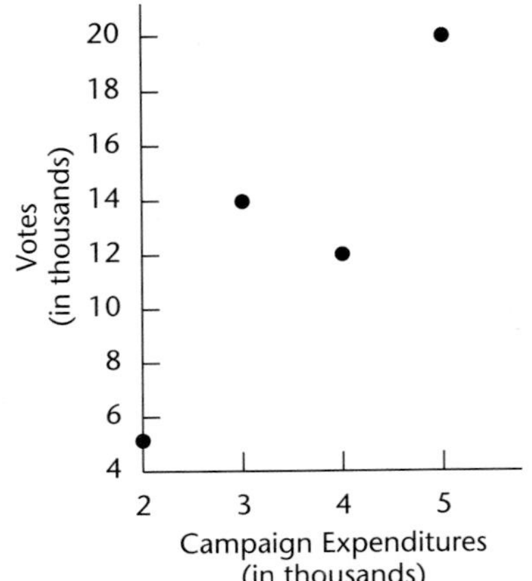

b.

$$b = \frac{4(200) - (14)(51)}{4(54) - (14)^2} = \frac{86}{20} = 4.3$$

$$a = \frac{51 - 4.3(14)}{4} = -2.3$$

$Y' = -2.3 + 4.3X$ (in thousands of votes)

c. For each additional \$1,000, the candidate could expect 4,300 votes.

14-4

$$s_{Y \cdot X} = \sqrt{\frac{\Sigma Y^2 - a\Sigma Y - b\Sigma XY}{n-2}}$$

$$= \sqrt{\frac{765 - (-2.3)(51) - 4.3(200)}{4-2}}$$

$$= \sqrt{\frac{765 + 117.3 - 860}{2}}$$

$$= 3.3$$

14-5 a. $Y' = -2.3 + 4.3(3.8) = 14.0$

Then:

$$14.0 \pm 2.920(3.3)\sqrt{\frac{1}{4} + \frac{(3.8 - 3.5)^2}{54 - \frac{(14)^2}{4}}}$$

$$= 9.0 \text{ and } 19.0$$

b. The probability is 0.90 (that is, the odds are 9 to 1) that a candidate who spends \$38,000 on his or her campaign will receive between 9,000 and 19,000 votes.

c.

$$14.0 \pm 2.920(3.3)\sqrt{1 + \frac{1}{4} + \frac{(3.8 - 3.5)}{54 - \frac{(14)^2}{4}}}$$

$$= 14.0 \pm 10.85$$

$$= 3.15 \text{ and } 24.85$$

d. The probability is 0.90 that if individual candidates spend \$38,000 on their campaigns they will receive between 3,150 and 24,850 votes.

e. The wider interval is the result of comparing individuals and comparing means.

CHAPTER 15

Multiple Regression and Correlation Analysis

OBJECTIVES

When you have completed this chapter, you will be able to

- describe the relationship between one dependent variable and two or more independent variables;
- interpret multiple regression and correlation output from a computer software package;
- conduct a test to determine if the coefficient of multiple determination is zero;
- conduct a test to determine if any of the regression coefficients are zero.

CHAPTER PROBLEM Predicting Prejudice

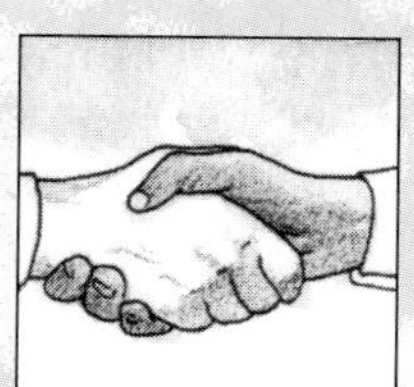

DR. SUSAN WELCH IS A NOTED SOCIOLOGIST DOING RESEARCH ON RACIAL PREJUDICE. SHE HAS DEVELOPED A TEST IN WHICH A RACIAL PREJUDICE "SCORE" IS GIVEN TO RESPONDENTS BASED ON THEIR ANSWERS TO A SERIES OF QUESTIONS. AN EQUALLY IMPORTANT PROBLEM FOR HER IS TO IDENTIFY FACTORS THAT ARE RELATED TO THIS SCORE. DR. WELCH WOULD LIKE TO BE ABLE TO PREDICT WHICH RESPONDENTS WILL HAVE LOW PREJUDICE SCORES AND WHICH WILL HAVE HIGH PREJUDICE SCORES. IS AGE A FACTOR? MARITAL STATUS? HOW ABOUT POLITICAL ORIENTATION—THAT IS, ARE LIBERALS MORE OR LESS LIKELY TO EXHIBIT RACIAL PREJUDICE THAN CONSERVATIVES? WHAT OTHER FACTORS SHOULD BE CONSIDERED?

INTRODUCTION

Our study of correlation and regression began in Chapter 13, where we showed that the relationship between two sets of interval-scaled measurements can be described by Pearson's product-moment coefficient of correlation. A coefficient of 0.91, for example, indicates that the relationship between the two variables is strong. A coefficient near 0, such as 0.17, would indicate a very weak relationship. If the relationship is linear, we could then use the equation $Y' = a + bX$ to predict the dependent variable Y based on the independent variable X.

In this chapter, we will continue our study of correlation and regression. However, instead of using only one independent variable to predict the dependent variable, we will consider *two or more* independent variables. The purpose of using several independent variables is to obtain a more accurate prediction or to explain more of the variation in the dependent variable. When more than one independent variable is used, we refer to the procedure as **multiple regression and correlation analysis.**

ASSUMPTIONS ABOUT MULTIPLE REGRESSION AND CORRELATION

Multiple regression and correlation analysis is based on five assumptions:

1. The independent variables and the dependent variable have a linear (straight-line) relationship.
2. The dependent variable is continuous and at least of interval scale.
3. The variation around the regression line is the same for all values of X. This means that Y varies the same amount when X is a low value as when it is a high value. Statisticians call this assumption **homoscedasticity.**
4. Successive observations of the dependent variable are uncorrelated. Violation of this assumption is called **autocorrelation.** Autocorrelation often occurs when data are collected over a period of time.
5. The independent variables are not highly correlated. When the independent variables are correlated, this is called a **multicollinearity.**

Even when one or more of these assumptions are violated, which often happens in practice, multiple regression and correlation techniques seem to work well, and the results are still adequate. The regression equation and measures of correlation provide the researcher with predictions that are better than any that could otherwise be made.

THE MULTIPLE REGRESSION EQUATION

Recall from Chapter 14 (formula 14–1) that for *one* independent variable, the linear regression equation has the form $Y' = a + bX$. The multiple regression case extends this equation to include other independent variables. For two independent variables, the regression equation is

$$Y' = a + b_1X_1 + b_2X_2 \tag{15-1}$$

where

X_1	is the first independent variable.
X_2	is the second independent variable.
a	is the point of intercept with the Y-axis.
b_1	is the rate of change in Y for each unit change in X_1, with X_2 held constant. It is called a **net regression coefficient.**
b_2	is the rate of change in Y for each unit change in X_2, with X_1 held constant. It too is called a net regression coefficient.

For any number of independent variables *(k)*, the general multiple regression equation is:

$$Y' = a + b_1X_1 + b_2X_2 + b_3X_3 + \cdots + b_kX_k \tag{15-2}$$

Problem The director of admissions at McLean University uses high-school grade point average (X_1) and IQ score (X_2) to predict first-quarter grade point averages (GPAs) of entering students. The director has determined the multiple regression equation to be:

$$Y' = 0.60 + 0.75X_1 + 0.001X_2$$

Predict the first-quarter GPA for a student whose high-school GPA is 3.0 and whose IQ is 100. Also predict the first-quarter GPA for an entering student with a high-school GPA of 4.0 and an IQ of 100. What do 0.60 and the regression coefficient 0.75 in the equation indicate?

Solution The predicted college GPA for a student with a 3.0 GPA in high school and with an IQ of 100 is 2.95, found by:

$$\begin{aligned} Y' &= 0.60 + 0.75X_1 + 0.001X_2 \\ &= 0.60 + (0.75)(3.0) + (0.001)(100) \\ &= 2.95 \end{aligned}$$

The predicted GPA for a student with a 4.0 GPA in high school and with an IQ of 100 is 3.70, found by:

$$Y' = 0.60 + (0.75)(4.0) + 0.001(100)$$

The 0.60 indicates that the regression equation crosses through the Y-axis at 0.60. This is also called the Y-intercept. The regression coefficient of 0.75 indicates that for each increase of 1 in the high school grade point average, the first-quarter GPA at McLean will increase 0.75 *regardless of the student's IQ.*

Self-Review 15-1

Answers to the Self-Review problems are at the end of the chapter.

An economist is studying a sample of households to determine the weekly amount saved by each. Three independent variables seem to hold promise as predictors of the amount saved. These are weekly income, weekly amount spent on food, and weekly amount spent on entertainment. The multiple regression equation was computed to be:

$$Y' = 20.0 + 0.50X_1 - 1.20X_2 - 1.05X_3$$

where

Y' is the weekly amount saved.
X_1 is the weekly income of the household.
X_2 is the weekly amount spent on food.
X_3 is the weekly amount spent on entertainment.

If a household had an income of \$300 per week and spent \$70 on food and \$50 on entertainment, how much would be saved per week?

Exercises

Answers to the even-numbered exercises are at the back of the book.

1. Refer to the following regression equation:

$$Y' = 4.00 + 5.00X_1 - 2.00X_2$$

   a. Compute the value of Y' if $X_1 = 10$ and $X_2 = 5$.
   b. How much does Y' change for a unit change in X_1 if X_2 is held constant?

2. Refer to the following regression equation:

$$Y' = 55.0 - 3.20X_1 + 2.3X_2$$

   a. Compute the value of Y' if $X_1 = 20$ and $X_2 = 30$.
   b. How much does Y' change for a unit change in X_1 if X_2 is held constant?

3. A medical researcher is studying the systolic blood pressure of business executives. Two independent variables are being used as predictors: income and age. The following multiple regression equation has been computed:

$$Y' = 130 + 0.32X_1 + 0.22X_2$$

where

Y' is the systolic pressure.
X_1 is the income (in thousands of dollars).
X_2 is the age.

   a. What is the expected systolic pressure reading of a 60-year-old executive earning \$80,000 a year?
   b. Regardless of age, how much does the systolic pressure increase for each \$10,000 additional annual income?

4. A sample of women was polled to determine the degree of satisfaction they felt in their marriages. All the women were between the ages of 35 and 50, and all had at least one child. An index of satisfaction was developed for the study. Three factors were used as predictor variables: the number of children (X_1), the number of years since the birth of the last child (X_2), and the mother's yearly earnings outside the home (X_3). The regression equation is:

$$Y' = 30 + 1.05X_1 + 6.5X_2 + 0.005X_3$$

   a. Predict the index of marriage satisfaction for a woman with three children who earns \$12,000 outside the home and had her last child eight years ago.
   b. Which would contribute more to marriage satisfaction: another year without a child or an additional \$5,000 in income?

STAT BYTE

Financial investors often use regression analysis to help estimate future trends in the stock market. For example, according to the Hemline Theory, a rise in the hemline of women's skirts will forecast an upturn in the Dow Jones Industrial Average. According to the Super Bowl Theory, if the team winning the Super Bowl was from the American Conference, the Dow Jones Industrial Average will be down the following year. If the winning team was from the National Conference, the index will be up. Incidentally, the Super Bowl Index was correct for 21 out of 23 years! Would you want to invest your money based on this association?

MEASURING THE STRENGTH OF THE ASSOCIATION

If the multiple regression equation fits the data perfectly, there is no error in the predicted value of Y. Of course, this seldom occurs in actual practice, so we need measures that can provide information on the strength of the association and the size of the error in our prediction. There are three such measures: (1) the coefficient of multiple correlation, (2) the coefficient of multiple determination, and (3) the multiple standard error of estimate. The first two are relative measures; the last is an absolute measure of error.

The **coefficient of multiple correlation** is the multi-dimensional analogue to Pearson's coefficient of correlation described in Chapter 13.

Multiple Coefficient of Correlation A measure of the strength of the association between the dependent variable and two or more independent variables.

As the following diagram shows, the multiple coefficient of correlation can have any value between 0 and +1.00, inclusive, and is designated R. Multiple R is always positive. A coefficient of 0.87 indicates a very strong association between the dependent and independent variables. A coefficient of 0.13 reveals a very weak correlation.

Weak correlation	Moderate correlation	Strong correlation

0 — No correlation; 0.50; 1.00 — Perfect correlation

The coefficient of multiple correlation permits us to say that the coefficient indicates a "very strong" association between the dependent and independent variables. A more precise measure of association is the **coefficient of multiple determination,** R^2. (Note that we find it by squaring the coefficient of multiple correlation.) Logically, if we know the coefficient of multiple determination R^2, we can find the coefficient of correlation R by taking the square root of R^2—that is, $\sqrt{R^2} = R$.

Coefficient of Multiple Determination The proportion (percent) of the total variation in the dependent variable Y that is explained by the set of independent variables.

The third measure of association, called the **multiple standard error of estimate,** is the natural extension to the standard error of estimate discussed in Chapter 14. It is based on the squared differences between Y and Y'. However, Y' is now based on *two or more* independent variables. It can be thought of as an average of the squared errors.

The multiple standard error of estimate is computed by

$$S_{Y\cdot 12} = \sqrt{\frac{\Sigma(Y - Y')^2}{n-(k+1)}} \qquad \textbf{15-3}$$

where

$S_{Y\cdot 12}$ is the multiple standard error of estimate. The subscript $Y\cdot 12$ indicates that Y is the dependent variable. The 12 indicates that there are two independent variables, labeled X_1 and X_2.

n is the number of observations in the sample.

k is the number of independent variables.

Problem An administrator at Valley General Hospital is studying patients' length of stay, in days, based on two variables: the patient's age (X_1) and sex (X_2). The administrator selects a random sample of nine patients who were in the hospital for at least one day during the last month. The results are as follows:

Age (in years)	Sex*	Stay (in days)
42	1	12
36	0	10
32	1	11
29	0	6
26	0	9
24	0	6
22	1	9
18	0	5
15	0	7

*A 1 indicates that the patient was female and a 0 that the patient was male.

a. Use a statistical software package to determine the regression equation.
b. Use this regression equation to estimate the length of stay for Ms. Kulla, a 42-year-old patient.
c. Identify and interpret the coefficient of determination.
d. Identify and interpret the multiple standard error of estimate.

Solution MINITAB was used to determine the regression equation and measures of association. The dependent variable is length of stay. The patient's age and sex are the independent variables.

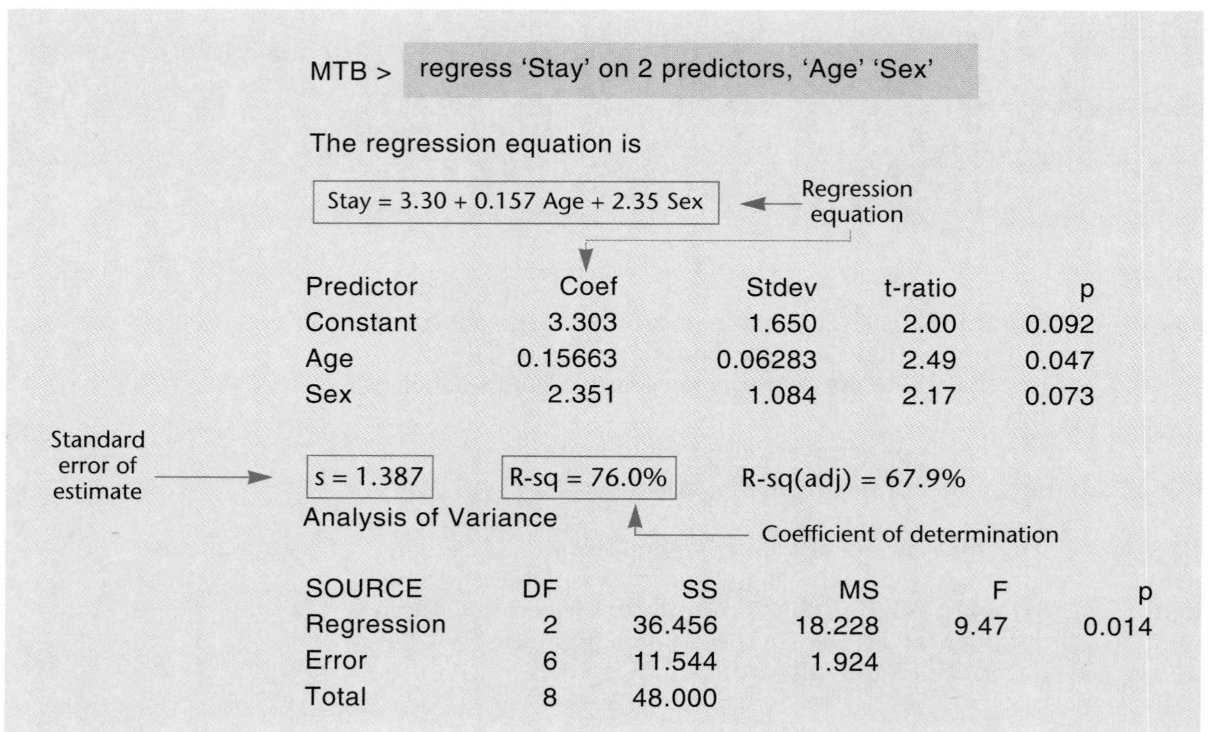

MTB > regress 'Stay' on 2 predictors, 'Age' 'Sex'

The regression equation is

Stay = 3.30 + 0.157 Age + 2.35 Sex

Predictor	Coef	Stdev	t-ratio	p
Constant	3.303	1.650	2.00	0.092
Age	0.15663	0.06283	2.49	0.047
Sex	2.351	1.084	2.17	0.073

s = 1.387 R-sq = 76.0% R-sq(adj) = 67.9%

Analysis of Variance

SOURCE	DF	SS	MS	F	p
Regression	2	36.456	18.228	9.47	0.014
Error	6	11.544	1.924		
Total	8	48.000			

a. The regression equation is:

$$Y' = 3.30 + 0.157X_1 + 2.35X_2$$

The intercept with the Y-axis is 3.30. This is the point at which both $X_1 = 0$ and $X_2 = 0$. An increase of one year in age—say, from 40 to 41 years—means an increase in the length of stay of 0.157 days. A 41-year-old patient can be expected to stay 0.157 days longer in the hospital than a 40-year-old patient. Similarly, a female patient can be expected to stay an additional 2.35 days. How is this so? If the sampled patient is male, the variable X_2 is coded as 0, so 2.35 is multiplied by 0 and this term drops out of the equation. If the sampled patient is female, then 2.35 is multiplied by 1, so the length of stay is increased by 2.35 days. There are only two conditions—male or female—with this variable.

b. Ms. Kulla's stay is estimated to be 12.244 days, found by:

$$\begin{aligned} Y' &= 3.30 + 0.157(42) + 2.35(1) \\ &= 12.244 \end{aligned}$$

c. The coefficient of determination is 0.760, which means that 76.0% of the variation in the length of a patient's stay can be explained by the variation in the patient's sex and age.

d. The multiple standard error of estimate is 1.387. This measure is interpreted much like the standard deviation. About 95% of the actual observations are within 2 standard errors of the regression equation. That is, we would expect to find most of the lengths of stay within 2.774 days of the regression equation, found by 2(1.387).

THE ANOVA TABLE

Included in many software programs is an ANOVA (analysis of variance) table similar to the one described in Chapter 12. In that chapter, the variation was divided into two components—that due to treatments, and that due to other causes, including random variation. Here the total variation is also divided into two categories—that explained by the regression (the independent variables), and the error (unexplained variation). These two categories (regression and error) are identified in the "SOURCE" column of the computer output. The column headed "DF" indicates the number of degrees of freedom. In the Valley General Hospital example, there were nine observations, so $n = 9$. The total number of degrees of freedom is $n - 1 = 9 - 1 = 8$. The degrees of freedom in the "Regression" row is equal to the number of independent variables. We use k to represent the number of independent variables, so $k = 2$. The degrees of freedom in the "error" row is $n - (k + 1) = 9 - (2 + 1) = 6$.

The term "*SS*" (middle column of the ANOVA table) refers to the sum of the squares. These terms are computed as follows:

$$\begin{aligned} \text{Total sum of squares} &= SS\text{ total} = \Sigma(Y - \overline{Y})^2 = 48 \\ \text{Error sum of squares} &= SSE = \Sigma(Y - Y')^2 - 11.544 \\ \text{Regression sum of squares} &= SSR = SS\text{ total} - SSE \\ &= 48.000 - 11.544 = 36.456 \end{aligned}$$

The column headed "*MS*," which refers to the mean square, is found by dividing *SS* by *df.* We obtain *MSR* by *SSR/k* and *MSE* by *SSE/*[*n* – (*k* + 1)]. Continuing, the "*F*" column is the ratio of the two mean squares, that is, *MSR/MSE*. The details of this calculation are discussed in the upcoming section on the global test of hypothesis. The *p* column refers to the *p*-value. As discussed in an earlier section, the *p*-value refers to the probability of obtaining the value of the test statistic this large or larger, if the null hypothesis is true.

The general format of the ANOVA table is:

Analysis of Variance

SOURCE	df	SS	MS	F	
Regression	k	SSR	MSR	MSR/MSE	p
Error	$n-(k+1)$	SSE	MSE		
Total	$n-1$	SS total			

We calculate the coefficient of determination R^2 using numbers in the ANOVA table. Recall from Chapter 13 that the coefficient of determination is the percent of the variation that is explained by the regression. In other words, the coefficient of determination is the ratio of the regression sum of squares to the total sum of squares. It is the variation explained by the regression divided by the total variation:

$$R^2 = \frac{SSR}{SS\text{ total}} \tag{15-4}$$

In the Valley General Hospital example, R^2 is 0.760, found by formula 15-4:

$$R^2 = \frac{SSR}{SS\text{ total}} = \frac{36.456}{48.000} = 0.760$$

The ANOVA table can also be used to arrive at the multiple standard error of estimate. The formula is:

$$s_{Y\cdot 12} = \sqrt{\frac{SSE}{n-(k+1)}} \tag{15-5}$$

Using this formula the standard error of estimate is:

$$s_{Y\bullet 12} = \sqrt{\frac{SSE}{n-(k+1)}} = \sqrt{\frac{11.544}{9-(2+1)}} = 1.387$$

This is the same result reported in the MINITAB output on page 540.

Exercises

5. Refer to the following ANOVA output:

SOURCE	DF	SS	MS	F
Regression	3	21.000	7.000	2.333
Error	15	45.000	3.000	
Total	18	66.000		

a. What is the sample size?
b. How many independent variables are there?
c. Compute the coefficient of determination.
d. Compute the standard error of estimate.

6. Refer to the following ANOVA output:

SOURCE	DF	SS	MS	F
Regression	5	60.000	12.000	1.714
Error	20	140.000	7.000	
Total	25	200.000		

a What is the sample size?
b. How many independent variables are there?
c. Compute the coefficient of determination.
d. Compute the standard error of estimate.

A Computer Application

Applied research often deals with a large number of independent variables. Frequently, however, just a few of them will account for most of the variation in the dependent variable. How can the researcher determine which independent variable or set of variables is most useful in the predicting equation? The following Problem will illustrate how several independent variables can be considered. The variables will be tested both individually and collectively to see if they have a significant connection to some dependent variable.

Problem Recall from the Chapter Problem that Dr. Welch, a noted sociologist, is studying factors related to racial prejudice. She has isolated four variables that she believes are related to racial prejudice:

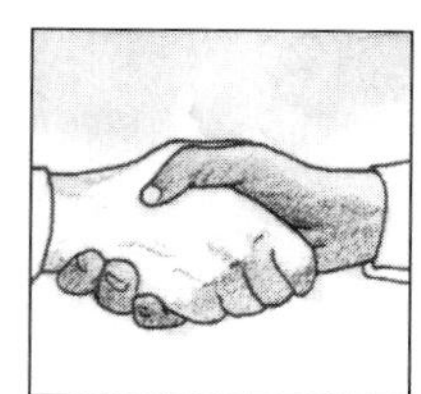

Mobility is a person's mobility, measured in number of months since his or her last change of address.

Age — is the person's age in years.

Status — is the person's "index of social and economic status." The norm for this index is 100. Someone very poor and with practically no social status might have an index score of, say, 10. A person with a high income and high social status might have a score of 192.

Pol — is the person's political orientation. Based on an interview, each person sampled is assigned a score from 1 (very conservative) to 10 (very liberal).

Prej — is the degree of racial prejudice as measured by a test. This is the dependent variable.

Dr. Welch gave her test to a sample of 21 people. The sample data are shown in Table 15–1. Using these sample data, determine the

Table 15-1 **Degree of Racial Prejudice (Based on Mobility, Age, Socioeconomic Status, and Political Orientation)**

Prejudice Score (Prej)	Mobility	Age	Status	Political Orientation (Pol)
80	38	23	135	2
86	44	30	170	6
48	28	38	82	4
40	35	30	77	5
66	26	19	72	3
64	38	23	122	2
94	33	17	159	3
32	42	70	68	7
48	14	51	129	3
32	23	31	67	1
88	38	21	129	6
76	24	22	126	4
52	33	33	53	8
40	25	31	75	6
76	43	22	123	4
14	22	65	43	2
62	18	26	153	2
40	49	38	79	8
72	24	22	116	3
60	34	29	91	3
76	24	52	133	5

multiple regression equation, and test the regression coefficients individually and as a group.

Solution There are four independent variables under consideration: mobility, age, status, and political orientation. To illustrate the relationships between the dependent variable, prejudice, and each of the four independent variables, the following scatter diagrams are drawn. It appears that mobility and status are both positively related to prejudice and that age and political orientation are negatively related. The strongest relationship appears to be between prejudice and status.

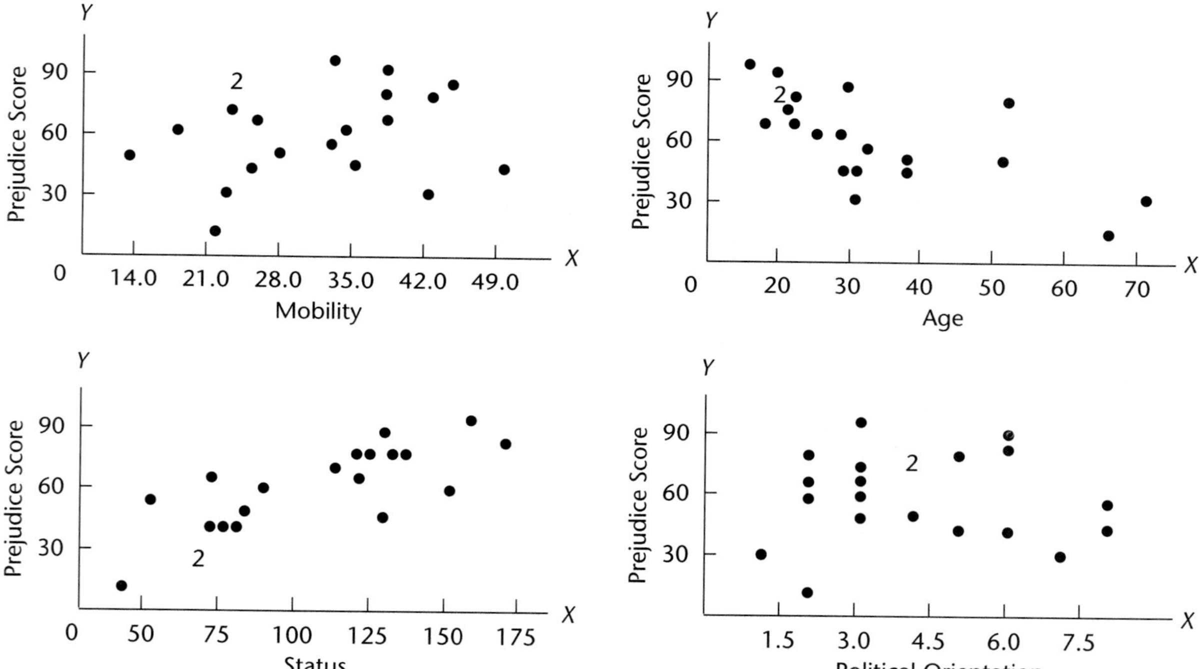

A **correlation matrix** is used to display all possible simple coefficients of correlation. That is, it shows the correlations between each independent variable and the dependent variable as well as the correlation among the independent variables. The following output was produced on MINITAB:

```
MTB > corr c1-c5

             Prej    Mobility      Age     Status
Mobility    0.207
Age        -0.662     -0.127
Status      0.825      0.052    -0.423
Pol        -0.052      0.523     0.224    -0.198
```

The strength and direction of the relationships noted in the scatter diagrams are confirmed in the correlation matrix. The correlation is strongest between prejudice and status, and that correlation is positive. The correlation matrix can also be used to look for **multicollinearity,** or correlation among the independent variables. Such correlation is undesirable because it can cause distortion in the standard error of estimate and can lead to incorrect conclusions about which of the independent variables have significant net regression coefficients. Here, the largest correlation among the independent variables—between political orientation and mobility—is 0.523, which is not large enough to cause a problem. How do we determine when the correlation is too large? One way is to conduct a test of hypothesis on the correlation coefficient as described in Chapter 13. A second method is to use the rule of thumb that the correlation must be stronger than –0.70 or +0.70 to be a problem.

Self-Review 15-2

The Skaff Appliance Company currently has 30 retail outlets across the United States. They retail name-brand electronic products, such as TVs, stereos, and VCRs. Skaff Appliance is considering opening several additional stores in other metropolitan areas. Paul Skaff, president, would like to study the relationship between his sales at the 30 existing locations and several factors regarding the existing store or its region. This information will be useful in selecting metropolitan areas in which to locate new stores. The following sample data were recorded:

Population (in thousands)	Percent Unemployed	Yearly Advertising Expense (in thousands of dollars)	Monthly Sales (in thousands of dollars)
7.500	5.1	$59.0	$ 5.170
8.710	6.3	62.5	5.780
10.000	4.7	61.0	4.840
7.450	5.4	61.0	6.000
8.670	5.4	6.1	6.000
11.000	7.2	12.5	6.120
13.180	5.8	35.8	6.400
13.810	5.8	59.9	7.100
14.430	6.2	57.2	8.500
10.000	5.5	35.8	7.500
13.210	6.8	27.9	9.300
17.100	6.2	24.1	8.800
15.120	6.3	27.7	9.960

Population (in thousands)	Percent Unemployed	Yearly Advertising Expense (in thousands of dollars)	Monthly Sales (in thousands of dollars)
18.700	5.0	24.0	9.830
20.200	5.5	57.2	10.120
15.000	5.8	44.3	10.700
17.600	7.1	49.2	10.450
19.800	7.5	23.0	11.320
14.400	8.2	62.7	11.870
20.350	7.8	55.8	11.910
18.900	6.2	50.0	12.600
21.600	7.1	47.6	12.600
25.250	4.0	43.5	14.240
27.500	4.2	55.9	14.410
21.000	7.0	51.2	13.730
19.700	6.4	76.6	13.730
24.150	5.0	63.0	13.800
17.650	8.5	68.1	14.920
22.300	7.1	74.4	15.280
24.000	8.0	70.1	14.410

The correlation matrix from the statistical software package is given below.

	Sales	Popul	unemp%
Popul	0.894		
unemp%	0.312	0.075	
Advexp	0.385	0.260	0.128

a. What is the correlation between population ("Popul") and the percent of unemployment ("unemp%")?

b. Which of the three independent variables has the strongest correlation with the dependent variable?

c. Does there appear to be any problem with correlation among the independent variables? What is the condition of strong correlation among the independent variables called?

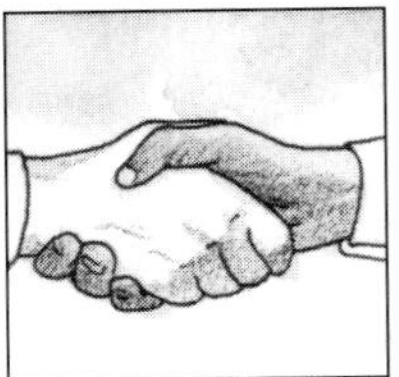

A GLOBAL TEST FOR THE MULTIPLE REGRESSION EQUATION

The overall ability of the set of independent variables to explain the variation of the dependent variable can be tested. To put it another way, can the hypothesis that all the regression coefficients are zero be rejected? A test of several regression coefficients at a time is referred to as a **global test.**

To relate the question to our Chapter Problem on racial prejudice, the null hypothesis is that the net regression coefficients for mobility, age, social status, and political orientation are all equal to zero. That is, none of these variables is useful in explaining the variation in prejudice. The following computer output gives the regression equation, the ANOVA table, and other statistics for the data in Table 15–1:

```
MTB >   regr 'Prej' on 4 predictors 'Mobility' 'Age' 'Status' 'Pol'

The regression equation is
Prej = 27.2 + 0.097 Mobility – 0.583 Age + 0.399 Status + 1.57 Pol

Predictor        Coef       Stdev     t-ratio        p
Constant        27.24       13.45        2.03    0.060
Mobility       0.0971      0.2909        0.33    0.743
Age           -0.5825      0.1726      – 3.38    0.004
Status        0.39854     0.06724        5.93    0.000
Pol             1.569       1.359        1.15    0.265

s = 9.910      R – sq = 82.9%      R – sq(adj) = 78.7%

Analysis of Variance

SOURCE        DF        SS        MS         F         p
Regression     4    7639.5    1909.9     19.45     0.000
Error         16    1571.2      98.2
Total         20    9210.7
```

The net regression coefficients of 0.097 for mobility, –0.583 for age, and so on are estimates of the population values β_1, β_2, β_3, and

β_4. We want to test whether these population values are zero or not. The null and alternate hypotheses are stated as follows:

$$H_0 : \beta_1 = \beta_2 = \beta_3 = \beta_4 = 0$$
$$H_a : \text{Not all the } \beta\text{s are 0.}$$

Not to reject the null hypothesis implies that the regression coefficients are all zero. Logically, this means that none of them is useful in explaining the variation in prejudice. Should this be the case, we would need to search for other independent variables.

To perform the test, we employ the F distribution, which, you will recall from Chapter 12, has the following major characteristics:

1. It is positively skewed.
2. It is based on two sets of degrees of freedom.
3. It cannot take on negative values.

The value of the test statistic is determined from the following formula:

$$F = \frac{SSR/k}{SSE/[n-(k+1)]} \qquad \textbf{15-6}$$

The degrees of freedom for the numerator of the F distribution are equal to k, the number of independent variables. In this example, there are four independent variables, so $k = 4$. The degrees of freedom in the denominator are equal to $n - (k + 1) = 21 - (4 + 1) = 16$, as reported in the "Analysis of Variance" section of the computer printout on page 548. The column headed "df" reports the degrees of freedom. Using the 0.05 significance level and Appendix E, we find the critical value of F to be 3.01. The null hypothesis is rejected if the computed value of F exceeds 3.01. The value of F is computed as follows:

$$F = \frac{SSR/k}{SSE/[n-(k+1)]} = \frac{7639.5/4}{1571.2/[21-(4+1)]} = 19.45$$

This information can also be found in the computer printout on page 548. The F value is 19.45.

Since the computed F value of 19.45 exceeds the critical value of 3.01, the null hypothesis is rejected, and we conclude that at least one of the regression coefficients is not equal to zero. The p-value is

0.000, so there is little likelihood that H_0 is actually true. At least one of the independent variables is useful in explaining the variation in prejudice. The next problem is to find which of the set of variables is useful.

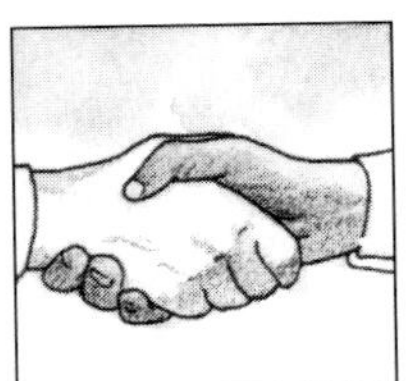

EVALUATING INDIVIDUAL REGRESSION COEFFICIENTS

If there are regression coefficients that could be equal to zero, we want to consider eliminating them from the regression equation. To do so, we will conduct a test of hypothesis for each regression coefficient individually.

For mobility	For age	For status	For political orientation
H_0: $\beta_1 = 0$	H_0: $\beta_2 = 0$	H_0: $\beta_3 = 0$	H_0: $\beta_4 = 0$
H_a: $\beta_1 \neq 0$	H_a: $\beta_2 \neq 0$	H_a: $\beta_3 \neq 0$	H_a: $\beta_4 \neq 0$

The test statistic is the t distribution with $n - (k + 1)$ degrees of freedom in each case. There were 21 people in the sample, so there are $n - (k + 1) = 21 - (4 + 1) = 16$ degrees of freedom in the test. Assuming a 0.05 significance level and using a two-tailed test, we find critical values of –2.120 and 2.120.

Refer again to the computer output on page 548. The column headed "Coef," which refers to the coefficient, reports the regression equation. In the next column, "Stdev," the standard deviation of the regression coefficient is given. How is this value interpreted? We selected a sample of 21 people and obtained the information on the dependent variable and each of the four independent variables for each person. If we were to select another random sample of 21 people and compute the regression equation on that sample, we would probably obtain similar, but slightly different, regression coefficients. If we were to repeat this process many times, we could obtain a sampling distribution for each of the regression coefficients. The value reported in the "Stdev" column reflects the variability in these coefficients.

STAT BYTE

In 1965, the first labels were placed on cigarette packs warning that smoking was harmful to one's health. Since 1969, tobacco companies have not been allowed to advertise on radio or TV. Many independent research projects, as well as the Office of the Surgeon General, have established a link between smoking and cancer. Critics of these reports contend that the research does show a correlation between smoking and cancer, but that the research does not establish a *causal* link. What do you think?

The sampling distribution of Coef/Stdev follows the t distribution with $n - (k + 1)$ degrees of freedom. Hence, by testing whether the corresponding net regression coefficient is significantly different from zero, we are able to determine whether an independent variable should be in the regression equation.

From the output on page 548, we see that the t-values for age and status exceed the critical values, and the t-values for mobility and political orientation do not. Therefore, the variables age and status are included in the regression and those for mobility and political orientation are dropped.

We run the regression computation again *using only the variables that are significant.* Note that 79.9% of the variation in prejudice is explained by the two variables age and status. So a reduction of two independent variables caused a loss of about three percentage points in explained variation—a worthwhile trade because the new model, with only two independent variables, is easier to understand. In addition, in the revised regression model, the two independent variables have significant net regression coefficients. The details of the test are not shown, but note that the *p*-values in the output are both less than 0.05, which indicates that they are significant at the 0.05 level.

```
MTB >   regress 'Prej' on 2 predictors 'Age' 'Status'

The regression equation is
Prej = 37.0 – 0.553 Age + 0.387 Status

Predictor       Coef       Stdev     t-ratio       p
Constant       36.96       10.98        3.37   0.003
Age          –0.5532      0.1690       –3.27   0.004
Status       0.38746     0.06805        5.69   0.000

s = 10.13      R – sq = 79.9%      R – sq(adj) = 77.7%

Analysis of Variance

SOURCE        DF        SS        MS        F        p
Regression     2    7363.5    3681.8    35.88    0.000
Error         18    1847.1     102.6
Total         20    9210.7
```

Caution should be exercised here. It is usually best to delete only one independent variable at a time. Why? If two independent variables are correlated, removing one will often make the other significant. In this case, both variables were deleted because their *p*-values were quite large (0.743 for mobility and 0.265 for political orientation), indicating that there was little likelihood these two variables would be significant under most conditions.

Self-Review 15-3

Refer to the Skaff Appliance example (Self-Review 15–2). The following multiple regression output was obtained. The regression equation is:

Sales = – 3.41 + 0.502 Popul + 0.663 unemp% + 0.0244 Advexp

Predictor	Coef	Stdev	t-ratio	p
Constant	–3.406	1.411	–2.41	0.023
Popul	0.50196	0.04253	11.80	0.000
unemp%	0.6633	0.1982	3.35	0.002
Advexp	0.02442	0.01287	1.90	0.069

s = 1.229 R – sq = 87.7% R – sq (adj) = 86.3%

Analysis of Variance

SOURCE	DF	SS	MS	F	p
Regression	3	279.921	93.307	61.82	0.000
Error	26	39.245	1.509		
Total	29	319.166			

a. What percent of the variation in sales is explained by the three independent variables?

b. In a region with a population of 25,000 and 5.0% unemployment, what sales would you estimate for a store with advertising expense of $30,000? (*Hint:* Remember that some of the data are in thousands.)

c. Conduct the global test of hypothesis. Can we conclude that any of the regression coefficients are not equal to zero? Use the 0.05 significance level.

d. Conduct a test of hypothesis on each of the regression coefficients. Which can we conclude are not equal to zero? Use the 0.05 significance level.

Exercises

7. Refer to the following computer output:

Predictor	Coef	Stdev
Constant	20.00	10.00
X_1	–1.00	0.25
X_2	12.00	8.00
X_3	–15.00	5.00

Analysis of Variance

SOURCE	DF	SS	MS	F
Regression	3	7500.0	?	?
Error	18	?	?	
Total	21	10000.0		

a. Complete the ANOVA table.
b. Conduct a global test of hypothesis. Can you conclude that any of the regression coefficients are not equal to zero? Use the 0.05 significance level.
c. Conduct a test of hypothesis on each of the regression coefficients to determine which are not equal to zero. Use the 0.05 significance level.

8. Refer to the following computer output:
a. Complete the ANOVA table.
b. Conduct a global test of hypothesis. Can you conclude that any of the regression coefficients are not equal to zero? Use the 0.05 significance level.
c. Conduct a test of hypothesis on each of the regression coefficients to determine which are not equal to zero. Use the 0.05 significance level.

Predictor	Coef	Stdev	t-ratio
Constant	−150.00	90.00	?
X_1	2000.00	500.00	?
X_2	−25.00	30.00	?
X_3	5.00	5.00	?
X_4	−300.00	100.00	?
X_5	0.60	0.15	?

Analysis of Variance

SOURCE	DF	SS	MS	F
Regression	5	1500.0	?	?
Error	15	?	?	
Total	20	2000.0		

SUMMARY

In Chapters 13 and 14, the discussion was limited to the relationship between a dependent variable and one independent variable. The techniques used in those two chapters have been extended to include the relationship between a dependent variable and two or more independent variables.

The multiple regression equation for two independent variables has the form

$$Y' = a + b_1 X_1 + b_2 X_2$$

where a is the point of intercept with the Y-axis, and b_1 and b_2 are the regression coefficients associated with X_1 and X_2. This equation can be expanded to include any number of independent variables. For four independent variables, it would be $Y' = a + b_1X_1 + b_2X_2 + b_3X_3 + b_4X_4$.

If the multiple regression equation fits the data perfectly, there is no error in predicting Y, the dependent variable. This rarely happens in actual research. The multiple standard error of estimate is available to measure the error in a prediction.

The strength of the relationship between the dependent and independent variables is measured by the coefficient of multiple correlation, R. It ranges from 0 to 1.00, with 0 indicating no relationship and 1.00 perfect relationship. The coefficient of multiple determination R^2 is a measure of the proportion of the variation in the dependent variable Y, which is explained by the multiple regression equation.

Five assumptions should be met in order for these techniques to be employed. (1) There should be a linear relationship between the dependent and independent variables. (2) The dependent variable must be continuous and at least of interval scale. (3) The variation around the regression line must be the same for all values of X. (4) Successive observations should not be correlated. (5) The set of independent variables should not be correlated. Of course, most research projects cannot fully satisfy all of these assumptions. However, despite such limitations, the use of these techniques is encouraged—but caution should be used in drawing conclusions about the results of regression and correlation analysis.

Two hypothesis-testing procedures were described. The first one tests (in a global fashion) whether any of the set of net regression coefficients do not equal zero. The second procedure tests the net regression coefficients individually, enabling the researcher to delete the nonsignificant independent variables.

Because of the many calculations involved in multiple regression, the use of a computer is virtually essential.

Exercises

9. A real estate agent is studying the selling prices of homes in a certain district of the city. The following equation has been developed:

$$Y' = 24.2 + 9.80X_1 + 3.50X_2$$

where

Y' is the selling price in thousands of dollars.
X_1 is the number of bedrooms.
X_2 is the number of bathrooms.

a. What do you estimate the selling price to be for a three-bedroom, two-bath home in the area?
b. Does the addition of a bedroom add more to the selling price than the addition of a bathroom?

10. From a study of the number of points scored by teams in the National Football League, the following multiple regression equation was developed:

$$Y' = 6.0 + 0.06X_1 + 0.05X_2 - 2.75X_3$$

where

Y' is the number of points scored.
X_1 is the number of yards gained running the ball.
X_2 is the number of yards gained passing the ball.
X_3 is the number of turnovers (the number of fumbles plus the number of passes intercepted).

a. If the Cleveland Browns gained 150 yards running the ball, gained 310 yards passing, and had three turnovers, how many points would you expect them to score?
b. How many points are lost for each turnover?

11. In the course of a study of the yearly amount spent on food, a sociologist found that annual income and number of persons in the family explained 86.3% of the variation in the yearly amount spent. The regression equation was computed to be

$$Y' = 500 + 0.023X_1 + 252X_2$$

where

X_1 is the annual family income.
X_2 is the number of family members.

a. How much does the food expenditure increase with each additional family member?
b. If a family's income increases by $1,000, how much would you expect the expenditure on food to increase?
c. Predict the amount to be spent on food by a family of five with an annual income of $20,000.

12. An insurance company is analyzing the profiles of its individual policyholders as they relate to the amount of life insurance coverage carried. Three independent variables are being considered: the policyholder's annual income (X_1), number of children under 21 (X_2), and age (X_3).

$$Y' = 2.43 + 0.32X_1 + 8.89X_2 + 0.14X_3$$

where Y' and X_1 are measured in thousands of dollars.
a. How much life insurance would a 40-year-old woman with an annual income of $50,000 and three children be expected to carry?
b. How much does the amount of coverage increase with each child?
c. If two policyholders were both 40 years old and each had five children, but one made $40,000 per year and the other $50,000, how much difference would you expect there to be in their respective insurance coverages?

13. A market research firm is studying the relationship between monthly income for a household (Y) and three factors: the mortgage payment for the household, the value of the family car, and the age of the husband. A random sample of 19 observations revealed

Monthly Income	Mortgage Payment	Car Value	Husband's Age
$3,552	$656	$ 7,800	36
2,520	568	5,100	37
3,384	416	10,500	30
4,020	504	9,500	34
3,168	736	6,260	39
2,676	584	4,380	34
1,956	432	3,760	31
3,684	544	7,350	41
3,540	784	6,580	38
4,152	896	7,900	37
3,816	504	9,450	40
4,020	608	12,600	35
3,920	648	10,630	33

Monthly Income	Mortgage Payment	Car Value	Husband's Age
2,544	568	5,340	32
2,736	416	4,690	36
2,556	536	1,200	39
3,120	440	9,600	42
3,720	736	8,800	36
3,192	608	11,400	33

a. The following correlation matrix was developed. What is the correlation between age and car? Does the fact that the correlation is negative surprise you? Explain.

	Income	Mortgage	Car
Mortgage	0.461		
Car	0.746	0.098	
Age	0.198	0.167	−0.119

b. Which of the set of independent variables has the strongest correlation with the dependent variable?

c. The following output was obtained. Compute the coefficient of determination and the multiple standard error of estimate.

The regression equation is
Income = −424 + 1.69 Mortgage + 0.157 Car + 42.5 Age

Predictor	Coef	Stdev	t-ratio	p
Constant	−424.3	928.8	−0.46	0.654
Mortgage	1.6863	0.6227	2.71	0.016
Car	0.15711	0.02741	5.73	0.000
Age	42.51	24.32	1.75	0.101

Analysis of Variance

SOURCE	DF	SS	MS	F	p
Regression	3	5430941	1810314	15.66	0.000
Error	15	1733741	115583		
Total	18	7164682			

d. Referring to the computer output in part c, conduct a global test of hypothesis to determine if any of the net regression coefficients are not equal to zero. Use the 0.05 significance level.

e. Referring to the computer output in part c, conduct an individual test of hypothesis on each independent variable to determine which of the net regression coefficients are not equal to zero. Use the 0.05 significance level.

14. The planning director for Dhont Automatic Door Company has been asked by the president of the company to prepare an analysis of the product sales as it relates to advertising expense and bonuses paid to the sales force. The planning director selected a sample of 15 sales territories and determined, for each, the sales in that region, the advertising expense, the total amount of sales, and bonuses paid to the sales force. The results were as follows:

Sales (in thousands of dollars)	Advertising Expense (in thousands of dollars)	Bonuses Paid (in thousands of dollars)
$158.75	$16.2	$10
186.25	22.5	11
132.50	17.1	7
203.75	21.6	14
127.50	13.5	7
225.00	23.4	17
201.25	22.5	14
160.00	14.4	12
173.75	15.3	12
180.00	20.7	12
198.75	19.8	14
172.50	13.5	15
155.00	12.6	8
130.30	13.5	9
160.80	11.6	14

a. The following correlation matrix was developed. What is the correlation between bonus and advertising expense? Would you expect this association to be positive or negative?

	Sales	Advexp
Advexp	0.780	
Bonus	0.849	0.447

b. Refer to the computer output in part a. Which of the set of independent variables has the strongest correlation with the dependent variable sales?

c. The following output was obtained. Compute the coefficient of determination and the standard error of estimate.

The regression equation is
Sales = 41.2 + 3.49 Advexp + 5.95 Bonus

Predictor	Coef	Stdev	t-ratio	p
Constant	41.17	11.29	3.65	0.003
Advexp	3.4925	0.6327	5.52	0.000
Bonus	5.9480	0.8614	6.90	0.000

Analysis of Variance

SOURCE	DF	SS	MS	F	p
Regression	2	10740.8	5370.4	70.18	0.000
Error	12	918.3	76.5		
Total	14	11659.1			

d. Referring to the computer output in part c, conduct a global test of hypothesis to determine if any of the net regression coefficients are different from zero. Use the 0.05 significance level.

e. Referring to the computer output in part c, conduct an individual test of hypothesis on each independent variable to determine which of the net regression coefficients are not equal to zero. Use the 0.05 significance level.

15. The production supervisor in a factory that assembles electric toasters wants to develop a multiple regression equation to predict the amount of the bonus employees can be expected to earn each week based on their productivity. He decides to use their score on a manual dexterity test and their years of experience with the company as independent variables. The following sample information is obtained:

Weekly Bonus Earned	Years of Experience	Manual Dexterity Score
$70	7	21
20	14	9
40	10	16
70	8	18
30	12	9
50	9	19
60	6	17
10	12	7
20	11	12
65	15	10
40	10	5
60	4	20
50	8	8
50	6	12

a. Develop a correlation matrix. Are the independent variables correlated? Does this appear to be a problem?

b. Conduct a global test to determine if at least one of the regression coefficients is not zero. Use the 0.05 significance level.

c. Conduct a test on the net regression coefficients individually. Use the 0.05 significance level. Are both variables significant?

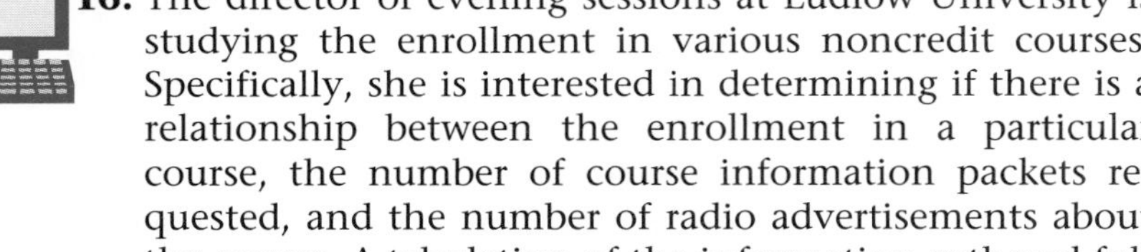

16. The director of evening sessions at Ludlow University is studying the enrollment in various noncredit courses. Specifically, she is interested in determining if there is a relationship between the enrollment in a particular course, the number of course information packets requested, and the number of radio advertisements about the course. A tabulation of the information gathered follows:

Enrollment	Information Packets	Radio Advertisements
53	85	3
44	80	2
30	60	1
63	150	4
60	84	2
57	108	2
60	210	3

Enrollment	Information Packets	Radio Advertisements
65	200	4
54	100	4
30	40	1
35	60	2
50	90	3
50	90	2
42	85	2
36	80	1

a. Develop a correlation matrix. Are the independent variables correlated? Will this cause a problem?
b. Conduct a global test of hypothesis to determine whether at least one regression coefficient is not zero.
c. Conduct a test of hypothesis on the net regression coefficients individually. Use the 0.05 significance level. Are both independent variables significant?

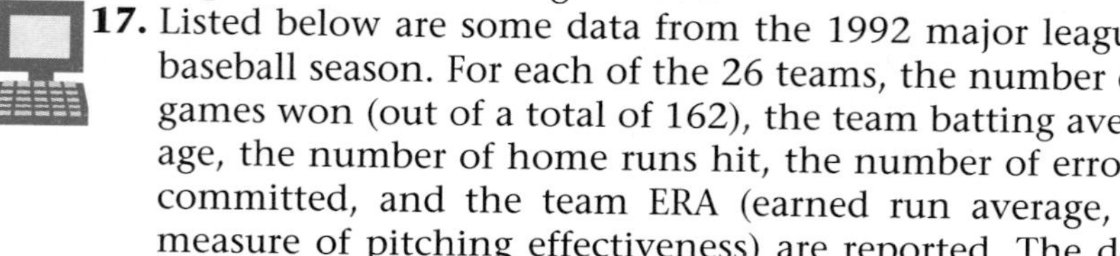

17. Listed below are some data from the 1992 major league baseball season. For each of the 26 teams, the number of games won (out of a total of 162), the team batting average, the number of home runs hit, the number of errors committed, and the team ERA (earned run average, a measure of pitching effectiveness) are reported. The dependent variable is the number of games won.

Team	Wins	Average	Home Runs	Errors	ERA
Toronto	96	263	163	93	3.91
Milwaukee	92	268	82	89	3.43
Baltimore	89	259	148	93	3.79
Cleveland	76	266	127	141	4.11
New York Yankees	76	261	163	114	4.21
Detroit	75	256	182	116	4.60
Boston	73	246	84	139	3.58
Oakland	96	258	142	125	3.73
Minnesota	90	277	104	95	3.70
Chicago White Sox	86	261	110	129	3.82
Texas	77	250	159	154	4.09
California	72	243	88	134	3.84
Kansas City	72	256	75	122	3.81
Seattle	64	263	149	112	4.55
Pittsburgh	96	255	106	101	3.35

Team	Wins	Average	Home Runs	Errors	ERA
Montreal	87	252	102	124	3.25
St. Louis	83	262	94	94	3.38
Chicago Cubs	78	254	104	114	3.39
New York Mets	72	235	93	116	3.66
Philadelphia	70	253	118	131	4.11
Atlanta	98	254	138	109	3.14
Cincinnati	90	260	99	96	3.46
San Diego	82	255	135	115	3.56
Houston	81	246	96	114	3.72
San Francisco	72	244	105	113	3.61
Los Angeles	63	248	72	174	3.41

a. Using a statistical software package, develop a correlation matrix showing the relationship between the dependent variable and each of the independent variables. Also show the correlations between all of the independent variables. Which independent variable has the strongest correlation with the dependent variable (number of games won)? Are there any potential problems with the correlations among the independent variables? If so, between which variables?

b. Determine the regression equation and write it out. Does it surprise you that the coefficients of the variables "error," and "ERA" are negative? Explain.

c. Estimate the number of wins for a team that had a team batting average of 260, hit 150 home runs, committed 100 errors, and had a team ERA of 3.50.

d. Find the coefficient of determination on your computer output. Comment on its value.

e. Conduct a global test of hypothesis on the set of independent variables. Use the .05 significance level. Is it reasonable to conclude that any of the independent variables have coefficients different from zero?

f. Conduct a test of hypothesis on each regression coefficient. Would you consider deleting any of the regression coefficients? Use the .05 significance level.

18. Select a dependent variable that you think can be predicted by several independent variables. One suggestion might involve predicting sales based on advertising expense, number of sales personnel, and amount spent on research and development. Another might involve the salary of executives based on the number of employees supervised, their age, and their number of years of experience. Question 17 gave information on major league baseball, but football fans might estimate the number of points scored in college or professional games, based on the number of yards gained running the ball, passing the ball, the number of fumbles, and the number of passes intercepted. For basketball, you might try points scored based on percent of field goals made, the number of foul shots attempted, and who was the home team. Car enthusiasts might estimate a car's mileage rating based on the engine size, weight, number of doors, and whether or not it was manufactured in the United States.

a. Collect the data and enter the information into the computer.

b. Analyze the computer printout and write a summary of your findings. Include in your discussion such measures as the coefficient of determination and an interpretation of the multiple regression equation. Also conduct global and individual tests of hypothesis.

Data Exercises

19. Refer to the Real Estate Data, which reports information on homes sold in Northwestern Ohio during 1992. Develop a multiple regression equation using selling price as the dependent variable and the area of the home in square feet, the number of bedrooms, the number of bathrooms, whether or not there is a fireplace, whether or not there is an attached garage, and the distance from the center of the city as independent variables.

a. Write out the regression equation. Interpret the regression coefficients. For example, how much does an extra bedroom add to the value of the home? How much is an attached garage or a fireplace worth to the value of the home? Does it surprise you that the regression coefficient for distance is negative? Why?

b. What is the coefficient of determination? Interpret.

c. Conduct a global test of hypothesis to determine if any of the set of independent variables have regression coefficients different from zero. Use $\alpha = .05$.

d. Conduct a test of hypothesis for each independent variable to determine which of the regression coefficients are different from zero. Would you consider deleting any of the independent variables? If so, which variables would you delete? Use $\alpha = .05$.

e. Rerun the regression equation deleting the variables from part d. Comment on the change in the regression equation. Was there much change in the coefficient of determination?

20. Refer to the Salary Data for middle managers in Sarasota, Florida. Develop a regression equation with salary as the dependent variable and age, years with the company, sex, and the number of employees supervised as independent variables.

a. Write out the regression equation. Interpret the regression coefficients. For example, how much does salary increase for each employee supervised? What is the difference in the salary for men and women?
b. What is the coefficient of determination? Interpret.
c. Conduct a global test of hypothesis to determine if any of the set of independent variables have regression coefficients different from zero. Use $\alpha = .05$.
d. Conduct a test of hypothesis for each independent variable to determine which of the regression coefficients are different from zero. Would you consider deleting any of the independent variables? If so, which variables would you delete? Use $\alpha = .05$.
e. Rerun the regression equation deleting the variables from part d. Comment on the change in the regression equation. Was there much change in the coefficient of determination?

HOW MUCH INFORMATION IS ENOUGH?

Mr. Ed Apple is a real estate tax appraiser for the city of Milan, Michigan. Ed is the only appraiser employed by the city, so he has a very large workload. In a recent continuing-education course, he was introduced to regression analysis, and he now would like to apply these techniques to automate the process of placing a value on each home in his taxing district. It is very expensive to gather and store information, so he cannot collect large amounts of data on each home. Listed below are data thought to be relevant to predicting market value for a sample of 20 homes sold within the last 30 days.

The first column is the actual market value of each home. The age of each building is expressed in years. Distance is the number of miles from the Milan courthouse, which is located in the center of the city. The number of square feet of living area and the number of bedrooms are shown in the last two columns, respectively.

Market Value (in thousands)	Age (in years)	Distance (in miles)	Size (in square feet)	Number of Bedrooms
$135.0	10	11.0	1,805	4
92.7	10	9.4	1,571	3
115.4	9	13.4	1,697	4
138.0	8	15.0	1,776	4
101.3	11	8.6	1,639	7
117.3	9	13.5	1,605	2

(continued)

Market Value (in thousands)	Age (in years)	Distance (in miles)	Size (in square feet)	Number of Bedrooms
119.0	12	6.9	1,686	4
149.9	4	25.1	1,779	3
116.6	10	10.0	1,729	3
107.5	7	14.7	1,645	3
126.9	8	12.4	1,719	4
137.5	5	19.6	1,732	4
119.6	9	11.4	1,629	4
134.2	9	12.4	1,719	7
158.5	9	15.6	1,904	7
127.3	12	8.7	1,920	4
150.8	4	22.6	1,866	6
105.8	10	9.6	1,561	5
135.4	4	22.2	1,578	5
112.0	7	14.5	1,650	3

What would you do if you were the real estate tax appraiser?

CHAPTER ACHIEVEMENT TEST

Answer all the questions. The answers are at the back of the book.

MULTIPLE-CHOICE QUESTIONS. Select the response that best answers each of the questions.

1. Which of the following are both measures of association?
   a. The regression equation and the standard error of estimate
   b. The coefficient of determination and the standard error of estimate
   c. The regression equation and the coefficient of determination
   d. The regression coefficient and the standard error of estimate

2. The variation around the regression line must be the same for all values of X. This requirement is called
   a. stepwise.
   b. autocorrelation.
   c. homoscedasticity.
   d. correlation.

3. If successive observations of the dependent variable are themselves correlated, then this is called
   a. autocorrelation.
   b. homoscedasticity.

c. regression.
d. stepwise.

4. Multiple regression analysis assumes that there is
a. a linear relationship between the variables.
b. no correlation in the population.
c. never any negative signs in the regression coefficients.
d. at least the nominal scale of measurement.

5. For the multiple regression technique to be employed, the dependent variable must be at least of which scale of measurement?
a. Interval
b. Ratio
c. Ordinal
d. Nominal

Questions 6–11 are based on the following information. A study has been undertaken to predict annual income based on number of years on the job, age, and years of education beyond eighth grade. This equation has been developed: $Y' = 10 + 0.2X_1 + 0.1X_2 + 2.5X_3$ (in thousands of dollars). Variable 1 is number of years on the job, variable 2 is age, and variable 3 is number of years of education beyond eighth grade.

6. The equation is called a
a. coefficient of regression equation.
b. simple regression equation.
c. multiple regression equation.
d. dependent variable equation.

7. How many dependent variables are there?
a. 0
b. 1
c. 2
d. 3

8. How many independent variables are there?
a. 0
b. 1
c. 2
d. 3
e. None of the above

9. What are the numbers 0.2, 0.1, and 2.5 called?
a. Partial correlation coefficients
b. Net regression coefficients
c. Coefficients of determination
d. Coefficients of standard errors
e. None of the above

10. An employee has been on the job 20 years, is 50 years old, and has eight years of education beyond eighth grade. What is the employee's predicted annual salary?
a. \$12,800

b. \$39,000
c. \$88,000
d. \$16,000
e. None of the above

11. For each additional year of education beyond eighth grade, annual income increases
a. \$2,500.
b. \$1,000.
c. \$1,280.
d. \$39,000.
e. None of the above

COMPUTATION PROBLEMS Refer to the following output:

Predictor	Coef	Stdev	t-ratio
Constant	5.00	2.00	2.50
X_1	2.00	1.00	2.00
X_2	0.50	0.15	3.33
X_3	0.30	0.10	3.00
X_4	1.00	0.25	4.00

s = 2.00 R – sq = 66.7%

Analysis of Variance

SOURCE	DF	SS	MS	F
Regression	4	200.0	50.0	12.50
Error	25	100.0	4.0	
Total	29	300.0		

12. Write the regression multiple equation.

13. Conduct a global test of hypothesis. Can you conclude that any of the net regression coefficients are not zero?

14. Test the net regression coefficients individually. Are there any that are not equal to zero? Should any of the regression equations be dropped?

ANSWERS TO SELF-REVIEW PROBLEMS

15-1 a. \$33.50 per week

$Y' = 20.0 + 0.50X_1 - 1.20X_2 - 1.05X_3$
$= 20.0 + 0.50(300) - 1.20(70)$
$- 1.05(50)$
$= 33.5$ (in dollars)

b. Households with a weekly income of \$300 that spend \$70 on food and \$50 on entertainment are expected to save \$33.50 per week. Note that as the amount earned increases, so does the amount to be saved. Similarly, as the amount spent on food and entertainment increases, the amount saved decreases.

15-2 a. 0.075

b. Population

c. No. The largest correlation is 0.260, between advertising expense and population. This is less than the rule-of-thumb limit of 0.70. The condition is called multicollinearity.

15-3 a. $R^2 = \dfrac{279.921}{319.166} = 0.877$

b. 13.187, found by:
$Y' = -3.41 + 0.502(25) + 0.663(5.0) + 0.0244(30)$

c. H_0: $\beta_1 = \beta_2 = \beta_3 = 0$; H_a: At least one coefficient is not zero.

H_0 is rejected if $F > 2.95$ (approximately).

$F = 93.307/1.509 = 61.82$

H_0 is rejected. At least one of the regression coefficients does not equal zero.

d. H_0: $\beta_1 = 0$; H_a: $\beta_1 \neq 0$
H_0: $\beta_2 = 0$; H_a: $\beta_2 \neq 0$
H_0: $\beta_3 = 0$; H_a: $\beta_3 \neq 0$
In each case, H_0 is rejected if $t < -2.056$ or $t > 2.056$.

H_0 is rejected for population and unemployment. Advertising expense is dropped from the analysis.

UNIT REVIEW

In the last three chapters, we introduced the fundamental concepts of correlation and regression analysis. We examined various measures used to describe the degree of relationship between a dependent variable and one or more independent variables. We also developed a mathematical equation that allows us to predict a dependent variable based on one or more independent variables.

Key Concepts

1. **Correlation analysis** allows us to assess the strength of the relationship between two or more variables and to determine what proportion of the total variation is explained by the independent variable(s).
   a. **Pearson's product-moment coefficient of correlation** (r). This measure reports the strength of the relationship between dependent and independent variables. Its use assumes that the data are of an interval scale. For two variables, r can have any value from –1.00 to +1.00, inclusive. The *strength* of the relationship is not dependent on the *direction* of the relationship. For example, correlation coefficients of –0.09 and +0.09 are equal in strength—both are very weak. For more than two variables, the coefficient of multiple correlation may assume a value from 0 to +1.00, inclusive.
   b. **Coefficient of determination.** This measure reports the proportion of the total variation that is explained by the independent variable (or variables). For two variables, it is the proportion of the total variation in one variable explained by the other variable. Likewise, for more than two independent variables, it is the proportion of the total variation in the dependent variable explained by the independent variables. It varies from 0 to +1.00. A coefficient of 0.80, for example, indicates that 80% of the total variation in the dependent variable is explained by the independent variable or variables.
   c. **Coefficient of nondetermination.** This measure reports the proportion of the total variation in the dependent variable that is *not* explained by the independent variable (or variables). We find it by subtracting the coefficient of determination from 1.00. Thus, it can assume any value from 0 to +1.00, inclusive. A coefficient of 0.20, for example, reveals that 20% of the total variation in the dependent variable is not explained by the independent variable or variables.

   d. **Spearman's rank-order correlation coefficient.** This statistic reports the relationship between two ordinal-level variables—that is, data that we can rank from high to low or vice versa. Spearman's coefficient can assume any value between –1.00 and +1.00.
2. **Regression analysis** is important because, by using a mathematical equation, we can estimate the value of one variable based on another variable.
   a. **Regression equation.** For one dependent and one independent variable, the equation has the form $Y' = a + bX$. For one dependent and three independent variables, it is $Y' = a + b_1X_1 + b_2X_2 + b_3X_3$.
   b. **Standard error of estimate** is a measure that allows us to assess the accuracy of regression.

Key Terms

Correlation analysis
Scatter diagram
Dependent variable
Independent variable
Pearson's product-moment correlation coefficient
Coefficient of determination
Coefficient of nondetermination
Spearman's rank-order correlation coefficient
Regression equation
Slope
Y-intercept
Least-squares principle
Standard error of estimate
Confidence interval
Prediction interval
Multiple regression and correlation analysis
Homoscedasticity
Regression coefficient
Autocorrelation
Multicollinearity
Multiple coefficient of correlation
Coefficient of multiple determination
Multiple standard error of estimate
Correlation matrix
Multicollinearity
Global test

Key Symbols

r	The coefficient of correlation for two variables.
r^2	The coefficient of determination.
R^2	The coefficient of multiple determination.
Y'	The dependent variable of a regression equation.
X	One of the independent variables.
$s_{Y \cdot X}$	The standard error of estimate.
r_s	Spearman's rank-order correlation coefficient.
a	The *Y*-intercept. The value of Y when $X = 0$.
b	The slope of the line. The variable b is the amount Y changes for a unit change in X.

Problems

1. Researchers at the Mississippi General Hospital Administrative Services Department are interested in determining the relationship between staff members' years of service and the number of sick leaves granted to employees in a year. A sample of the personnel files reveals the following data:

Number of Years of Service X	Number of Sick Leaves Y
11	16
8	12
20	27
2	3
4	7
5	4
14	22

   a. Draw a scatter diagram.
   b. Calculate Pearson's coefficient of correlation, the coefficient of determination, and the coefficient of nondetermination.
   c. Determine the line of regression.
   d. Plot the line of regression on the scatter diagram.
   e. Estimate the number of sick leaves per year for a person who has 12 years of service.
   f. Interpret your findings.

2. The relationship between the annual birth rate and the suicide rate for United Nations countries is to be investigated. A sample of 10 countries yielded the following figures:

Country	Birth Rate (per 1,000 population)	Suicide Rate (per 1,000 population)
Italy	12.5	5.8
Poland	19.0	12.1
United States	15.3	12.7
Australia	15.7	11.1
Finland	13.5	25.1
Germany	13.9	30.5
Mexico	35.3	2.1
Spain	17.2	4.1
Czechoslovakia	18.4	21.9
Singapore	17.0	11.3

a. Draw a scatter diagram using suicide rate as the dependent variable and birth rate as the independent variable.
b. Compute the regression equation.
c. For a country with a birth rate of 15.0 per 1,000 population, what is the predicted suicide rate?
d. Determine the coefficients of correlation, determination, and nondetermination.
e. Interpret your findings.

3. An educator is examining the relationship between number of hours of classroom study per day and students' ability to speak a foreign language. Each member of a large group of students is randomly assigned to one of five classrooms. The number of hours of study varies in each. After 10 weeks, a proficiency examination is given and the score determined:

Hours of Study	Score
1	55
1	51
1	43
2	67
2	59
2	63
3	85
3	80
3	72
4	89
4	80
4	84
4	74

a. Compute the regression equation.
b. What score would you estimate for a student studying four hours per week?
c. Compute the coefficient of correlation.
d. Test to determine if there is correlation in the population.

4. Eight physicians scored an experimental drug on a scale of 1 to 20 with respect to its effectiveness and its aftereffect on the patient.

Physician	Effectiveness Score	Aftereffect Score
Otto	8	2
Bono	16	17

Physician	Effectiveness Score	Aftereffect Score
Smythe	3	9
Apple	20	10
Butz	3	2
Archer	9	16
Marchand	6	4
Sabino	13	7

Calculate Spearman's rank-order correlation coefficient to measure the strength of the relationship between the effectiveness of the experimental drug and its aftereffect.

5. Ten professional rodeo contestants have competed in last season's California Roundup and Wyoming Roundup. Their scores are:

Contestant	California Roundup Scores	Wyoming Roundup Scores
Best	69	70
Asner	86	79
Jones	81	82
Belk	96	91
Sampson	72	67
Sawicki	69	62
Aey	70	70
Damon	92	98
Bardi	57	70
Aslo	84	80

A trainer is interested in evaluating the ranking of the contestants at the two roundups.

a. Rank the points scored in each rodeo (watch ties).

b. Calculate Spearman's rank- order correlation coefficient to evaluate the consistency of each contestant, conduct the appropriate test of hypothesis, and interpret your findings.

6. The salary structure of the Evergreen School System is being analyzed. Two variables are thought to relate to a teacher's salary—number of years with the system and whether or not the individual has obtained a master's degree. A sample of 12 teachers revealed the following:

Salary (in thousands of dollars)	Experience (in years)	Master's (yes = 1, no = 0)
$22.5	4	1
23.6	8	1
24.1	10	1
25.3	16	1
28.0	22	1
27.0	18	1
21.2	3	0
22.2	8	0
23.7	11	0
24.4	13	0
24.8	19	0
25.0	20	0

a. Compute the regression equation.
b. What percent of the variation in salary is explained by the two independent variables?
c. Estimate the salary for a teacher with a master's degree and 10 years of experience.
d. If we hold experience constant, how much does a master's degree add to salary?
e. Conduct a global test of hypothesis to determine if there is a significant correlation in the population between the dependent variable (salary) and the two independent variables (experience and possessing a master's degree).
f. Determine if both experience and a master's degree are significant independent variables.

7. An agricultural economist is studying the relationship between the per capita income in a county, the percent of the population that is employed in agriculture, and the mean number of years of education for people over 25 years of age. Twenty nonurban Midwestern counties are selected. From the data given in the following table, answer these questions:
a. Determine the correlation matrix and the regression equation.
b. What percent of the variation in per capita income is explained by education and percent employed in agriculture?
c. Conduct a global test to determine if the two independent variables are significant predictors.
d. Individually, are both independent variables significant?

Income (in thousands of dollars)	Percent Employed in Agriculture	Education (in years)
$ 9.8	10.2	10.1
9.7	13.4	11.9
8.0	10.2	10.1
9.9	10.8	7.9
10.9	10.3	10.0
9.1	13.3	11.0
9.1	11.3	11.0
10.7	10.3	10.1
7.6	12.7	7.9
12.4	8.5	10.1
10.7	12.6	11.0
9.6	12.8	10.4
13.2	9.7	8.7
12.9	9.5	9.0
10.1	10.5	10.2
7.6	13.0	7.8
9.9	10.6	8.0
12.4	8.4	10.2
13.3	9.6	8.8

Using Statistics

Situation. A pharmaceutical company obtained a random sample of 30 chronic arthritis sufferers. The 30 sufferers were randomly assigned to five different groups. Each of the five groups was then administered doses, ranging from 20 to 180 mg per kilogram of body weight, of a new drug that reduces pain. The time until the patient sensed a change in his or her condition is noted. The sample data follows:

Dosage	Response Time (in minutes)					
20	4.3	4.9	5.0	5.9	6.4	7.2
40	2.5	2.8	3.0	3.1	3.3	3.8
60	1.5	1.8	1.9	2.2	2.3	2.8
120	0.8	1.1	1.3	1.4	1.5	1.7
180	0.8	0.9	1.0	1.1	1.2	1.6

The manufacturer would like to develop an equation that expresses the relationship between dosage and response time. Plot the data and summarize the results.

Discussion. The dependent variable is the response time—the time until relief is obtained. The independent variable is the dosage. The scatter diagram shown below clearly reveals that the relationship between dosage and response time is not linear (straight-line). This violates one of the basic regression assumptions. In addition, the coefficient of determination is 60.1%. This is lower than the company would like.

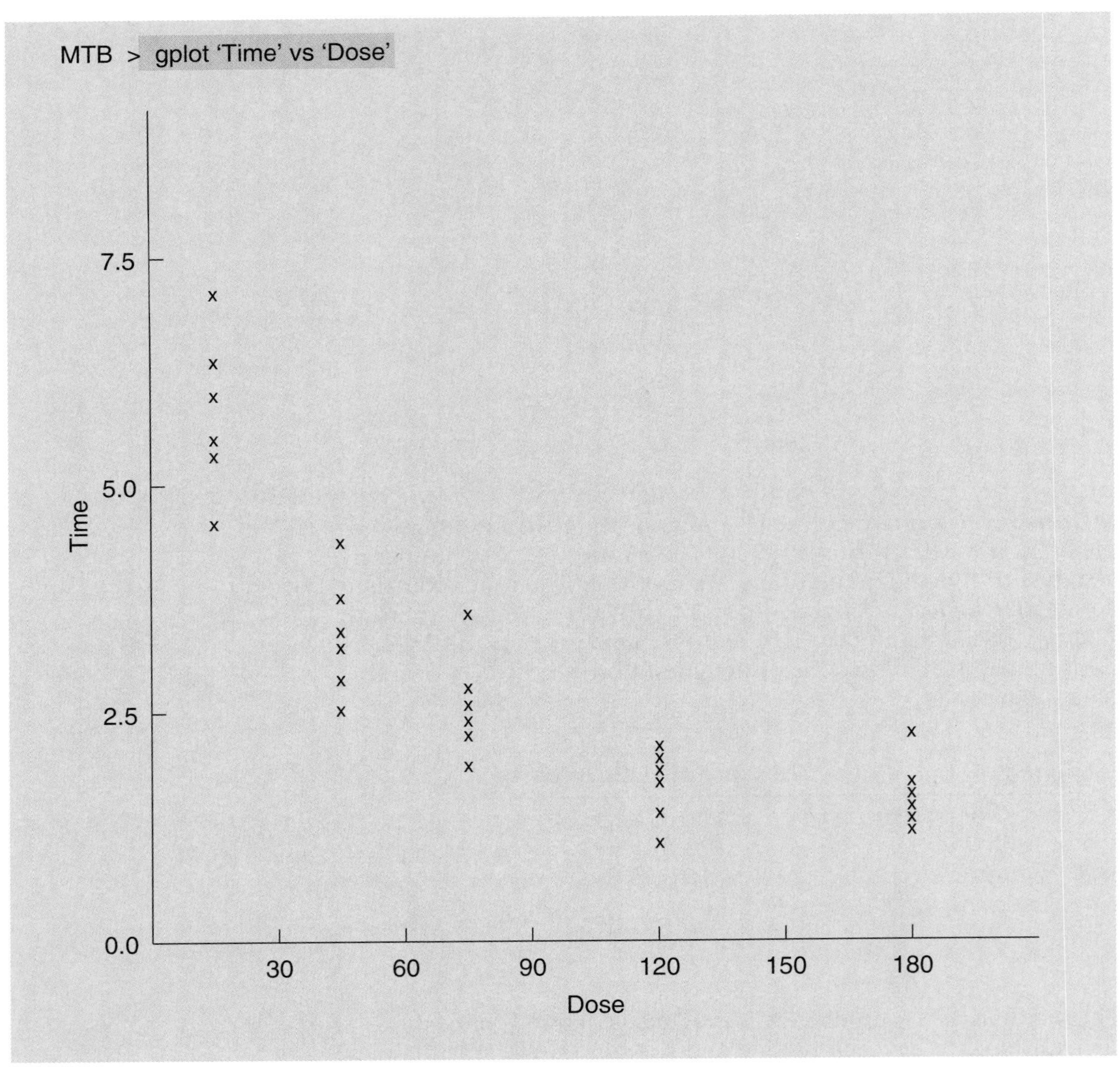

```
MTB >  regr 'Time' 1 'Dose'

The regression equation is
Time = 4.56 – 0.0229 Dose

Predictor        Coef        Stdev     t-ratio       p
Constant       4.5636       0.3616      12.62     0.000
Dose        -0.022940     0.003533      -6.49     0.000

s = 1.132    R – sq = 60.1%        R – sq (adj) = 58.7%

Analysis of Variance

SOURCE      DF       SS      MS       F       p
Regression   1   54.056  54.056   42.17   0.000
Error       28   35.894   1.282
Total       29   89.950
```

How can this be improved? Note from the scatter diagram that the response time is very short for large dosages, but quite long for small doses. In fact, it appears that the response time is proportional to the rate at which the drug works, rather than to the dose itself. We compute the rate by taking the reciprocal of the dosage. That is, the rate at which a 20-mg dose works is 1/20 = 0.05 minutes. The second scatter diagram, on page 578, shows the association between the rate at which the drug works and the response time.

Note from this second scatter diagram that the relationship now appears to be linear. In addition, in the output on page 579, the value of R-sq is increased to 90.2%. The least squares equation (rounded to two places) is $Y' = 0.46 + 102.98X$. The intercept value of 0.46 implies a minimum response time of 0.46 minutes, regardless of the dose. We determine the additional time until the patient senses relief by dividing the amount of the dose into 102.98. So, a small dose, such as 20 mg, will take 102.98/20 = 5.149 minutes to work after the initial response. A larger dose, such as 180 mg, will take less time to work after the initial response (102.98/180 = 0.572 minutes).

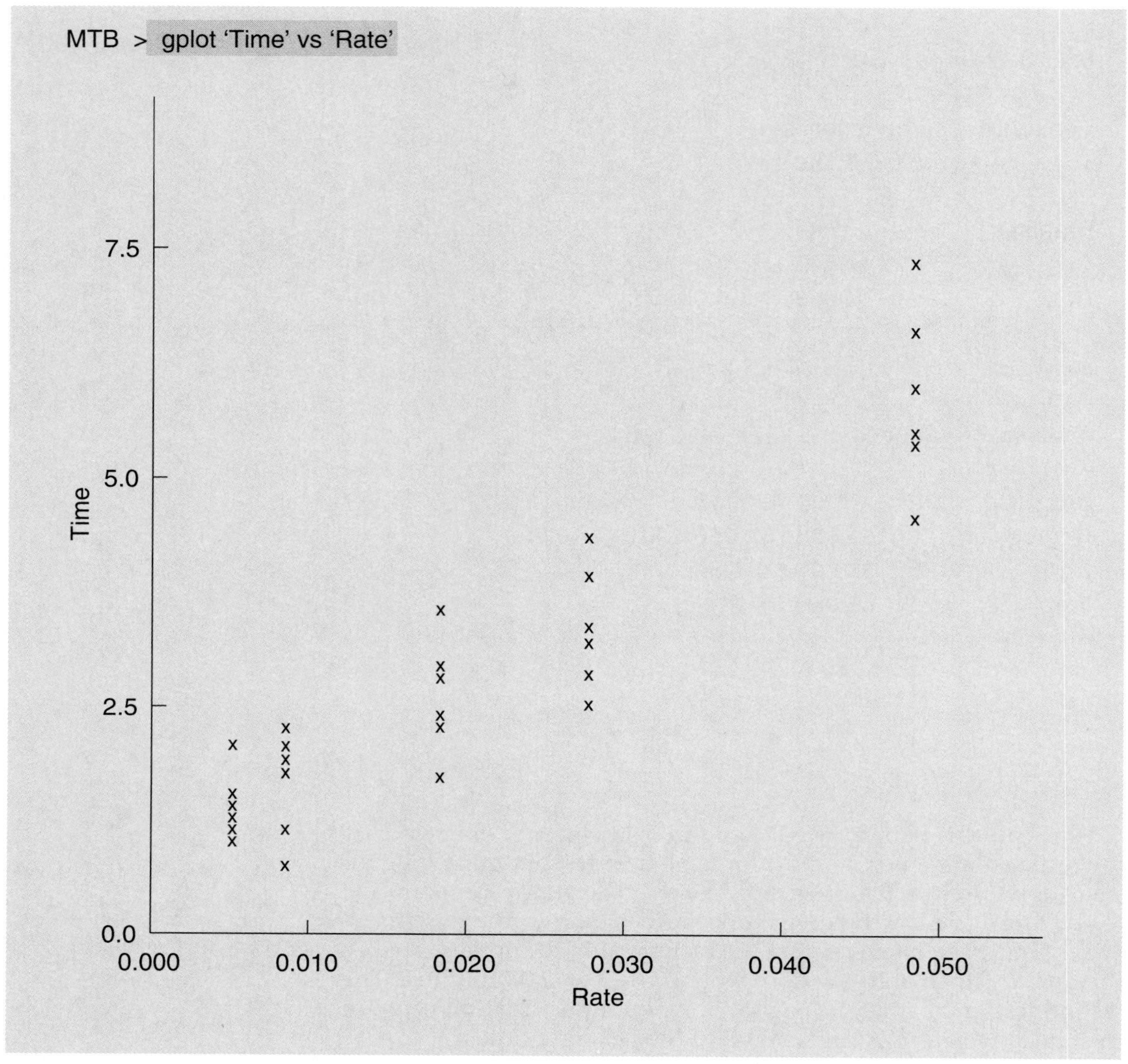
MTB > gplot 'Time' vs 'Rate'
Time
7.5
5.0
2.5
0.0
0.000
0.010
0.020
0.030
0.040
0.050
Rate

```
MTB >  regr 'Time' 1 'Rate'

The regression equation is
Time = 0.463 + 103 Rate

Predictor        Coef       Stdev    t-ratio       p
Constant       0.4626      0.1701       2.72   0.011
Rate          102.981       6.425      16.03   0.000

s = 0.5619       R - sq = 90.2%       R - sq (adj) = 89.8%

Analysis of Variance

SOURCE      DF        SS        MS        F       p
Regression   1    81.109    81.109   256.88   0.000
Error       28     8.841     0.316
Total       29    89.950
```

In summary we began the analysis assuming that response time was a linear function of the dose level. However, examination of the scatter diagram and further analysis lead us to conclude that response time is more accurately described as a linear function of the reciprocal of the dose level. There may be other times in statistics when transforming or redefining a variable will be very helpful.

CHAPTER 16

Analysis of Nominal-Level Data: The Chi-Square Distribution

OBJECTIVES

When you have completed this chapter, you will be able to

- explain the characteristics of the chi-square distribution;
- test a hypothesis regarding the difference between an observed set of frequencies and a corresponding expected set of frequencies;
- test whether two criteria of classification are related.

CHAPTER PROBLEM What Melts in Your Mouth, Not in Your Hands?

HAVE YOU EVER WONDERED ABOUT THE DISTRIBUTION OF COLORS IN A BAG OF M&M'S CHOCOLATE PEANUT CANDIES? AFTER CHECKING WITH THE MANUFACTURER, THE *LANSING STATE JOURNAL* REPORTED THAT 30% OF THE CANDIES ARE BROWN, 20% GREEN, 20% RED, 20% YELLOW, AND 10% ORANGE. TO CHECK THIS REPORT, YOU PURCHASE A ONE-POUND BAG OF M&M'S CHOCOLATE PEANUT CANDIES AT THE K-MART STORE. THERE ARE A TOTAL OF 188 CANDIES IN THE BAG—67 BROWN, 24 GREEN, 51 RED, 22 YELLOW, AND 24 ORANGE. DOES YOUR BAG OF CANDIES AGREE WITH THE NEWSPAPER'S FINDINGS?

INTRODUCTION

The hypothesis tests we examined in Chapters 10–12 dealt with the interval level of measurement for problems in which the population was assumed to be normal. This chapter begins our study of statistical tests for *nominal* and *ordinal* scales of measurement or for cases in which no assumptions can be made about the shape of the population. Recall from Chapter 1 that the nominal level of data is the most "primitive," or the "lowest," type of measurement. Nominal-level information, such as male or female, can be classified only into categories. The ordinal level of measurement assumes that one category is ranked higher than the next one. To illustrate, each staff therapist in a clinic might be rated as being superior, good, fair, or poor. A rating of "superior" is higher than a "good" rating, a "good" rating is higher than a "fair" rating, and so on.

Tests involving nominal or ordinal levels of measurements are called **nonparametric** or **distribution-free** tests. In this chapter, we will examine two tests that employ the **chi-square distribution** as the test statistic.

THE CHI-SQUARE (χ^2) DISTRIBUTION

The chi-square distribution is appropriate for both nominal-level and ordinal-level data. It is designated χ^2, and because it involves squared observations *it is always positive*. Like the t and F distributions discussed earlier, there are many chi-square distributions, each with a different shape that depends on the number of degrees of freedom. Figure 16–1 shows the shape of various chi-square distributions for selected degrees of freedom (df). Note that as the number of degrees of freedom increases, the distribution approaches a symmetric distribution and the peak moves to the right.

The number of degrees of freedom is determined by the number of categories minus one, that is, $k - 1$, and not by the size of the sample. For example, if a sample of 150 undergraduate students were classified as freshmen, sophomores, juniors, or seniors, there would be $k - 1 = 4 - 1 = 3$ degrees of freedom.

There are three major properties of a chi-square distribution:

1. Chi-square is nonnegative—that is, it is either zero or positive.
2. A chi-square distribution is not symmetrical. Its skewness is positive. However, as the number of degrees of freedom increases, chi-square approaches a symmetric distribution.
3. There is a family of chi-square distributions. There is a particular distribution for each degree of freedom.

THE GOODNESS-OF-FIT TEST: EQUAL EXPECTED FREQUENCIES

We use the chi-square distribution as a test statistic to determine how well a set of observations "fits" a theoretical, or expected, set of

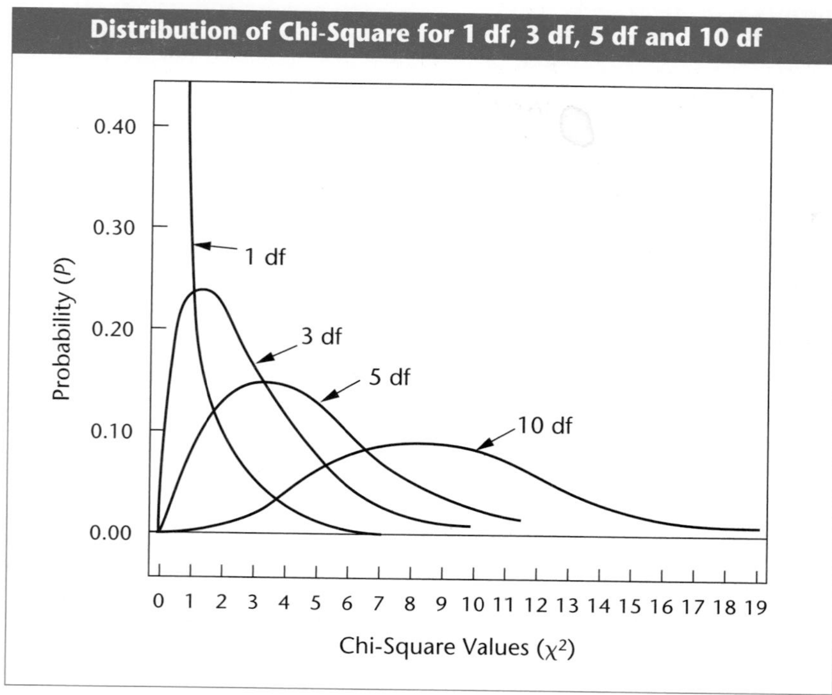

Figure 16-1

observations. To put it another way, the objective is to find out how well an *observed* set of frequencies, abbreviated f_o, fits an *expected* set of frequencies, f_e. This test is called the **goodness-of-fit test.**

The general nature of the goodness-of-fit test can be shown by a specific application. The following Problem illustrates the procedure when the expected frequencies, f_e, are all the same.

Problem A psychologist is interested in determining whether mentally retarded children, given four choices of colors, prefer one over the other three. The researcher conjectures that color preference may have some effect on behavior. Eighty mentally retarded children are given a choice of a brown, orange, yellow, or green T-shirt. Following is a tally of their selections:

Color	Number
Brown	25
Orange	18
Yellow	19
Green	18
Total	80

Do the children have a color preference or not?

STAT BYTE

In 1992, there were 264 major league baseball players who earned more than $1 million, including incentives. There are 650 major league roster positions, so more than 40% of the active players made over a million dollars. Broken down by American and National League:

League	Earnings (in millions of dollars) $1–2	$2–3	$3–4	More Than $4	Total
American	53	54	29	11	147
National	40	49	16	12	117
Total	93	103	45	23	264

Are the salaries of $1 million or more related to the league?

Solution Use the five-step hypothesis-testing procedure.

STEP 1. State the null hypothesis and alternate hypothesis. The null hypothesis, H_0, is that there is no preference among the four colors. The alternate hypothesis, H_a, is that there is a preference among the four colors.

STEP 2. Select a level of significance. The 0.05 level has been chosen. Recall that this is the probability of making a Type I error. It means that the probability is 0.05 of incorrectly rejecting a true null hypothesis.

STEP 3. Choose an appropriate test statistic. Here it is chi-square, defined as

$$\chi^2 = \sum \frac{(f_o - f_e)^2}{f_e} \qquad \textbf{16-1}$$

where

f_o is an observed frequency in a category (color in the problem).
f_e is an expected frequency in a particular category.

STEP 4. Formulate a decision rule. As in other hypothesis-testing situations, we look at the sampling distribution of the test statistic (chi-square, in this case) in order to arrive at a *critical value*. Recall that the critical value is the number that separates the region in which H_0 is not rejected from the region in which it is rejected. The shape of the chi-square distribution depends on the number of degrees of freedom. In a goodness-of-fit test, there are $k - 1$ degrees of freedom, where k stands for the number of categories. There are four categories (colors); therefore, $k - 1 = 4 - 1 = 3$ degrees of freedom. The critical value of chi-square is found in Appendix F, where the number of degrees of freedom is shown in the left margin. The various column headings, such as 0.10 and 0.05, represent the area, or probability, to the right of the particular χ^2 value. In this example, go down the left-hand column to 3 degrees of freedom (df). Then move across that row to the column headed 0.05, the level of significance selected for this problem. The critical value is 7.815. A portion of Appendix F is reproduced as Table 16–1. The decision rule is: If the computed value of chi-square is greater than 7.815, reject the null hypothesis.

Table 16-1

Critical Values of the Chi-Square Distribution

Degrees of Freedom df	Possible Values of χ^2 Right-Tail Area 0.10	0.05	0.02	0.01
1	2.706	3.841	5.412	6.635
2	4.605	5.991	7.824	9.210
3	6.251	7.815	9.837	11.345
4	7.779	9.488	11.668	13.277
5	9.236	11.070	13.388	15.086

The decision rule is shown graphically.

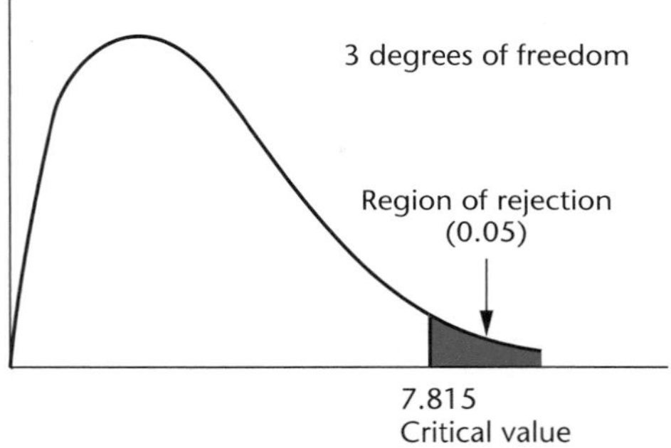

A large computed value of chi-square (over 7.815 in this problem) occurs when there is a *substantial* difference between observed and expected frequencies. If that happens, we reject the null hypothesis that there is no difference between the two distributions. If the differences between f_o and f_e are small, then the computed value of χ^2 is small (7.815 or less in this Problem), indicating that the difference between the two distributions *occurred by chance.*

STEP 5. Compute chi-square and make a decision. Recall that out of the 80 children, 25 chose a brown T-shirt, 18 chose an orange one, 19 chose a yellow one, and 18 chose a green one. These are the *observed* frequencies, denoted by f_o. How are the *expected* frequencies determined? As noted in Step 1, the null hypothesis states that there is no preference with respect to the four colors. If the null hypothesis is true, we would expect one-fourth of the 80 children, or 20, to select orange, and so on. Thus, in this Problem, all the expected frequencies (f_e) are equal (20).

The observed and expected frequencies and the calculations needed for chi-square are in Table 16–2. The procedure is as follows:

1. Subtract each expected frequency from each observed frequency; that is, $f_o - f_e$.
2. Square the difference: $(f_o - f_e)^2$.

Table 16-2 **Calculations for Chi-Square**

Color	f_o	f_e	$f_o - f_e$	$(f_o - f_e)^2$	$\frac{(f_o - f_e)^2}{f_e}$
Brown	25	20	5	25	25/20 = 1.25
Orange	18	20	−2	4	4/20 = 0.20
Yellow	19	20	−1	1	1/20 = 0.05
Green	18	20	−2	4	4/20 = 0.20
Total	80	80	0		1.70
	Must be equal	Must be equal	Must be 0		Computed χ^2

3. Divide each squared difference by the corresponding expected frequency:

$$\chi^2 = \sum \frac{(f_o - f_e)^2}{f_e}$$

4. Sum these quantities to obtain chi-square, as shown in formula 16–1:

$$\chi^2 = \sum \frac{(f_o - f_e)^2}{f_e}$$

Note that in the table the computed value of χ^2 is 1.70.

The computed value of 1.70 is less than the critical value of 7.815. Therefore, we fail to reject the null hypothesis. The mentally retarded children have no preference among the four colors.

Self-Review 16-1

Answers to the Self-Review problems are at the end of the chapter.

A toothpaste manufacturer hopes to market one or more toothpastes with these unusual flavors: cranberry, root beer, lime, vanilla, and orange. The manufacturer wonders if consumers have a preference among the flavors. Small tubes of each flavor are given to 200 consumers, and each consumer is asked to state his or her preference. The 0.01 level of significance is to be used.

a. What are the null hypothesis and the alternate hypothesis?
b. What is the decision rule?

The preferences of the 200 in the sample are:

Flavor	Number Preferring Flavor
Cranberry	32
Root beer	30
Lime	28
Vanilla	58
Orange	52
Total	200

c. Calculate the value of chi-square.
d. What is your decision with respect to the null hypothesis?

Exercises

Answers to the even-numbered exercises are at the back of the book. Use the five-step hypothesis-testing procedure for each of the following exercises.

1. A city has three television stations, each with its own evening news program from 6:00 p.m. to 6:30 p.m. The Acklin Survey Group is hired to determine if there is a preference among the viewing audience for any station. A random sample of 150 viewers revealed that 53 watched the evening news on WNAE-TV, 64 on WMWM-TV, and 33 on WRRN-TV. At the 0.05 level of significance, is there sufficient evidence to show that the three stations do not have equal shares of the evening news audience?
2. It is suspected that a particular six-sided die is "loaded," making it not a "true" die. As an experiment, this die is rolled 120 times. The results are:

Face	1	2	3	4	5	6	**Total**
Observed Frequency	15	29	14	17	28	17	120

Using the five-step procedure and the 0.05 significance level, test the null hypothesis that there is no difference between the distribution of observed frequencies and the distribution of expected frequencies.

3. A sociologist is studying retired people living in Florida. A sample of 150 who had lived in the Northeast revealed that they were from the following metropolitan regions:

Region	Observed Frequency
North Jersey–New York	60
South Jersey–Philadelphia	40
Buffalo	25
Pittsburgh	25

At the 0.05 significance level, can we conclude that an equal number of retirees were from the four regions?

4. In the Michigan lottery "Daily Game," officials select numbers by having plastic balls with digits on them blown at random, mechanically, through a plastic tube. It is argued that each of the 10 digits, 0 through 9, has the same chance of occurrence. To test for potential bias, researchers recorded the number of times each number was selected during 200 trials.

a. State the null and alternate hypotheses.

b. Determine the expected frequencies for each digit.

c. For a test at the 0.01 level of significance, what is the critical value?

d. Here are the results:

Digit	Frequency
0	10
1	23
2	15
3	24
4	21
5	23
6	19
7	18
8	25
9	22

Compute the chi-square test statistic.

e. Make a decision about the potential bias. Interpret your results.

5. In a study of work patterns, an organizational behavior specialist collected the data below on absenteeism by day of the week.

Day	Frequency
Monday	124
Tuesday	74
Wednesday	104
Thursday	98
Friday	120

a. If 520 absences were recorded over a five-day work week, what is the null hypothesis and what is the expected number of absences each day?

b. What is the critical value for a test at the 0.05 level of significance?

c. Conduct a test using chi-square and arrive at a decision.

THE GOODNESS-OF-FIT TEST: UNEQUAL EXPECTED FREQUENCIES

The expected frequencies in the preceding problems were all equal. For the color-preference problem involving a sample of 80 mentally retarded children, we expected 20 to select brown, 20 to select orange, and so on. What if the expected frequencies are not equal? The chi-square distribution can still be applied. The following Problem will serve as an illustration.

Problem Recall from the Chapter Problem the *Lansing State Journal* reported that in a bag of M&M's Chocolate Peanut Candies there are 30% brown, 20% green, 20% red, 20% yellow, and 10% orange. You purchase a one-pound bag at the K-Mart. There are a total of 188 candies in the bag—67 brown, 24 green, 51 red, 22 yellow, and 24 orange. At the .01 significance level, does your bag of M&M's Chocolate Peanut Candies agree with the distribution suggested by the newspaper?

Solution The usual hypothesis-testing procedure is used:

H_0: The distribution is as updated by the newspaper.
H_a: The distribution is not as updated by the newspaper.

If the null hypothesis is true, we would expect 30% of the 188 candies to be brown. Thus, .30 (188) = 56.4. Likewise, we expect 37.6 of the candies to be green, found by 188(.20).

The observed and expected frequencies are as follows:

Color	f_o	f_e	f_e found by
Brown	67	56.4	.30(188)
Green	24	37.6	.20(188)
Red	51	37.6	.20(188)
Yellow	22	37.6	.20(188)
Orange	24	18.8	.10(188)
Total	188	188.0	

The computed value of chi-square is determined using formula 16–1:

$$\chi^2 = \Sigma \frac{(f_o - f_e)^2}{f_e}$$

$$= \frac{(67-56.4)^2}{56.4} + \frac{(24-37.6)^2}{37.6} + \frac{(51-37.6)^2}{37.6} + \frac{(22-37.6)^2}{37.6} + \frac{(24-18.8)^2}{18.8}$$

$$= 19.598$$

STAT BYTE

In 1991, 28% of Americans over the age of 24 who had never been married had completed college. Of those who had never been married, 23% had not completed college. Is whether or not a person has been married related to completing college?

Because there are five categories, there are four degrees of freedom, found by $k - 1 = 5 - 1 = 4$. The critical value at the 0.01 significance level is 13.277. The null hypothesis is rejected if the computed value of chi-square exceeds 13.277 (from Appendix F). Because the computed value of 19.598 is larger than the critical value, the null hypothesis is rejected in favor of the alternate hypothesis. The distribution of colors found in the bag of M&M's Chocolate Peanut Candies purchased at the K-Mart is not from a population as described in the *Lansing State Journal*. The *p*-value is less than 0.01 (from the software program Electronic Tables it is estimated to be 0.0006).

The versatility of the chi-square distribution is further demonstrated in the following Problem.

Problem A national study revealed that, within five years of their release from prison, 20% of criminals had not been arrested again, 38% had been arrested once, and so on. Table 16–3 shows the complete distribution.

A social agency in a large city has developed a guidance program for former prisoners who settle there. Anxious to compare local results with the national figures in Table 16–3, the director of the social agency selected at random former prisoners who were in the guidance program. The distributions of the frequencies observed and the national experience are shown in Table 16–4.

How would the director of the social agency use chi-square to compare the local experience with the national experience? Use the 0.01 level of significance.

Solution The *number* in each category resulting from local experience cannot be directly compared with the *percent* from the national study. The national percentages can, however, be converted to expected frequencies (f_e). Logically, if there is no difference be-

Table 16-3 Number of Arrests and Percent of the Total

Number of Arrests after Release from Prison	Percent of Total
0	20.0
1	38.0
2	18.0
3	13.5
4 or more	10.5
Total	100.0

Table 16-4

Comparison of Local and National Distributions

Number of Arrests after Release from Prison	Local Experience (number)	National Experience (percent of total)
0	58	20.0
1	62	38.0
2	28	18.0
3	16	13.5
4 or more	36	10.5
	200	100.0

tween the local experience and the national experience, 20% of the 200 sampled, or 40, would never be arrested again after being released from prison. Likewise, 38% of the 200, or 76, would be arrested once, and so on. For a complete set of observed and expected frequencies, see Table 16–5.

The null hypothesis (H_0) is that there is no difference between the local experience and the national experience. That is, any differences between the observed and the expected frequencies are due to chance (sampling). The alternate hypothesis (H_a) is that there is a difference between the local experience and the national experience. There are five categories in Table 16–5, so there are $k - 1 = 5 - 1 = 4$ degrees of freedom. The critical value of χ^2 from Appendix F is 13.277. The decision rule is shown graphically at the top of the following page.

Table 16-5

Comparison of Local and National Frequencies

Number of Arrests after Release from Prison	Local Experience (number)	National Experience (percent of total)
0	58	40
1	62	76
2	28	36
3	16	27
4 or more	36	21
	200	200

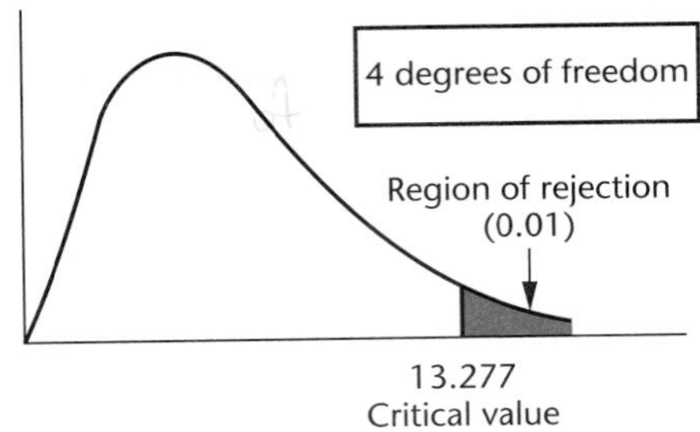

The computed value of chi-square is 27.653 (see Table 16–6).

The computed value of 27.653 is in the region beyond the critical value of 13.277. At the 0.01 level of significance, therefore, the null hypothesis is rejected and the alternate hypothesis is accepted. The conclusion is that there is a difference between the local and national experiences. The director now has evidence that his program results in significantly fewer arrests.

Self-Review 16-2

National figures revealed that 42% of vacationers who travel outside the United States go to Europe, 20% to the Far East, 16% to South and Central America, 6% to the Middle East, 12% to the South Pacific, and 4% go elsewhere. A local travel agency wondered if its customers differ significantly from this breakdown with respect to their travel destinations. A sample of the files of 200 of its customers revealed the following:

Destination	Number of Vacationers
Europe	80
Far East	44
South and Central America	34
Middle East	16
South Pacific	20
All others	6

a. What level of measurement is involved?
b. State both the null hypothesis and the alternate hypothesis.
c. Show the decision rule graphically. Use the 0.05 level.
d. Compute χ^2 and arrive at a decision.

Exercises

6. The occupations of parents are thought to influence the choice of occupation of their children. The percent distribution of ca-

Table 16-6

Computation of Chi-Square

Number of Arrests after release from prison	Local Experience f_o	National Experience f_e	$f_o - f_e$	$(f_o - f_e)^2$	$\frac{(f_o - f_e)^2}{f_o}$
0	58	40	+18	324	324/40 = 8.100
1	62	76	−14	196	196/76 = 2.579
2	28	36	−8	64	64/36 = 1.778
3	16	27	−11	121	121/27 = 4.482
4 or more	36	21	+15	225	225/21 = 10.714
	200	200	0		27.653
			Must be 0		χ^2

reer employment by categories of professional, technical, and service for a population of parents is shown here:

Occupation	Percent
Professional	26
Technical	64
Service	10

A sample of 360 children revealed the following about their occupations:

Occupation	Number
Professional	117
Technical	206
Service	37

Use the 0.01 level of significance to test the hypothesis that there has been no change in the distribution of occupations.

7. According to Mendel's theory of heredity, if plants with wrinkled, green seeds are crossbred with plants whose seeds are round and yellow, the offspring will follow this probability distribution:

Type of Skin	Probability
Round yellow	9/16
Wrinkled yellow	3/16
Round green	3/16
Wrinkled green	1/16

This indicates that out of every 16 seeds that germinate, 9 will have round yellow skin, and so on. Use the 0.05 level of significance to check whether Mendel's theory is contradicted by an experiment that yields the results shown:

Type of Skin	Number
Round yellow	334
Wrinkled yellow	123
Round green	120
Wrinkled green	31
	608

8. According to figures previously released by the Census Bureau, health insurance coverage for American citizens was distributed as follows:

Insurance	Percent
Private	65
Medicare/Medicaid	19
Uninsured	16

A random sample of 235 U.S. citizens conducted several years later revealed the following distribution:

Insurance	Number
Private	137
Medicare/Medicaid	53
Uninsured	45

Use the 0.05 level of significance to test the hypothesis that there has been no change in the distribution.

9. Aldini Medical Services studied a sample of 3,459 abortions. The number of previous abortions reported by these patients was:

Abortions	Patients
None	2,441
One	784
Two	184
Three or more	50

Are these data consistent with national data that indicate that 67% have never had a previous abortion, 27% have had one previous abortion, 5% have had two previous abortions, and 1% have had three or more previous abortions? Use the 0.05 level of significance.

10. Professor Graham teaches at least one section of sophomore-level statistics each quarter. He selected 50 students from his previous class and found 30 sophomores, 10 freshmen, and 10 juniors. Can Professor Graham conclude that there are twice as many sophomores as freshmen and juniors? Use the 0.05 significance level.

11. According to national sources, the size distribution of farms in the United States is as follows:

Size of Farm (in acres)	Percent of All Farms
Under 500	32
500–999	14
1,000–1,999	11
2,000 and over	43

An independent survey is made of 250 farms in Alabama.

a. What is the expected frequency for the survey results?
b. At the 0.01 level of significance, what is the critical value?
c. The number of farms in each size category is 80, 27, 33, and 110, respectively. Compute the test statistic.
d. Is Alabama's distribution significantly different from the national one? Discuss your decision.

12. The law guarantees a defendant the right to be tried before a jury of his or her peers. However, many students of American law complain that the wealthy and powerful are easily excused from jury duty. As part of a study of jury compositions, the following data were developed for Logan County. The population is 80% white, 15% black, and 5% Hispanic. Three hundred persons who recently served on juries were found to fall into those categories as follows:

White	Black	Hispanic
233	53	14

a. What are the null and alternate hypotheses?
b. Find the expected frequencies for each.
c. Determine the critical value for a test at the 0.01 level of significance.
d. What is the value of your test statistic?
e. What conclusion do you draw? Discuss your decision.

13. When three "fair" coins are tossed, probability theory predicts that the number of heads is 0, 1, 2, or 3, with probabilities 1/8, 3/8, 3/8, and 1/8, respectively. A set of three coins is tossed 160 times, and you wish to check for "fairness."
a. State the null and alternate hypotheses.
b. Determine the expected frequencies for each outcome.
c. For a 0.01 level of significance, what is the critical value?
d. On the toss of three coins 160 times, you observed that no heads appear on 25 tosses, only one head 55 times, two heads on 70 tosses, and 3 heads on 10 tosses. What is the value of the test statistic?
e. Can you reject the null hypothesis?

CONTINGENCY-TABLE ANALYSIS

The goodness-of-fit test discussed in the previous section was concerned with only a *single* trait, such as the color of a T-shirt offered to mentally retarded children. What testing procedure is followed if we are interested in the relationship between two characteristics, such as a person's adjustment to retirement and whether or not that person moved to a retirement community? By classifying adjustment as excellent, good, fair, or poor, the research data may be tallied into a table:

	Adjustment to Retirement			
Status	Excellent	Good	Fair	Poor
Moved to retirement community	卌 卌 卌 卌 卌 |||	卌 卌 卌 ||	卌 卌 卌 卌	卌 卌 卌
Did not move to retirement community	卌 卌 卌 |||	卌 卌 卌 ||	卌 卌 卌 |	卌 卌 卌 卌 卌 ||

This table is called a **contingency table.** It is also popularly referred to as cross-tabulated data.

> **Contingency Table** Frequency data resulting from the simultaneous classification of *more than one* variable or trait of the observed item.

STAT BYTE

In 1969, the Selective Service Administration conducted a lottery to determine the sequence in which young men would be drafted into the armed services. Many claimed that the results were not random. Those with birthdays at the end of the year seemed to have a predominance of the lower draft numbers. The results are shown below by quarter of birth and identify whether the draft number was low (1–183) or high (184–366).

Quarter	Low	High
First	34	57
Second	39	52
Third	50	42
Fourth	60	32

Is draft order related to the quarter of one's birth?

We can also use the chi-square statistic to determine whether two traits—home status and adjustment to retirement—are related. The null hypothesis is that the two traits are independent (not related). This is consistent with our earlier use of the null hypothesis to represent the case in which there was no change. We always took as the H_0 that $\mu_1 = \mu_2$, or that μ was equal to some particular value. Like the goodness-of-fit test, the contingency-table analysis uses observed frequencies (f_o) and expected frequencies (f_e). The observed frequencies are recorded in the preceding contingency table in the form of tallies. The expected frequencies must be computed.

Problem A metropolitan law enforcement agency classifies crimes committed within its jurisdiction as either "violent" or "nonviolent." An investigation has been ordered to find out whether the type of crime (violent or nonviolent) depends on the age of the person who committed it. A sample of 100 crimes was selected at random from the police files. The results are cross-classified in Table 16–7.

Does it appear that there is any relationship between the age of a criminal and the nature of the crime? Use the 0.05 level of significance.

Solution As usual, the initial step is to state the null hypothesis H_0 and the alternate hypothesis H_a:

H_0: There is no relationship between the type of crime committed and the age of the criminal.
H_a: There is a relationship between the type of crime committed and the age of the criminal.

The observed frequencies are shown in Table 16–7. To determine the expected frequencies, note first the number of crimes being studied (100). In the marginal totals on the right side of the contingency table, observe that 55 of the 100 crimes, or 55%, were violent. If the null hypothesis is true (that there is no relationship between type of crime and age), logically it can be expected that 11 of the 20 criminals under age 25 (55%) committed a violent crime, found by 0.55×20. Likewise, if the null hypothesis is true, one would expect

Classification of Crimes by Type and by Age of Person — Table 16-7

	Age			
Type of Crime	Under 25	25–49	50 and Over	**Total**
Violent	15	30	10	55
Nonviolent	5	30	10	45
Total	20	60	20	100

33 of the criminals in the 25–49 age bracket, or 55%, to have committed a violent crime. The expected frequency in the 50-and-over group is the same as the under-25 group.

The same logic can be followed to find the expected frequencies for the nonviolent crimes. Again, from the right-hand marginal totals, 45/100, or 45%, were nonviolent crimes. Then:

$45/100 \times 20 = 9$ expected frequencies for the under-25 age group
$45/100 \times 60 = 27$ expected frequencies in the 25–49 age group

By now, no doubt you will have noticed that under the assumption of independence an expected frequency, f_e can be computed by:

$$\text{Expected frequency for a cell} = \frac{\text{(Row total)(Column total)}}{\text{Grand total}}$$

For the nonviolent, 25–49 age cell—just computed to be 27 expected frequencies—the formula would be

$$f_e = \frac{\text{(Row total)(Column total)}}{\text{Grand total}} = \frac{(45)(60)}{100}$$
$$= 27, \text{ the same answer as the one computed earlier}$$

The observed frequencies and the corresponding expected frequencies are shown in the form of a contingency table (see Table 16–8).

We formulate a decision rule by first determining the number of degrees of freedom. For a contingency table, we find the number of degrees of freedom by multiplying number of rows minus one by number of columns minus one. This contingency table has two rows and three columns. To find the degrees of freedom, we compute as follows: (rows – 1)(columns – 1) = (2 – 1)(3 – 1) = 2 *df*. The critical value can now be located in Appendix F. Recall that the 0.05 level of significance had been selected. Go down the left margin in Appendix F to 2 df and read the critical value under the 0.05 column. It is 5.991. Thus, the decision rule is: Reject the null hypothesis if the computed value of χ^2 is greater than 5.991.

As before, chi-square is computed by formula 16–1:

$$\chi^2 = \Sigma \frac{(f_o - f_e)^2}{f_e}$$

Inserting the observed frequencies f_o and the expected frequencies f_e from Table 16–8 in the formula for χ^2, we obtain:

$$\chi^2 = \frac{(15-11)^2}{11} + \frac{(5-9)^2}{9} + \frac{(30-33)^2}{33} + \frac{(30-27)^2}{27}$$

$$+\frac{(10-11)^2}{11}+\frac{(10-9)^2}{9}$$
$$=1.45+1.78+0.27+0.33+0.09+0.11$$
$$=4.03$$

Table 16-8

Observed Frequencies and Expected Frequencies

	Age							
	Under 25		25–49		50 and Over		**Total**	
Type of Crime	f_o	f_e	f_o	f_e	f_o	f_e	f_o	f_e
Violent	15	11	30	33	10	11	55	55
Nonviolent	5	9	30	27	10	9	45	45
Total	20	20	60	60	20	20	100	100

Must be equal (Under 25 totals); $\frac{(45)(60)}{100}$ (25–49 Nonviolent f_e); Must be equal (Total column totals)

The computed chi-square value of 4.03 is less than the critical value of 5.991. Therefore, we fail to reject the null hypothesis. The law enforcement agency cannot conclude that there is any relationship between the age of the criminal and the degree of violence of the crime. To put it another way, the age of the criminal is independent of the degree of violence.

CAUTION. In the formula for χ^2, note that the expected frequencies, f_e, are in the denominator. If the expected frequency for any cell is quite small, the corresponding, $(f_o - f_e)^2/f_e$ value for that cell may be disproportionately large. In turn, this one large cell value might result in a computed χ^2 greater than the critical value, causing the null hypothesis to be rejected. If the expected frequency for that one cell were larger, the null hypothesis would probably not be rejected. To avoid this problem, a useful rule of thumb is to look for an expected frequency of at least 5 in each cell. Whenever the number is smaller than that, you may want to consider combining several adjacent cells. The obvious alternative, of course, would be to increase the size of the sample.

Here is the MINITAB output for the same example:

```
MTB >  chisquare c1-c3

Expected counts are printed below observed counts
            C1        C2        C3     Total
    1       15        30        10        55
         11.00     33.00     11.00

    2        5        30        10        45
          9.00     27.00      9.00
Total       20        60        20       100

ChiSq =  1.455 + 0.273 + 0.091 +
         1.778 + 0.333 + 0.111 = 4.040
df = 2                             ↑
                              Computed χ²
```

Self-Review 16-3

In a city with a maximum security prison, the residents have been polled to determine if a relationship exists between marital status and a resident's stand on capital punishment.

a. What are the null hypothesis and the alternate hypothesis? A random sample of 200 residents was asked for their opinion and for their marital status. The results were cross-classified into the following table:

Stand on Capital Punishment	Marital Status: Married	Marital Status: Not Married	Total
Favor	100	20	120
Oppose	50	30	80
Total	150	50	200

b. What is this table called?
c. How many cells does the table contain?
d. How many degrees of freedom are there?
e. For the 0.01 level of significance, what is the critical value of chi-square?
f. What is the computed value of chi-square?
g. What is your decision regarding the null hypothesis?

Exercises

14. Five hundred persons were divided into two groups to be sampled—one group consisting of people with religious affiliation and the other of people without religious affiliation. Each person was given a test to determine his or her degree of racial prejudice. The question to be explored is whether religious status and racial prejudice are related.

a. State the null hypothesis and the alternate hypothesis. The results of the survey were cross-classified into the following contingency table:

	Degree of Racial Prejudice		
Religious Status	Highly Prejudiced	Somewhat Prejudiced	Not Prejudiced
Church-affiliated	70	160	170
Not church-affiliated	20	50	30

b. Using the 0.05 level, determine the critical value of chi-square.

c. Compute χ^2 and arrive at a decision.

15. New employees hired for staff positions in a large social service agency immediately go into a six-week orientation and training program. At the end of the six weeks, each employee is rated as a below-average prospect, an average prospect, an above-average prospect, or an outstanding prospect. Each employee is then assigned to a supervisor. After a period of two months, the supervisor rates the employee as poor, fair, good, or superior. The question to be explored is whether an employee's performance in the training program is correlated with the supervisor's ratings. A sample of the personnel files revealed the following:

- After two months, of the 30 employees who had been rated below-average prospects, 11 were rated poor by the supervisor, 8 were rated fair, 6 good, and 5 superior.
- Of the 40 employees who had been rated average, 9 were rated poor, 18 fair, 8 good, and 5 superior.
- Of the 60 employees who had been rated above average, 7 were rated poor, 11 fair, 28 good, and 14 superior.
- Of the 70 employees who had been rated outstanding, 5 were rated poor, 10 fair, 22 good, and 33 superior.

Using the 0.01 level of significance and the five-step hypothesis-testing procedure, organize the counts into a contingency table and reach a decision.

16. A sample of people was asked by the Gallup organization to rate the honesty and ethical standards of various groups. Some of the results are shown below.

	High	Average	Low
Medical doctors	53	35	10
Lawyers	24	43	27
Business executives	18	55	20
Members of Congress	14	43	38

Use the 0.10 significance level to determine if the selected professions are related with respect to honesty and ethical standards.

17. A random sample of 780 registered voters was asked to evaluate President Clinton's handling of the abortion issue. They were cross-classified by party affiliation into the following table:

	Democrat	Republican	Independent
Approved	200	125	70
Disapproved	130	230	25

Does the sample support the hypothesis that party affiliation is independent of opinion of President Clinton's handling of the abortion issue? Use the 0.05 level of significance.

18. The table below shows the distribution of hair color for a sample of 422 people. You wish to see if hair color and sex are related.

Hair Color	Men	Women
Black	56	32
Blonde	37	66
Brown	84	90
Red	19	38
Total	196	226

Use the 0.01 level of significance to test whether these distributions are significantly different. Explain your findings.

19. A random sample of voter registrations for several adjacent precincts led to the following data:

	Precinct 16	Precinct 23	Precinct 27
Democrat	300	420	270
Independent	25	17	10
Republican	150	300	200

Test whether the party affiliation distribution is the same in all three precincts. Use the 0.01 level of significance. Explain your findings.

SUMMARY

This chapter dealt with data of the nominal scale—that is, data that were classified into categories. Two types of hypothesis-testing problems involving the use of the chi-square distribution as the test statistic were discussed. If the objective is to determine whether a set of observed frequencies "fits" a set of expected or assumed frequencies, a goodness-of-fit test is applied. The five-step hypothesis-testing procedure is followed: (1) Both the null hypothesis and the alternate hypothesis are stated. (2) A level of significance is chosen. (3) The appropriate test statistic is decided upon. It is chi-square in this type of problem. (4) A decision rule is formulated that allows the researcher to reject or fail to reject the null hypothesis stated in Step 1. (5) A sample is selected, the observed frequency, f_o, in each category is matched with the expected frequency, f_e, and χ^2 is computed. Based on the decision rule and on the computed value of chi-square, a decision regarding the null hypothesis is made.

The second type of problem discussed in this chapter dealt with data cross-classified into a contingency table. The null hypothesis is that the two criteria of classification are unrelated. Based on the marginal totals of the rows and the columns, an expected frequency, f_e, is computed for each cell. Chi-square is calculated and, given a predetermined level of significance, a decision is made regarding the null hypothesis.

The chi-square distribution has the following major characteristics:

1. Its value is always positive.
2. There is a family of chi-square distributions; for each number of degrees of freedom, there is a particular distribution.
3. It is positively skewed but approaches a symmetrical distribution as the number of degrees of freedom increases.

Exercises

20. The number of military recruits from each of four regions of the United States is listed below:

Region	Number
I	37
II	36
III	37
IV	50

Is there sufficient evidence at the 5% level of significance to demonstrate a difference in recruitment among the four regions?

21. A group of 385 mental patients has been classified according to parental social class, with the following results:

Social Class	Frequency
Upper	18
Upper-middle	31
Middle	46
Lower-middle	126
Lower	164

Test at the 0.05 level of significance to verify that the data are consistent with the assumption that all social classes are equally likely to be represented.

22. Refer to Exercise 21. Test the hypothesis that 30% of the patients are lower class, 30% are lower-middle class, 20% are middle class, 10% are upper-middle class, and 10% are upper class. Use the 0.05 significance level.

23. Each human being has blood of type A, B, O, or AB. The proportion of the general population having each type follows:

Type	Proportion
A	0.39
B	0.11
O	0.46
AB	0.04
	1.00

A sample of 700 people revealed that 250 have type A, 70 have type B, 340 have type O, and 40 have type AB. At the 0.05 significance level, can we conclude that this group of people differs from the general population?

24. A coin has been altered so that two heads appear face up for every one tail. Find the maximum number of heads occurring that will allow acceptance of the hypothesis of $P(\text{Head}) = 2/3$ if 30 tosses are performed. Use the 0.10 significance level.

25. As part of a study of air traffic at the Express Airport, a record has been kept of the number of aircraft arrivals during any one half-hour period. The table that follows shows the number of half-hour periods in which there were 0, 1, 2, 3, or 4 or more arrivals. Test the hypothesis that the number of such arrivals follows the expected frequencies shown. Use the 0.05 significance level.

Number of Arrivals	Observed Frequency	Expected Frequency
0	19	14
1	23	27
2	28	27
3	18	18
4 or more	12	14
Total	100	100

26. The Department of Education reports that the educational level of adults in the United States conforms to the distribution that follows. The Phoenix, Arizona, Chamber of Commerce wishes to compare a sample of 500 residents against this norm. What conclusion should be drawn from the results of this analysis? Use the 0.05 significance level.

Years of Education	Expected	Observed
1–8	100	77
9–12	150	198
13–16	200	178
Over 16	50	47

27. Students claim not to like morning classes. As a test, a college statistics department offered sections of a basic statistics course at various times during the day. No limit was set on class size. The following table shows the number of students who selected each class. At the 0.01 level of significance, is there sufficient evidence to show that students have a time preference?

Time	Number of Students
Early a.m.	40
Late a.m.	62
Early p.m.	60
Late p.m.	35
Early evening	58
Late evening	45
Total	300

28. A group of executives was classified according to total income and age. Test the hypothesis, at the 0.01 level, that age is not related to level of income.

Age	Income: Less Than $100,000	$100,000–$399,999	$400,000 or more
Under 40	6	9	5
40–54	18	19	8
55 or older	11	12	17

29. A manufacturer of women's apparel wants to know if a woman's age is a factor in determining whether she would buy a particular garment. Accordingly, the firm surveyed three age groups and asked each woman to rate the garment as excellent, average, or poor. The results follow. Test the hypothesis, at the 0.05 level, that rating is not related to age group.

Rating	Age Group: 15–24	25–39	40–55	Total
Excellent	40	47	46	133
Average	51	74	57	182
Poor	29	19	37	85
Total	120	140	140	400

30. Four coins are tossed 100 times; the number of heads for each trial is shown. At the 0.01 level, test the hypothesis that the coins are fair. (*Hint:* Use the binomial distribution to determine the expected frequencies.)

Number of Heads	Observed Frequency
0	8
1	30
2	29
3	23
4	10
	100

31. The following results were reported in Japan's parliamentary elections for the lower house:

	1983	1990
Liberal Democratic Party	250	300
Socialist Party	112	85
Clean Government Party	58	56
Democratic Socialist Party	38	26
Communist Party	26	26
New Liberal Club	8	6
Independents	19	13

Use the 0.01 level of significance to test if there has been a significant change in the distribution.

32. A study was conducted of 400 families, each having exactly three children, to determine whether or not boy and girl births were equally likely. The results were as given. What is the value of χ^2? Are results significantly different from those expected at the 0.05 significance level?

Number of Boys: 0, 1, 2, 3
Number of Cases: 65, 140, 148, 47

(*Hint:* Use the binomial distribution to find the expected frequencies.)

33. Five different brands of canned tuna were examined for quality. A total of 24 cans of each brand was checked. The number of acceptable and unacceptable cans are as follows:

	Brands				
	A	B	C	D	E
Acceptable	3	10	5	3	9
Unacceptable	21	14	19	21	15

At the 0.05 significance level, can we conclude that the five brands are of the same quality?

34. A manufacturer claims that a certain drug is effective in treating arthritis. In an experiment, 176 people were given the drug and 176 people were given sugar pills. The results are shown below. At the 0.05 significance level, is there a difference in effectiveness of the two drugs?

	Better	**No Effect**	**Worse**	**Total**
Drug	88	50	38	176
Sugar	60	65	51	176
Total	148	115	89	352

Data Exercises

35. Refer to Real Estate Data, which reports information on homes sold in Northwest Ohio during 1992.

Tally the selling prices of the homes, in thousands of dollars, into three classes: \$0–\$179.9, \$180–\$189.9, and \$190.0 or more.

(1) Create a table that shows the grouped selling prices and whether or not the home had an attached garage. At the

0.05 significance level, can we conclude that selling price is related to whether or not there is an attached garage?

(2) Create a table that shows the grouped selling prices and whether or not the home had a fireplace. At the 0.05 significance level, can we conclude that selling price is related to whether or not the home had a fireplace?

(3) Tally the distance into three categories: 0–17 miles, 18–24 miles, and 25 or more miles. Create a table that shows the grouped selling prices and the grouped distances. Can we conclude that the homes nearer the center of the city had the lower selling price? Use the 0.05 significance level.

36. Refer to Salary Data, which refers to a sample of middle managers in Sarasota, Florida.

a. Conduct a test of hypothesis to determine if the four regions are equally represented in the sample. Use the .05 significance level.

b. Conduct a test of hypothesis to determine if the types of industry are equally represented. Use the 0.05 significance level.

c. Tally the salaries of the middle managers, in thousands of dollars, into three groups: $0–$59.9, $60–$64.9, and $65.0 or more.

(1) Create a table that shows both the grouped salaries and the sex of the manager. At the 0.05 significance level, can we conclude that sex and salary are related?

(2) Create a table that shows both the grouped salaries and the type of the industry in which the manager is employed. At the 0.05 significance level, can we conclude that type of industry and salary are related?

CASE: RETIRE AND RELOCATE?

The Department of Health and Human Services is preparing an informational booklet on several retirement issues. One of the topics is whether retirees are more content if they continue to reside in their preretirement homes. To explore this question further, the department sent a detailed questionnaire to a sample of retirees who remain in their preretirement homes and to a sample of those who have relocated. Of those who stayed in the same community, 28 made an excellent adjustment to retirement, 41 a good one, 19 fair, 9 poor, and 1 very poor. Of those who moved to other locations, 23 made an excellent adjustment to retirement, 36 a good one, 23 fair, 2 poor, and 8 very poor. Does it appear that moving affects adjustment to retirement?

CHAPTER ACHIEVEMENT TEST

Answer all the questions. The answers are at the back of the book.

MULTIPLE-CHOICE QUESTIONS. Select the response that best answers each of the questions.

1. For a study to determine if region of birth (four categories) is related to age (six categories), the number of degrees of freedom in the test statistic would be
  a. 10.
  b. 24.
  c. 15.
  d. 2.

2. If the chi-square test is to be applied to data cross-classified in a contingency table,
  a. the expected frequencies cannot be determined.
  b. the observed frequencies must total 100.
  c. there must be at least ten expected frequencies in each cell.
  d. None of the above

3. The chi-square distribution
  a. is positively skewed.
  b. never assumes a negative value.
  c. changes when the number of degrees of freedom changes.
  d. All of the above

4. A sample of 100 students consists of 60 females and 40 males. The purpose of the study is to determine if the sex of the student is related to passing or failing the first test. Out of the 100 students in the sample, 70 students passed. The expected number of males passing is
  a. 70.
  b. 40.
  c. 28.
  d. 12.
  e. None of the above

Questions 5–10 are based on the following problem. A supervisor wants to determine if unexcused absences in her department are distributed evenly throughout the workweek. A random sample of personnel files revealed the following:

Day	Frequency
Monday	20
Tuesday	16
Wednesday	17
Thursday	21
Friday	26
	100

5. How many degrees of freedom are there?
  a. 4
  b. 99
  c. 5
  d. 25
  e. None of the above

6. If the null hypothesis is true, what is the expected number of absences on Friday?
  a. 25
  b. 20
  c. 26
  d. 16
  e. None of the above

7. At the 0.01 significance level, the critical value of chi-square is
  a. 13.277.
  b. 15.086.
  c. 9.488.
  d. 11.070.
  e. None of the above

8. The computed value of χ^2 is
  a. 9.
  b. 3.1.
  c. 62.
  d. None of the above

9. The correct decision is:
  a. Reject H_0.
  b. Do not reject H_0.
  c. Not enough information is available to make a determination.

10. The correct statistical conclusion is that
  a. the unexcused absences are evenly distributed throughout the workweek.
  b. the unexcused absences are not evenly distributed throughout the workweek.
  c. Neither is a correct conclusion.

COMPUTATION PROBLEMS

11. Ninety congresspersons were surveyed to determine if there is a relationship between their party affiliation and their position on a proposed Social Security tax increase. At the 0.05 significance level, test the null hypothesis that there is no relationship. The survey results were:

	Favor	Oppose	Undecided
Democrats	22	13	28
Republicans	8	7	12

12. An analysis of last year's automotive sales revealed that for every one full-sized automobile purchased, two medium-sized, three compact, and four subcompact cars were purchased. A sample of purchases made during the last month shows car sales of 38 full-sized, 62 medium-sized, 41 compact, and 59 subcompact cars. Test at the 0.01 significance level to determine whether a change has occurred in the type of automobile purchased.

ANSWERS TO SELF-REVIEW PROBLEMS

16-1 a. H_0: There is no preference.
H_a: There is a preference.

b. Reject H_0 if χ^2 is greater than 13.277. Otherwise, reject it. ($k - 1 = 5 - 1 = 4$ df)

c.

$f_o - f_e$	$(f_o - f_e)^2$	$\frac{(f_o - f_e)^2}{f_e}$
32–40	$(-8)^2$	64/40 = 1.6
30–40	$(-10)^2$	100/40 = 2.5
28–40	$(-12)^2$	144/40 = 3.6
58–40	$(18)^2$	324/40 = 8.1
52–40	$(12)^2$	144/40 = 3.6
		$\chi^2 = 19.4$

d. Reject H_0 because 19.4 is in the region beyond 13.277; accept H_a. There is a difference with respect to taste preference.

16-2 a. Nominal.

b. H_0: There is a no difference between local and national vacation destinations.
H_a: There is a difference between local and national destinations.

c.

d.

f_o	f_e	$\frac{(f_o - f_e)^2}{f_e}$
80	84	16/84 = 0.190
44	40	16/40 = 0.400
34	32	4/32 = 0.125
16	12	16/12 = 1.333
20	24	16/24 = 0.667
6	8	4/8 = 0.500
		$\chi^2 = 3.215$

Do not reject the null hypothesis, because 3.215 is less than 11.070.

16-3 a. H_0: There is no relationship between a person's stand on capital punishment and his or her marital status.
H_a: There is a relationship between a person's stand on capital punishment and his or her marital status.

b. A 2-by-2 contingency table
c. 4, found by 2×2
d. 1, found by $(2 - 1)(2 - 1)$
e. 6.635, from Appendix F

f.
$$\chi^2 = \frac{(100-90)^2}{90} + \frac{(20-30)^2}{30} + \frac{(50-60)^2}{60} + \frac{(30-20)^2}{20} = 11.11$$

g. Reject the null hypothesis. Conclusion: At the 0.01 level, there is a relationship between a person's stand on capital punishment and his or her marital status.

CHAPTER 17

Nonparametric Methods: Analysis of Ranked Data

OBJECTIVES

When you have completed this chapter, you will be able to

- perform the sign test and describe its applications;
- calculate the Wilcoxon signed-rank test and describe its applications;
- describe the Wilcoxon rank-sum test;
- compute the Kruskal–Wallis analysis of variance by ranks test.

CHAPTER PROBLEM **Mirror, Mirror on the Wall . . .**

THE SHELBYVILLE CAREER TRAINING CENTER SPECIALIZES IN HELPING RECENTLY WIDOWED HOMEMAKERS REENTER THE JOB MARKET. OF PARTICULAR CONCERN IS THE DEGREE OF SELF-CONFIDENCE EXHIBITED. A SAMPLE OF 13 WIDOWS WAS SELECTED AT RANDOM. BEFORE THE START OF THE TRAINING PROGRAM, EACH OF THE WIDOWS WAS RATED BY A STAFF OF PSYCHIATRISTS AS HAVING NO SELF-CONFIDENCE, SOME SELF-CONFIDENCE, OR GREAT SELF-CONFIDENCE. AFTER THE TRAINING PROGRAM, THE SAME STAFF RATED EACH WIDOW'S DEGREE OF SELF-CONFIDENCE ON THE SAME SCALE. DO THE DATA IN TABLE 17-1 DEMONSTRATE AN INCREASE IN SELF-CONFIDENCE?

STAT BYTE

There is a simple, but time-consuming, way to convince a group of people that you are a genius. Start with a list of, say, 512, people. Contact half of them and tell them it is going to rain next Thursday. Tell the other half it is not going to rain. After next Thursday, there will be 256 people for whom you correctly predicted the weather. Now contact half of this smaller group and tell them it is going to rain on the following Monday; tell the other half it is not going to rain on the following Monday. When Monday arrives, there will be 128 people for whom you have now made two correct predictions. Repeat this process again, contacting only those for whom you provided "correct" predictions. After, say, five predictions, there will be 16 people for whom you have been right five times in a row. From their point of view, you will have done something truly unusual!

INTRODUCTION

In Chapter 16, we began examining nonparametric hypothesis tests, of which the chi-square goodness-of-fit test is one example. For that particular test, the only requirement is that the data have a nominal level of measurement. We noted that such tests are especially useful if no assumptions—such as normality—can be made about the shape of the population.

All four nonparametric tests in *this* chapter—namely, the sign test, the Wilcoxon signed-rank test, the Wilcoxon rank-sum test, and the Kruskal–Wallis analysis of variance by ranks test—assume that the data are at least of ordinal level of measurement, which is to say that the observations can be ordered (ranked) from lowest to highest. For example, if families were to be classified according to socioeconomic status (lower, lower-middle, middle, upper-middle, or upper class), the level of measurement would be ordinal scale.

THE SIGN TEST

The **sign test** is one of the more widely used nonparametric tests. As the name implies, it is based on the sign of the difference in paired observations. The pairs frequently represent "before" and "after" measurements taken on the same individual or item. Before the experiment, a person is given a rating, asked for an opinion, or asked to perform a specific task, the results of which are recorded. Then the experiment or treatment takes place, after which another observation is obtained and any changes are recorded. An increase may be given a " + " sign and a decrease a " – " sign. The underlying assumption is that no change occurs (the null hypothesis). If the null hypothesis is true and the treatment had no effect, the distribution of positive differences will be binomial, with $\pi = 0.50$. If the treatment had *some* impact, the number of positive differences will be either significantly large or significantly small.

The test statistic is the number of plus signs. When the null hypothesis is true, it follows the binomial distribution discussed in Chapter 6. Recall the principal features of the binomial: (1) the probability of a success remains constant from trial to trial, (2) successive trials are independent, and (3) we count the number of successes in a given number of trials. These traits are generally consistent with the sign test. The following Problem illustrates the details of a one-tailed sign test.

Problem Recall from the Chapter Problem that the Shelbyville Career Training Center specializes in programs designed to increase the level of self-confidence among recently widowed homemakers. The sample information is reported in Table 17–1. At the 0.10 sig-

Table 17-1

Self-Confidence of Widows Before and After a Training Program

	Self-Confidence		
Widow	Before	After	**Sign of Difference**
1	None	Some	+
2	Some	Great	+
3	None	Great	+
4	Some	Great	+
5	Some	None	–
6	None	Some	+
7	Some	Great	+
8	None	Great	+
9	None	Some	+
10	None	Some	+
11	Great	None	–
12	None	Great	+
~~13~~	~~None~~	~~None~~	~~0~~

Excluded from further analysis ← 13

nificance level, can we conclude that there has been an increase in the level of self-confidence?

Solution A "+" sign indicates that the special training program was a *success*—that is, the widow had more confidence after the program. Note, for example, that widow 1 had no confidence in herself before the training program, but showed some confidence after the program. Therefore, the sign of the difference for her is "+." A "–" sign indicates a *failure*—that is, the widow was less self-confident about reentering the job market after the training program than before it. If a subject felt *no change* in self-confidence, she was omitted from further analysis. Widow 13 had no confidence in her chances in the job market either before or after the program. Thus, she was excluded and the size of the sample was reduced to 12.

It does seem logical that if the special training program were *not* effective in developing self-confidence among the widows, half of them would show increased self-confidence, the other half reduced self-confidence. That is, there would be an equal proportion of pluses and minuses, and the *probability* of a success in the experiment would be $\pi = 1/2 = 0.50$. Therefore, the null hypothesis to be tested is $\pi \leq 0.50$. The null and alternate hypotheses are:

H_0: There is no change in a widow's degree of self-confidence as a result of the training program ($\pi \leq 0.50$).
H_a: A widow's confidence is increased ($\pi > 0.50$).

The 0.10 level of significance is chosen. The appropriate test statistic is the binomial distribution, with π, the probability of a success, equal to 0.50 and n, the size of the sample, equal to 12.

The alternate hypothesis, stated earlier, is $\pi > 0.50$. Recall from previous hypothesis-testing problems that if a direction is predicted (such as that π is greater than 0.50), a *one-tailed* test is used. And because only the tails of the sampling distribution are used in determining the critical region, only the right tail of the binomial distribution will be needed in this problem (because it is stated that π is *greater* than 0.50).

We locate the region of rejection by accumulating the probabilities starting with 12 successes, then 11 successes, and so on, until *we come as close as possible to the level of significance of 0.10 without exceeding it.* In this case, the probability of 12 successes is 0.000, the probability of 11 successes is 0.003, and so on, as shown in the middle column of Table 17–2. (The probabilities are found in Appendix A for a π of 0.50 and an n of 12.)

The probabilities of success are cumulated starting with 12 successes. The cumulative probability for 10 or more successes, for example, is found by 0.003 + 0.016. Refer to the cumulative probabilities in the right column of Table 17–2. A decision rule can now be formulated. The cumulative probability of 0.073 is as close as possible to the significance level of 0.10 without exceeding it. Thus, if nine or more pluses (increases in self-confidence among the widows) appear in the sample of 12 usable pairs of ratings, the null hypothesis will be rejected. The rejection region is shown in Figure 17–1. The probabilities on the Y-axis are from Appendix A.

A count of the pluses and minuses in Table 17–1 on the previous page shows that there are 10 pluses (successes) and two minuses (failures). The 10 successes are in the region of rejection. The null hypothesis that $\pi \leq 0.50$ is rejected; the alternate hypothesis that $\pi > 0.50$ is accepted. This means the special training program was a success. Out of 12 widows, 10 had more self-confidence after the

Table 17-2 **Cumulative Probabilities for 12, 11, 10, 9, and 8 Successes**

Number of Success	Probability of Success		Cumulative Probability
⋮	⋮		⋮
8	0.121		0.194 = 0.073 + 0.121
9	0.054	↑	0.073 = 0.019 + 0.054
10	0.016		0.019 = 0.003 + 0.016
11	0.003	Add	0.003
12	0.000	up	0.000

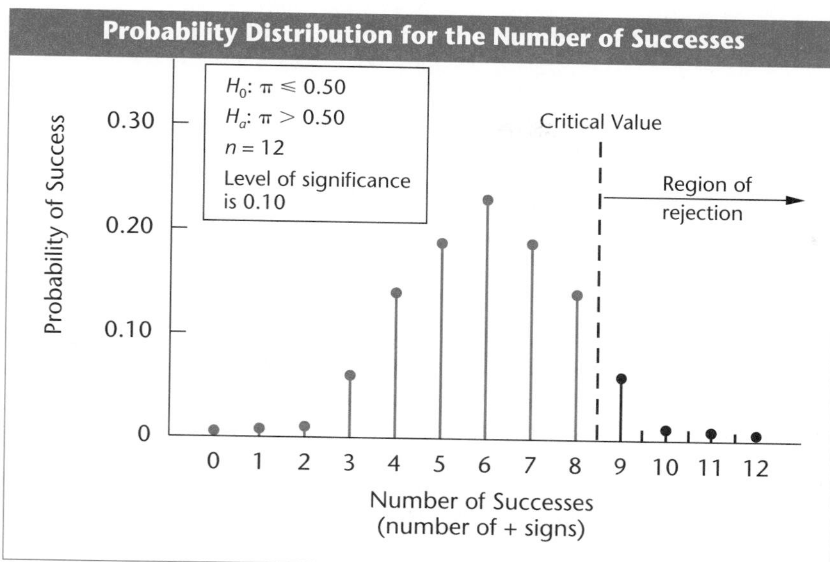

Figure 17-1

program than before it. In rejecting the null hypothesis, we are saying that it is highly unlikely that such a large number of successes (10 successes out of 12 in the sample) *could be due to chance.* It is reasonable to assume, therefore, that this large number of successes is due to the effects of the special training program.

Self-Review 17-1

Answers to the Self-Review problems are at the end of the chapter.

A physical-fitness instructor believes that his planned exercise program is effective in weight reduction. To investigate this claim, a random sample of seven persons enrolled in his program was selected. All were weighed before they started the program and again after completing it. These were the results:

Student	Before	After
Dave	192	185
Rick	198	185
Jim	170	162
Kelly	130	118
Stacey	113	118
Rob	165	159
Andy	168	164

The sign test is to be applied to determine if there is evidence to demonstrate that the weight-reduction program is effective. The 0.10 significance level is used. (Note: In solving the problem, we consider a *reduction* in a person's weight a "success" and designate it by a plus sign. An increase in a person's weight is considered a "failure" and is designated by a minus sign.)

a. State the null hypothesis and the alternate hypothesis.
b. State the decision rule.
c. Should the null hypothesis be rejected? Explain.
d. Interpret your findings.

The Problem that follows illustrates the use of the sign test in formulating a decision rule for a *two-tailed* hypothesis test.

Problem Twenty police officers are paired into 10 teams that include one black and one white officer each. The "partners" are then observed and rated as to their degree of mutual trust. A high score indicates a great deal of trust. After this observation period, the 10 pairs are subjected to a "wilderness experience" and then retested on their degree of mutual trust. Do the data in Table 17–3 suggest any significant changes attributable to the experience?

Table 17-3 **Degree of Trust Before and After a Wilderness Experience**

Partners	Rating Before Experience	Rating After Experience	Sign of Difference
A	68	32	–
B	55	55	0
C	49	55	+
D	40	75	+
E	20	50	+
F	18	25	+
G	30	23	–
H	70	49	–
I	52	62	+
J	50	41	–

Solution Use the five-step hypothesis-testing procedure.

STEP 1. The null hypothesis is that there is no difference in the ratings before and after the wilderness experience. This will be tested against the alternate hypothesis that there is a difference. Note that this is a two-sided alternative. We are looking for *any* significant change.

Again, it seems logical that if the wilderness experience had *not* affected mutual trust among the police officers, half of the pairs would show increased trust and the other half less trust.

That is, there would be an equal proportion of pluses and minuses. Thus, if the wilderness experience has no impact, the probability of a success is $\pi = 0.50$. The null and alternate hypotheses are:

$$H_0 : \pi = 0.50$$
$$H_a : \pi \neq 0.50$$

STEP 2. For this problem, the researcher selected a 20% level of significance.

STEP 3. The decision rule is based on the binomial distribution. Note that because the rating for partners B was neither an increase nor a decrease, their pair is removed from further calculations. Thus, nine partner pairs remain. As in the Chapter Problem, we construct the sampling distribution applicable for a true null hypothesis by referring to Appendix A, as shown in Table 17–4.

STEP 4. The test statistic is the number of plus (+) signs. When the null hypothesis is true, it follows the binomial distribution. In this illustration the number of + signs is 5 (from Table 17–3).

STEP 5. Because this is a two-sided test with a 20% level of significance, each tail of the sampling distribution should contain a probability of 10%. We can readily see the probability of 0.090 that there will be fewer than three " + " signs and, by symmetry, the probability of 0.090 that there will be more than six " + " signs. Therefore, the decision rule for a significance level of approximately 20% is to reject the null hypothesis if there are fewer than three or

Table 17-4

Binomial Probability Distribution ($n = 9$, $\pi = 0.50$)

Number of Successes	Probability of Success		Cumulative Probability	
0	.002	Add down ↓	.002	
1	.018		.020	
2	.070		.090	
				Critical value
3	.164		.254	
4	.246		.500	
5	.246		.500	
6	.164		.254	
				Critical value
7	.070	Add up ↑	.090	
8	.018		.020	
9	.002		.002	

more than six successes (plus signs). Otherwise, do not reject the null hypothesis. This decision rule is illustrated in Figure 17–2. The probabilities are based on Table 17–4.

Figure 17-2

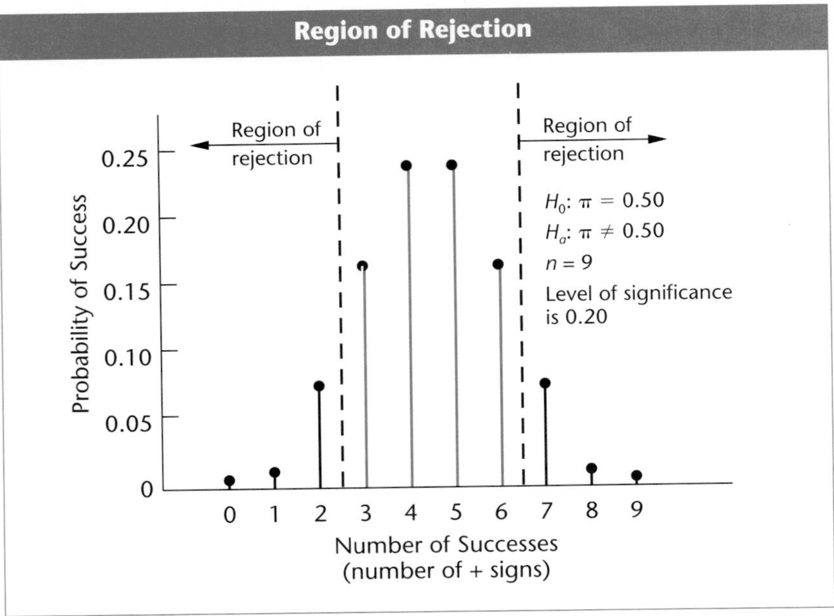

Referring to Table 17–3, note again that there are five " + " signs. This is fewer than six successes and therefore is not in the region of rejection. We conclude, therefore, that the wilderness experience did *not* change the degree of mutual trust between the partners. Essentially, we are saying that five successes could reasonably be attributed to chance.

Self-Review 17-2

Two groups, Calorie Counters and Fitness Fanatics, take different approaches to weight loss. Our interest is in determining if there is a difference in the percentage of weight loss experienced by the members of the two groups. A sample of 15 members from each of the two groups is selected and matched on the basis of age, current weight, and several other factors. The percentage of weight lost by each pair is recorded. Based on this information (see p. 621), would you conclude that the two weight loss methods differ in their success rates? Use the 0.05 significance level.

a. What is the null hypothesis? The alternate hypothesis?
b. What is the sample size?
c. What is the decision rule?
d. What is your decision?

Pair	Calorie Counters	Fitness Fanatics	Sign
A	12	4	+
B	8	17	–
C	26	5	+
D	19	6	+
E	7	12	–
F	2	24	–
G	15	18	–
H	23	4	+
I	5	15	–
J	8	21	–
K	3	25	–
L	18	14	+
M	4	15	–
N	17	8	+
O	5	21	–

Exercises

Answers to the even-numbered exercises are at the back of the book.

1. A random sample of eight college students and their parent of the same sex were administered a questionnaire that yields a score on their attitudes toward premarital sex. Seven of the parents were classified as more conservative than their offspring and one as less conservative. Use the sign test to determine if there is sufficient evidence at the 10% level of significance to claim that the attitudes of the two groups are significantly different.

2. The salaries paid six women and six men in similar positions are paired as shown. Use the sign test to see if there is a significant difference in salary between the males and females at the 5% level of significance.

Pair	Males	Females
1	$33,963	$28,673
2	37,271	28,697
3	26,130	18,180
4	33,016	38,265
5	60,718	57,653
6	17,957	13,214

3. A course was offered to help students improve their reading comprehension. A sample of 12 students was tested before and after this course, and each student's comprehension percentage was recorded:

Student	Before	After
Tom	62	65
Alice	70	68
Greg	73	70
Alan	59	71
Cindy	67	68
Jackie	79	73
Scott	46	52
Janet	80	76
Chris	59	64
Dan	67	72
Sara	69	66
Norm	75	80

Use the sign test and the 0.05 significance level to determine if students' comprehension increased after they took the course.

4. The numbers of emergency patients at two emergency care facilities in the same city are to be compared. The number of cases at each facility is obtained for the first 15 days of November. At the 0.05 significance level, can it be concluded that St. Vincent handles more emergency patients than St. Rose?

Date	St. Rose	St. Vincent
1	15	21
2	18	18
3	19	22
4	16	20
5	19	21
6	24	19
7	16	20
8	19	15
9	18	30
10	16	32
11	12	17
12	18	21
13	23	25
14	17	20
15	32	22

5. A new type of radar is installed at air traffic control towers on the West Coast, and there is concern that controllers may not be as accurate as they were on the East Coast. The number of "near misses" is recorded each day during the first two weeks of October. They are:

Date	East Coast	West Coast
1	4	15
2	8	6
3	3	7
4	0	7
5	2	5
6	6	1
7	8	9
8	5	26
9	3	6
10	1	6
11	5	8
12	3	6
13	5	9
14	9	0

Use the sign test and the 0.05 level of significance to test if there were significantly more "near misses" on the West Coast than the East Coast.

THE SIGN TEST WITH LARGE SAMPLES

As discussed in Chapter 7, the binomial probability distribution closely approximates the normal probability distribution when both $n\pi$ and $n(1-\pi)$ exceed 5. The mean and standard deviation of the binomial distribution are computed by:

$$\text{Mean}: \quad \mu = n\pi$$
$$\text{Standard deviation}: \quad \sigma = \sqrt{n\pi(1-\pi)}$$

These formulas were introduced in Chapter 6.

The test statistic, z, is given by the formula

$$z = \frac{X - n\pi}{\sqrt{n\pi(1-\pi)}} \tag{17-1}$$

where

X is the number of plus signs or minus signs.

π is the probability of a plus sign, which is 0.50 if the null hypothesis is true.

n is the number of paired observations.

z is the test statistic.

Because $\pi = 0.50$, these formulas for the mean and standard deviation reduce to $\mu = 0.50n$ and $\sigma = 0.50\sqrt{n}$. To meet the requirement that $n\pi$ is at least 5, n must be at least 10. Hence, the large sample methods can be employed when the sample size is at least 10. Another consideration remains—the continuity correction factor (described in Chapter 7). Recall that a continuity correction factor of 0.50 is used when a continuous distribution approximates a discrete distribution. In this case, the binomial is approximated by the normal. The test statistic simplifies to

$$z = \frac{(X \pm 0.50) - .50n}{0.50\sqrt{n}} \qquad \text{17-2}$$

where X is the number of plus or minus signs, whichever is being tested or counted. $X + 0.50$ is used if X is less than $n/2$, and $X - 0.50$ is used if X is more than $n/2$. The following illustrative example will help to clarify these points further.

Problem Recall the Chapter Problem about the self-confidence level of widows before and after a training program (see Table 17–1, p. 615). Did the training program increase the widows' self-confidence? As before, use the 0.10 significance level.

Solution We can simplify the calculations if we use the normal approximation to the binomial. Since $n\pi = 12(0.5) = 6.0$ and $n(1 - \pi) = 12(0.5) = 6.0$, the qualifications for using the normal approximation to the binomial are met. The null and alternate hypotheses are as before:

$$H_0: \pi \geq 0.50$$
$$H_a: \pi > 0.50$$

We determine the decision rule by using the normal probability distribution table (Appendix C), with 0.10 in the upper tail of the curve. The decision rule is to reject H_0 if z is greater than 1.28; otherwise, we do not reject H_0.

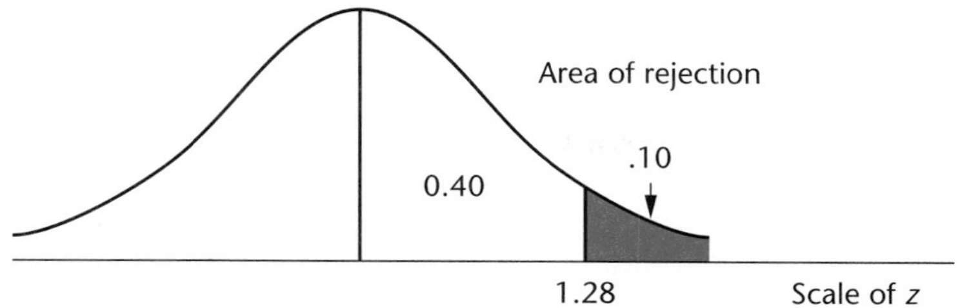

Reviewing the components in this problem, we find:

$X = 10$ the number of widows exhibiting increased self-confidence after the training program.

$n = 12$ the size of the sample.

Substituting these values into formula 17–2,

$$z = \frac{(X \pm 0.50) - 0.50n}{0.50\sqrt{n}}$$

we get:

$$z = \frac{(10 - 0.50) - 0.50(12)}{0.50\sqrt{12}}$$

$$= 2.02$$

Note that 0.50 is subtracted from 10 because 10, the number of "successes," exceeds half the sample size. The steps are shown in the following diagram:

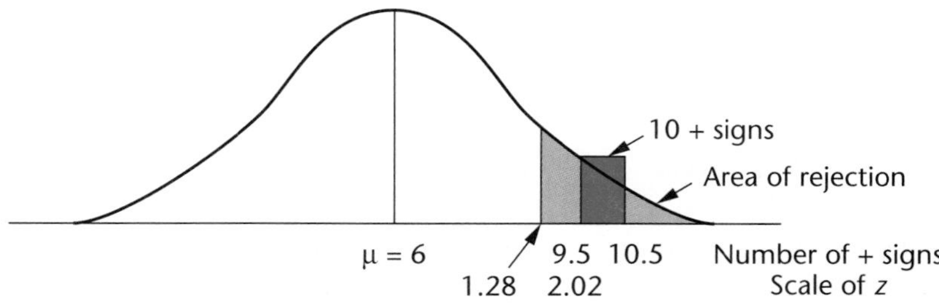

The computed value of 2.02 is in the rejection region beyond 1.28. Therefore, we reject H_0 and accept H_a. Again, we conclude that the training program does enhance the widows' self-confidence.

Self-Review 17-3

Refer to Self-Review 17–2, in which 15 members from each of two health clubs were matched on the basis of similar characteristics. The percent of weight loss for each person was determined. Six plus signs were noted. Use the sign test, the normal approximation to the binomial, and the 0.05 significance level to determine if there is a difference in the effectiveness of the two weight-control programs.

Exercises

6. A gerontologist examined 20 elderly couples and rated both the husband and the wife on a scale designed to measure various degrees of optimism. On this scale, 0 represents absolutely no optimism and 100 extreme optimism. The following data show the paired ratings. Use the sign test to determine if there is a significant difference between husband and wife with respect to optimism. Use the 0.05 significance level.

Husband	70	85	73	75	65	50	80	71	80	51	72	76	79	65	59	72	84	90	56	57
Wife	65	41	45	80	84	50	71	52	42	78	62	38	80	65	54	67	87	90	38	43
Difference	5	44	28	–5	–19	0	9	19	38	–27	10	38	–1	0	5	5	–3	0	18	14

7. Refer to Exercise 1, but assume the sample size to be 28, with 20 of the parents classified as more conservative than their offspring and eight as less conservative. Use the 0.05 significance level and apply the sign test. Can the normal approximation be used? If so, solve the problem using the normal approximation.

8. Refer to Exercise 3. Use the normal approximation to the sign test to solve the problem.

9. Refer to Exercise 4. Use the normal approximation to the sign test to solve the problem.

10. Refer to Exercise 5. Use the normal approximation to the sign test to solve the problem.

THE WILCOXON SIGNED-RANK TEST

The sign test, discussed in the previous section, considers only the sign of the difference between two paired observations. This appears to waste information because it ignores the value of the difference between the two paired observations. The **Wilcoxon signed-rank test,** developed by Frank Wilcoxon in 1945, considers not only the sign of the difference but also the magnitude (amount) of the difference between two paired observations. Thus, it is considered more efficient than the sign test. The Wilcoxon signed-rank test is applied only to paired interval-level data. It is a replacement for the paired t test discussed in Chapter 11. If you suspect that the distribution of differences between the paired observations is not normal, then you should avoid the paired t test and use the Wilcoxon signed-rank test in its place.

To apply the Wilcoxon signed-rank test, follow these steps:

1. Compute the difference between each pair of observations. Whenever the difference between the paired values is zero, discard that pair of observations.
2. Rank these differences from lowest to highest, without regard to their signs. That is, rank the smallest absolute difference a 1, the next larger absolute difference a 2, and so on. If ties occur, average the ranks involved and award each tied observation the average value.
3. Put the ranks with a positive difference in one column and those with a negative difference in another.
4. Determine the sums of the positive (R^+) and negative (R^-) ranks.
5. Compare the smaller of the two sums (R^+ or R^-) with the critical values found in Appendix G. Based on this comparison, make your decision regarding the null hypothesis.

Problem Consider again the earlier problem dealing with the wilderness experience of paired police officers. The before-and-after ratings are repeated in Table 17–5. The numerical differences between each paired rating, the ranks, and the ranks' signs have been added. Use the Wilcoxon signed-rank test to determine if there is a difference in the degree of trust before and after the police officers' wilderness experience.

Table 17-5

Degree of Trust Before and After a Wilderness Experience

	Rating					
Partners	Before Experience	After Experience	**Difference**	**Rank**	**Positive Ranks**	**Negative Ranks**
A	68	32	−36	9		9
B	55	55	0			
C	49	55	+6	1	1	
D	40	75	+35	8	8	
E	20	50	+30	7	7	
F	18	25	+7	2.5	2.5	
G	30	23	−7	2.5		2.5
H	70	49	−21	6		6
I	52	62	+10	5	5	
J	50	41	−9	4		4
					23.5	21.5

Solution The null and alternate hypotheses are:

H_0: The trust ratings are the same after the wilderness experience as before it.

H_a: The trust ratings are not the same after the wilderness experience as before it.

If the null hypothesis is true, then it is to be expected that the total of the positive ranks is about equal to the total of the negative ranks. A large difference in the two sums is an indication that the null hypothesis is false. The Wilcoxon test serves as a basis for determining how large a difference between the positive and negative sums can reasonably be attributed to chance.

For a two-tailed alternative hypothesis, the smaller of R^+ and R^- is compared with the critical value. Appendix G is used to formulate the decision rule. Suppose the 0.05 significance level is to be used; to locate the critical value, find the column headed by 0.05. Next, move down that column to the row corresponding to the sample size. In this case, find the two-tailed probability 0.05 and move down to $n = 9$. (Recall that we started out with 10 pairs, but because one pair—pair B—showed no difference, that pair was omitted, leaving a sample of only nine.) The critical value from Appendix G is 5. The decision rule is to reject H_0 if the smaller of R^+ and R^- is less than or equal to 5. Referring to Table 17–5, we see that the one sum is 23.5, the other 21.5. The smaller of the two (21.5) is greater than 5; hence, the null hypothesis cannot be rejected. Thus, we conclude that the trust ratings are the same after the wilderness experience as before it.

The following Problem illustrates a case where a one-tailed test is appropriate.

Problem A study is to be conducted of the difference in ages between husbands and wives at their first marriages. The question to be explored: Are husbands older than their wives?

Solution Use the five-step hypothesis-testing procedure.

STEP 1. The null hypothesis is that there is no significant difference between the two ages. Put another way, the null hypothesis states that the sum of the positive ranks equals the sum of the negative ranks ($\Sigma R^+ = \Sigma R^-$). It is suspected that in first marriages the husbands are older than their wives; this is the *alternate hypothesis.* Because we are interested in determining if the husbands are *older* than their wives (only one direction), a one-tailed test is indicated.

STEP 2. A 10% level of significance has been chosen.

STEP 3. The Wilcoxon signed-rank test is an appropriate test to determine if there is a significant difference between two paired observations.

Table 17-6

Comparison of Husbands' and Wives' Ages

Couple	Husband's Age	Wife's Age	Difference	Rank	R^+	R^-
A	24	21	3	6	6	
B	25	23	2	4	4	
C	27	19	8	8	8	
D	22	24	−2	4		4
E	20	18	2	4	4	
F	30	25	5	7	7	
G	18	17	1	1.5	1.5	
H	21	22	−1	1.5		1.5
					30.5	5.5

STEP 4. Eight married couples are randomly selected for inclusion in the study. We formulate the decision rule by referring to Appendix G. Go down the *n* column until you locate the number in the sample (8). Next, find the critical one-tail value in the column headed 0.10; it is 8. Hence, the decision rule is as follows: If the negative sum of the negative ranks is 8 or less, reject the null hypothesis and accept the alternate hypothesis. Otherwise, do not reject H_0.

STEP 5. The ages of both husbands and wives are listed in Table 17–6. The age differences are shown in column 4, and are *ranked* in column 5, where we disregarded the sign of the difference. There is a tie for first position: Couple G had a difference of 1, couple H, – 1 (see column 3). When the differences were ranked, we resolved this tie by giving each of the two couples a rank of 1.5, which we found by adding ranks 1 and 2, then dividing by 2. There is also a three-way tie—for third place (couples B, D, and E). We resolved this tie by adding ranks 3, 4, and 5 and dividing by 3. Each of these three couples was assigned a rank of 4.

Note that the sum of the negative ranks is 5.5, less than the critical value of 8. Based on the decision rule in Step 4, the null hypothesis is rejected. It is reasoned that such a large difference between the sum of the positive signed ranks (30.5) and the sum of the negative signed ranks (5.5) *could not be due to chance alone.* The alternate hypothesis—that in first marriages husbands are older than their wives—is accepted. We conclude that, in first marriages, a preponderance of husbands are older than their wives.

Self-Review 17-4

The burglary rates for 1985 and last year for a sample of 10 major cities are shown in the following tabulation:

1985	Last Year
10.1	20.4
10.6	22.1
8.2	10.2
4.9	9.8
11.5	13.7
17.3	24.7
12.4	15.4
11.1	12.7
8.6	13.3
10.0	18.4

Is there sufficient evidence, at the 0.05 level of significance, to show that the rates have increased between 1985 and last year?

a. State the null and alternative hypotheses.
b. State the decision rule.
c. State your conclusion and interpret.

Exercises

11. Ten sets of identical twins were separated at the age of six. One twin in each set was educated at a private school while the other twin attended a public school. At the end of eight years, each of the twins' academic achievement was measured. Is there sufficient evidence, at the 0.05 level of significance, to claim that private schools generally produce a higher level of educational achievement?

Twin Pair	Public-School Education	Private-School Education
A	12	19
B	14	13
C	8	6
D	11	24
E	14	12
F	12	16
G	13	10
H	15	18
I	18	24
J	17	22

a. State the null and alternate hypotheses.
b. Using the sign test, state the decision rule.
c. What conclusion would you reach using the sign test?
d. If the Wilcoxon signed-rank test were used, what would the decision rule be?
e. What conclusion would be reached from the Wilcoxon signed-rank test?

12. Apply the Wilcoxon signed-rank test to the data on weight reduction presented in Self-Review 17–2. Use the 0.05 significance level.

13. A random sample of seven young professionals and their parents was asked the size of their homes (expressed in square feet of usable space).

City	Young Professional	Parent
Atlanta	1,725	1,275
Baltimore	1,310	1,130
Chicago	1,647	1,476
Dallas/Ft. Worth	1,672	1,762
New York	1,520	1,250
San Francisco	1,886	1,688
Washington, D.C.	1,672	1,762

Use the Wilcoxon signed-rank test to determine if there is statistical evidence at the 10% level to claim that "yuppies" live in larger homes than do their parents.

14. Refer to Exercise 4. Use the signed-rank test to determine if St. Vincent Hospital handles more emergency patients than St. Rose.

THE WILCOXON RANK-SUM TEST

One test specifically designed to determine whether two independent samples come from the same population is the **Wilcoxon rank-sum test.** It is an alternative to the two-sample *t* test described in Chapter 11. Recall that the *t* test required that the two populations be normally distributed and have equal variances. These conditions need not be met in order for the Wilcoxon rank-sum test to apply.

The test is based on the sum of the ranks. The data from the two samples are ranked as if they were part of a single sample. If the null hypothesis is true, then the ranks will be nearly evenly distributed between the two samples. That is, the low, medium, and high ranks should be about equally divided between the two samples. If the alternate hypothesis is true, one of the samples will have more of the lower ranks and thus a smaller total. The other sample will have the higher ranks and a larger total. If both samples are eight or larger, then the normal distribution may be used as the test statistic. The formula is

$$z = \frac{W - n_1 \dfrac{(n_1 + n_2 + 1)}{2}}{\sqrt{\dfrac{n_1 \cdot n_2 (n_1 + n_2 + 1)}{12}}} \qquad \textbf{17-3}$$

where

n_1 is the number of samples from the first population.
n_2 is the number of samples from the second population.
W is the sum of ranks of the first population.
z is the standard normal deviate.

Problem An insurance company separates its hospital claims into two categories: short-term (less than 3 days) and long-term (3 days or more). Analysts have suggested that, among the long-term patients, male patients are confined longer than females. To investigate the validity of this suggestion, the insurance company randomly selects nine men and eight women from its files and records the length of their hospital confinement. That information is presented in Table 17–7. Use the 0.05 significance level.

Solution If long-term confinement is the same for men and women, then one can expect the totals of the ranks to be about the same.

Note that there are two men and two women with confinements of 9 and 11 days. As before, when there is a tie the ranks involved are averaged. That is, for the cases involving nine days' confinement, the ranks involved are 2 and 3. They are averaged and a rank of 2.5 is assigned to each of the two cases. The confinements of 11 days are handled similarly (see Table 17–8).

The insurance company believes that men are confined longer. Therefore, a one-tailed test will be employed, with the region of rejection in the upper tail. The two hypotheses are as follows:

H_0: The length of hospital confinement among extended-care patients is the same for men and women.
H_a: The length of confinement is longer for men.

The test statistic follows the standard normal distribution. Using the 0.05 significance level, we find that from Appendix C the criti-

Table 17-7

Length of Hospital Confinement (in days) for a Sample of Men and Women

Men	Women
13	11
15	14
9	10
18	8
11	16
20	9
24	17
22	21
25	

Table 17-8

Ranked Length of Hospital Confinement

Men		Women	
Length	Rank	Length	Rank
13	7	11	5.5
15	9	14	8
9	2.5	10	4
18	12	8	1
11	5.5	16	10
20	13	9	2.5
24	16	17	11
22	15	21	14
25	17		
	97		56

cal value of z is 1.65. Therefore, the null hypothesis will be rejected if the computed value of z is greater than 1.65.

The alternate hypothesis concerns the claim that men are confined to the hospital longer. Therefore, the value of W is calculated for that group. W is 97, which we find by totaling the ranks of the men in the combined sample.

Computing z from formula 17–3 gives:

$$z = \frac{W - \frac{n_1(n_1 + n_2 + 1)}{2}}{\sqrt{\frac{n_1 \cdot n_2(n_1 + n_2 + 1)}{12}}}$$

$$z = \frac{97 - \frac{9(9+8+1)}{2}}{\sqrt{\frac{9 \cdot 8(9+8+1)}{12}}} = 1.54$$

Because the value of z is less than 1.65, the null hypothesis is not rejected. The evidence is not sufficient to show that, among long-term patients, men are confined longer than women.

Note that, in using the Wilcoxon rank-sum test, you may number the two populations in either order. However, once you have made a choice, W *must* be the sum of the ranks from the population identified as number 1. If, in this example, the population of women had been identified as population number 1, then the test statistic would be:

$$z = \frac{56 - \frac{8(8+9+1)}{2}}{\sqrt{\frac{(8)(9)(8+9+1)}{12}}}$$

$$= \frac{56 - 72}{10.4} = -1.54$$

This is the same z-value you found earlier, except that the sign has changed because the populations were renumbered.

Self-Review 17-5

The research director for a golf ball manufacturer wants to know if there is any difference in the distance two of the company's golf balls travel. Eight of their Dino brand and seven of their Maxi brand were hit by an automatic driving machine. The results (distances in yards traveled) were as follows:

Dino: 252, 263, 279, 273, 271, 265, 257, 280
Maxi: 262, 242, 256, 260, 258, 243, 239, 265

Using the 0.05 level and the Wilcoxon rank-sum test,

a. State the null and alternate hypotheses.
b. State the decision rule.

c. Compute z.
d. State and interpret your decision.

Exercises

15. A study dealing with the social mobility of women gathered data on two groups of women, one group composed of women who had never married, the second group consisting of married women. Each group was rated on an upward-mobility scale from 0 to 100, with a higher score denoting greater mobility. At the 0.10 level of significance, is there a difference between these two groups with respect to social mobility?

Never married: 42, 49, 53, 54, 55, 56, 58, 60, 65, 68, 70, 75
Married: 30, 35, 42, 45, 55, 60, 60, 70

16. Samples of eight socially active students and 10 socially nonactive students were randomly selected to determine if there was a difference between the two groups in a measure of social prestige. Each student was evaluated on a 15-point scale in which a low score denotes lack of prestige. The findings are recorded below. Test the null hypothesis of no difference between the two populations using a 0.10 level of significance.

Active students: 11, 13, 14, 15, 8, 10, 5, 6
Nonactive students: 8, 9, 10, 16, 17, 1, 7, 2, 5, 4

17. A supervisor in the Hamtramck, Michigan, auto plant examined the quality of the work of several male and female employees, with the following results (higher scores represent better quality):

Males: 68, 72, 78, 83, 84, 85, 86, 90, 96, 99
Females: 71, 74, 79, 80, 82, 85, 86, 88

Is there any difference between these two age groups? Use the 0.10 level of significance.

18. The time until a light bulb failed for each of two brands is obtained (in hours). At the 0.05 significance level, can we conclude there is a difference in the two bulbs?

Longlast: 372, 283, 712, 849, 623, 382, 427, 821
Everlast: 645, 682, 913, 742, 691, 689, 842, 751

19. *Consumer Reports* magazine wants to compare the fuel efficiency of two similar makes of golf carts. Each is driven on a random basis over a 17-day period and the mileages are recorded:

Day	Make	Mileage
1	Go-Go	24.1
2	Go-Go	25.0
3	Putt-Putt	24.8
4	Putt-Putt	24.3
5	Go-Go	24.2
6	Putt-Putt	25.3
7	Putt-Putt	24.2
8	Putt-Putt	23.6
9	Go-Go	24.5
10	Go-Go	24.4
11	Go-Go	24.5
12	Putt-Putt	23.2
13	Putt-Putt	24.0
14	Putt-Putt	23.8
15	Go-Go	23.8
16	Go-Go	23.0
17	Go-Go	22.9

Using the 0.05 level of significance and the Wilcoxon rank-sum test, can we conclude that there is a difference in the two makes?

20. Heights (in inches) of children of comparable ages from developed and underdeveloped countries are shown below. Is there a significant difference in the two distributions? Use the rank-sum test with a significance level of 0.05.

Developed: 30.2, 34.6, 37.8, 40.8, 43.4, 45.9, 39.2, 40.4, 40.0
Undeveloped: 29.4, 33.8, 37.5, 40.7, 43.4, 31.2, 32.1, 33.4

STAT BYTE

A quality control inspector at MJP Electronics was told to select three items out of a box of 30, test them, and accept the box if all three were satisfactory. He grabbed three items off the top, found them to be acceptable, and sent the box on. Much to his dismay, the person at the first workstation found that the box contained 25 defective items. How could the quality control inspector and the employee at the first workstation reach such different conclusions? Is it really possible that 25 out of 30 items in the box were defective? If so, how could this happen?

ANALYSIS OF VARIANCE BY RANKS: THE KRUSKAL–WALLIS TEST

As noted at the beginning of Chapter 12, if we want to apply the ANOVA method to test if the means of several populations are equal, it is assumed that (1) the populations are normally distributed, (2) the standard deviations of the populations are equal, and (3) the populations are independent. If these assumptions cannot be met—or if the level of measurement is of ordinal scale (ranked)—then we can use an analysis of variance technique developed in

1952 by W. H. Kruskal and W. A. Wallis. It is referred to as the **Kruskal–Wallis one-way analysis of variance by ranks.**

The procedure for the Kruskal–Wallis test is to substitute the *rankings* of the items in each sample for the actual values. Also, instead of working with the means of each treatment—as in ANOVA—we use the sum of the ranks for each treatment. The test statistic, designated as H, is

$$H = \frac{12}{n(n+1)}\left(\frac{S_1^2}{n_1} + \frac{S_2^2}{n_2} + \cdots + \frac{S_k^2}{n_k}\right) - 3(n+1) \qquad \text{17-4}$$

where

S_1 is the sum of the ranks for the sample designated 1, S_2 is the sum of the ranks for the sample designated 2, and so on.

n_1 is the number in the sample designated 1, n_2 is the number in the sample designated 2, and so on.

n is the combined number of observations for all samples.

k is the number of populations.

In practice, this distribution is approximated by the chi-square distribution with $k - 1$ degrees of freedom, where k is the number of populations. This approximation is accurate whenever each sample consists of at least five observations. Should any one of the samples include fewer than five observations, a special table of critical values for the analysis of variance by ranks (Kruskal–Wallis test) will have to be consulted. Many advanced texts contain such tables; for our purposes, an analysis of samples greater than or equal to five will suffice.

Problem Five persons from each of three different ethnic groups were randomly selected and their perceptions of their own social mobility were recorded. The groups are identified as Black, Asian, and Hispanic. Each person's response was recorded on a scale on which 100 indicates extreme mobility and 0 indicates no mobility. The actual responses by group were:

Black: 73, 51, 55, 64, 71
Asian: 63, 74, 53, 92, 84
Hispanic: 81, 93, 84, 91, 84

The hypothesis to be tested is whether the perception scores of the three groups are the same. The scores for each group are presented in Table 17–9.

Table 17-9

Ranking of Perceived Mobility

Black (S_1)		**Asian (S_2)**		**Hispanic (S_3)**	
Score	Rank	Score	Rank	Score	Rank
73	7	63	4	81	9
51	1	74	8	93	15
55	3	53	2	84	11
64	5	92	14	91	13
71	6	84	11	84	11
	22		39		59

Solution The responses were ranked without regard to ethnic background, and the sum of each of the three ranks was determined (see Table 17–9). Note that there was a tie for the eleventh rank—three responses of 84. To break the tie and still have the entire sample ordered from 1 through 15, we assigned the averaged rank [(10 + 11 + 12) ÷ 3 = 11] to each of the three tied values.

The sums of the ranks for the samples are $S_1 = 22$, $S_2 = 39$, and $S_3 = 59$. These values are inserted into formula 17–4 and the value of H is computed:

$$H = \frac{12}{n(n+1)}\left(\frac{S_1^2}{n_1} + \frac{S_2^2}{n_2} + \frac{S_3^2}{n_3}\right) - 3(n+1)$$

$$= \frac{12}{15(15+1)}\left(\frac{(22)^2}{5} + \frac{(39)^2}{5} + \frac{(59)^2}{5}\right) - 3(15+1)$$

$$= 54.86 - 48$$

$$= 6.86$$

The critical value of H is located in Appendix F. Recall that there are $k - 1$ degrees of freedom, where k is the number of populations. There are three populations (Black, Asian, and Hispanic), so $k - 1 = 3 - 1 = 2$ degrees of freedom. The critical value for 2 degrees of freedom at the 0.05 significance level, as found in Appendix F, is 5.991. We conclude that at least one population perceives its social mobility in a manner significantly different from that of the others.

The Kruskal–Wallis test is available on several computer software programs. One computer output follows. The first column, labeled "Score," is the mobility ranking for the 15 subjects. The second column, labeled "Group," identifies the minority group. The Blacks are labeled "1," the Asians "2," and the Hispanics "3." The computed value of H is 6.86, the same value as determined above.

```
MTB >  print 'Score' 'Group'

Score      Group

   73          1
   51          1
   55          1
   64          1
   71          1
   63          2
   74          2
   53          2
   92          2
   84          2
   81          3
   93          3
   84          3
   91          3
   84          3

MTB >  Kruskal 'Score' 'Group'

LEVEL     NOBS     MEDIAN     AVE. RANK     z VALUE
    1        5      64.00           4.4       -2.20
    2        5      74.00           7.8       -0.12
    3        5      84.00          11.8        2.33
OVERALL     15                      8.0

H = 6.860
H(ADJ. FOR TIES) = 6.909
```

Self-Review 17-6

In a study of the effect of packaging on sales of fresh produce, 24 supermarkets in a nearby metropolitan area were each assigned randomly to one of three groups. The first group of stores sells its produce in bulk; customers select their own produce and place it in paper bags. The second group of stores places produce in a mesh bag, visible to the customer. The third group's produce is prepackaged in kraft bags, through which the customer cannot see it. The amount of produce (in pounds) sold in one week by each group is shown in the table that follows:

Produce Sold Each Week (in pounds)		
In Kraft Bags	In Mesh Bags	In Paper Bags
576	464	272
640	480	356
752	720	224
784	756	335
368	208	279
596	512	304
608	288	192
448	496	
	672	

Use the Kruskal–Wallis test and the 0.05 significance level to determine if the different packaging methods result in the same weekly produce sales.

Exercises

21. The Orange County Sheriff's Department is planning to purchase a large number of patrol cars. Four different models are being considered. Samples of all four makes were driven 10,000 miles each, and separate records were kept of the expenditures for repairs. At the 0.05 significance level, is there a difference in the amounts spent on repairs for the four models?

Model A	Model B	Model C	Model D
\$220	\$120	\$180	\$168
290	250	215	247
256	173	231	241
273	235	200	210
265	179	190	276
	249	258	193
	236		273
	147		

22. An insurance company is preparing a new advertising campaign aimed at professionals and wants to know the amount of term insurance carried by lawyers, physicians, veterinarians, and college professors. The amount of term insurance (in thousands of dollars) carried for a sample taken from each of these four categories is shown in the following table. Is there a difference in the amount of insurance carried by the four groups? Use the 0.01 significance level.

Amount of Insurance Carried (in thousands of dollars)

By Lawyers	By Physicians	By Veterinarians	By College Professors
$200	$120	$ 20	$ 30
120	190	105	25
75	200	90	120
65	195	70	190
140	130	200	45
135		125	
		175	

23. In an attempt to justify budget requests for more officers in some precincts, a police commissioner collected the following data on the number of crimes committed weekly in several randomly selected weeks and precincts:

Precinct	Number of Crimes
Central	73, 68, 64, 65, 80
North	42, 39, 51, 38, 39, 40
South	49, 60, 52, 51, 54

At the 0.01 significance level, is there a significant difference in the number of crimes in the three precincts?

24. A drug company tests two drugs by administering them, along with a placebo, to three different treatment groups. The data below are reaction times in minutes.

Drug	Time
Drug A	21, 25, 24, 28, 32
Drug B	19, 22, 21, 25, 23
Placebo	28, 28, 26, 35, 30

Is there a significant difference at the 0.10 level of significance?

25. To test four types of baseball bats, a professional player with the Florida Marlins used each for one week in rotation during

20 weeks of play. He recorded his batting average each week with each bat:

Bat	Average				
A	.251	.262	.237	.228	.242
B	.232	.255	.282	.248	.256
C	.230	.225	.209	.216	.236
D	.259	.222	.277	.250	.238

At the 0.01 significance level, is there a difference among the bats?

SUMMARY

This chapter presented four of the many nonparametric tests—namely, (1) the sign test, (2) the Wilcoxon signed-rank test, (3) the Wilcoxon rank-sum test, and (4) the Kruskal–Wallis analysis of variance by ranks. These four tests illustrate the general nature of "distribution-free" hypothesis tests. With nonparametric tests, we need not assume that the populations in question are normally distributed. Usually, such tests are also easier to apply than the corresponding parametric tests and have wider applications because they require only ordinal-level measurements.

The sign test employs the sign of the difference of each paired observation. The sign of the difference is either positive (+) or negative (–). If there is no difference, the pair is dropped from the study. The sign test actually builds on the binomial distribution because there are only two possible outcomes.

The Wilcoxon signed-rank test, also based on dependent samples, considers not only the sign, but also the magnitude of the differences. The test is based on the sum of the ranks.

To test two or more unrelated or independent samples, the Wilcoxon rank-sum test and the Kruskal-Wallis analysis of variance by ranks can be used. Both of these tests rank all the data together as if the observations had come from a common population. To perform the actual test, we must compute the sum of the ranks for a particular group. The Wilcoxon rank-sum test can be applied to only two populations. When the null hypothesis is true, its test statistic is approximately normal. The Kruskal-Wallis analysis of variance by ranks, on the other hand, is applied to two or more independent populations. The distribution of the test statistic for the null hypothesis approximates that of chi-square.

Exercises

26. The work-methods department of an electronics manufacturer has suggested that soft background music might increase productivity. A stereo system has been set up, and the productivity of a random sample of eight workers is checked against their productivity before music was piped in. Can we conclude that the workers are more productive with the background music on? Use the 0.05 significance level and the sign test.

Worker	With Music	Without Music
Smith	34	32
Clark	34	37
Craig	41	28
Berry	34	33
Roberts	31	28
Nardelli	36	33
Palmer	29	27
Tang	39	38

27. A study to determine the effect of nursing intervention on behavior modification of patients over a two-month period produced the following test results. Use a sign test to determine whether nursing intervention produced a significant change in behavior, as indicated by an increased score. Use the 0.05 level of significance.

Subject	Initial Score	Final Score
1	31	32
2	40	39
3	41	42
4	118	134
5	57	73
6	12	22
7	120	132
8	44	60
9	47	29
10	53	72

28. Refer to Exercise 27. Examine the effect of nursing intervention using the normal approximation to the sign test. Use the 0.05 significance level.

29. Apply the Wilcoxon signed-rank test to the data regarding the effect of nursing intervention. Use the 0.05 significance level.

30. In an attempt to judge the effectiveness of an antismoking campaign, researchers exposed 12 smokers to a series of presentations. The number of cigarettes the subjects smoked the week before the presentations and the week after the presentations were obtained as follows:

	Subject											
	1	2	3	4	5	6	7	8	9	10	11	12
Before	23	23	21	24	21	27	22	19	19	23	26	27
After	23	12	21	15	19	15	17	21	18	14	18	19

Use the sign test to determine whether the antismoking campaign was effective at the 0.05 level of significance.

31. Refer to Exercise 30. Examine the antismoking campaign using the normal approximation to the sign test. Use the 0.05 significance level.

32. Apply the Wilcoxon signed-rank test to the data in Exercise 30 regarding the antismoking campaign. Use the 0.05 significance level.

33. Two major league baseball fans were comparing the relative strengths of the National and American Leagues. One contended that the National League home-run leader had more home runs than the American League leader. A 17-year record of home runs hit by the leader in each of the two leagues follows. Use the sign test to determine whether the difference found is significant at the 0.05 level.

American	National	American	National
32	38	44	39
36	38	49	44
32	36	32	52
32	44	49	47
37	40	45	44
33	48	48	49
44	45	61	46
49	45	40	41
44	36		

34. The number of traffic citations given by two police officers is to be compared. The following is a 25-day sample of citations. At the 0.01 level of significance, can it be concluded that officer A gives more citations than officer B? Use the sign test.

Date	A	B	Date	A	B	Date	A	B	Date	A	B	Date	A	B
5–1	10	7	5–8	7	3	5–15	11	8	5–22	6	5	5–29	10	8
5–2	11	5	5–9	4	2	5–16	12	10	5–23	7	9	5–30	9	7
5–3	7	4	5–10	5	8	5–17	11	7	5–24	9	3	5–31	8	4
5–4	8	8	5–11	8	1	5–18	7	2	5–25	12	8	6–1	5	2
5–5	8	9	5–12	7	2	5–19	5	6	5–26	10	7	6–2	6	7

35. A study conducted by the National Weather Service on the change in temperature following a thunderstorm found that after 40 storms the temperature increased on five occasions, stayed the same on 10 occasions, and decreased on 25 occasions. Is this sufficient evidence, at the 0.05 significance level, to conclude that there is a significant decrease in temperature after a thunderstorm?

36. Refer to Exercise 26. Use the Wilcoxon signed-rank test to determine if workers are more productive with background music.

37. Refer to Exercise 33. Use the Wilcoxon signed-rank test to determine if there is a difference in the number of home runs hit by the leaders in the American and National Leagues.

38. Refer to Exercise 34. Use the Wilcoxon signed-rank test to determine if police officer A gives more citations.

39. Seventeen randomly selected married men with incomes greater than $100,000 are asked whether they consider their marriages to be satisfactory. Each man is also asked to report the length of his marriage in years. The results of the study follow. At the 0.01 significance level, use the Wilcoxon rank-sum test to determine if there is a difference in years between the satisfied and the dissatisfied groups.

Length of Marriage	
Satisfied	Not Satisfied
15	8
3	5
9	9
12	12
23	10
29	13
14	4
6	7
16	

40. Two judges normally preside over the family court where all divorce hearings are held. Ms. Kerger, a young lawyer new to the area, represents the husband in an upcoming case. She gathers the following data on the monthly alimony payments (in dollars per month) recently awarded by the two judges. At the 0.01 significance level, use the Wilcoxon rank-sum test to determine if there is a difference in the alimony settlements awarded by the two judges. The samples for each group have been arranged from low to high.

Judge Cain	Judge Stevens
$ 80	$120
160	140
220	192
520	204
580	252
640	440
840	500
920	560
1,200	
1,360	
2,000	

41. The owner of G-Mart, a large retailer, is concerned about the effectiveness of the store's advertising. G-Mart advertises three different special articles in three different media and observes sales over the next few weeks. The data are from five or six randomly selected weeks and are not necessarily the same weekly sales for all media. Here are the results. Is there a significant difference among the media at the 0.01 level? (Sales are in thousands of dollars.)

Yellow Pages	Television	Newspaper
$14	$46	$39
21	49	48
36	59	51
12	52	50
18	50	49
	51	

42. A study has been made of the murder rates (per 1,000 population) for a sample of medium-sized cities in the Northeast, South, and Far West. Use the 0.05 significance level to determine if there is a difference in the murder rates of different geographical areas. The sample for each region has been arranged from low to high.

Northeast	South	Far West
2.3	1.6	1.7
4.5	1.9	3.7
6.7	3.0	3.8
7.8	6.5	4.3

Northeast	South	Far West
9.5	7.2	5.9
12.7	11.6	6.2
13.1	13.1	7.9
	14.5	8.4
	15.1	

Data Exercises

43. Refer to Real Estate Data, which reports information on homes sold in Northwest Ohio during 1992.

a. Consider only the homes with an attached garage. Use a nonparametric test to determine if there is a difference in the distributions of the selling prices of the homes with a fireplace and those without. Use the 0.05 significance level.

b. Consider only the homes without a fireplace. Use a nonparametric test to determine if there is a difference in the distributions of the selling prices of the homes with a garage and those without. Use the 0.05 significance level.

44. Refer to Salary Data, which refers to a sample of middle managers in Sarasota, Florida.

a. Use an appropriate nonparametric test to determine if there is a difference in the distributions of salaries in the four regions. Use the 0.05 significance level.

b. Use an appropriate nonparametric test to determine if there is a difference in the distributions of salaries for the various types of managers. Use the 0.05 significance level.

c. Consider only the men. They are coded 1 in the survey. Use an appropriate nonparametric test to determine if there is a difference in the distributions of their salaries by region. Use the 0.05 significance level.

DO ATHLETES NEED HELP?

CASE

The number of classes athletes missed before and after they were selected for a sports travel team are shown on page 648. Some people claim that the number of absences increases in such a situation and that the athletes need additional academic help, such as tutoring. Others claim that there is really no change in the absence pattern. What do you think?

Student	Absences Before	Absences After
Balbier	2	12
Bernhoffer	8	8
Comes	9	10
Davis	15	12
Glowacki	10	10
Haut	8	14
Knowltin	19	18
Langerman	3	2
Lewis	4	10
Mikolas	16	14
Murawa	8	12
Parker	17	20
Pier	8	17
Proshek	3	7
Ritchart	3	14

CHAPTER ACHIEVEMENT TEST

Answer all the questions. The answers are at the back of the book.

MULTIPLE-CHOICE QUESTIONS. Select the response that best answers each of the questions.

1. Which of the following is *not* a nonparametric test?
   a. Sign test
   b. *t* test
   c. Wilcoxon signed-rank test
   d. Wilcoxon rank-sum test
   e. All are nonparametric tests.
2. The sign test should be employed instead of the paired *t* test when
   a. the sample is greater than 30.
   b. the population from which the sample is drawn is known to be nonnormal.
   c. both $n\pi$ and $n(1 - \pi)$ are less than 5.
   d. All of the above
   e. None of the above
3. When the sign test is employed, the null hypothesis is that the probability of a positive difference in a paired observation is 0.5 and the sampling distribution for testing follows a binomial distribution.

a. True
b. False

4. In comparing the sign test and the Wilcoxon signed-rank test, we find that the Wilcoxon signed-rank test
   a. considers the magnitude of the difference between paired observations.
   b. considers only the sign of the difference.
   c. is actually the same as the sign test.
   d. assumes a normal distribution.
   e. must always have 30 or more observations.

5. In comparing the sign test and the Wilcoxon signed-rank test to the Wilcoxon ranked-sum test, we find that the Wilcoxon ranked-sum test assumes
   a. a normal distribution.
   b. a positive difference.
   c. dependent samples.
   d. independent samples.

6. If we are using the Wilcoxon signed-rank test on a set of data with a sample size of 15 and we want a two-tailed test with a 0.01 significance level, then the null hypothesis will be rejected if
   a. R^+ is greater than 19.
   b. R^- is greater than 19.
   c. the smaller of R^+ and R^- is less than or equal to 15.
   d. the smaller of R^+ and R^- is less than or equal to 25.

7. In the Wilcoxon signed-rank test, if the absolute differences between paired observations are equal (tied) and greater than zero, the suggested procedure is to
   a. discard the tied observations.
   b. use the smaller value of the corresponding ranks.
   c. use the larger of the corresponding ranks.
   d. assign the average value of the corresponding ranks.
   e. None of the above

8. In the Wilcoxon rank-sum test, the differences between pairs of scores are ranked.
   a. True
   b. False

9. A principal advantage of the nonparametric tests is that the underlying assumptions are often less restrictive.
   a. True
   b. False

10. The region of rejection for the Wilcoxon rank-sum test can be in
    a. the upper tail only.
    b. the lower tail only.
    c. the lower tail or the upper tail or both tails.
    d. None of the above

COMPUTATION PROBLEMS

11. A medical researcher wishes to determine if the lung capacity of nonsmokers is greater than that of persons who smoke. A random sample of 15 matched pairs has been selected and tested. The lung capacity of the nonsmokers has been subtracted from that of the smoking partners. A total of 13 positive differences was obtained (two were negative). Is this sufficient evidence to show that the lung capacity of nonsmokers is greater? Use a 0.01 level of significance and the sign test.

12. Several married couples were asked to estimate the age at which their partner achieved emotional maturity. Use the Wilcoxon signed-rank test at the 0.01 level of significance to test for a difference between the perceptions of husbands and wives.

Couple	Husbands	Wives
A	43	31
B	28	32
C	36	34
D	39	39
E	43	35
F	28	22
G	31	28
H	23	24
I	29	34
J	39	39
K	26	19
L	34	29
M	35	35

13. A study compared the achievement scores of a group of randomly selected inner-city sixth-graders to the scores of a similar group of students from suburban and rural areas within the same school district. Use a 0.05 significance level and the Kruskal-Wallis test to determine if a difference exists between the two groups. Note that the data have been ordered.

Inner-city
students: 95, 100, 100, 106, 108, 115, 120, 125, 130
Suburban
students: 108, 114, 118, 122, 124, 126, 130, 136, 142, 150
Rural
students: 102, 109, 112, 120, 127, 129, 140

ANSWERS TO SELF-REVIEW PROBLEMS

17-1 a. H_0: There is no change as a result of the planned exercise program.

H_a: The program is successful in reducing weight.

b. Refer to Appendix A for a π of 0.50 and an n of 7. The probability of seven successes is 0.008, of six successes 0.055. Adding gives 0.063, which is as close as possible to 0.10 without exceeding it. Therefore, reject the null hypothesis if there are six or seven pluses.

c. Yes; there are six pluses.

d. The planned exercise program is successful in reducing a person's weight.

17–2 a. H_0: $\pi = 0.50$

H_a: $\pi \neq 0.50$

b. $n = 15$

c. The probability of three or fewer plus signs is 0.017. The probability of 12 or more plus signs is also 0.017; hence, the probability of 3 or fewer or 12 or more plus signs is 0.034. H_0 is rejected if the number of plus signs is not between 4 and 11.

d. Since there are 6 plus signs, H_0 cannot be rejected.

17-3 Since $n\pi = 15(0.5) = 7.5$ and $n(1 - \pi) = 15(0.5) = 7.5$, the normal approximation may be used.

$$H_0: \pi = 0.50$$
$$H_a: \pi \neq 0.50$$

The null hypothesis is rejected if z is less than −1.96 or greater than 1.96. Six plus signs were noted.

$$z = \frac{6 + 0.5 - 0.5(15)}{0.5\sqrt{15}} = -\,0.52$$

H_0 is not rejected. There is no difference in the two groups. Results are the same as those we obtained earlier using the sign test.

17-4 a. H_0: The sum of the positive ranks (R^+) equals the sum of the negative ranks (R^-).

H_a: R^- is less than R^+.

b. Reject the null hypothesis if R^- is 10 or less.

c. $R^- = 0$; hence, we reject the null hypothesis. There has been a significant increase in the burglary rates during the period.

17-5 a. H_0: There is no difference in the distances traveled by Dino and by Maxi.

H_a: There is a difference in the distances traveled by Dino and by Maxi.

b. Do not reject H_0 if the computed z is between 1.96 and −1.96 (from Appendix C); otherwise, reject H_0 and accept H_a.

c. $n_1 = 8$, the number of observations in the first sample

Dino		Maxi	
Distance	Rank	Distance	Rank
252	4	262	9
263	10	242	2
279	15	256	5
273	14	260	8
271	13	258	7
265	11.5	243	3
257	6	239	1
280	16	265	11.5
Total	89.5		46.5

$W = 89.5$

$$z = \frac{89.5 - \dfrac{8(8+8+1)}{2}}{\sqrt{\dfrac{(8)(8)(8+8+1)}{12}}}$$
$$= \frac{21.5}{9.52} = 2.26$$

d. Reject H_0; accept H_a. There is a difference in the distances traveled by the two golf balls.

17-6 H_0: The sale of produce is the same.
H_a: The sale of produce is not the same.

Rankings of Produce Sold Weekly

Kraft Rank	Mesh Rank	Paper Rank
16	12	4
19	13	9
22	21	3
24	23	8
10	2	5
17	15	7
18	6	1
11	14	
	20	
137	126	37

$$H = \frac{12}{24(24+1)}\left[\frac{(137)^2}{8} + \frac{(126)^2}{9} + \frac{(37)^2}{7}\right] - 3(24+1)$$

$$= 86.11 - 75 = 11.11$$

Since the computed value of H is greater than 5.991, H_0 is rejected and H_a is accepted. The conclusion is that the sale of produce is not the same under the various packaging conditions.

UNIT REVIEW

Chapters 16 and 17 presented some of the basic concepts of hypothesis testing that apply when nominal and ordinal levels of measurement are involved. We discussed several tests, including the chi-square goodness-of-fit test, the sign test, the Wilcoxon signed-rank test, the Wilcoxon rank-sum test, and the Kruskal–Wallis analysis of variance by ranks.

Key Concepts

1. **Nominal level of measurement.** The nominal level of measurement is considered the "lowest" level of measurement. Data on this level can be classified only into categories; there is no particular order for the categories, and the categories are mutually exclusive. Recall that "mutually exclusive" means a respondent can be placed in one and only one category.
2. **Ordinal level of measurement** is the next-higher level of measurement. For ordinal-scaled data, one category is rated higher than the previous one. The categories are also mutually exclusive.
3. **Nonparametric (distribution-free) tests.** Hypothesis tests applied to nominal-level and ordinal-level data are often referred to as nonparametric tests, or distribution-free tests. These tests do not make any assumptions about the distribution of the population from which the sample or samples are selected. That is, to apply these nonparametric tests we need not have a normally distributed population.
4. **Chi-square goodness-of-fit test.** This test requires only nominal-scaled data. It is concerned with a single trait—the religious affiliation of the respondent, for example. The purpose of the test is to find out how well an observed set of data compares to an expected set of data.
5. **Contingency tables.** This application of the chi-square statistic is concerned with the simultaneous classification of two traits. That is, are the two traits related or not? The chi-square statistic compares observed frequencies with expected frequencies.
6. The **chi-square distribution** has the following major characteristics:
   - **a.** Its value is nonnegative.
   - **b.** There is a different chi-square distribution for each number of degrees of freedom.
   - **c.** It is positively skewed, but approaches a symmetrical distribution as the number of degrees of freedom increases.

7. The **sign test** is used to investigate changes in paired or related observations. The sign of the difference may be either positive or negative, and the binomial distribution is used as the test statistic.
8. The **Wilcoxon signed-rank test** is an extension of the sign test. It considers not only the sign of the difference in related observations, but also the magnitude of the differences.
9. The **Wilcoxon rank-sum test** concerns random samples taken from two independent populations. It is an alternative to the *t* test, but the assumption of a normal population is not required. If both samples are at least of eight items, the normal distribution may be used as the test statistic.
10. The **Kruskal–Wallis test** concerns random samples obtained from more than two independent samples. It is a nonparametric alternative to the ANOVA test, but the normality assumption is not required. The chi-square distribution is used as the test statistic.

Key Terms

Nonparametric tests
Distribution-free tests
Chi-square distribution
Observed frequency
Expected frequency
Goodness-of-fit test
Contingency table
Nominal level of data
Ordinal level of data
Sign test
Wilcoxon signed-rank test
Wilcoxon rank-sum test
Kruskal–Wallis one-way analysis of variance by ranks

Key Symbols

f_e Frequency expected.
f_o Frequency observed.
k Number of categories.
χ^2 Chi-square.
R^+ Sum of the positive signed differences.
R^- Sum of the negative signed differences.
W Test statistic for the Wilcoxon-rank sum test.
H Test statistic for the Kruskal–Wallis test.

Problems

1. A traffic engineer is studying the traffic pattern on a four-lane inbound expressway. For an hour, he counts the number of vehicles using each lane. The results are as follows:

Lane	Frequency
1	100
2	80
3	60
4	60
Total	300

At the 0.05 significance level, test the hypothesis that the traffic uses all four lanes equally.

2. Historically, 25.2% of the used cars purchased in a particular geographical area are subcompacts, 52.9% are mid-sized, and 21.9% are full-sized cars. A sample of 500 sales during the first quarter revealed that 129 of the used cars sold were subcompacts, 258 were mid-sized, and 113 were full-sized. At the 0.05 level, test the hypothesis that there has been no change in the buying habits with respect to used cars.

3. There is interest in finding out if there is a relationship between the political affiliation of registered voters and their reaction to a proposed ban on the buildup of nuclear armaments. The responses to a questionnaire were tabulated and are shown in the following table:

Political Affiliation	Favor Ban	Oppose Ban	No Opinion
Democrat	295	82	73
Republican	191	69	62
Independent	99	17	41
All others	45	12	14

Test the hypothesis, at the 0.05 significance level, that political affiliation is independent of the reaction to a nuclear armament reduction.

4. A sample of 200 cola drinkers is obtained. They are classified by sex and whether they prefer regular or diet cola:

Cola	Male	Female	Total
Regular	45	40	85
Diet	55	60	115
Total	100	100	200

At the 0.01 significance level, can we conclude that there is a difference in the cola preferences of men and women?

5. You want to use a simple "before-and-after" test to find out whether or not the experience of playing softball on a team made up of both boys and girls is successful in reducing sexism

among boys. You test the boys for sexism both before league play begins and at the end of the season. The scores are:

Boy	**Score**	
	Before Season	After Season
Carter	22	20
Jones	30	26
Ford	28	28
Yamamoto	38	30
Cosell	12	11
West	40	41
Archer	37	32
Oreon	32	29
Giles	50	40
Raggi	42	36
Nice	36	29
Cork	33	21
Mueller	36	21

Use the 0.05 significance level to test whether the softball experience has been effective in reducing sexism among boys.

6. An oil company advertises that customers will obtain higher gas mileage using "Super" lead-free gasoline than using their regular lead-free. To investigate this claim, a sample of 14 cars of various makes and models was selected. The 14 cars are driven the same distances over the same roads with both the "Super" lead-free and the regular lead-free gasolines. The results, in miles per gallon, follow:

Car	**Super**	**Regular**
1	21.5	20.9
2	32.5	32.9
3	29.7	30.9
4	35.4	35.7
5	26.2	26.0
6	40.4	41.7
7	25.9	25.3
8	25.1	25.0
9	24.0	23.8
10	18.9	18.5
11	23.4	23.2
12	30.2	28.5
13	26.8	26.2
14	27.6	27.0

At the 0.01 significance level, can it be concluded that the lead-free mileage is greater? Use the sign test.

7. Refer to Problem 5. Use the Wilcoxon signed-rank test to investigate whether the softball experience reduced sexism among boys. Use the 0.05 significance level.

8. Refer to Problem 6. Use the Wilcoxon signed-rank test to investigate whether Super lead-free gasoline yields higher mileage. Use the 0.01 significance level.

9. The amount of money spent for lunches by working men and women is being compared. The amount spent by a sample of 10 men and 12 women on a particular day is tabulated; the results are shown in the table that follows. Can it be concluded that men spend more than women? Use the 0.01 significance level.

Men	Women
$3.73	$3.85
4.40	4.05
4.72	4.10
3.95	4.15
5.10	3.90
3.95	3.65
4.00	3.55
4.35	3.20
4.45	3.05
3.85	3.55
	2.90
	4.62

10. A study of the amount of overpayment to welfare recipients is made in three regions. Is there a difference in the amount of overpayment in the three regions? Use the 0.05 level of significance.

West	Northwest	Midwest
$20	$24	$33
18	27	32
26	26	40
14	29	23
25	30	28
19	31	35
22	21	38
17	28	30

Using Statistics

Situation. In 1969, the Selective Service Administration conducted a lottery to determine the sequence in which young men would be drafted into the armed services. Officials claimed that they selected the birthdates randomly and assigned a sequence number by mixing small capsules containing the dates. Several people objected, however, that the drawing was not random. Here are the results of that experience, where the numbers indicate priority from high (1) to low (366) corresponding to each birthdate. All those men with birthdays on September 14 were drafted first, followed by those with birthdays on April 24, and so on. Do you think this set of numbers is random?

Date	Jan.	Feb.	Mar.	Apr.	May	Jun.	Jul.	Aug.	Sep.	Oct.	Nov.	Dec.
1	305	086	108	032	330	249	093	111	225	359	019	129
2	159	144	029	271	298	228	350	045	161	125	034	328
3	251	297	267	083	040	301	115	261	049	244	348	157
4	215	210	275	081	276	020	279	145	232	202	266	165
5	101	214	293	269	364	028	188	054	082	024	310	056
6	224	347	139	253	155	110	327	114	006	087	076	010
7	306	091	122	147	035	085	050	168	008	234	051	012
8	199	181	213	312	321	366	013	048	184	283	097	105
9	194	338	317	219	197	335	277	106	263	342	080	043
10	325	216	323	218	065	206	284	021	071	220	282	041
11	329	150	136	014	037	134	248	324	158	237	046	039
12	221	068	300	346	133	272	015	142	242	072	066	314
13	318	152	259	124	295	069	042	307	175	138	126	163
14	238	004	354	231	178	356	331	198	001	294	127	026
15	017	089	169	273	130	180	322	102	113	171	131	320
16	121	212	166	148	055	274	120	044	207	254	107	096
17	235	189	033	260	112	073	098	154	255	288	143	304
18	140	292	332	090	278	341	190	141	246	005	146	128
19	058	025	200	336	075	104	227	311	177	241	203	240
20	280	302	239	345	183	360	187	344	063	192	185	135
21	186	363	334	062	250	060	027	291	204	243	156	070
22	337	290	265	316	326	247	153	339	160	117	009	053
23	118	057	256	252	319	109	172	116	119	201	182	162
24	059	236	258	002	031	358	023	036	195	196	230	095
25	052	179	343	351	361	137	067	286	149	176	132	084
26	092	365	170	340	357	022	303	245	018	007	309	173
27	355	205	268	074	296	064	289	352	233	264	047	078
28	077	299	223	262	308	222	088	167	257	094	281	123

Date	Jan.	Feb.	Mar.	Apr.	May	Jun.	Jul.	Aug.	Sep.	Oct.	Nov.	Dec.
29	349	285	362	191	226	353	270	061	151	229	099	016
30	164		217	208	103	209	287	333	315	038	174	003
31	211		030		313		193	011		079		100

Discussion. The most powerful test we can apply to the ranked data is the Kruskal–Wallis test. The null hypothesis is that the sum of the ranks is the same for each of the 12 months. The alternate hypothesis is that the sum of the ranks is not the same. The sums of the monthly ranks are:

Jan.	Feb.	Mar.	Apr.	May	Jun.	Jul.	Aug.	Sep.	Oct.	Nov.	Dec.
6236	5886	7000	6110	6447	5872	5628	5377	4719	5656	4462	3768

We then obtain

$$H = \frac{12}{366 \times 367}\left[\frac{(6236)^2}{31} + \cdots + \frac{(3768)^2}{31}\right] - 3(367)$$
$$= 25.96$$

We find the critical value for the 0.01 significance level from the chi-square table, with 11 degrees of freedom. It is 24.725. Since H exceeds this value, we can reject the null hypothesis of equality of the ranks and conclude that the draft numbers were *not* randomly selected.

An alternate analysis based on the chi-square two-way classification of the data is to group the ranks as either high (184–366) or low (1–183) and aggregate the months into quarters. In this case, the data would appear as:

	Rank	
Quarter	Low	High
First	34	57
Second	39	52
Third	50	42
Fourth	60	32

A MINITAB printout of the chi-squared test follows:

```
MTB >  chisquare c1 c2

Expected counts are printed below observed counts

            C1       C2    Total
    1       34       57       91
         45.50    45.50
    2       39       52       91
         45.50    45.50
    3       50       42       92
         46.00    46.00
    4       60       32       92
         46.00    46.00

Total      183      183      366

ChiSq =  2.907 + 2.907 +
         0.929 + 0.929 +
         0.348 + 0.348 +
         4.261 + 4.261 = 16.888

df = 3
```

The test statistic is 16.888, which *exceeds* the critical value of 11.3. So the null hypothesis that the ranks are independent of the quarter of the year can be rejected. You can again conclude that there is some bias in the draft priority. It may be of interest that two of the authors' birthdates fall in the later part of the year and hence were particularly sensitive to this apparently nonrandom lottery.

APPENDIX TABLES

A—The Binomial Probability Distribution

N= 1
PROBABILITY

X	.05	.1	.2	.3	.4	.5	.6	.7	.8	.9	.95
0	.950	.900	.800	.700	.600	.500	.400	.300	.200	.100	.050
1	.050	.100	.200	.300	.400	.500	.600	.700	.800	.900	.950

N= 2
PROBABILITY

X	.05	.1	.2	.3	.4	.5	.6	.7	.8	.9	.95
0	.903	.810	.640	.490	.360	.250	.160	.090	.040	.010	.003
1	.095	.180	.320	.420	.480	.500	.480	.420	.320	.180	.095
2	.003	.010	.040	.090	.160	.250	.360	.490	.640	.810	.903

N= 3
PROBABILITY

X	.05	.1	.2	.3	.4	.5	.6	.7	.8	.9	.95
0	.857	.729	.512	.343	.216	.125	.064	.027	.008	.001	.000
1	.135	.243	.384	.441	.432	.375	.288	.189	.096	.027	.007
2	.007	.027	.096	.189	.288	.375	.432	.441	.384	.243	.135
3	.000	.001	.008	.027	.064	.125	.216	.343	.512	.729	.857

N= 4
PROBABILITY

X	.05	.1	.2	.3	.4	.5	.6	.7	.8	.9	.95
0	.815	.656	.410	.240	.130	.062	.026	.008	.002	.000	.000
1	.171	.292	.410	.412	.346	.250	.154	.076	.026	.004	.000
2	.014	.049	.154	.265	.346	.375	.346	.265	.154	.049	.014
3	.000	.004	.026	.076	.154	.250	.346	.412	.410	.292	.171
4	.000	.000	.002	.008	.026	.062	.130	.240	.410	.656	.815

N= 5
PROBABILITY

X	.05	.1	.2	.3	.4	.5	.6	.7	.8	.9	.95
0	.774	.590	.328	.168	.078	.031	.010	.002	.000	.000	.000
1	.204	.328	.410	.360	.259	.156	.077	.028	.006	.000	.000
2	.021	.073	.205	.309	.346	.312	.230	.132	.051	.008	.001
3	.001	.008	.051	.132	.230	.312	.346	.309	.205	.073	.021
4	.000	.000	.006	.028	.077	.156	.259	.360	.410	.328	.204
5	.000	.000	.000	.002	.010	.031	.078	.168	.328	.590	.774

N= 6
PROBABILITY

X	.05	.1	.2	.3	.4	.5	.6	.7	.8	.9	.95
0	.735	.531	.262	.118	.047	.016	.004	.001	.000	.000	.000
1	.232	.354	.393	.303	.187	.094	.037	.010	.002	.000	.000
2	.031	.098	.246	.324	.311	.234	.138	.060	.015	.001	.000
3	.002	.015	.082	.185	.276	.312	.276	.185	.082	.015	.002
4	.000	.001	.015	.060	.138	.234	.311	.324	.246	.098	.031
5	.000	.000	.002	.010	.037	.094	.187	.303	.393	.354	.232
6	.000	.000	.000	.001	.004	.016	.047	.118	.262	.531	.735

N= 7
PROBABILITY

X	.05	.1	.2	.3	.4	.5	.6	.7	.8	.9	.95
0	.698	.478	.210	.082	.028	.008	.002	.000	.000	.000	.000
1	.257	.372	.367	.247	.131	.055	.017	.004	.000	.000	.000
2	.041	.124	.275	.318	.261	.164	.077	.025	.004	.000	.000
3	.004	.023	.115	.227	.290	.273	.194	.097	.029	.003	.000
4	.000	.003	.029	.097	.194	.273	.290	.227	.115	.023	.004
5	.000	.000	.004	.025	.077	.164	.261	.318	.275	.124	.041
6	.000	.000	.000	.004	.017	.055	.131	.247	.367	.372	.257
7	.000	.000	.000	.000	.002	.008	.028	.082	.210	.478	.698

N= 8
PROBABILITY

X	.05	.1	.2	.3	.4	.5	.6	.7	.8	.9	.95
0	.663	.430	.168	.058	.017	.004	.001	.000	.000	.000	.000
1	.279	.383	.336	.198	.090	.031	.008	.001	.000	.000	.000
2	.051	.149	.294	.296	.209	.109	.041	.010	.001	.000	.000
3	.005	.033	.147	.254	.279	.219	.124	.047	.009	.000	.000
4	.000	.005	.046	.136	.232	.273	.232	.136	.046	.005	.000
5	.000	.000	.009	.047	.124	.219	.279	.254	.147	.033	.005
6	.000	.000	.001	.010	.041	.109	.209	.296	.294	.149	.051
7	.000	.000	.000	.001	.008	.031	.090	.198	.336	.383	.279
8	.000	.000	.000	.000	.001	.004	.017	.058	.168	.430	.663

N= 9
PROBABILITY

X	.05	.1	.2	.3	.4	.5	.6	.7	.8	.9	.95
0	.630	.387	.134	.040	.010	.002	.000	.000	.000	.000	.000
1	.299	.387	.302	.156	.060	.018	.004	.000	.000	.000	.000
2	.063	.172	.302	.267	.161	.070	.021	.004	.000	.000	.000
3	.008	.045	.176	.267	.251	.164	.074	.021	.003	.000	.000
4	.001	.007	.066	.172	.251	.246	.167	.074	.017	.001	.000
5	.000	.001	.017	.074	.167	.246	.251	.172	.066	.007	.001
6	.000	.000	.003	.021	.074	.164	.251	.267	.176	.045	.008
7	.000	.000	.000	.004	.021	.070	.161	.267	.302	.172	.063
8	.000	.000	.000	.000	.004	.018	.060	.156	.302	.387	.299
9	.000	.000	.000	.000	.000	.002	.010	.040	.134	.387	.630

N= 10
PROBABILITY

X	.05	.1	.2	.3	.4	.5	.6	.7	.8	.9	.95
0	.599	.349	.107	.028	.006	.001	.000	.000	.000	.000	.000
1	.315	.387	.268	.121	.040	.010	.002	.000	.000	.000	.000
2	.075	.194	.302	.233	.121	.044	.011	.001	.000	.000	.000
3	.010	.057	.201	.267	.215	.117	.042	.009	.001	.000	.000
4	.001	.011	.088	.200	.251	.205	.111	.037	.006	.000	.000
5	.000	.001	.026	.103	.201	.246	.201	.103	.026	.001	.000
6	.000	.000	.006	.037	.111	.205	.251	.200	.088	.011	.001
7	.000	.000	.001	.009	.042	.117	.215	.267	.201	.057	.010
8	.000	.000	.000	.001	.011	.044	.121	.233	.302	.194	.075
9	.000	.000	.000	.000	.002	.010	.040	.121	.268	.387	.315
10	.000	.000	.000	.000	.000	.001	.006	.028	.107	.349	.599

N= 11
PROBABILITY

X	.05	.1	.2	.3	.4	.5	.6	.7	.8	.9	.95
0	.569	.314	.086	.020	.004	.000	.000	.000	.000	.000	.000
1	.329	.384	.236	.093	.027	.005	.001	.000	.000	.000	.000
2	.087	.213	.295	.200	.089	.027	.005	.001	.000	.000	.000
3	.014	.071	.221	.257	.177	.081	.023	.004	.000	.000	.000
4	.001	.016	.111	.220	.236	.161	.070	.017	.002	.000	.000
5	.000	.002	.039	.132	.221	.226	.147	.057	.010	.000	.000
6	.000	.000	.010	.057	.147	.226	.221	.132	.039	.002	.000
7	.000	.000	.002	.017	.070	.161	.236	.220	.111	.016	.001
8	.000	.000	.000	.004	.023	.081	.177	.257	.221	.071	.014
9	.000	.000	.000	.001	.005	.027	.089	.200	.295	.213	.087
10	.000	.000	.000	.000	.001	.005	.027	.093	.236	.384	.329
11	.000	.000	.000	.000	.000	.000	.004	.020	.086	.314	.569

N= 12
PROBABILITY

X	.05	.1	.2	.3	.4	.5	.6	.7	.8	.9	.95
0	.540	.282	.069	.014	.002	.000	.000	.000	.000	.000	.000
1	.341	.377	.206	.071	.017	.003	.000	.000	.000	.000	.000
2	.099	.230	.283	.168	.064	.016	.002	.000	.000	.000	.000
3	.017	.085	.236	.240	.142	.054	.012	.001	.000	.000	.000
4	.002	.021	.133	.231	.213	.121	.042	.008	.001	.000	.000
5	.000	.004	.053	.158	.227	.193	.101	.029	.003	.000	.000
6	.000	.000	.016	.079	.177	.226	.177	.079	.016	.000	.000
7	.000	.000	.003	.029	.101	.193	.227	.158	.053	.004	.000
8	.000	.000	.001	.008	.042	.121	.213	.231	.133	.021	.002
9	.000	.000	.000	.001	.012	.054	.142	.240	.236	.085	.017
10	.000	.000	.000	.000	.002	.016	.064	.168	.283	.230	.099
11	.000	.000	.000	.000	.000	.003	.017	.071	.206	.377	.341
12	.000	.000	.000	.000	.000	.000	.002	.014	.069	.282	.540

N= 13
PROBABILITY

X	.05	.1	.2	.3	.4	.5	.6	.7	.8	.9	.95
0	.513	.254	.055	.010	.001	.000	.000	.000	.000	.000	.000
1	.351	.367	.179	.054	.011	.002	.000	.000	.000	.000	.000
2	.111	.245	.268	.139	.045	.010	.001	.000	.000	.000	.000
3	.021	.100	.246	.218	.111	.035	.006	.001	.000	.000	.000
4	.003	.028	.154	.234	.184	.087	.024	.003	.000	.000	.000
5	.000	.006	.069	.180	.221	.157	.066	.014	.001	.000	.000
6	.000	.001	.023	.103	.197	.209	.131	.044	.006	.000	.000
7	.000	.000	.006	.044	.131	.209	.197	.103	.023	.001	.000
8	.000	.000	.001	.014	.066	.157	.221	.180	.069	.006	.000
9	.000	.000	.000	.003	.024	.087	.184	.234	.154	.028	.003
10	.000	.000	.000	.001	.006	.035	.111	.218	.246	.100	.021
11	.000	.000	.000	.000	.001	.010	.045	.139	.268	.245	.111
12	.000	.000	.000	.000	.000	.002	.011	.054	.179	.367	.351
13	.000	.000	.000	.000	.000	.000	.001	.010	.055	.254	.513

N= 14
PROBABILITY

X	.05	.1	.2	.3	.4	.5	.6	.7	.8	.9	.95
0	.488	.229	.044	.007	.001	.000	.000	.000	.000	.000	.000
1	.359	.356	.154	.041	.007	.001	.000	.000	.000	.000	.000
2	.123	.257	.250	.113	.032	.006	.001	.000	.000	.000	.000
3	.026	.114	.250	.194	.085	.022	.003	.000	.000	.000	.000
4	.004	.035	.172	.229	.155	.061	.014	.001	.000	.000	.000
5	.000	.008	.086	.196	.207	.122	.041	.007	.000	.000	.000
6	.000	.001	.032	.126	.207	.183	.092	.023	.002	.000	.000
7	.000	.000	.009	.062	.157	.209	.157	.062	.009	.000	.000
8	.000	.000	.002	.023	.092	.183	.207	.126	.032	.001	.000
9	.000	.000	.000	.007	.041	.122	.207	.196	.086	.008	.000
10	.000	.000	.000	.001	.014	.061	.155	.229	.172	.035	.004
11	.000	.000	.000	.000	.003	.022	.085	.194	.250	.114	.026
12	.000	.000	.000	.000	.001	.006	.032	.113	.250	.257	.123
13	.000	.000	.000	.000	.000	.001	.007	.041	.154	.356	.359
14	.000	.000	.000	.000	.000	.000	.001	.007	.044	.229	.488

N= 15
PROBABILITY

X	.05	.1	.2	.3	.4	.5	.6	.7	.8	.9	.95
0	.463	.206	.035	.005	.000	.000	.000	.000	.000	.000	.000
1	.366	.343	.132	.031	.005	.000	.000	.000	.000	.000	.000
2	.135	.267	.231	.092	.022	.003	.000	.000	.000	.000	.000
3	.031	.129	.250	.170	.063	.014	.002	.000	.000	.000	.000
4	.005	.043	.188	.219	.127	.042	.007	.001	.000	.000	.000
5	.001	.010	.103	.206	.186	.092	.024	.003	.000	.000	.000
6	.000	.002	.043	.147	.207	.153	.061	.012	.001	.000	.000
7	.000	.000	.014	.081	.177	.196	.118	.035	.003	.000	.000
8	.000	.000	.003	.035	.118	.196	.177	.081	.014	.000	.000
9	.000	.000	.001	.012	.061	.153	.207	.147	.043	.002	.000
10	.000	.000	.000	.003	.024	.092	.186	.206	.103	.010	.001
11	.000	.000	.000	.001	.007	.042	.127	.219	.188	.043	.005
12	.000	.000	.000	.000	.002	.014	.063	.170	.250	.129	.031
13	.000	.000	.000	.000	.000	.003	.022	.092	.231	.267	.135
14	.000	.000	.000	.000	.000	.000	.005	.031	.132	.343	.366
15	.000	.000	.000	.000	.000	.000	.000	.005	.035	.206	.463

N= 16
PROBABILITY

X	.05	.1	.2	.3	.4	.5	.6	.7	.8	.9	.95
0	.440	.185	.028	.003	.000	.000	.000	.000	.000	.000	.000
1	.371	.329	.113	.023	.003	.000	.000	.000	.000	.000	.000
2	.146	.275	.211	.073	.015	.002	.000	.000	.000	.000	.000
3	.036	.142	.246	.146	.047	.009	.001	.000	.000	.000	.000
4	.006	.051	.200	.204	.101	.028	.004	.000	.000	.000	.000
5	.001	.014	.120	.210	.162	.067	.014	.001	.000	.000	.000
6	.000	.003	.055	.165	.198	.122	.039	.006	.000	.000	.000
7	.000	.000	.020	.101	.189	.175	.084	.019	.001	.000	.000
8	.000	.000	.006	.049	.142	.196	.142	.049	.006	.000	.000
9	.000	.000	.001	.019	.084	.175	.189	.101	.020	.000	.000
10	.000	.000	.000	.006	.039	.122	.198	.165	.055	.003	.000
11	.000	.000	.000	.001	.014	.067	.162	.210	.120	.014	.001
12	.000	.000	.000	.000	.004	.028	.101	.204	.200	.051	.006
13	.000	.000	.000	.000	.001	.009	.047	.146	.246	.142	.036
14	.000	.000	.000	.000	.000	.002	.015	.073	.211	.275	.146
15	.000	.000	.000	.000	.000	.000	.003	.023	.113	.329	.371
16	.000	.000	.000	.000	.000	.000	.000	.003	.028	.185	.440

N= 17
PROBABILITY

X	.05	.1	.2	.3	.4	.5	.6	.7	.8	.9	.95
0	.418	.167	.023	.002	.000	.000	.000	.000	.000	.000	.000
1	.374	.315	.096	.017	.002	.000	.000	.000	.000	.000	.000
2	.158	.280	.191	.058	.010	.001	.000	.000	.000	.000	.000
3	.041	.156	.239	.125	.034	.005	.000	.000	.000	.000	.000
4	.008	.060	.209	.187	.080	.018	.002	.000	.000	.000	.000
5	.001	.017	.136	.208	.138	.047	.008	.001	.000	.000	.000
6	.000	.004	.068	.178	.184	.094	.024	.003	.000	.000	.000
7	.000	.001	.027	.120	.193	.148	.057	.009	.000	.000	.000
8	.000	.000	.008	.064	.161	.185	.107	.028	.002	.000	.000
9	.000	.000	.002	.028	.107	.185	.161	.064	.008	.000	.000
10	.000	.000	.000	.009	.057	.148	.193	.120	.027	.001	.000
11	.000	.000	.000	.003	.024	.094	.184	.178	.068	.004	.000
12	.000	.000	.000	.001	.008	.047	.138	.208	.136	.017	.001
13	.000	.000	.000	.000	.002	.018	.080	.187	.209	.060	.008
14	.000	.000	.000	.000	.000	.005	.034	.125	.239	.156	.041
15	.000	.000	.000	.000	.000	.001	.010	.058	.191	.280	.158
16	.000	.000	.000	.000	.000	.000	.002	.017	.096	.315	.374
17	.000	.000	.000	.000	.000	.000	.000	.002	.023	.167	.418

N= 18
PROBABILITY

X	.05	.1	.2	.3	.4	.5	.6	.7	.8	.9	.95
0	.397	.150	.018	.002	.000	.000	.000	.000	.000	.000	.000
1	.376	.300	.081	.013	.001	.000	.000	.000	.000	.000	.000
2	.168	.284	.172	.046	.007	.001	.000	.000	.000	.000	.000
3	.047	.168	.230	.105	.025	.003	.000	.000	.000	.000	.000
4	.009	.070	.215	.168	.061	.012	.001	.000	.000	.000	.000
5	.001	.022	.151	.202	.115	.033	.004	.000	.000	.000	.000
6	.000	.005	.082	.187	.166	.071	.015	.001	.000	.000	.000
7	.000	.001	.035	.138	.189	.121	.037	.005	.000	.000	.000
8	.000	.000	.012	.081	.173	.167	.077	.015	.001	.000	.000
9	.000	.000	.003	.039	.128	.185	.128	.039	.003	.000	.000
10	.000	.000	.001	.015	.077	.167	.173	.081	.012	.000	.000
11	.000	.000	.000	.005	.037	.121	.189	.138	.035	.001	.000
12	.000	.000	.000	.001	.015	.071	.166	.187	.082	.005	.000
13	.000	.000	.000	.000	.004	.033	.115	.202	.151	.022	.001
14	.000	.000	.000	.000	.001	.012	.061	.168	.215	.070	.009
15	.000	.000	.000	.000	.000	.003	.025	.105	.230	.168	.047
16	.000	.000	.000	.000	.000	.001	.007	.046	.172	.284	.168
17	.000	.000	.000	.000	.000	.000	.001	.013	.081	.300	.376
18	.000	.000	.000	.000	.000	.000	.000	.002	.018	.150	.397

N= 19
PROBABILITY

X	.05	.1	.2	.3	.4	.5	.6	.7	.8	.9	.95
0	.377	.135	.014	.001	.000	.000	.000	.000	.000	.000	.000
1	.377	.285	.068	.009	.001	.000	.000	.000	.000	.000	.000
2	.179	.285	.154	.036	.005	.000	.000	.000	.000	.000	.000
3	.053	.180	.218	.087	.017	.002	.000	.000	.000	.000	.000
4	.011	.080	.218	.149	.047	.007	.001	.000	.000	.000	.000
5	.002	.027	.164	.192	.093	.022	.002	.000	.000	.000	.000
6	.000	.007	.095	.192	.145	.052	.008	.001	.000	.000	.000
7	.000	.001	.044	.153	.180	.096	.024	.002	.000	.000	.000
8	.000	.000	.017	.098	.180	.144	.053	.008	.000	.000	.000
9	.000	.000	.005	.051	.146	.176	.098	.022	.001	.000	.000
10	.000	.000	.001	.022	.098	.176	.146	.051	.005	.000	.000
11	.000	.000	.000	.008	.053	.144	.180	.098	.017	.000	.000
12	.000	.000	.000	.002	.024	.096	.180	.153	.044	.001	.000
13	.000	.000	.000	.001	.008	.052	.145	.192	.095	.007	.000
14	.000	.000	.000	.000	.002	.022	.093	.192	.164	.027	.002
15	.000	.000	.000	.000	.001	.007	.047	.149	.218	.080	.011
16	.000	.000	.000	.000	.000	.002	.017	.087	.218	.180	.053
17	.000	.000	.000	.000	.000	.000	.005	.036	.154	.285	.179
18	.000	.000	.000	.000	.000	.000	.001	.009	.068	.285	.377
19	.000	.000	.000	.000	.000	.000	.000	.001	.014	.135	.377

N= 20
PROBABILITY

X	.05	.1	.2	.3	.4	.5	.6	.7	.8	.9	.95
0	.358	.122	.012	.001	.000	.000	.000	.000	.000	.000	.000
1	.377	.270	.058	.007	.000	.000	.000	.000	.000	.000	.000
2	.189	.285	.137	.028	.003	.000	.000	.000	.000	.000	.000
3	.060	.190	.205	.072	.012	.001	.000	.000	.000	.000	.000
4	.013	.090	.218	.130	.035	.005	.000	.000	.000	.000	.000
5	.002	.032	.175	.179	.075	.015	.001	.000	.000	.000	.000
6	.000	.009	.109	.192	.124	.037	.005	.000	.000	.000	.000
7	.000	.002	.055	.164	.166	.074	.015	.001	.000	.000	.000
8	.000	.000	.022	.114	.180	.120	.035	.004	.000	.000	.000
9	.000	.000	.007	.065	.160	.160	.071	.012	.000	.000	.000
10	.000	.000	.002	.031	.117	.176	.117	.031	.002	.000	.000
11	.000	.000	.000	.012	.071	.160	.160	.065	.007	.000	.000
12	.000	.000	.000	.004	.035	.120	.180	.114	.022	.000	.000
13	.000	.000	.000	.001	.015	.074	.166	.164	.055	.002	.000
14	.000	.000	.000	.000	.005	.037	.124	.192	.109	.009	.000
15	.000	.000	.000	.000	.001	.015	.075	.179	.175	.032	.002
16	.000	.000	.000	.000	.000	.005	.035	.130	.218	.090	.013
17	.000	.000	.000	.000	.000	.001	.012	.072	.205	.190	.060
18	.000	.000	.000	.000	.000	.000	.003	.028	.137	.285	.189
19	.000	.000	.000	.000	.000	.000	.000	.007	.058	.270	.377
20	.000	.000	.000	.000	.000	.000	.000	.001	.012	.122	.358

B—Poisson Distribution: Probability of Exactly *x* Occurrences

	μ								
x	**0.1**	**0.2**	**0.3**	**0.4**	**0.5**	**0.6**	**0.7**	**0.8**	**0.9**
0	0.9048	0.8187	0.7408	0.6703	0.6065	0.5488	0.4966	0.4493	0.4066
1	0.0905	0.1637	0.2222	0.2681	0.3033	0.3293	0.3476	0.3595	0.3659
2	0.0045	0.0164	0.0333	0.0536	0.0758	0.0988	0.1217	0.1438	0.1647
3	0.0002	0.0011	0.0033	0.0072	0.0126	0.0198	0.0284	0.0383	0.0494
4		0.0001	0.0003	0.0007	0.0016	0.0030	0.0050	0.0077	0.0111
5					0.0002	0.0004	0.0007	0.0012	0.0020
6							0.0001	0.0002	0.0003

	μ								
x	**1.0**	**2.0**	**3.0**	**4.0**	**5.0**	**6.0**	**7.0**	**8.0**	**9.0**
0	0.3679	0.1353	0.0498	0.0183	0.0067	0.0025	0.0009	0.0003	0.0001
1	0.3679	0.2707	0.1494	0.0733	0.0337	0.0149	0.0064	0.0027	0.0011
2	0.1839	0.2707	0.2240	0.1465	0.0842	0.0446	0.0223	0.0107	0.0050
3	0.0613	0.1804	0.2240	0.1954	0.1404	0.0892	0.0521	0.0286	0.0150
4	0.0153	0.0902	0.1680	0.1954	0.1755	0.1339	0.0912	0.0573	0.0337
5	0.0031	0.0361	0.1008	0.1563	0.1755	0.1606	0.1277	0.0916	0.0607
6	0.0005	0.0120	0.0504	0.1042	0.1462	0.1606	0.1490	0.1221	0.0911
7	0.0001	0.0034	0.0216	0.0595	0.1044	0.1377	0.1490	0.1396	0.1171
8		0.0009	0.0081	0.0298	0.0653	0.1033	0.1304	0.1396	0.1318
9		0.0002	0.0027	0.0132	0.0363	0.0688	0.1014	0.1241	0.1318
10			0.0008	0.0053	0.0181	0.0413	0.0710	0.0993	0.1186
11			0.0002	0.0019	0.0082	0.0225	0.0452	0.0722	0.0970
12			0.0001	0.0006	0.0034	0.0113	0.0264	0.0481	0.0728
13				0.0002	0.0013	0.0052	0.0142	0.0296	0.0504
14				0.0001	0.0005	0.0022	0.0071	0.0169	0.0324
15					0.0002	0.0009	0.0033	0.0090	0.0194
16						0.0003	0.0014	0.0045	0.0109
17						0.0001	0.0006	0.0021	0.0058
18							0.0002	0.0009	0.0029
19							0.0001	0.0004	0.0014
20								0.0002	0.0006
21								0.0001	0.0003
22									0.0001

C—The Normal Probability Distribution

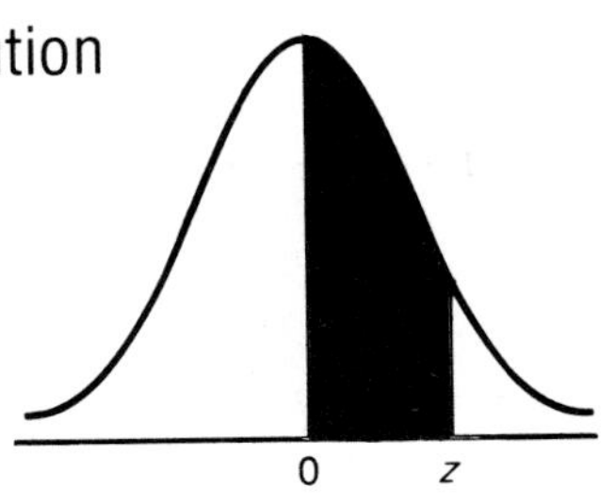

z	0	1	2	3	4	5	6	7	8	9
	SECOND DECIMAL PLACE OF z									
0.0	.0000	.0040	.0080	.0120	.0160	.0199	.0239	.0279	.0319	.0359
0.1	.0398	.0438	.0478	.0517	.0557	.0596	.0636	.0675	.0714	.0753
0.2	.0793	.0832	.0871	.0910	.0948	.0987	.1028	.1064	.1103	.1141
0.3	.1179	.1217	.1255	.1293	.1331	.1368	.1406	.1443	.1480	.1517
0.4	.1554	.1591	.1628	.1664	.1700	.1736	.1772	.1808	.1844	.1879
0.5	.1915	.1950	.1985	.2019	.2054	.2088	.2123	.2157	.2190	.2224
0.6	.2257	.2291	.2324	.2357	.2389	.2422	.2454	.2486	.2517	.2549
0.7	.2580	.2611	.2642	.2673	.2704	.2734	.2764	.2794	.2823	.2852
0.8	.2881	.2910	.2939	.2967	.2995	.3023	.3051	.3078	.3106	.3133
0.9	.3159	.3186	.3212	.3238	.3264	.3289	.3315	.3340	.3365	.3389
1.0	.3413	.3438	.3461	.3485	.3508	.3531	.3554	.3577	.3599	.3621
1.1	.3643	.3665	.3686	.3708	.3729	.3749	.3770	.3790	.3810	.3830
1.2	.3849	.3869	.3888	.3907	.3925	.3944	.3962	.3980	.3997	.4015
1.3	.4032	.4049	.4066	.4082	.4099	.4115	.4131	.4147	.4162	.4177
1.4	.4192	.4207	.4222	.4236	.4251	.4265	.4279	.4292	.4306	.4319
1.5	.4332	.4345	.4357	.4370	.4382	.4394	.4406	.4418	.4429	.4441
1.6	.4452	.4463	.4474	.4484	.4495	.4505	.4515	.4525	.4535	.4545
1.7	.4554	.4564	.4573	.4582	.4591	.4599	.4608	.4616	.4625	.4633
1.8	.4641	.4649	.4656	.4664	.4671	.4678	.4686	.4693	.4699	.4706
1.9	.4713	.4719	.4726	.4732	.4738	.4744	.4750	.4756	.4761	.4767
2.0	.4772	.4778	.4783	.4788	.4793	.4798	.4803	.4808	.4812	.4817
2.1	.4821	.4826	.4830	.4834	.4838	.4842	.4846	.4850	.4854	.4857
2.2	.4861	.4864	.4868	.4871	.4875	.4878	.4881	.4884	.4887	.4890
2.3	.4893	.4896	.4898	.4901	.4904	.4906	.4909	.4911	.4913	.4916
2.4	.4918	.4920	.4922	.4925	.4927	.4929	.4931	.4932	.4934	.4936
2.5	.4938	.4940	.4941	.4943	.4945	.4946	.4948	.4949	.4951	.4952
2.6	.4953	.4955	.4956	.4957	.4959	.4960	.4961	.4962	.4963	.4964
2.7	.4965	.4966	.4967	.4968	.4969	.4970	.4971	.4972	.4973	.4974
2.8	.4974	.4975	.4976	.4977	.4977	.4978	.4979	.4979	.4980	.4981
2.9	.4981	.4982	.4982	.4983	.4984	.4984	.4985	.4985	.4986	.4986
3.0	.4987									
3.5	.4998									
4.0	.49997									
4.5	.499997									
5.0	.4999997									

The entries in this table are the proportion of the observations from a *normal* distribution that have z scores between 0 and z. This proportion is represented by the colored area under the curve in the figure.

Reprinted with permission from *CRC Standard Mathematical Tables*, Fifteenth Edition (West Palm Beach. Fla.: Copyright The Chemical Rubber Co., CRC Press, Inc.).

D—Critical Values of Student's *t* Distribution

One-tailed value

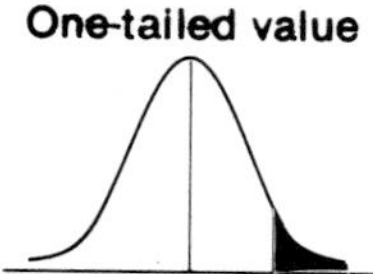

Two-tailed value

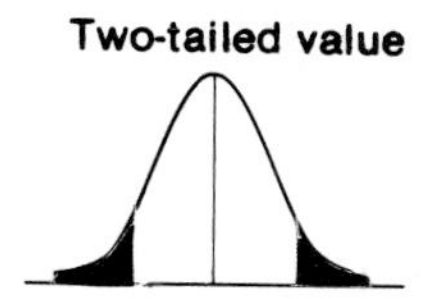

DEGREES OF FREEDOM	ONE-TAILED VALUE					
	0.25	0.10	0.05	0.025	0.01	0.005
	TWO-TAILED VALUE					
	0.50	0.20	0.10	0.05	0.02	0.01
1	1.000	3.078	6.314	12.706	31.821	63.657
2	0.816	1.886	2.920	4.303	6.965	9.925
3	.765	1.638	2.353	3.182	4.541	5.841
4	.741	1.533	2.132	2.776	3.747	4.604
5	.727	1.476	2.015	2.571	3.365	4.032
6	.718	1.440	1.943	2.447	3.143	3.707
7	.711	1.415	1.895	2.365	2.998	3.499
8	.706	1.397	1.860	2.306	2.896	3.355
9	.703	1.383	1.833	2.262	2.821	3.250
10	.700	1.372	1.812	2.228	2.764	3.169
11	.697	1.363	1.796	2.201	2.718	3.106
12	.695	1.356	1.782	2.179	2.681	3.055
13	.694	1.350	1.771	2.160	2.650	3.012
14	.692	1.345	1.761	2.145	2.626	2.977
15	.691	1.341	1.753	2.131	2.602	2.947
16	.690	1.337	1.746	2.120	2.583	2.921
17	.689	1.333	1.740	2.110	2.567	2.898
18	.688	1.330	1.734	2.101	2.552	2.878
19	.688	1.328	1.729	2.093	2.539	2.861
20	.687	1.325	1.725	2.086	2.528	2.845
21	.686	1.323	1.721	2.080	2.518	2.831
22	.686	1.321	1.717	2.074	2.508	2.819
23	.685	1.319	1.714	2.069	2.500	2.807
24	.685	1.318	1.711	2.064	2.492	2.797
25	.684	1.316	1.708	2.060	2.485	2.787
26	.684	1.315	1.706	2.056	2.479	2.779
27	.684	1.314	1.703	2.052	2.473	2.771
28	.683	1.313	1.701	2.048	2.467	2.763
29	.683	1.311	1.699	2.045	2.462	2.756
30	.683	1.310	1.697	2.042	2.457	2.750
35	.682	1.306	1.690	2.030	2.438	2.724
40	.681	1.303	1.684	2.021	2.423	2.704
45	.680	1.301	1.680	2.014	2.412	2.690
50	.680	1.299	1.676	2.008	2.403	2.678
55	.679	1.297	1.673	2.004	2.396	2.669
60	.679	1.296	1.671	2.000	2.390	2.660
70	.678	1.294	1.667	1.994	2.381	2.648
80	.678	1.293	1.665	1.989	2.374	2.638
90	.678	1.291	1.662	1.986	2.368	2.631
100	.677	1.290	1.661	1.982	2.364	2.625
120	.677	1.289	1.658	1.980	2.358	2.617
∞	.674	1.282	1.645	1.960	2.326	2.576

E—Critical Values of the F Statistic (0.05 level of significance)

Degrees of Freedom in Numerator

Degrees of Freedom in Denominator

	1	2	3	4	5	6	7	8	9	10	12	15	20	24	30	40	60	120	∞
1	161	200	216	225	230	234	237	239	241	242	244	246	248	249	250	251	252	253	254
2	18.5	19.0	19.2	19.2	19.3	19.3	19.4	19.4	19.4	19.4	19.4	19.4	19.4	19.5	19.5	19.5	19.5	19.5	19.5
3	10.1	9.55	9.28	9.12	9.01	8.94	8.89	8.85	8.81	8.79	8.74	8.70	8.66	8.64	8.62	8.59	8.57	8.55	8.53
4	7.71	6.94	6.59	6.39	6.26	6.16	6.09	6.04	6.00	5.96	5.91	5.86	5.80	5.77	5.75	5.72	5.69	5.66	5.63
5	6.61	5.79	5.41	5.19	5.05	4.95	4.88	4.82	4.77	4.74	4.68	4.62	4.56	4.53	4.50	4.46	4.43	4.40	4.37
6	5.99	5.14	4.76	4.53	4.39	4.28	4.21	4.15	4.10	4.06	4.00	3.94	3.87	3.84	3.81	3.77	3.74	3.70	3.67
7	5.59	4.74	4.35	4.12	3.97	3.87	3.79	3.73	3.68	3.64	3.57	3.51	3.44	3.41	3.38	3.34	3.30	3.27	3.23
8	5.32	4.46	4.07	3.84	3.69	3.58	3.50	3.44	3.39	3.35	3.28	3.22	3.15	3.12	3.08	3.04	3.01	2.97	2.93
9	5.12	4.26	3.86	3.63	3.48	3.37	3.29	3.23	3.18	3.14	3.07	3.01	2.94	2.90	2.86	2.83	2.79	2.75	2.71
10	4.96	4.10	3.71	3.48	3.33	3.22	3.14	3.07	3.02	2.98	2.91	2.85	2.77	2.74	2.70	2.66	2.62	2.58	2.54
11	4.84	3.98	3.59	3.36	3.20	3.09	3.01	2.95	2.90	2.85	2.79	2.72	2.65	2.61	2.57	2.53	2.49	2.45	2.40
12	4.75	3.89	3.49	3.26	3.11	3.00	2.91	2.85	2.80	2.75	2.69	2.62	2.54	2.51	2.47	2.43	2.38	2.34	2.30
13	4.67	3.81	3.41	3.18	3.03	2.92	2.83	2.77	2.71	2.67	2.60	2.53	2.46	2.42	2.38	2.34	2.30	2.25	2.21
14	4.60	3.74	3.34	3.11	2.96	2.85	2.76	2.70	2.65	2.60	2.53	2.46	2.39	2.35	2.31	2.27	2.22	2.18	2.13
15	4.54	3.68	3.29	3.06	2.90	2.79	2.71	2.64	2.59	2.54	2.48	2.40	2.33	2.29	2.25	2.20	2.16	2.11	2.07
16	4.49	3.63	3.24	3.01	2.85	2.74	2.66	2.59	2.54	2.49	2.42	2.35	2.28	2.24	2.19	2.15	2.11	2.06	2.01
17	4.45	3.59	3.20	2.96	2.81	2.70	2.61	2.55	2.49	2.45	2.38	2.31	2.23	2.19	2.15	2.10	2.06	2.01	1.96
18	4.41	3.55	3.16	2.93	2.77	2.66	2.58	2.51	2.46	2.41	2.34	2.27	2.19	2.15	2.11	2.06	2.02	1.97	1.92
19	4.38	3.52	3.13	2.90	2.74	2.63	2.54	2.48	2.42	2.38	2.31	2.23	2.16	2.11	2.07	2.03	1.98	1.93	1.88
20	4.35	3.49	3.10	2.87	2.71	2.60	2.51	2.45	2.39	2.35	2.28	2.20	2.12	2.08	2.04	1.99	1.95	1.90	1.84
21	4.32	3.47	3.07	2.84	2.68	2.57	2.49	2.42	2.37	2.32	2.25	2.18	2.10	2.05	2.01	1.96	1.92	1.87	1.81
22	4.30	3.44	3.05	2.82	2.66	2.55	2.46	2.40	2.34	2.30	2.23	2.15	2.07	2.03	1.98	1.94	1.89	1.84	1.78
23	4.28	3.42	3.03	2.80	2.64	2.53	2.44	2.37	2.32	2.27	2.20	2.13	2.05	2.01	1.96	1.91	1.86	1.81	1.76
24	4.26	3.40	3.01	2.78	2.62	2.51	2.42	2.36	2.30	2.25	2.18	2.11	2.03	1.98	1.94	1.89	1.84	1.79	1.73
25	4.24	3.39	2.99	2.76	2.60	2.49	2.40	2.34	2.28	2.24	2.16	2.09	2.01	1.96	1.92	1.87	1.82	1.77	1.71
30	4.17	3.32	2.92	2.69	2.53	2.42	2.33	2.27	2.21	2.16	2.09	2.01	1.93	1.89	1.84	1.79	1.74	1.68	1.62
40	4.08	3.23	2.84	2.61	2.45	2.34	2.25	2.18	2.12	2.08	2.00	1.92	1.84	1.79	1.74	1.69	1.64	1.58	1.51
60	4.00	3.15	2.76	2.53	2.37	2.25	2.17	2.10	2.04	1.99	1.92	1.84	1.75	1.70	1.65	1.59	1.53	1.47	1.39
120	3.92	3.07	2.68	2.45	2.29	2.18	2.09	2.02	1.96	1.91	1.83	1.75	1.66	1.61	1.55	1.50	1.43	1.35	1.25
∞	3.84	3.00	2.60	2.37	2.21	2.10	2.01	1.94	1.88	1.83	1.75	1.67	1.57	1.52	1.46	1.39	1.32	1.22	1.00

Source: This table is reproduced from M. Merrington and C.M. Thompson, "Tables of Percentage Points of the Inverted Beta (F) Distribution," *Biometrika*, vol. 33 (1943), by permission of the Biometrika trustees.

E—Critical Values of the *F* Statistic (0.01 level of significance)

Degrees of Freedom in Numerator (columns); *Degrees of Freedom in Denominator* (rows)

	1	2	3	4	5	6	7	8	9	10	12	15	20	24	30	40	60	120	∞
1	4,052	5,000	5,403	5,625	5,764	5,859	5,928	5,982	6,023	6,056	6,106	6,157	6,209	6,235	6,261	6,287	6,313	6,339	6,366
2	98.5	99.0	99.2	99.2	99.3	99.3	99.4	99.4	99.4	99.4	99.4	99.4	99.4	99.5	99.5	99.5	99.5	99.5	99.5
3	34.1	30.8	29.5	28.7	28.2	27.9	27.7	27.5	27.3	27.2	27.1	26.9	26.7	26.6	26.5	26.4	26.3	26.2	26.1
4	21.2	18.0	16.7	16.0	15.5	15.2	15.0	14.8	14.7	14.5	14.4	14.2	14.0	13.9	13.8	13.7	13.7	13.6	13.5
5	16.3	13.3	12.1	11.4	11.0	10.7	10.5	10.3	10.2	10.1	9.89	9.72	9.55	9.47	9.38	9.29	9.20	9.11	9.02
6	13.7	10.9	9.78	9.15	8.75	8.47	8.26	8.10	7.98	7.87	7.72	7.56	7.40	7.31	7.23	7.14	7.06	6.97	6.88
7	12.2	9.55	8.45	7.85	7.46	7.19	6.99	6.84	6.72	6.62	6.47	6.31	6.16	6.07	5.99	5.91	5.82	5.74	5.65
8	11.3	8.65	7.59	7.01	6.63	6.37	6.18	6.03	5.91	5.81	5.67	5.52	5.36	5.28	5.20	5.12	5.03	4.95	4.86
9	10.6	8.02	6.99	6.42	6.06	5.80	5.61	5.47	5.35	5.26	5.11	4.96	4.81	4.73	4.65	4.57	4.48	4.40	4.31
10	10.0	7.56	6.55	5.99	5.64	5.39	5.20	5.06	4.94	4.85	4.71	4.56	4.41	4.33	4.25	4.17	4.08	4.00	3.91
11	9.65	7.21	6.22	5.67	5.32	5.07	4.89	4.74	4.63	4.54	4.40	4.25	4.10	4.02	3.94	3.86	3.78	3.69	3.60
12	9.33	6.93	5.95	5.41	5.06	4.82	4.64	4.50	4.39	4.30	4.16	4.01	3.86	3.78	3.70	3.62	3.54	3.45	3.36
13	9.07	6.70	5.74	5.21	4.86	4.62	4.44	4.30	4.19	4.10	3.96	3.82	3.66	3.59	3.51	3.43	3.34	3.25	3.17
14	8.86	6.51	5.56	5.04	4.70	4.46	4.28	4.14	4.03	3.94	3.80	3.66	3.51	3.43	3.35	3.27	3.18	3.09	3.00
15	8.68	6.36	5.42	4.89	4.56	4.32	4.14	4.00	3.89	3.80	3.67	3.52	3.37	3.29	3.21	3.13	3.05	2.96	2.87
16	8.53	6.23	5.29	4.77	4.44	4.20	4.03	3.89	3.78	3.69	3.55	3.41	3.26	3.18	3.10	3.02	2.93	2.84	2.75
17	8.40	6.11	5.19	4.67	4.34	4.10	3.93	3.79	3.68	3.59	3.46	3.31	3.16	3.08	3.00	2.92	2.83	2.75	2.65
18	8.29	6.01	5.09	4.58	4.25	4.01	3.84	3.71	3.60	3.51	3.37	3.23	3.08	3.00	2.92	2.84	2.75	2.66	2.57
19	8.19	5.93	5.01	4.50	4.17	3.94	3.77	3.63	3.52	3.43	3.30	3.15	3.00	2.92	2.84	2.76	2.67	2.58	2.49
20	8.10	5.85	4.94	4.43	4.10	3.87	3.70	3.56	3.46	3.37	3.23	3.09	2.94	2.86	2.78	2.69	2.61	2.52	2.42
21	8.02	5.78	4.87	4.37	4.04	3.81	3.64	3.51	3.40	3.31	3.17	3.03	2.88	2.80	2.72	2.64	2.55	2.46	2.36
22	7.95	5.72	4.82	4.31	3.99	3.76	3.59	3.45	3.35	3.26	3.12	2.98	2.83	2.75	2.67	2.58	2.50	2.40	2.31
23	7.88	5.66	4.76	4.26	3.94	3.71	3.54	3.41	3.30	3.21	3.07	2.93	2.78	2.70	2.62	2.54	2.45	2.35	2.26
24	7.82	5.61	4.72	4.22	3.90	3.67	3.50	3.36	3.26	3.17	3.03	2.89	2.74	2.66	2.58	2.49	2.40	2.31	2.21
25	7.77	5.57	4.68	4.18	3.86	3.63	3.46	3.32	3.22	3.13	2.99	2.85	2.70	2.62	2.53	2.45	2.36	2.27	2.17
30	7.56	5.39	4.51	4.02	3.70	3.47	3.30	3.17	3.07	2.98	2.84	2.70	2.55	2.47	2.39	2.30	2.21	2.11	2.01
40	7.31	5.18	4.31	3.83	3.51	3.29	3.12	2.99	2.89	2.80	2.66	2.52	2.37	2.29	2.20	2.11	2.02	1.92	1.80
60	7.08	4.98	4.13	3.65	3.34	3.12	2.95	2.82	2.72	2.63	2.50	2.35	2.20	2.12	2.03	1.94	1.84	1.73	1.60
120	6.85	4.79	3.95	3.48	3.17	2.96	2.79	2.66	2.56	2.47	2.34	2.19	2.03	1.95	1.86	1.76	1.66	1.53	1.38
∞	6.63	4.61	3.78	3.32	3.02	2.80	2.64	2.51	2.41	2.32	2.18	2.04	1.88	1.79	1.70	1.59	1.47	1.32	1.00

F—Critical Values of the Chi-square Statistic

Possible Values of χ^2

DEGREES OF FREEDOM df	RIGHT-TAIL AREA 0.10	0.05	0.02	0.01
1	2.706	3.841	5.412	6.635
2	4.605	5.991	7.824	9.210
3	6.251	7.815	9.837	11.345
4	7.779	9.488	11.668	13.277
5	9.236	11.070	13.388	15.086
6	10.645	12.592	15.033	16.812
7	12.017	14.067	16.622	18.475
8	13.362	15.507	18.168	20.090
9	14.684	16.919	19.679	21.666
10	15.987	18.307	21.161	23.209
11	17.275	19.675	22.618	24.725
12	18.549	21.026	24.054	26.217
13	19.812	22.362	25.472	27.688
14	21.064	23.685	26.873	29.141
15	22.307	24.996	28.259	30.578
16	23.542	26.296	29.633	32.000
17	24.769	27.587	30.995	33.409
18	25.989	28.869	32.346	34.805
19	27.204	30.144	33.687	36.191
20	28.412	31.410	35.020	37.566
21	29.615	32.671	36.343	38.932
22	30.813	33.924	37.659	40.289
23	32.007	35.172	38.968	41.638
24	33.196	36.415	40.270	42.980
25	34.382	37.652	41.566	44.314
26	35.563	38.885	42.856	45.642
27	36.741	40.113	44.140	46.963
28	37.916	41.337	45.419	48.278
29	39.087	42.557	46.693	49.588
30	40.256	43.773	47.962	50.892

This table contains the values of χ^2 that correspond to a specific right-tail area and specific numbers of degrees of freedom df.

Source: From Table IV of Fisher & Yates: *Statistical Tables for Biological, Agricultural and Medical Research*, published by Longman Group Ltd., London (previously published by Oliver and Boyd Ltd., Edinburgh), by permission of the authors and publishers.

G—Critical Values of the Wilcoxon Signed-rank Statistic

ONE-TAILED VALUE	0.10	0.05	0.025	0.01	0.005
TWO-TAILED VALUE	0.20	0.10	0.05	0.02	0.01
n					
5	2	0	—	—	—
6	3	2	0	—	—
7	5	3	2	0	—
8	8	5	3	1	0
9	10	8	5	3	1
10	14	10	8	5	3
11	17	13	10	7	5
12	21	17	13	9	7
13	26	21	17	12	9
14	31	25	21	15	12
15	36	30	25	19	15
16	42	35	29	23	19
17	48	41	34	27	23
18	55	47	40	32	27
19	62	53	46	37	32
20	69	60	52	43	37

Each entry represents the largest value of W with a value less than or equal to the value in the column heading.

Adapted from Table II in F. Wilcoxon, S.K. Katti, and Roberta A. Wilcox. *Critical Values and Probability Levels for the Wilcoxon Rank Sum Test and the Wilcoxon Signed Rank Test*, American Cyanamid Company (Lederle Laboratories Division, Pearl River, N.Y.) and The Florida State University (Department of Statistics, Tallahassee, Fla.), August 1963. Used with the permission of American Cyanamid Company and The Florida State University.

H—A Table of Random Numbers

77940	44905	86088	90164	80699	77739	58053	81156	89580
86271	61188	47642	07467	98286	72475	49211	64105	20356
56715	00377	23817	82971	16648	88965	08600	02426	20965
19828	18730	60358	40177	96759	10825	98714	12025	21242
47741	75379	54318	35263	05766	54137	78724	12778	96585
78119	12051	42769	05619	80166	76597	46083	41167	26281
64957	77143	39063	15832	04167	69414	83671	81740	84627
89999	28946	86251	06464	35832	17041	99677	03431	67128
85161	12260	58820	18492	23814	01986	10094	48170	85204
37402	52782	05495	39887	99531	85691	64388	37482	55257
80963	39222	86055	73573	71093	90753	18059	56221	39904
37249	53066	84141	30378	88793	61097	30628	78210	56070
18213	51159	59577	28931	52027	93081	43111	11060	16095
32503	03871	37063	41887	88399	81743	90260	63412	01383
57004	69626	13747	29240	69608	75632	47153	18464	83417
73164	97693	13084	39802	81823	12056	59392	29799	55653
47180	61957	93044	31395	37482	71890	37471	54890	28745
94850	63606	95604	05380	79694	00575	16644	52639	64217
25931	96025	15138	59941	66729	18006	22565	51626	25736
59738	07828	34746	67218	66996	79781	32849	87964	87373
42631	99979	06075	17008	63053	16384	60468	57887	13243
39489	48511	67616	79479	46650	50653	43311	05217	72018
04907	32917	19307	54424	64966	74716	32464	58546	52792
94745	34952	29910	92610	26144	78419	95089	59549	38100
35317	64037	50984	16500	41011	93496	99673	94182	31842
49477	68247	16214	59292	89048	18337	93732	05948	08074
20579	31251	10466	38375	55795	35458	63302	94647	30949
49392	85585	59029	49301	26029	40077	98764	01797	30775
97406	03597	33277	21516	73334	90685	14704	90822	27159
85730	12403	77927	90317	91187	48191	07049	24849	26954
81372	99901	35002	41279	52363	17914	68038	67622	21102
11718	47809	27996	44228	42535	91947	26786	35908	56264
13107	06911	81068	65713	35385	07260	89789	67565	51864
14315	74663	80814	15140	63881	45676	00240	94143	62116
89863	95501	88552	32314	97593	64693	86754	26258	97026
62388	44931	57781	95594	84173	38109	25399	23721	40811
85145	30748	26602	13859	77904	26177	94500	80918	90943
30780	55982	86406	92656	93776	21143	75102	88168	44051
99573	50192	93637	01853	17172	81428	15281	49381	79499
27455	13662	64072	98108	72338	78389	65179	02713	64049
09256	75292	69064	01805	72467	63812	12321	06450	57805
45861	31907	71084	70716	05161	38953	89274	69079	53817
43626	93477	73706	54347	32253	40376	67022	09641	84813
21639	51765	74863	10244	82732	23345	62589	54117	44796
94531	45676	14881	02100	72025	20099	28209	93363	34633
76301	99636	94834	41479	94100	75088	38863	83404	65304
69180	74282	25833	31438	60025	65978	28610	78712	52570
72657	29881	74560	31153	14183	56353	83250	38294	18201
98621	84064	47415	95665	93820	83278	83169	27831	98405
18328	30120	87629	70058	78387	75388	01665	96022	96974
74339	84573	33771	49295	44655	67986	47967	65168	55103
48605	10226	89975	96608	62499	87411	17786	26172	51980
33518	14979	81644	74439	05423	57926	61494	20932	39230
51594	16410	14971	95952	27475	17701	06534	03898	56064
76758	78109	16635	92762	08013	06500	05015	66938	32416
15998	24339	07648	33168	18224	69085	10474	29263	02381
77459	98146	57121	75169	48567	85150	52773	69026	01845
16363	02719	22198	20662	42787	38567	17495	53670	02033
44092	05751	66307	45292	19334	08828	52746	86944	99757
90372	97120	45699	99889	47445	52929	54901	69435	18271
61835	67670	60336	26054	79324	91775	24997	75208	17646
25213	20209	12528	62499	02435	67974	02113	99980	77052
17373	93101	44764	93063	95469	38547	87416	49835	07386
24004	21302	07243	16405	53033	67377	05320	97912	63779
51205	88783	19446	90353	20185	90748	32495	29063	86533
09931	71773	51215	15456	44771	53316	39425	95154	29705
84794	79070	88377	96671	61516	47024	49223	07032	95299
56439	44950	58721	88768	38361	90893	45666	15275	65793
05133	00035	20035	33952	88667	94896	08728	23770	98016

09271	14121	72138	46183	80606	27366	64281	09268	91687
26014	21851	13183	42368	42640	05439	27749	41974	65176
91564	43821	79359	61772	65333	80059	05310	45707	35031
15870	23835	00596	77829	48670	68908	49106	52076	49988
45966	89965	65797	01425	32255	11240	43490	02836	58560
55785	68397	28224	81887	39831	90537	88485	23419	68558
25716	57223	09434	25663	45629	92289	77125	51992	55082
16173	81330	53130	05431	08301	66631	24941	91385	94820
67375	92665	93498	26646	66068	29986	08004	76321	42363
73174	07833	96387	87631	14247	96055	10231	81766	47764
88788	47107	19403	97531	54644	67066	93256	78881	00124
92351	34851	60579	73815	32873	35824	16473	06558	35381
67301	54071	88766	65327	33815	21769	13281	91295	87940
12016	01553	44669	06535	68984	74464	10470	03201	44164
14197	96247	65313	79877	31240	21509	08961	83075	43913
46608	06132	32172	72646	70801	10053	71191	95739	94286
19940	78844	19424	65159	55326	98343	41223	23080	25925
83673	66882	23876	73828	78551	64134	87630	35698	75159
28121	88665	36571	51052	52874	65117	18587	40729	43639
16321	46418	07337	62203	59318	67395	18863	38108	57817
00726	94505	45754	42513	04274	22718	30001	27431	50377
16119	89218	27219	88613	13316	85674	77057	73205	65766
24332	15244	76139	52918	63831	12993	75468	01335	89271
96889	87155	96210	60509	93278	77961	33374	50595	31377
72709	72773	67810	12529	38271	93242	78257	24606	00258
85108	22417	93383	02512	95727	17801	47808	88766	33535
44544	12092	66656	87003	06146	17421	92785	43188	88277
80586	77372	73473	36513	79823	17854	27079	59619	19581
82004	37139	13473	61937	24675	09656	51300	59897	63976
85498	56102	70765	97085	88532	11250	80752	95475	75066
43848	21931	85931	30414	75018	17463	09198	85268	60836
55532	43497	34954	02847	13541	36903	96290	13112	73483
50693	34796	86999	42357	64927	81431	48734	15435	41875
07169	51679	77409	85545	56960	31012	00060	86255	10395
54615	36912	86515	08564	99535	75780	48283	19689	94222
50191	47693	31598	53627	80818	31902	72011	00837	48576
54105	47807	88871	23772	27689	26181	44898	40946	88565
78721	47619	12932	60587	70959	99451	64505	06009	97613
80666	54708	09279	05651	00861	16469	76014	04691	59494
71365	56598	41482	32666	82410	75280	78673	45695	56963
73096	69899	72701	44459	06563	21450	38041	21727	17726
24881	13830	32041	86367	20436	11110	26302	63801	63854
28022	58835	80954	00141	19869	34375	97847	80024	16822
57893	85057	13002	75427	40254	26617	88895	00056	61951
14058	76454	16544	47941	86903	81315	16287	06463	37793
33327	99153	56129	61558	66546	62734	24259	70591	12369
52883	14700	99358	76401	96405	75370	88502	33893	60475
94231	90228	61464	31781	54900	52396	65982	19774	00783
29109	49886	73740	94847	45344	93850	94092	24420	72825
37290	61291	48075	96043	17399	39445	88389	83035	97943
73751	56771	99869	03652	36968	07040	66382	84774	99326
78515	80860	58541	18263	84417	85001	46923	04443	06163
21570	12411	70755	13033	06144	46189	75605	57093	65725
86600	90902	50451	64652	12407	67311	93642	80618	31107
10961	44418	53433	66411	21083	19522	73564	00436	61706
68738	41608	67496	48059	21250	98716	66540	45959	38826
05718	26266	54091	35154	31922	41873	07925	53557	22489
78469	42051	40057	62558	09206	81372	94627	76128	41416
42128	17579	14569	72507	67577	69620	09011	36029	42809
48944	11900	94350	71696	09349	56039	29803	69263	17083
17581	85707	86695	64250	11310	84514	70706	49899	45257
96459	92341	19729	48855	26314	75117	16815	99611	93242
83560	49089	41509	43566	08623	06343	32840	11146	74632
06541	04092	35923	29980	65908	65007	54308	61528	21683
69075	02792	48727	44129	95957	01158	31965	32690	72583
11269	99090	02161	85201	11590	90354	25531	25885	60830
99955	96090	19180	34698	63605	59189	23715	82102	80560
15816	13021	24063	11403	75737	92894	12338	84391	91062
32493	59419	64387	80167	40259	22665	70669	58646	19156
73523	90085	24101	37380	10769	68389	77612	73799	60231
41423	68907	76067	15046	95215	38187	96503	72843	00511
61584	81842	70720	00080	49732	05470	10361	38390	04685

79480	01759	80315	81115	82938	79467	52140	89328	55936
84681	38650	68625	10126	89096	00322	70714	98093	21531
91944	06239	04284	69820	66464	32687	67819	90649	51047
47318	61100	61508	87044	90432	76083	05674	82251	70993
82670	80830	95370	52422	54513	98909	77672	91942	25461
49145	81955	07254	94787	56740	73487	94410	49677	16742
87633	53864	52530	03781	60322	39639	02109	93619	12066
48178	32650	25806	04113	91624	54058	60836	51178	15677
26441	59454	68363	06205	79890	45781	79923	85579	21950
05948	67387	48967	92465	84091	16373	65014	43113	26720
47552	77499	00364	50359	60428	82415	82848	82247	07012
47841	85655	51756	91752	98607	15656	29298	28272	28683
54987	04672	43622	13146	46388	62945	86570	82120	21587
49522	15111	11570	88806	20414	25904	95538	39111	67068
91796	45118	81536	21379	83599	96297	16895	18884	48968
44681	27319	43120	20267	86169	00195	11280	91729	20201
83737	93204	64448	29959	75734	03684	11507	06692	87685

GLOSSARY

alternate hypothesis A claim about the population that is accepted if the null hypothesis is rejected.

analysis of variance (ANOVA) A statistical technique used to determine whether more than two populations have the same mean.

arithmetic mean The sum of the values divided by the total number of values.

average A single value that is representative of the set of data.

autocorrelation A condition in regression in which successive values of the dependent variable are related.

central limit theorem The distribution of the sample means approaches the normal probability distribution as the sample size increases, regardless of the shape of the population.

central tendency A measure that describes the middle of a distribution.

Chebyshev's Theorem The proportion of observations lying within k standard deviations of the mean is at least $1 - 1/k^2$.

chi-square distribution A positively skewed continuous probability distribution based on the number of degrees of freedom. It is nonnegative and approaches a symmetric distribution as the number of degrees of freedom increases.

class An interval within which data are tallied.

class frequency The number of observations, or tallies, that occur in each class.

classical probability Number of favorable outcomes divided by total number of possible outcomes.

coefficient of correlation A measure of the strength of the association between two interval-scaled variables. It may range from -1.0 to 1.0 with -1.0 and 1.0 indicating perfect correlation and 0 indicating the absence of correlation.

coefficient of determination The proportion of the total variation in one variable that is explained by the variation in the other variable.

coefficient of *non*determination The proportion of the variation in one variable *not* explained by the variation in another variable.

coefficient of skewness A measure of the lack of symmetry in a distribution.

coefficient of variation The standard deviation divided by the mean. A measure of the relative dispersion of a data set.

combination One particular group of objects or persons selected from a larger group.

Complement Rule The probability of an event not happening. It is found by subtracting the probability of it happening from 1.0, written $P(\text{not } A) = 1 - P(A)$.

conditional probability The likelihood that an event will occur, assuming that another event has already occurred.

confidence interval A range within which the population parameter is expected to fall for a preselected level of confidence.

contingency table Frequency data from the simultaneous classification of more than one variable or trait of the observed item, often tallied in one table.

continuous probability distribution A probability distribution that assumes an infinite number of values within a specific range of values.

correlation analysis The statistical techniques used to determine the strength of the relationship between two variables. The basic objective of correlation analysis is to determine the degree of correlation (relationship) between variables, from zero (no correlation) to perfect (complete) correlation.

critical value The value(s) that separate the region of rejection from the remaining values.

cumulative A result of adding or collecting from one extreme value.

decision rule A statement of the condition or conditions under which the null hypothesis is rejected.

degrees of freedom The number of items in a sample that are free to vary.
dependent samples Samples that are paired or related in some fashion.
dependent variable The variable that is being predicted or estimated.
descriptive statistics The methods used to describe the data that have been collected.
discrete probability distribution A distribution that can assume only certain values. It is usually the result of counting the number of favorable outcomes of an experiment.

event A collection of one or more outcomes of an experiment.
exhaustive Each person, object, or item must be classified in at least one category.
experiment The observation of some activity, or the act of taking some type of measurement.

***F* distribution** A continuous probability distribution where the value of F is always positive. The distribution is always positively skewed.
finite correction factor A term that reduces (corrects) the standard error estimate when samples are taken from a finite population.
first quartile The point below which one-fourth of the observations occur.
frequency distribution An arrangement of the data that shows the frequency of occurrence of the values of interest.
frequency polygon A chart using straight lines to graphically portray the frequency distribution.

general rule of addition If two events are combined that are not mutually exclusive, the probability that one or the other will occur is their sum minus the probability of the joint occurrence, written $P(A \text{ or } B) = P(A) + P(B) - P(A \text{ and } B)$.
general rule of multiplication The probability of the events A and B occurring is the probability of event A times the probability of event B, given that event A has occurred, written $P(A \text{ and } B) = P(A) \cdot P(B \mid A)$.
geometric mean The nth root of the product of n measurements.
goodness-of-fit test A test to determine if an observed set of frequencies could have been obtained from a population with a given distribution.

histogram A chart using bars to graphically portray the frequency distribution.
homoscedasticity A condition in regression in which the dependent variable has the same variance for all values of the independent variable.

independent events The occurrence of one event does not affect the probability that the other event will take place.
independent samples Samples that are not related in any way.
independent variable A variable that provides the basis for estimation. It is the predictor variable and is usually designated X.
inferential statistics A decision, estimate, prediction, or generalization about a population based on a sample.
interquartile range The difference between the third quartile and the first quartile.
interval estimate A range of values within which we have some confidence that the population parameter lies.
interval scale The distance between numbers is a known, constant size, but the zero value is arbitrary.

joint probability A measure of the likelihood that two or more events will happen concurrently.

Kruskal–Wallis test A test based on ranks to determine whether more than two samples come from the same population.

least-squares method A method used to determine the regression equation by minimizing the sum of the squares of the distances between the actual Y values and the predicted values of Y.
level of significance The probability of rejecting the null hypothesis when it is true.

mean deviation The average of the deviations between all the observations in a set of data and its mean.
median The midpoint of the values after they have been arranged from the smallest to the largest (or the largest to the smallest).
midpoint A point that divides a class into two equal parts.
mode The value of the observation that appears most often.

multicollinearity A condition that occurs in multiple regression analysis if the "independent" variables are highly correlated.
mutually exclusive An individual or item that, by virtue of being included in one category, must be excluded from any other category.

negatively skewed distribution A distribution having a mean smaller than the median and mode. The long tail of the distribution is to the left, or in a negative direction.
nominal scale Data that are organized into categories the order of which is not important.
nonparametric tests Tests where assumptions regarding the shape of the population are not required. Such tests are applicable for nominal or ordinal scale of measurement.
nonprobability sampling Items whose inclusion in the sample is based on the judgment of the person selecting the sample.
normal distribution A particular "bell-shaped" probability distribution that is completely described by μ and σ.
null hypothesis A claim about the value of a population parameter.

one-tailed test A hypothesis test in which the rejection region is in one tail of the sampling distribution.
ordinal scale Data or categories that can be ranked; that is, one category is higher than another.
outcome A particular result of an experiment.

***p*-value** The probability that a test statistic in a hypothesis test is at least as extreme as the one obtained.
parameter One measurable characteristic of a population.
partial correlation A measure that shows the relationship between the dependent variable and an independent variable not yet considered in the multiple regression equation when the other independent variables in the equation are considered but held constant.
permutation An ordered arrangement of a group of objects.
point estimate The value, computed from a sample, used to estimate a population parameter.
Poisson distribution A discrete probability distribution that describes a purely random process.
pooled variance estimate An estimate of the population variance obtained by combining two or more sample variances.
population A collection of all possible members of a set of individuals, objects, or measurements.
positively skewed distribution A distribution having a mean larger than the median or mode. The long tail of the distribution is to the right, or in a positive direction.
probability A fraction that measures the likelihood that a particular event will occur.
probability distribution A listing of the outcomes that may occur and of their corresponding probabilities.
probability sampling A method of sampling in which each member of the population of interest has a known likelihood of being included in the sample.
proportion A fraction, ratio, percent, or probability that indicates what part of the sample or population has a particular trait.

quartile deviation One-half of the interquartile range. To compute the quartile deviation, the first quartile is subtracted from the third quartile and the difference is divided by two.

random sample A sample chosen so that each member of the population has the same chance of being selected.
random variable A quantity that assumes one and only one numerical value as a result of the outcome of an experiment.
range The difference between the highest and lowest observations in a set of data.
rank-order correlation coefficient A measure of the strength of association between two ordinal-scaled sets of variables. It may range from -1.0 to 1.0.
ratio scale Data possessing a natural zero point and organized into measures for which differences are meaningful.
raw data Numerical information presented in an ungrouped form.
regression equation A mathematical equation that defines the relationship between two variables.
relative frequency The number of times a particular event occurred in the past divided by the total number of observations.

sample A part or portion of the population.
sampling distribution of the mean A probability distribution of all possible sample means of a given size selected from a population.
sampling distribution of the sample proportion A probability distribution of all possible proportions of a given size sample selected from a population.

sampling error The difference between the value of the population parameter and its corresponding sample statistic.
scatter diagram A graphic tool that visually portrays the relationship between two variables.
sign test A test based on the sign of the differences in a set of paired observations.
special rule of addition If two mutually exclusive events A and B are combined, the probability that one or the other will happen is their sum, written $P(A \text{ or } B) = P(A) + P(B)$.
special rule of multiplication A rule for combining two or more independent events. The probability of the joint occurrence of two events is $P(A \text{ and } B) = P(A) \cdot P(B)$.
standard deviation The square root of the arithmetic mean of the squared deviations from the mean.
standard error of estimate A measure of the variability of the observed values around the regression line.
standard error of a proportion The standard deviation of the sampling distribution of the sample proportion.
standard error of the mean The standard deviation of the sampling distribution of the sample means.
standard normal distribution A special normal distribution with a mean of zero and a standard deviation of one unit.
stated limits The actual boundaries for a particular class in a frequency distribution.
statistic One measurable characteristic of a sample.
statistics The body of techniques used to facilitate the collection, organization, presentation, analysis, and interpretation of data for the purpose of making better decisions.
stem-and-leaf chart A histogram in which the tallies are replaced by digits in order to present data.
stratified random sample After the population of interest is divided into logical strata, a sample is drawn from each stratum or subgroup. The manner in which the sample is gathered may be either nonproportional or proportional to the total number of members in each stratum.
Student's *t* distribution A continuous probability distribution with a mean of zero. It is flatter at the apex and more spread out than the standard normal distribution.
subjective probability The likelihood of an event assigned on the basis of whatever information is available.
symmetric distribution A distribution that has the same shape on both sides of the median.
systematic random sample The members of the population are arranged in some fashion. A random starting point is selected. Then every kth element is chosen for the sample.

test statistic A quantity, calculated from the sample information, used as a basis for deciding whether or not to reject the null hypothesis.
third quartile The point below which three-fourths of the observations occur.
treatment A specific source or cause of variation in a set of data.
true limits The real boundaries of a class, given continuous data.
two-tailed test A hypothesis test in which the rejection region is divided equally between the two tails of the sampling distribution.
Type I error An error that occurs when a true null hypothesis is rejected.
Type II error An error that occurs when a false null hypothesis is not rejected.

weighted mean The values of the observations are weighted by the frequency of occurrence.
Wilcoxon rank-sum test A test based on the sum of the ranks to determine whether two samples came from the same population.
Wilcoxon signed-rank test A test applied to paired data. It is a replacement for the paired t test.

***Y*-intercept** The point at which the regression equation crosses the Y-axis.

***z*-value or *z*-score** A unit of measure with respect to the standard normal distribution. It measures the distance from the mean of a normal distribution in terms of the number of standard deviations.

ANSWERS TO CHAPTER ACHIEVEMENT TESTS

CHAPTER ONE

1. False, ordinal **2.** True **3.** False, nominal **4.** True
5. False, inferential statistics **6.** False, sample **7.** True **8.** True

CHAPTER TWO

1. b **2.** b **3.** a **4.** b **5.** a **6.** b **7.** a **8.** c **9.** a
10. c **11.**

Time	Patients	
1–5	///	3
6–10	~~////~~ /	6
11–15	~~////~~	5
16–20	~~////~~ /	6
21–25	~~////~~	5
		25

12.

Tread Depth ($\frac{1}{32}$")	Frequency	Percent Frequency	Less Than Cummulative Frequency
0–3	4	7	4
4–7	15	29	19
8–11	25	48	44
12–15	5	10	49
16–19	3	6	52
	52	100	

a.

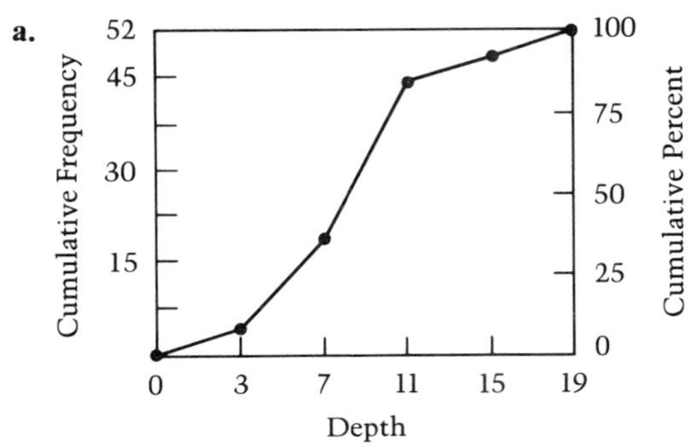

b. 19 out of 52, or 37%, have less than $\frac{7}{32}$″ Tread Depth.
c. 40% have less than $\frac{8}{32}$″ Tread Depth.

13.

Expense	Amount	Percent of Total
Housing	$ 400	38%
Utilities	140	13
Medical	25	2
Food	190	18
Transportation	150	14
Clothing	50	5
Savings	50	5
Miscellaneous	50	5
	$1,055	

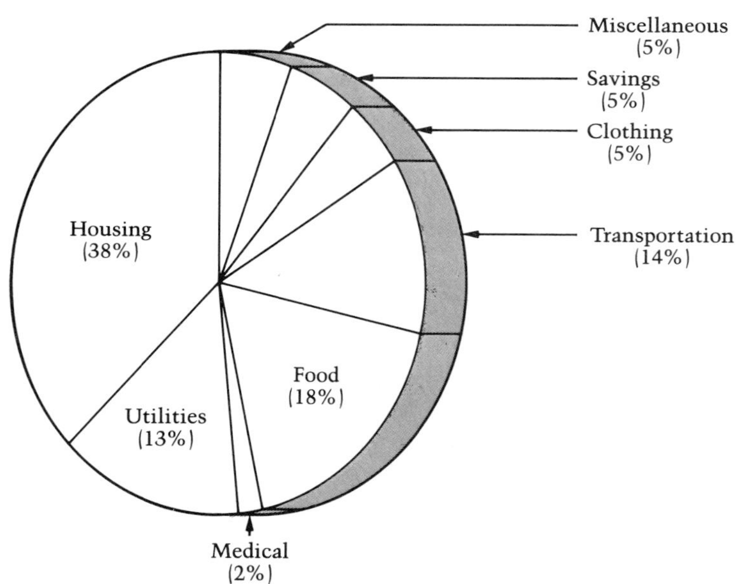

CHAPTER THREE

1. a **2.** c **3.** c **4.** b **5.** a **6.** b **7.** c **8.** b **9.** d

10. a **11. a.** $\frac{1860}{4} = \$465$ **b.** $432.50 **12.** $Md = \$76.10$

13.

Pounds	f	X	fX	CF
0.0 to 0.4	3	0.2	0.6	3
0.5 to 0.9	8	0.7	5.6	11
1.0 to 1.4	20	1.2	24.0	31
1.5 to 1.9	10	1.7	17.0	41
2.0 to 2.4	6	2.2	13.2	47
2.5 to 2.9	3	2.7	8.1	50
	50		68.5	

$$\bar{X} = \frac{\Sigma fX}{n} = \frac{68.5}{50} = 1.37 \text{ pounds}$$

14. $\text{Median} = L + \frac{(n/2) - CF}{f}(i) = 0.95 + \frac{(50/2) - 11}{20}(0.5) = 1.30 \text{ pounds}$

15. Mode = 1.20 pounds

CHAPTER FOUR

1. c **2.** c **3.** c **4.** d **5.** b **6.** d **7.** b **8.** b

9. b **10.** a

11. a. range = 6 − 2 = 4

b. $\bar{X} = 20/5 = 4$; $MD = \frac{2 + 2 + 1 + 0 + 1}{5} = 1.2$

c. Sample variance: $s^2 = \frac{90 - \frac{(20)^2}{5}}{5 - 1} = 2.5$

d. $s = \sqrt{2.5} = 1.58$

12. a. 79 − 20 = 59

b. $Q_1 = 39.5 + \frac{30 - 23}{20}(10) = 43.0$

$Q_3 = 49.5 + \frac{90 - 43}{50}(10) = 58.9$

$QD = \frac{58.9 - 43.0}{2} = 7.95$

c. $s^2 = \frac{\Sigma fX^2 - \frac{(\Sigma fX)^2}{n}}{n - 1} = \frac{[335{,}610 - (6160)^2/120]}{119} = 163.0$

$s = \sqrt{163} = 12.8$

13. a. $CV = \frac{9}{87}(100) = 10.3$ in the Sanford district versus $\frac{12.8}{51.3}(100) = 24.9$ in the Jefferson district.

b. $sk = \frac{3(\$87{,}000 - \$84{,}000)}{\$9{,}000} = 1.00$ in the Sanford district

$sk = \frac{3(\$51{,}333 - \$52{,}900)}{\$12{,}800} = -0.37$ in the Jefferson district

A moderate positive skew in Sanford versus a slight negative skew in Jefferson.

CHAPTER FIVE

1. c **2.** c **3.** d **4.** c **5.** a **6.** b **7.** b **8.** b

9. b **10.** c **11.** b **12.** a **13.** c **14.** d **15.** b

16. $\frac{325}{500} = 0.65$

17. a.

	Men	Women	Totals
In State	5	18	23
Out of State	5	2	7
Totals	10	20	30

b. $\frac{5}{30} = 0.167$

c. $\frac{5}{7} = 0.714$

d. $\frac{12}{30} = 0.4$

18. $\frac{6}{10} \cdot \frac{5}{9} = \frac{1}{3} = 0.333$ **19.** $(0.35)(0.25) = 0.0875$

20. 0.60 **21.** $5/8 = 0.625$

22. a. 0.33 **b.** $0.67 + 0.23 - 0.12 = 0.78$

23. ${}_{30}P_3 = 30 \cdot 29 \cdot 28 = 24{,}360$

24. ${}_{15}C_3 = \frac{15!}{12!\,3!} = \frac{15 \cdot 14 \cdot 13}{3.2} = 455$

25. $\frac{1}{{}_{30}C_5} = \frac{5!\,25!}{30!} = \frac{5 \cdot 4 \cdot 3 \cdot 2}{30 \cdot 29 \cdot 28 \cdot 27 \cdot 26} = 0.000007$

CHAPTER SIX

1. a **2.** b **3.** b **4.** c **5.** a **6.** c **7.** b **8.** a

9. b **10.** c

11. $\pi = 0.9, n = 8$

a. $P(X = 6) = \frac{8!}{6!\,2!}(0.9)^6(0.1)^2 = 0.1488$

b. $P(X \geq 6) = 0.149 + 0.383 + 0.430 = 0.962$

c. $P(3) + P(4) + P(5) = 0.000 + 0.005 + 0.033 = 0.038$

12. $\mu = 150(1/50) = 3$

a. $P(0) = \frac{3^0e^{-3}}{0!} = 0.0498$

b. $P(2) = \frac{3^2e^{-3}}{2!} = 0.2240$

c. $P(X < 4) = P(0) + P(1) + P(2) + P(3)$

$P(1) = \frac{3^1e^{-3}}{1!} = 0.1494$ $\qquad P(2) = \frac{3^2e^{-3}}{2}$

$P(3) = \frac{3^3e^{-3}}{3!} = 0.2240$

$P(X < 4) = 0.0498 + 0.1494 + 0.2240 + 0.2240 = 0.6472$

d. $1 - P(0) = 0.9502$

13.

X	*P(X)*	*X · P(X)*	$(X - \mu)^2$	$(X - \mu)^2P(X)$
0	0.05	0.00	5.5225	0.2761
1	0.15	0.15	1.8225	0.2734
2	0.30	0.60	0.1225	0.0368
3	0.40	1.20	0.4225	0.1690
4	0.10	0.40	2.7225	0.2723
		2.35		1.0275

$\mu = 2.35 \qquad \sigma = \sqrt{1.0275} = 1.0137$

CHAPTER SEVEN

1. a **2.** d **3.** b **4.** d **5.** c **6.** b **7.** a **8.** a

9. b **10.** d

11. $\mu = 3.00, \sigma = 1.00$

	Answer
a. $z = (4.5 - 3.00)/1.00 = 1.50$	0.4332
b. $z = (4.00 - 3.00)/1.00 = 1.00$	$0.5000 - 0.3413 = 0.1587$
c. $z = (2.00 - 3.00)/1.00 = -1.00$	
$z = (3.50 - 3.00)/1.00 = 0.50$	$0.3413 + 0.1915 = 0.5328$
d. $z = (1.00 - 3.00)/1.00 = -2.00$	$0.5000 - 0.4772 = 0.0228$
e. $1.28 = (x - 3.00)/1.00$	
$x = 3.00 + 1.28(1.00) = 4.28$	

12. $\mu = n\pi = 80(0.40) = 32$ $\qquad \sigma^2 = n\pi(1 - \pi) = 80(0.40)(0.60) = 19.2$

$\sigma = \sqrt{19.2} = 4.38$

a. $z = (25.5 - 32)/4.38 = -1.48$ $\qquad 0.4306 + 0.5000 = 0.9306$

b. $z = (40.5 - 32)/4.38 = 1.94$ $\quad 0.5000 - 0.4738 = 0.0262$
c. $z = (34.5 - 32)/4.38 = 0.57$ $\quad 0.5000 + 0.2157 = 0.7157$
d. $z = (29.5 - 32)/4.38 = -0.57$
$z = (36.5 - 32)/4.38 = 1.03$ $\quad 0.2157 + 0.3485 = 0.5642$

13. a. 0.0838 **b.** 0.7148

CHAPTER EIGHT

1. d **2.** b **3.** c **4.** c **5.** c **6.** c **7.** c **8.** d
9. b **10.** b

11. a. $\sigma_{\bar{x}} = \dfrac{\sigma}{\sqrt{n}} = \dfrac{\$70}{\sqrt{70}} = \$8.37$

b. $X \pm z\sigma_{\bar{x}} = \$510 \pm 1.96(8.37)$
$= \$510 \pm 16.40$; from \$493.60 up to \$526.40

12. a. $\dfrac{10}{\sqrt{50}} = 1.41$ years

b. $38 \pm 2.58(1.41) = 38 \pm 3.6$; from 34.4 up to 41.6 years

13. a. $\sigma_P = \sqrt{\dfrac{\pi(1-\pi)}{n}} = \sqrt{\dfrac{(0.7)(0.3)}{20}} = 0.10$

b. $0.65 \pm 2.58(0.10) = 0.65 \pm 0.26$; from 39 up to 91%

14. a. $\sqrt{\dfrac{(0.75)(0.25)}{500}} = 0.019$

b. $0.75 \pm 1.65(0.019) = 0.75 \pm 0.03$; from 0.72 up to 0.78

15. $n = \left(\dfrac{zs}{E}\right)^2 = \left[\dfrac{(1.96)(3000)}{200}\right]^2 = 865$

CHAPTER NINE

1. a **2.** b **3.** c **4.** d **5.** b **6.** c **7.** b **8.** b
9. c **10.** b

11. a. $H_0: \mu_1 = \mu_2 \qquad H_a: \mu_1 \neq \mu_2$

b. If z is not between -1.81 and 1.81, reject H_0.

c. $z = \dfrac{5.21 - 4.65}{\sqrt{\dfrac{(2.3)^2}{40} + \dfrac{(1.9)^2}{30}}} = \dfrac{0.56}{0.50} = 1.11$

d. There is *no* difference in the mean length of stay in the two cities.

12. a. $H_0: \mu = \$19{,}500 \qquad H_a: \mu < \$19{,}500$

b. If $z < -1.28$, reject H_0.

c. $z = \dfrac{\bar{x} - \mu}{\sigma/\sqrt{n}} = \dfrac{18{,}750 - 19{,}500}{1581/\sqrt{50}} = \dfrac{-750}{224} = -3.35$

d. Reject H_0. The Spartan accountants *do* average less than the national mean.

CHAPTER TEN

1. c **2.** c **3.** c **4.** d **5.** b

6. a. $H_0: \pi_1 = \pi_2$ $H_a: \pi_1 < \pi_2$

b. If $z < -2.33$, reject H_0.

c. $\bar{p} = \dfrac{9 + 11}{316 + 214} = 0.0377$

$$z = \frac{\dfrac{9}{316} - \dfrac{11}{214}}{\sqrt{(0.0377)(0.9623)\left(\dfrac{1}{316} + \dfrac{1}{214}\right)}} = \frac{-0.023}{0.017} = -1.36$$

d. The null hypothesis cannot be rejected. The data prove nothing.

7. a. $H_0: \pi = 0.38$ $H_a: \pi \neq 0.38$

b. If z is not between -2.33 and 2.33, reject H_0.

c. $z = \dfrac{\dfrac{10}{25} - 0.38}{\sqrt{\dfrac{(0.38)(0.62)}{25}}} = \dfrac{0.02}{0.097} = 0.21$

d. The null hypothesis should *not* be rejected. The data do not contradict Congresswoman Jones.

8. $H_0: \pi = 0.50$ $H_a: \pi \neq 0.50$

If z is not between -1.96 and 1.96, reject H_0.

$$z = \frac{\dfrac{125}{240} - 0.50}{\sqrt{\dfrac{(0.5)(0.5)}{240}}} = \frac{0.02}{0.032} = 0.62$$

The mix seems to contain the advertised proportion.

9. $H_0: \pi_1 = \pi_2$ $H_a: \pi_1 \neq \pi_2$

If z is not between -2.58 and 2.58, reject H_0.

$\bar{p} = \dfrac{10 + 30}{18 + 72} = 0.444$

$$z = \frac{\dfrac{10}{18} - \dfrac{30}{72}}{\sqrt{(0.444)(0.556)\left(\dfrac{1}{18} + \dfrac{1}{72}\right)}} = \frac{0.139}{0.13} = 1.06$$

We fail to reject the null hypothesis. These data show no significant difference.

CHAPTER ELEVEN

1. d **2.** d **3.** b **4.** b **5.** b **6.** b **7.** d **8.** a

9. b **10.** a **11.** c

12. $H_0: \mu \geq 98.6^\circ$ F $H_a: \mu < 98.6^\circ$ F $\alpha = 0.05$

df $= n - 1 = 25 - 1 = 24$

Reject H_0 if t is less than -1.711; otherwise, fail to reject.

$$t = \frac{\bar{X} - \mu}{\frac{s}{\sqrt{n}}} = \frac{98.3 - 98.6}{\frac{0.64}{\sqrt{25}}} = -2.34$$

H_0 is rejected and H_a accepted. The evidence indicates unusually low body temperatures.

13. $H_0\colon \mu_d \leq 0$ $\quad H_a\colon \mu_d > 0$ $\quad \alpha = 0.01$ $\quad \text{df} = n - 1 = 7 - 1 = 6$
Reject H_0 if t is greater than 3.143; otherwise, fail to reject H_0.

Before	After	d	d^2
180	160	20	400
156	164	−8	64
188	172	16	256
132	130	2	4
208	200	8	64
196	190	6	36
190	184	6	36
		50	860

$\bar{d} = 50/7 = 7.14$

$$s_d = \sqrt{\frac{860 - \frac{(50)^2}{7}}{6}} = 9.15$$

$$t = \frac{7.14}{\frac{9.15}{\sqrt{7}}} = 2.06$$

A significant reduction is not shown by these data.

14. $H_0\colon \mu_1 = \mu_2$, $H_a\colon \mu_1 \neq \mu_2$, $\alpha = 0.05$, $\text{df} = n + n - 2 = 5 + 8 - 2 = 11$
Reject H_0 if t is less than -2.201 or more than 2.201; otherwise, fail to reject H_0.

$$s_1^2 = \sqrt{\frac{7{,}734 - \frac{(178)^2}{5}}{5 - 1}} = 18.69 \qquad s_2^2 = \sqrt{\frac{7{,}102 - \frac{(208)^2}{8}}{8 - 1}} = 15.56$$

$$s_p = \sqrt{\frac{4(18.69)^2 + 7(15.56)^2}{11}} = 16.77 \qquad t = \frac{35.6 - 26}{16.77\sqrt{\frac{1}{5} + \frac{1}{8}}} = 1.00$$

There is no evidence that one location is significantly better than the other.

CHAPTER TWELVE

1. a **2.** b **3.** a **4.** c **5.** c **6.** a **7.** a **8.** d

9. b **10.** a

11. a. $H_0\colon \mu_1 = \mu_2 = \mu_3$ $\qquad H_a$: At least one is different.

b.

Source	Sum of Squares	df	Mean Square
Treatments	0.45	2	0.225
Within	25.80	9	2.867
Total	26.25		

$$SST = \frac{(18)^2}{3} + \frac{(31)^2}{5} + \frac{(26)^2}{4} - \frac{(75)^2}{12}$$
$$= 0.45$$
$$SSE = 495 - \left[\frac{(18)^2}{3} + \frac{(31)^2}{5} + \frac{(26)^2}{4}\right]$$
$$= 25.80$$

c. The critical value is 4.26.

d. Because the test statistic is only 0.08, we cannot reject H_0. The means all appear to be equal.

CHAPTER THIRTEEN

1. c **2.** d **3.** c **4.** c **5.** a **6.** b **7.** c **8.** a
9. d **10.** c

11. a.

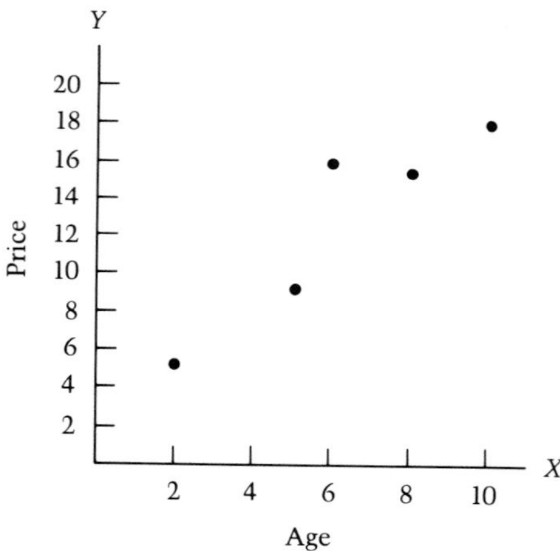

b.

X	Y	XY	X^2	Y^2	
2	5	10	4	25	
6	16	96	36	256	
10	18	180	100	324	
5	9	45	25	81	
8	15	120	64	225	
$\Sigma X = 31$	$\Sigma Y = 63$	$\Sigma XY = 451$	$\Sigma X^2 = 229$	$\Sigma Y^2 = 911$	$n = 5$

$$r = \frac{5(451) - (31)(63)}{\sqrt{[5(229) - (31)^2][5(911) - (63)^2]}} = 0.92$$

c. $r^2 = (0.92)^2 = 0.85$
$1 - r^2 = 1 - 0.85 = 0.15$

d. $H_0: \rho = 0 \qquad H_a: \rho \neq 0$
With 0.05 level of significance and 3 df, the critical value is 3.182.
$$t = \frac{0.92\sqrt{5-2}}{\sqrt{1-(.92)^2}} = 4.07 > 3.182. \text{ Reject } H_0.$$

e. The price of wine does increase with age. The price and age are positively correlated with 85% of variaton in one accounted for by variation of the other.

12. a.

Coaches	Sportswriters	d	d^2
1	1	0	0
2	5	−3	9
3	4	−1	1
4	6	−2	4
5	2	3	9
6	3	3	9
7	10	−3	9
8	11	−3	9
9	7	2	4
10	12	−2	4
11	8	3	9
12	9	3	9
			$\Sigma d^2 = 76$

$$r_s = 1 - \frac{6(76)}{12(12^2 - 1)} = 0.73$$

b. A positive correlation does exist between ranking among coaches and ranking among sportswriters.

c. $t = 0.73\sqrt{\frac{(12-2)}{1-(.73)^2}} = 3.38$

H_0: No correlation in population.
H_a: Positive rank correlation in population critical value is 1.812.
$t > 1.812$. Reject H_0. We can conclude that a significant correlation exists.

CHAPTER FOURTEEN

1. a **2.** c **3.** d **4.** a **5.** c **6.** a **7.** b **8.** c **9.** e

10. a.

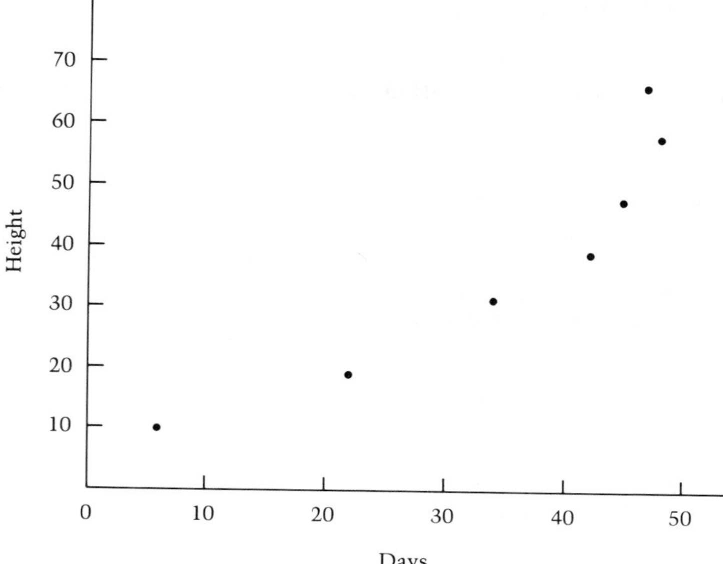

b.

Days X	Height Y	X^2	Y^2	XY
6	10	36	100	60
22	19	484	361	418
34	31	1,156	961	1,054
42	39	1,764	1,521	1,638
45	47	2,025	2,209	2,115
48	58	2,304	3,364	2,784
47	66	2,209	4,356	3,102
244	270	9,978	12,872	11,171

$$b = \frac{7(11{,}171) - 244(270)}{7(9{,}978) - (244)^2} = \frac{12{,}317}{10{,}310} = 1.195$$

$$a = \frac{270 - 1.195(244)}{7} = \frac{-21.58}{7} = -3.083 \qquad Y' = -3.083 + 1.195X$$

c. $$s_{Y \cdot X} = \sqrt{\frac{12{,}872 - (-3.083)(270) - (1.195)(11{,}171)}{7 - 2}} = 8.43$$

d. $Y' = -3.083 + 1.195(25) = 26.792$

$$26.792 \pm (2.015)(8.43)\sqrt{\frac{1}{7} + \frac{(25 - 244/7)^2}{9{,}978 - (244)^2/7}}$$

$26.792 \pm 7.762 = 19.030$ to 34.544

CHAPTER FIFTEEN

1. b **2.** c **3.** a **4.** b **5.** a **6.** c **7.** b **8.** d **9.** b

10. b **11.** a **12.** $Y' = 5.00 + 2.00X_1 + 0.50X_2 + 0.30X_3 + 1.00X_4$

13. $F = \frac{50}{4} = 12.50$, which is significant even at the 0.01 level (4.18 critical value). So at least one of the net regression coefficients is not zero.

14. The test statistic corresponding to X_1 is only 2.0. That value may not be significant and could be dropped. For a two-sided test with a 0.05 level of significance, the critical values are ± 2.06. At this level each of the other three variables appears significant.

CHAPTER SIXTEEN

1. c **2.** d **3.** d **4.** c **5.** a **6.** b **7.** a **8.** b

9. b **10.** a

11. H_0: There is no relationship H_a: There is a relationship.
df = (2 − 1)(3 − 1) = 2
If $\chi^2 > 5.991$, reject H_0.

	Favor		Oppose		Undecided		
	f_o	f_e	f_o	f_e	f_o	f_e	Totals
Democrats	22	21	13	14	28	28	63
Republicans	8	9	7	6	12	12	27
	30		20		40		90

$$\chi^2 = \frac{(22-21)^2}{21} + \frac{(13-14)^2}{14} + \frac{(28-28)^2}{28} + \frac{(8-9)^2}{9} + \frac{(7-6)^2}{6} + \frac{(12-12)^2}{12} = 0.40$$

We fail to reject H_0. There is no relationship.

12. $H_0: \pi_1 = 1/10, \pi_2 = 2/10, \pi_3 = 3/10, \pi_4 = 4/10$
H_a: At least one is different.

	f_o	f_e	$f_o - f_e$	$(f_o - f_e)^2/f_e$
Full-sized	38	20	18	16.2
Medium-sized	62	40	22	12.1
Compact	41	60	−19	6.0
Subcompact	59	80	−21	5.5
	200			39.8

The critical value is only 11.345, so something has changed!

CHAPTER SEVENTEEN

1. b **2.** b **3.** a **4.** a **5.** d **6.** c **7.** d **8.** b

9. a **10.** c

11. Consult a binomial table with $n = 15$ and $\pi = 0.5$. The probability of 13 or more "+" signs is only 0.003, which is smaller than 0.01. So the researcher could reject the null hypothesis of no difference. The lung capacity of nonsmokers is greater.

12.

Couple	d	R^+	R^-
A	12	10	
B	−4		4
C	2	2	
D	0		
E	8	9	
F	6	7	
G	3	3	
H	−1		1
I	−5		5.5
J	0		
K	7	8	
L	5	5.5	
M	0		
		44.5	10.5

The critical value is 3, so we fail to reject H_0. There is no significant difference between the sexes.

13.

Inner City		Suburban		Rural	
95	1	108	6.5	102	4
100	2.5	114	10	109	8
100	2.5	118	12	112	9
106	5	122	15	120	13.5
108	6.5	124	16	127	19
115	11	126	18	129	20
120	13.5	130	21.5	140	24
125	17	136	23		
130	21.5	142	25		
		150	26		
	80.5		173		97.5

$$H = \frac{12}{(26)(27)}\left[\frac{(80.5)^2}{9} + \frac{(173)^2}{10} + \frac{(97.5)^2}{7}\right] - 3(27)$$

$H = 5.68$

This is less than the critical value of 5.991, so we cannot conclude that a difference exists among the three groups.

Answers to Even-Numbered Chapter Exercises

CHAPTER ONE

2. **a.** Yes. The order of the categories could be changed. Race, for example, could be listed first followed by the sex of the smoker.
   b. Yes. Each smoker is listed in one category but excluded from the other categories.
   c. Yes. Every smoker who responded appears in at least one category.
   d. The percentage of male smokers dropped from about 50 percent in 1965 to about 32 percent in 1991. Likewise the percentage of whites who smoked decreased from about 40 percent of the population in 1965 to about 28 percent in 1991. And the percentage of blacks smoking declined from 43 percent to 33 percent in 1991.

4. **a.** Each degree is ranked higher than the previous one. Example: a doctor's degree is ranked higher than a master's degree. That assumes "first professional" is higher than master's. If not, the data is not ordinal.
   b. Yes. Each graduate is placed in one category but excluded from any other category.
   c. Yes. Each graduate appears in at least one category.
   d. In 1992 the largest number of degrees awarded was the bachelor's degree; doctorates the smallest number. There doesn't appear to be significant changes in each category from 1992 to 2000.

6. An overwhelming number of consumers tested (400/500, or 80%) believe this toothpaste is excellent. Based on these findings, we can expect a great majority of all consumers to feel the same way.

8. Answers will vary.

10. True.

CHAPTER TWO

2.

Number of Rolls	Class Frequencies
0–2	10
3–5	23
6–8	10
9–11	3
12–14	3
15–17	1
	Total 50

Based on the frequency distribution, the number of rolls ranges from 0 to 17. The largest concentration is in the 3–5 class.

4.

Number of Patrons	Class Frequencies
25–34	3
35–44	5
45–54	6
55–64	5
65–74	4
75–84	3
85–94	4
	Total 30

Based on the frequency distribution, the number of patrons using the library ranged from 25 to 94. The largest concentration is in the 45–54 class.

6.

Leading Digit	Trailing Digit
1	2479
2	112679
3	23456789
4	01235
5	37

8.

Leading Digit	Trailing Digit
0	678
1	2344889
2	01238
3	2468
4	0

Or, using groups of 5, the solution would be

Leading Digit	Trailing Digit
(0–4) 0	
(5–9) 0	678
(10–14) 1	2344
(15–19) 1	889
(20–24) 2	0123
(25–29) 2	8
(30–34) 3	24
(35–39) 3	68
(40–44) 4	0
(45–49) 4	

10. 

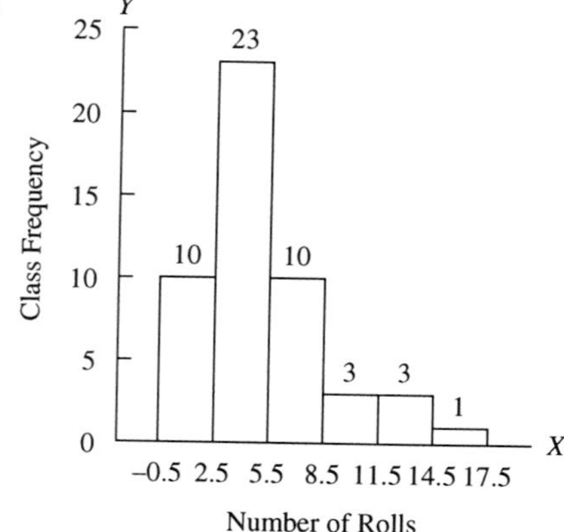

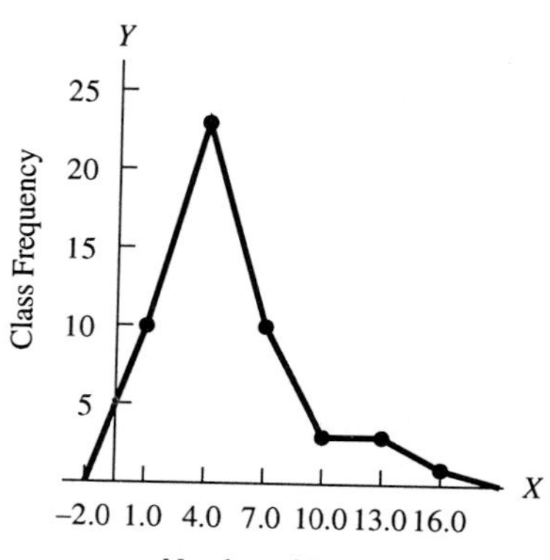

12.

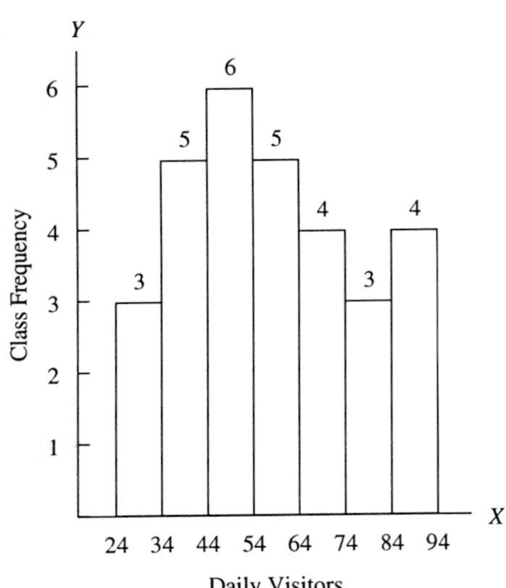

14. a.

Seconds	Cumulative Frequency
2.0–2.9	200
3.0–3.9	197
4.0–4.9	190
5.0–5.9	175
6.0–6.9	146
7.0–7.9	65
8.0–8.9	15
9.0–9.9	5

b. More-Than Cumulative Frequency Distribution:

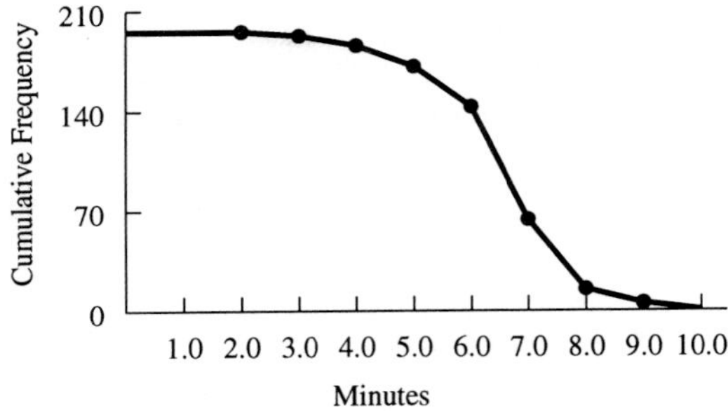

c. About 90% of the mice take more than 4.67 seconds to complete the maze.

16.

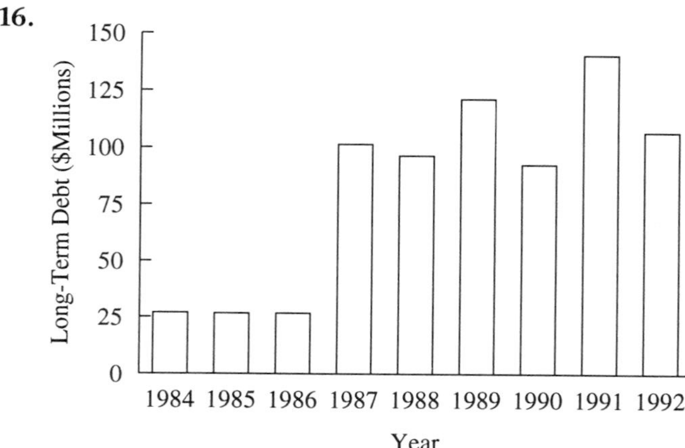

18.

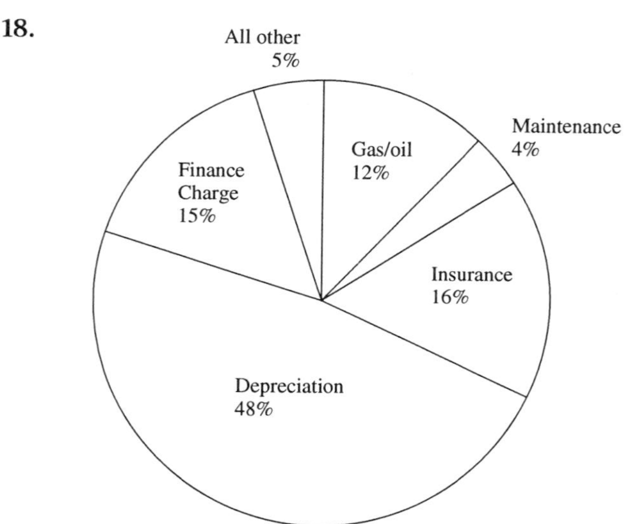

20. a.

Account Balance	Tallies	Number of Accounts	More-than Cumulative Frequencies
$ 0–99	𝍸 ////	9	40
100–199	𝍸 /	6	31
200–299	𝍸 /	6	25
300–399	𝍸 /	6	19
400–499	𝍸	5	13
500–599	//	2	8
600–699	/	1	6
700–799	///	3	5
800–899	/	1	2
900–999	/	1	1

b.

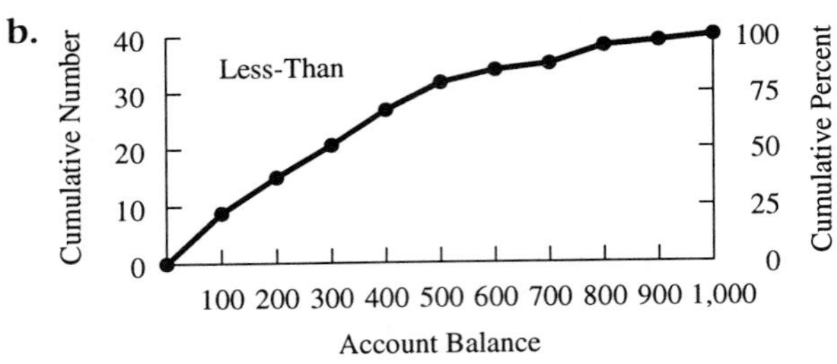

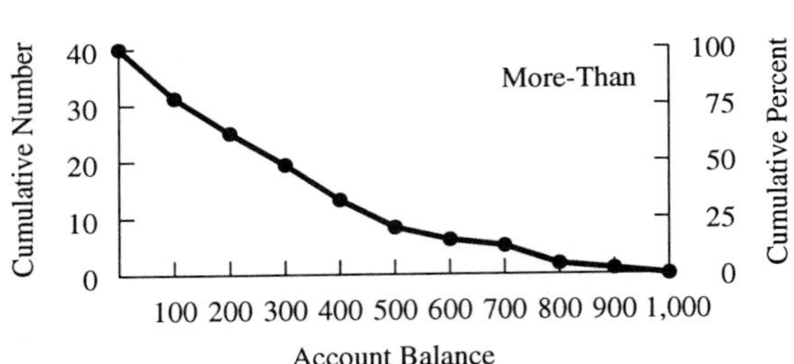

c. 20.0%

d. $400

22. a.

	Tallies	Total
4.5–5.4	~~////~~ ////	9
5.5–6.4	~~////~~ /	6
6.5–7.4	~~////~~ ~~////~~ ~~////~~ //	17
7.5–8.4	~~////~~ ///	8
8.5–9.4	///	3

b.

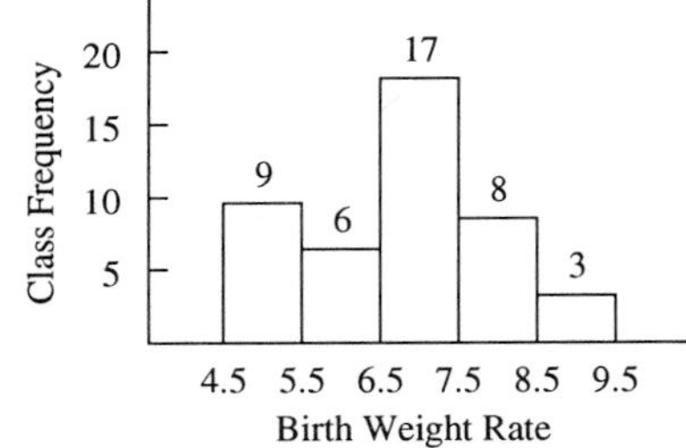

c.

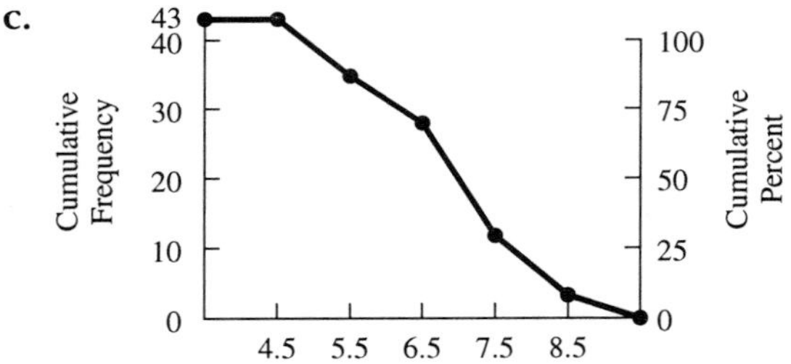

d. Approximately 20.

24. a.

Classes	Frequencies
Annual Percent Change	*Number of Countries*
0–4.9	5
5.0–9.9	11
10.0–14.9	4
15.0–19.9	0
20.0–24.9	2
25.0–29.9	0
30.0–34.9	0
35.0–39.9	1
	Total 23

b. Based on the frequency distribution, the annual average percent change in consumer prices during the past five years ranged from 0% to 39.9%. Practically all countries had annual percent increases between 0 and 14.9%, with the highest concentration between 5.0 and 9.9%.

c. Using an increment of 5 and MINITAB, we obtain the following stem-and-leaf display. (Note that the first row is 0   23444. The first leaf is 2, which represents 2.7%. The 7 was omitted. The second leaf is 3, representing 3.9%. The 9 was omitted by MINITAB.

```
MTB > stem c1;
SUBC> increment 5.

Stem-and-Leaf of C1          N = 23
Leaf Unit = 1.0

            STEM      LEAF
  (0-4)       0       23444
  (5-9)       0       56777788999
 (10-14)      1       2223
 (15-19)      1
 (20-24)      2       03
 (25-29)      2
 (30-34)      3
 (35-39)      3       7
```

d.

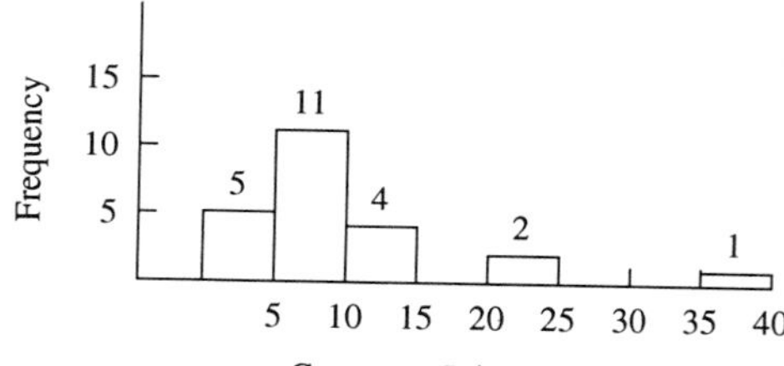

e.

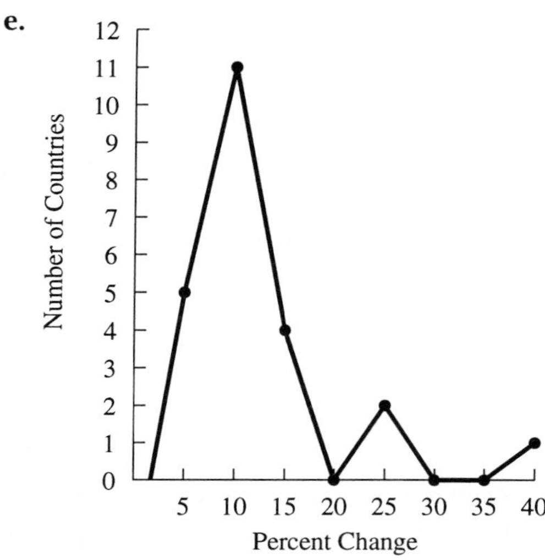

26. a. Using a class interval of $1,000 and starting at $15,000, we obtain the following:

Classes	Frequencies
Average Annual Pay	*Number of States*
$15,000–15,999	1
16,000–16,999	3
17,000–17,999	5
18,000–18,999	7
19,000–19,999	2
20,000–20,999	1
	Total 19

b. The concentration of annual pay is between $17,000 and $19,000. The lowest pay is about $15,000, the highest about $21,000.

c. The stem-and-leaf display for the average annual pay follows. Notice that the stem includes the "ten thousand" and "thousand" digits. The leaf is the "hundred" digit, and the "ten" and "unit" digits are omitted. Also note that values are truncated, not rounded.

```
Stem        Leaf
 15       4
 16       156
 17       33667
 18       1347789
 19       69
 20       2
```

d. Histogram using the stated class limits:

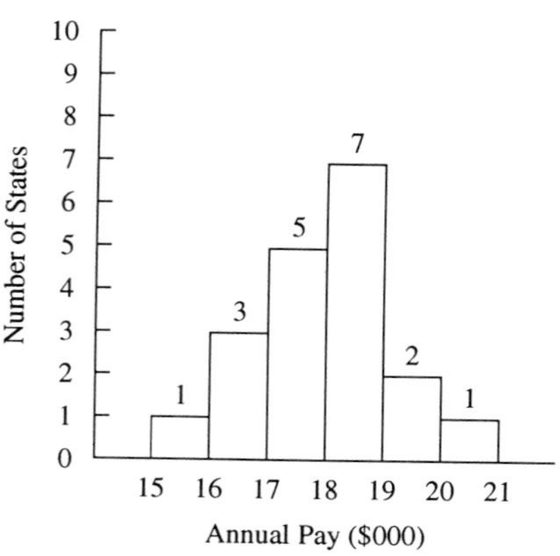

e. Frequency polygon using the class midpoints:

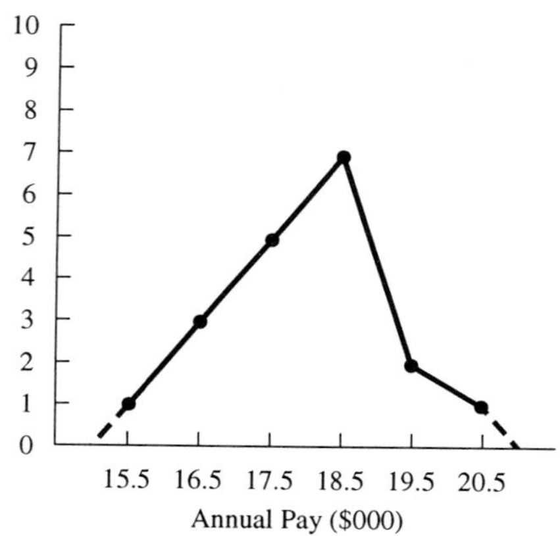

28. a.

Classes	Frequency
40–49	4
50–59	6
60–69	10
70–79	6
80–89	4
90–99	2

b.

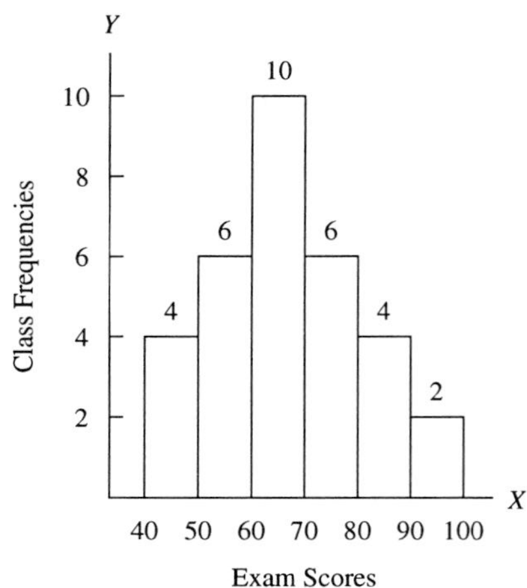

c.

Leading Digit		Trailing Digit
(40–44)	4	12
(45–49)	4	79
(50–54)	5	02
(55–59)	5	5679
(60–64)	6	013
(65–69)	6	5556889
(70–74)	7	2244
(75–79)	7	89
(80–84)	8	1
(85–89)	8	578
(90–94)	9	0
(95–99)	9	5

30.

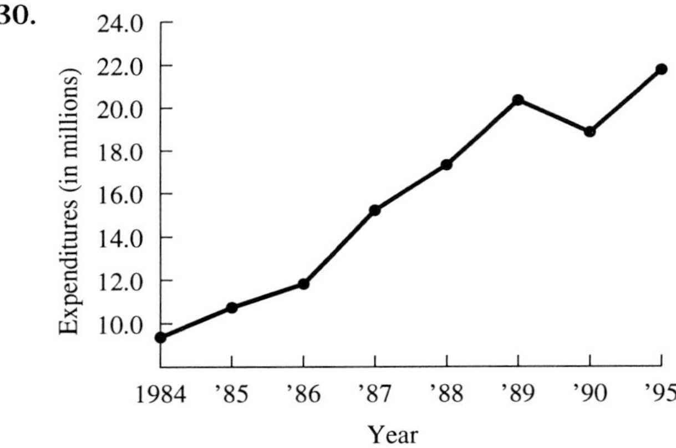

32.

Age	Cumulative Frequency	Frequency
1–3	35	35
4–6	60	25
7–9	75	15
10–12	78	3
13–15	80	2

34.

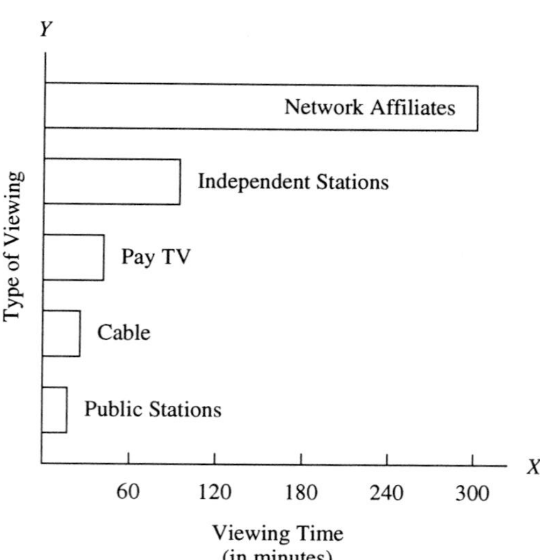

36. a.

Class	*f*	*CF*
$0.97–0.99	2	2
1.00–1.02	13	15
1.03–1.05	15	30
1.06–1.08	8	38
1.09–1.11	4	42
	42	

b. From $1.03 to 1.05.

c.

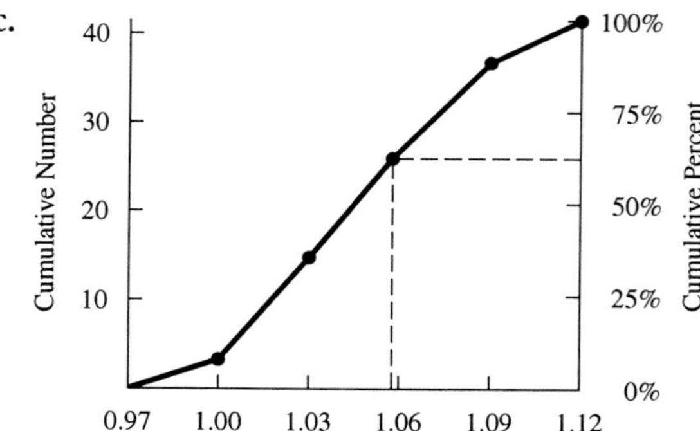

d. About $1.058.

38.

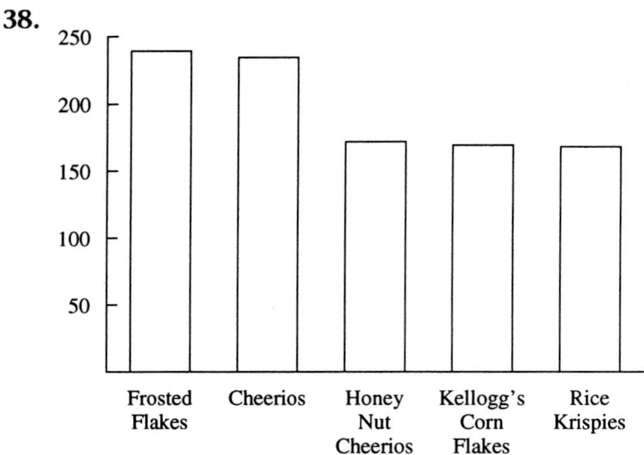

40.

Trash	Percent	Cumulative Percent
Paper	40	40
Yard	18	58
Metals	9	67
Glass	8	75
Rubber	8	83
Food Wastes	8	91
Plastic	7	98
Other	2	100

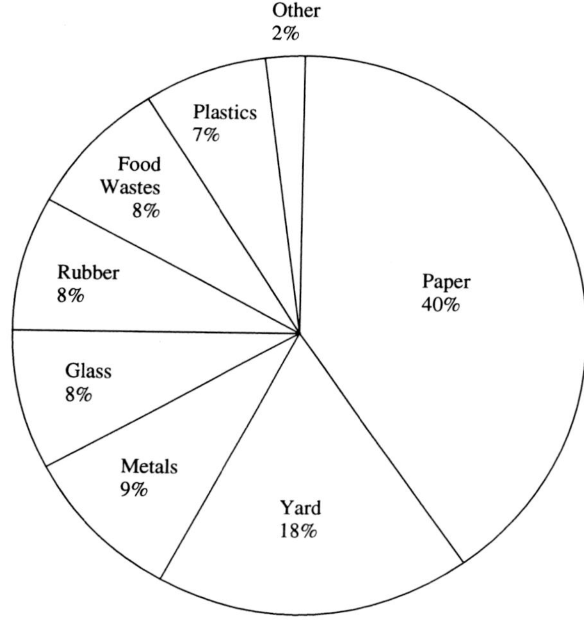

42. a. Ratio level.

b.

Grade Point Average	Class Frequencies
1.70–1.89	4
1.90–2.09	11
2.10–2.29	26
2.30–2.49	21
2.50–2.69	17
2.70–2.89	11
2.90–3.09	7
3.10–3.29	6
3.30–3.49	7
3.50–3.69	3
3.70–3.89	5
3.90–4.10	2

c.

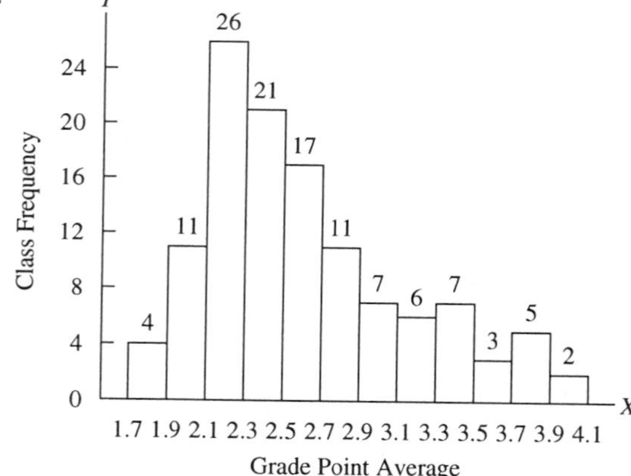

d.

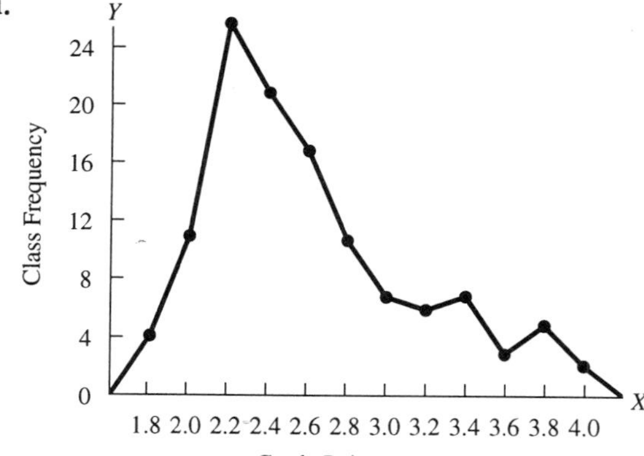

e.

Upper Class Limits	Cumulative Frequencies
1.90	4
2.10	15
2.30	41
2.50	62
2.70	79
2.90	90
3.10	97
3.30	103
3.50	110
3.70	113
3.90	118
4.00	120

f.

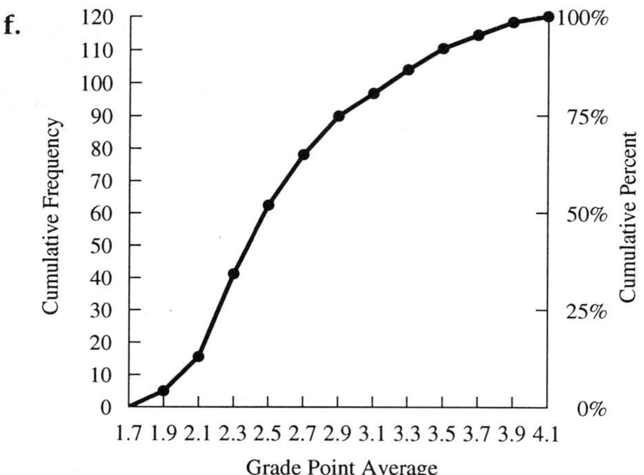

g.

Lower Class Limits	Cumulative Frequencies
1.70	120
1.90	116
2.10	105
2.30	79
2.50	58
2.70	41
2.90	30
3.10	23
3.30	17
3.50	10
3.70	7
3.90	2
4.10	0

h.

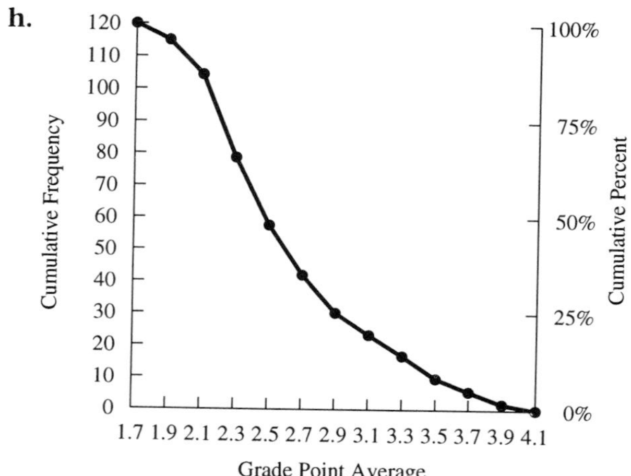

i. Based on the frequency distributions and the frequency polygons, (1) the grade point averages range from about 1.7 to about 4.1 (actually 4.0); (2) the major concentration is between 2.1 and 2.7; (3) a typical grade point average is 2.2; and (4) the grade point averages taper off rapidly after 2.9.

44. a.

Selling Price ($000)		*f*	*CF*
170.0	179.0	7	7
180.0	189.0	21	28
190.0	199.0	17	45
200.0	209.0	4	49
210.0	220.0	1	50
		50	

b. About $185,000.
c. About 92% sell for less than $200,000.
d. About 40% sell for less than about $185,000.
e. Mound shape.
f. A typical home sold for $185,000 to $190,000, with the selling prices ranging from about $170,000 up to more than $210,000.

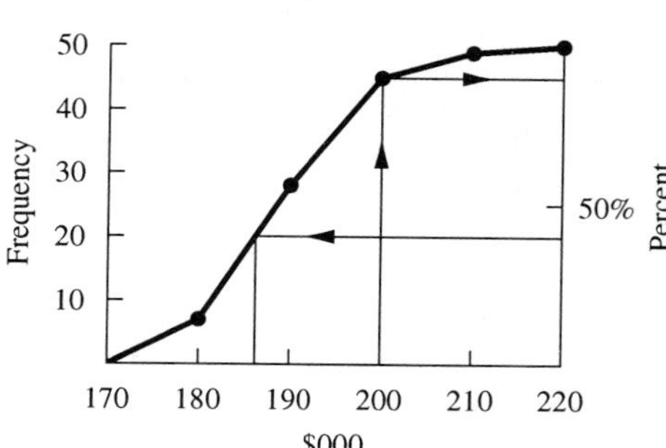

CHAPTER THREE

2. $\overline{X} = 69.2/8 = 8.65$ gallons.

4. White: $\overline{X} = 59.6/3 = 19.87$. Black: $\overline{X} = 29.8/3 = 9.93$. The rate for whites is nearly double that for blacks.

6. **a.** $\overline{X} = 33/6 = 5.5$ miles.
   b. $(1 - 5.5) + (4 - 5.5) + (9 - 5.5) + (8 - 5.5) + (5 - 5.5) + (6 - 5.5) = 0$.

8. **a.** $\overline{X} = \$523/8 = \$65.375$.
   b. $(\$46 - \$65.375) + (\$42 - \$65.375) \ldots + (46 - 65.375) = 0$.

10. Weighted mean = 28.2 per 100,000, found by 225.3/8.

12. $(2 + 10)/2 = 6$.

14. 41 students.

16. 3.38%, found by $(3.25 + 3.51)/2$. Half of instruments had savings rates below 3.38%, the other half above 3.38%.

18.

Selling Price ($000)	Class Frequencies f	Midpoint X	Frequency × Midpoint fX
40–49	3	44.5	133.5
50–59	6	54.5	327.0
60–69	19	64.5	1,225.5
70–79	23	74.5	1,713.5
80–89	9	84.5	760.5
	$\Sigma f = n = 60$		$\Sigma fX = 4,160.0$

$\overline{X} = 4,160/60 = 69.33 = \$69,330$ mean selling price.

20.

Age	Class Frequencies f	Midpoint X	fX
15–19	10	17	170
20–24	12	22	264
25–29	14	27	378
30–34	9	32	288
35–39	5	37	185
	50		1,285

$\overline{X} = \frac{1285}{50} = 25.7$ years.

22. a. 70–79 class **b.** 69.5 **c.** 10 **d.** 28

e. Median $= \$69.5 + \frac{(60/2) - 28}{23}(\$10) = \$70.37.$ **f.** Mode = \$74.50.

24.

Age	Class Frequencies	Cumulative Frequencies
Under 20	7	7
20 but less than 30	13	20
30 but less than 40	26	46
40 but less than 50	15	61
50 but less than 60	6	67
60 or over	3	70
Total	70	

a. Median $= 30.0 + \frac{(70/2) - 20}{26}(10) = 35.77$ years.

b. Mode = 35, the midpoint of the 30 but less than 40 class.
c. You cannot find a mid-point for an open-ended class.

26. a. A population.

b. Juneau: $\mu = 483.1/12 = 40.26$ degrees
San Juan: $\mu = 943.6/12 = 78.63$ degrees

c. Juneau: 40.35 degrees
San Juan: 78.95 degrees

28. 615,000,000,000/54,000,000 = 11,388.9 annually
11,388.9/365 = 31.2 daily

30.

Hours	×	Points	=	Total
3	×	2	=	6
5	×	3	=	15
4	×	4	=	16
12				37

37/12 = 3.08 Weighted Grade Point

32.

Miles per Gallon	Class Frequencies f	Midpoints X	fX	CF
20–24	2	22	44	2
25–29	7	27	189	9
30–34	15	32	480	24
35–39	8	37	296	32
40–44	3	42	126	35
	35		1,135	

a. $\overline{X} = 1{,}135/35 = 32.43$ mpg. **b.** $\text{Median} = 29.5 + 5\left(\dfrac{17.5 - 9}{15}\right) = 32.33.$

34. **a.** $\overline{X} = 5.51/7 = \$.79$ per pint Median = \$.79 Mode = \$.83
b. The mode is inferior because it represents only two observations.

36.

Time	f(percent)	X	fX
4 A.M.–5 A.M.	7.0	4.5	31.5
5 A.M.–6 A.M.	65.0	5.5	357.5
6 A.M.–7 A.M.	24.0	6.5	156.0
7 A.M.–8 A.M.	4.0	7.5	30.0
	100.0		575.0

$\overline{X} = 575/100 = 5.75$
$= 5{:}45$ A.M.

38.

Number of Nails	Number of Packages	CF
20 or less	5	5
21	8	13
22	11	24 ← median in this class
23	10	34
24	5	39
25 or more	1	40

$\text{Median} = 21.5 + \dfrac{20 - 13}{11}(1) = 22.14.$

40. a.

Number of Gallons	Frequency	X	fX	CF
90–99	1	95	95	1
100–109	5	105	525	6
110–119	12	115	1380	18
120–129	13	125	1625	31
130–139	10	135	1350	41
140–149	4	145	580	45
	45		5555	

b. $\overline{X} = 5555/45 = 123.44$.

c. $\text{Median} = 119.5 + \dfrac{22.5 - 18}{13}(10) = 122.96$.

42. a. The mean salary is $64,548 and the median salary is $63,900, so there is a difference of about $650. The distribution is somewhat positively skewed.

b. The typical manager is about 47 years old, has nearly 15 years of experience, and supervises 11 employees.

c.

Sex	Salary	Frequency
Women	63,486	49
Men	66,550	26

d. There were more women in the study by almost 2 to 1, but the men earn an average of nearly $3,000 more.

CHAPTER FOUR

2. a. Pond A: $\overline{X} = 14.2$ inches Pond B: $\overline{X} = 15.9$ inches

b. Pond A: Range $= 16.5 - 12.0 = 4.5$ inches
Pond B: Range $= 20.5 - 12.0 = 8.5$ inches
The range in the length of the trout in pond B is greater than those in pond A.

c. Pond A: $MD = 10.4/10 = 1.04$ inches
Pond B: $MD = 27.8/10 = 2.78$ inches

4. a. Range $= 9.75 - 3.30 = 6.45$

b. $\bar{X} = 61/10 = 6.1\%$

X	$\lvert X - \bar{X} \rvert$
9.75	3.65
3.75	2.35
5.50	0.60
7.30	1.20
6.00	0.10
7.30	1.20
7.25	1.15
4.25	1.85
3.30	2.80
6.60	0.50
61.00	15.4

$MD = 15.4/10 = 1.54$

On the average, the interest rate deviated 1.54 percentage points from the mean rate of 6.1 percent.

6. a. Black: Range = 49.1 − 19.6 = 29.5. Arithmetic mean = 252.2/10 = 25.22. Mean deviation = 6.972, found by 23.88 + 10.98 + 1.92 . . . = 69.72/10.
White: Range = 26.0 − 9.6 = 16.4. Arithmetic mean = 132.7/10 = 13.27. Mean deviation = 12.73 + 6.73 + .97 . . . = 38.92/10 = 3.892.

b. Yes. More spread in mortality rate for black males because 29.5 > 16.4 and 6.972 > 3.892.

8. a.

Pond A		Pond B	
$X - \bar{X}$	$(X - \bar{X})^2$	$X - \bar{X}$	$(X - \bar{X})^2$
−1.7	2.89	2.1	4.41
−0.2	0.04	4.1	16.81
−0.7	0.49	−3.9	15.21
0.3	0.09	−1.4	1.96
0.8	0.64	3.1	9.61
2.3	5.29	−2.4	5.76
1.8	3.24	−1.4	1.96
−2.2	4.84	−2.9	8.41
−0.2	0.04	4.6	21.16
−0.2	0.04	−1.9	3.61
	17.60		88.90

Pond A: $s^2 = 17.60/(10 - 1) = 1.96$
Pond B: $s^2 = 88.90/(10 - 1) = 9.88$

b. Pond A: $s = \sqrt{1.96} = 1.40$ inches
Pond B: $s = \sqrt{9.88} = 3.14$ inches

c. The variation in the lengths of the trout in pond B (3.14 inches) is greater than for pond A (1.40 inches).

10.

X	$(X - \bar{X})$	$(X - \bar{X})^2$
9.75	3.65	13.32
3.75	−2.35	5.52
5.50	−0.60	0.36
7.30	1.20	1.44
6.00	−0.10	0.01
7.30	1.20	1.44
7.25	1.15	1.32
4.25	−1.85	3.42
3.30	−2.80	7.84
6.60	0.50	0.25
	0	34.92

$\bar{X} = 6.1$
$n = 10$
$s^2 = 34.92/9 = 3.88$
$s = \sqrt{3.88} = 1.97$

12. $\sigma^2 = (259231.34 - (1402.8)^2/10)/10 = 6244.6556$. σ is equal to 79.023. Note answer is in thousands of dollars.

14. Pond A: $s^2 = \dfrac{2{,}034 - (142)^2/10}{10 - 1} = 1.96$

$s = \sqrt{1.96} = 1.40$ inches (same answer)

Pond B: $s^2 = \dfrac{2{,}617 - (159)^2/10}{10 - 1} = 9.88$

$s = \sqrt{9.88} = 3.14$ inches (same answer)

16.

X	X^2
9.75	95.06
3.75	14.06
5.50	30.25
7.30	53.29
6.00	36.00
7.30	53.29
7.25	52.56
4.25	18.06
3.30	10.89
6.60	43.56
61.00	407.02

$$s^2 = \frac{407.02 - (61)^2/10}{9} = \frac{407.02 - 372.10}{9} = 3.88$$

$s = 1.97$

18. a. Monday:

$$Q_1 = 11.5 + \frac{18 - 5}{15}(4) = 14.97$$

$$Q_3 = 19.5 + \frac{54 - 46}{16}(4) = 21.50$$

$$QD = \frac{21.50 - 14.97}{2} = 3.27$$

Friday:

$$Q_1 = 11.5 + \frac{16 - 5}{21}(4) = 13.60$$

$$Q_3 = 19.5 + \frac{48 - 48}{13}(4) = 19.50$$

$$QD = (19.50 - 13.60)/2 = 2.95$$

b. The dispersion between the quartiles is greater on Monday than on Friday because 3.27 is greater than 2.95.

20. a. $Q_1 = 70.00 + \frac{(92/4) - 17}{23}(10) = 72.60$

$$Q_1 = 90.00 + \frac{3(92)/4 - 68}{17}(10) = 90.60$$

b. $\frac{90.60 - 72.60}{2} = 9.00$

22. a. Considering these as all the admissions on Monday and Friday (that is, populations):

Monday			
X	f	fX	fX^2
5.5	1	5.5	30.25
9.5	4	38.0	361.00
13.5	15	202.5	2,733.75
17.5	26	455.0	7,962.50
21.5	16	344.0	7,396.00
25.5	7	178.5	4,551.75
29.5	3	88.5	2,610.75
	72	1,312.0	25,646.00

$$\sigma^2 = \frac{25{,}646 - (1{,}312)^2/72}{72} = 24.15$$

Standard Deviation $\sigma = \sqrt{24.15} = 4.91$

Friday: $\sigma^2 = \frac{18{,}916 - (1{,}068)^2/64}{64} = 17.09$

Standard Deviation $\sigma = \sqrt{17.09} = 4.13$

b. Since 4.91 is greater than 4.13, the dispersion on Monday is greater than on Friday.

24.

f	X	fX	fX^2
5	55	275	15,125
12	65	780	50,700
23	75	1,725	129,375
28	85	2,380	202,300
17	95	1,615	153,425
7	105	735	77,175
92		7,510	628,100

$$s = \sqrt{\frac{628{,}100 - (7{,}510)^2/92}{92 - 1}}$$

$$= \sqrt{\frac{628{,}100 - 613{,}044.57}{91}}$$

$$= \sqrt{165.44} = 12.86$$

26. Salary: $CV = \dfrac{\$3{,}000}{\$31{,}000}(100) = 9.7\%$

Length of employment: $CV = \dfrac{4}{15}(100) = 26.7\%$

There is relatively more dispersion in their lengths of employment.

28. Homeowner claims: $CV = \dfrac{\$425}{\$1{,}260}(100) = 34\%$

Auto Policy: $CV = \dfrac{\$300}{\$875}(100) = 34\%$

The relative dispersion is the same for the two policies.

30. $sk = \dfrac{3(11.5 - 11.95)}{4.5} = -0.3$ This distribution has a slight negative skewness.

32. $sk = \dfrac{3(\$376 - \$406)}{\$120} = -0.75$ Negative skewness.

34. **a.** $390 - 134 = 256$

b. $MD = \dfrac{3.4 + 10.4 + 51.4 + 79.4 + 176.6 + 29.4 + 2.4}{7} = 50.4$

c. $\sigma^2 = \dfrac{359{,}986 - (1{,}494)^2/7}{7} = 5{,}874.82$ **d.** $\sigma = \sqrt{5{,}874.82} = 76.6$

e. $sk = \dfrac{3(213.4 - 203)}{76.65} = 0.407$

36. **a.** Range $= 7 - 0 = 7$ **b.** $s^2 = \dfrac{365 - (95)^2/40}{40 - 1} = 3.57$

c. $s = \sqrt{3.57} = 1.89$ **d.** $QD = \dfrac{3.5 - 0.95}{2} = 1.27$

e. $sk = \dfrac{3(2.375 - 1.944)}{1.89} = 0.68$

38. a. Range = 29 − 0 = 29

b.

Weight	CF
0–4	10
5–9	47
10–14	97
15–19	165
20–24	195
25–29	200

$$Q_1 = 9.5 + \frac{50 - 47}{50}(5) = 9.8$$

$$Q_3 = 14.5 + \frac{150 - 97}{68}(5) = 18.4$$

Interquartile range = 18.4 − 9.8 = 8.6.

$$QD = \frac{18.4 - 9.8}{2} = 4.3$$

c.

X	f	fX	fX^2
2	10	20	40
7	37	259	1,813
12	50	600	7,200
17	68	1,156	19,652
22	30	660	14,520
27	5	135	3,645
	200	2,830	46,870

Considering this as all persons enrolled (a population),

$$\sigma^2 = \frac{46{,}870 - (2{,}830)^2/200}{200}$$

$$= \frac{46{,}870 - 39{,}564.845}{200} = 34.1275$$

$$\sigma = \sqrt{34.1275} = 5.84$$

40.

Mileage	Class Frequencies f	Cumulative Frequencies	Midpoints X	$f \cdot X$	fX^2
11–13	3	3	12	36	432
14–16	5	8	15	75	1,125
17–19	12	20	18	216	3,888
20–22	8	28	21	168	3,528
23–25	4	32	24	96	2,304
	32			591	11,277

a. $$Q_1 = 13.5 + \frac{(32/4) - 3}{5}(3) = 16.5$$

$$Q_3 = 19.5 + \frac{32(3)/4 - 20}{8}(3) = 21.00$$

b. Quartile Deviation $$= \frac{21.00 - 16.50}{2} = 2.25$$

c. $s = \sqrt{\dfrac{11{,}277 - (591)^2/32}{32 - 1}} = 3.42$

42.

Selling Price	f	X	fX	fX^2
40–49	3	44.5	133.5	5,941
50–59	6	54.5	327.0	17,822
60–69	19	64.5	1,225.5	79,045
70–79	23	74.5	1,713.5	127,656
80–89	9	84.5	760.5	64,262
	60		4,160.0	294,725

$\sigma = \sqrt{\dfrac{294{,}725 - (4160)^2/60}{60}}$

$= 10.25$

44. a. *Food*

$\overline{X} = 69.3/12 = 5.775 \qquad \sigma = \sqrt{\dfrac{497.51 - \dfrac{(69.3)^2}{12}}{12}} = 2.85$

$CV = \dfrac{2.85}{5.775}(100) = 49.35$

$sk = \dfrac{3(5.775 - 4.950)}{2.85} = 0.87$

New Cars

$\overline{X} = 56.4/12 = 4.70 \qquad \sigma = \sqrt{\dfrac{321.16 - \dfrac{(56.4)^2}{12}}{12}} = 2.16$

$CV = \dfrac{2.16}{5.775}(100) = 37.40$

$sk = \dfrac{3(4.70 - 4.05)}{2.16} = 0.90$

Medical Care

$\overline{X} = 103.9/12 = 8.658 \qquad \sigma = \sqrt{\dfrac{935.89 - \dfrac{(103.9)^2}{12}}{12}} = 1.74$

$CV = \dfrac{1.74}{8.658}(100) = 20.01$

$sk = \dfrac{3(8.658 - 8.600)}{1.74} = 0.10$

b. Over the 12-year period, the mean percent change was the largest in medical care and the variation was the largest in medical care also. The relative dispersion was the smallest in medical care. There was some positive skewness in all three data sets.

46. In comparing the mean percent using Marijuana/Hashish and LSD, we note that over the period the mean percent that have tried Marijuana/Hashish is 52.48 percent versus 8.63. The standard deviation is also larger for Marijuana/Hashish. Note also that the percent using LSD has remained fairly constant over the ten-year period, but the use of Marijuana has declined since 1980, and in 1989 was less than it was ten years earlier.

48.

Age	Number	*CF*
10–14	9,907	9,907
15–19	312,499	322,406
20–24	350,905	673,311
25–29	196,365	869,676
30–34	94,874	964,550
35–39	34,408	998,958
40–44	6,341	1,005,299
	1,005,299	

a. The distance between Q_1 and Q_3 is 8.19 years.

b. 18.36 years, found by $14.5 + \frac{251{,}324.75 - 9{,}907}{312{,}499}(5)$.

c. 26.55 years, found by $24.5 + \frac{753{,}974.25 - 673{,}311}{196{,}365}(5)$.

d. 4.095 years, found by $(26.55 - 18.36)/2$.

e. 34 years, found by $44 - 10$. The difference between the youngest and oldest unmarried mother is 34 years.

50. The following output was obtained from MINITAB:

```
                 N     MEAN    MEDIAN    TRMEAN    STDEV    SEMEAN
Distance        50   20.840    22.000    20.841    3.066     0.434
Selling         50   188.44    188.35    188.17     8.17      1.15

               MIN      MAX        Q1        Q3
Distance    15.000   26.000    18.000    23.000
Selling     174.60   211.20    182.32    193.07
```

a. $s = 8.17$. The $CV = 8.17/188.44 = 0.043$.
$sk = 3(188.44 - 188.35)/8.17 = 0.03$.

b. $\overline{X} = 20.84$, median $= 22.0$, and $s = 3.066$.
$CV = 3.066/20.84 = .147$. $sk = 3(20.84 - 22.00)/3.066 = -1.135$.
There is a moderate negative skew to the data.

CHAPTER FIVE

2. **a.** The election **b.** Democrat, Republican, Independent
c. Male or female

4. $P(X \text{ or } Y) = 0.05 + 0.10 = 0.15$

6. **a.** $P(\text{Two primary}) = P(\text{Lung}) + P(\text{Prostate})$
$= 12{,}226/37{,}555 + 10{,}835/37{,}555 = 0.614$
b. $P(\text{Stomach or Pancreas}) = P(\text{Stomach}) + P(\text{Pancreas})$
$= 3{,}037/37{,}555 + 3{,}031/37{,}555 = 0.162$
c. $P(\text{Not lung}) = P(\text{Prostate}) + P(\text{Colon}) + P(\text{Stomach}) + P(\text{Pancreas})$
$= 10{,}835/37{,}555 + 8{,}426/37{,}555 + 3{,}037/37{,}555 + 3{,}031/37{,}555$
$= 25{,}329/37{,}555 = 0.674$

8. $P(X \text{ or } Y) = 0.55 + 0.35 - 0.20 = 0.70$

10. Let W = Women members, U = University persons
$P(W \text{ or } U) = P(W) + P(U) - P(W \text{ and } U) = 0.20 + 0.05 - 0.02 = 0.23$

12. $P(S \text{ or } F) = P(S) + P(F) - P(S \text{ and } F) = 0.30 + 0.70 - 0.20 = 0.80$

14. **a.** $P(Y) = 4/10$
b. $P(M \mid Y) = 1/4$
c. $P(M \text{ and } Y) = 1/10$

16. **a.** $50/200 = 0.25$ **b.** $50/200 = 0.25$
c.

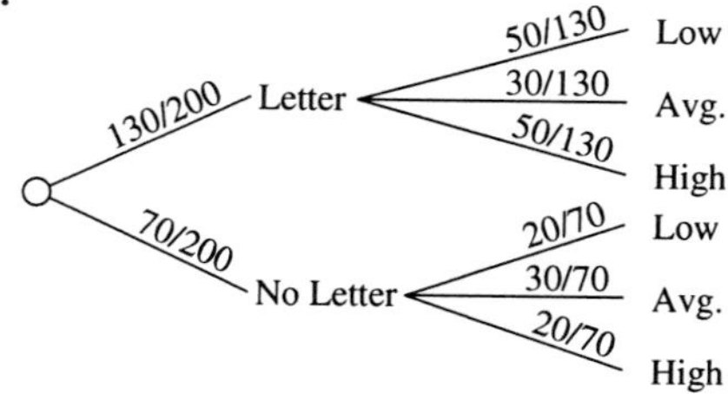

18.

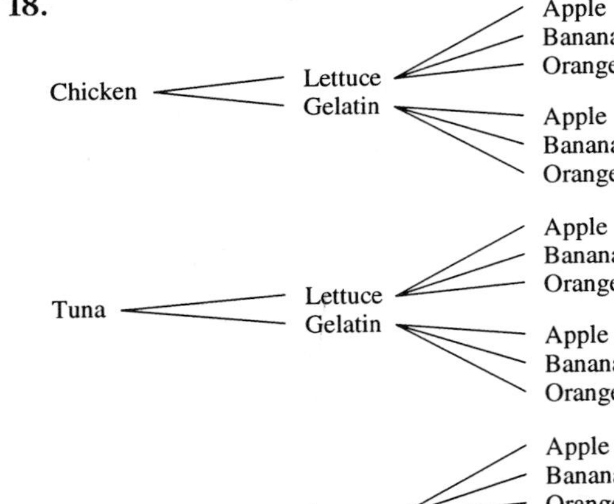

a. (3)(2)(3) = 18
b. 1/18

20. **a.** 0.25 **b.** 1/3 **c.** 2/24

22. **a.** P(Both finish) = P(First) · P(Second) = 60/100 · 60/100 = 0.36
b. Yes **c.** Special Rule of Multiplication **d.** Relative Frequency
e. P(Neither) = 40/100 · 40/100 = 0.16

24. **a.** $(0.3)^3 = 0.027$ **b.** $(0.7)^3 = 0.343$

26. $P(\sim Y) = 1 - P(Y) = 1 - \frac{6}{24} = 0.75$

28. (0.80)(0.80) = 0.64

30. (5)(4)(6) = 120

32. $(10)^4 = 10{,}000$

34. $4! = 4 \cdot 3 \cdot 2 \cdot 1 = 24$

36. ${}_8P_5 = \frac{8!}{3!} = 8 \cdot 7 \cdot 6 \cdot 5 \cdot 4 = 6{,}720$

38. ${}_{20}C_8 = \frac{20!}{(20 - 8)!8!} = \frac{20!}{12!8!} = \frac{20 \cdot 19 \cdot 18 \cdot 17 \cdot 16 \cdot 15 \cdot 14 \cdot 13}{8 \cdot 7 \cdot 6 \cdot 5 \cdot 4 \cdot 3 \cdot 2 \cdot 1} = 125{,}970$

40. ${}_6C_4 = \frac{6!}{2!4!} = \frac{6 \cdot 5}{2} = 15$

42. Observing the type of car is the experiment. The possible events are selling a subcompact, selling a compact, or selling a luxury. The possible outcomes with respect to doors are two-door or four-door.

44. $P(\text{all}) = 0.25 + 0.20 + 0.10 = 0.55$
$P(\text{Sporting event}) = P(\text{Basketball}) + P(\text{Horse race}) = 0.25 + 0.10 = 0.35$

46. a. $P(\text{Larger than 4}) = \dfrac{5}{10}$

b. $P(\text{Odd}) = \dfrac{5}{10}$

c. $P(\text{Odd and larger than 4}) = \dfrac{3}{10}$

48. Let D = Disney World and B = Busch Gardens
$P(D \text{ or } B) = P(D) + P(B) - P(D \text{ and } B) = 0.70 + 0.50 - 0.40 = 0.80$

50. Let S = Snow and P = Profit
$P(S \text{ and } P) = P(S) \cdot P(P|S) = (0.80)(0.85) = 0.68$

52. a. $\dfrac{13.3}{29.3} = 0.454$ **b.** $\dfrac{4.0}{29.3} = 0.137$ **c.** $\dfrac{14.9}{29.3} = 0.509$ **d.** $\dfrac{2.4}{13.3} = 0.180$

54. $P(H \text{ and } W) = P(H) \cdot P(W) = (0.5)(0.7) = 0.35$

56. $10 \cdot 10 \cdot 10 \cdot 10 = 10{,}000$

58. $_8C_3 = \dfrac{8!}{(8-3)!3!} = \dfrac{8!}{5!3!} = \dfrac{8 \cdot 7 \cdot 6}{3 \cdot 2} = 56$

60. $P(\text{All three}) = P(\text{First}) \cdot P(\text{Second}) \cdot P(\text{Third}) = (0.90)(0.90)(0.90) = 0.729$
Let D = Detected and ND = Not detected
$$\begin{aligned} P(\text{Two out of three detected}) &= P(D) \cdot P(D) \cdot P(ND) + P(D) \cdot P(ND) \cdot P(D) + P(ND) \cdot P(D) \cdot P(D) \\ &= (0.9)(0.9)(0.1) + (0.9)(0.1)(0.9) + (0.1)(0.9)(0.9) \\ &= 0.243 \end{aligned}$$

62. a. $(40)^3 = 64{,}000$
b. $1/64{,}000 = 0.000016$

c. $\left(\dfrac{13}{40}\right)^3 = 0.034$

d. $\left(\dfrac{1}{2}\right)\left(\dfrac{1}{2}\right)\left(\dfrac{1}{4}\right) = 0.0625$

e. $\left(\dfrac{3}{40}\right)\left(\dfrac{3}{40}\right)\left(\dfrac{1}{2}\right) = 0.0028$

64. a. $P(\text{Neither}) = \left(\frac{8}{9}\right)\left(\frac{10}{11}\right) = 0.808$

b. $1 - P(\text{Neither}) = 1 - 0.808 = 0.192$

66. a. 450/670 = .672
b. 70/670 = .104
c. 450/670 + 70/670 − 50/670 = .701
d. 50/670 = .075
e. 50/70 = .714

68. a.

	Illiterate	Not Illiterate	Total
Male	.0980	.3920	0.49
Female	.0867	.4233	0.51
	.1847	.8153	1.00

$P(\text{Male and Illiterate}) = P(M) \cdot P(I/M) = (.49)(.20) = .098$
$P(\text{Female and Illiterate}) = P(F) \cdot P(I/F) = (.51)(.17) = .0867$

b. .1847
c. $P(M/I) = .098/.1847 = .531$

70. a.

	Salary			
	$0 up to $59.9	$60.0 up to $64.9	$65.0 or More	Total
Female	10	24	15	49
Male	1	9	16	26
	11	33	31	75

(1) 49/75 = .653
(2) 15/49 = .306
(3) 15/75 = .20

(4) $\frac{31}{75} + \frac{49}{75} - \frac{15}{75} = \frac{65}{75} = .867$

b.

Region	0 up to 59.9	60.0 up to 64.9	65.0 or More	Total
1	1	4	7	12
2	4	11	8	23
3	3	11	14	28
4	3	7	2	12
	11	33	31	75

(1) $\frac{28}{75} = .373$

(2) $14/28 = .50$
(3) $14/75 = .187$

(4) $\frac{31}{75} + \frac{28}{75} - \frac{14}{75} = \frac{45}{75} = 0.60$

(5) $\left(\frac{44}{75}\right)\left(\frac{43}{74}\right)\left(\frac{42}{73}\right) = .196$

CHAPTER SIX

2.

X	$P(X)$	$X \cdot P(X)$	$(X - \mu)^2$	$(X - \mu)^2 \cdot P(X)$
25	0.10	2.50	342.25	34.23
35	0.25	8.75	72.25	18.06
45	0.40	18.00	2.25	0.90
55	0.20	11.00	132.25	26.45
65	0.05	3.25	462.25	23.11
		$\mu = 43.50$		$102.75 = \sigma^2$

$\sigma = 10.14$

4.

X	$P(X)$	$X \cdot P(X)$	$(X - \mu)^2 \cdot P(X)$
0	0.2	0	0.578
1	0.2	0.2	0.098
2	0.3	0.6	0.027
3	0.3	0.9	0.507
		$\mu = 1.7$	1.21

$\sigma = \sqrt{1.21} = 1.10$

6. $P(3) = \frac{5!}{3!(5-3)!}(0.40)^3(0.60)^2 = 0.23$

8. $P(0) = \frac{5!}{0!5!}(0.25)^0(0.75)^5 = 0.2373$

10. a. $P(2) = 0.073$ b. $P(4) = 0.200$ c. $P(6) = 0.041$

12. a. $P(3) = \frac{5!}{3!2!}(0.5)^3(0.5)^2 = 0.313$

b. $P(5) = \frac{5!}{5!0!}(0.5)^5(0.5)^0 = 0.031$

c. $P(0) = \frac{5!}{0!5!}(0.5)^0(0.5)^5 = 0.031$

$P(\text{At least one girl}) = P(X \geq 1) = 1 - P(0) = 1 - 0.031 = 0.969$

14. a. 0.201 b. $0.201 + 0.111 + 0.042 + 0.011 + 0.002 = 0.367$
c. $0.042 + 0.011 + 0.002 = 0.055$ d. $1 - 0.367 = 0.633$

16. $\mu = 60(0.05) = 3$ $P(0) = \frac{3^0 e^{-3}}{0!} = 0.0498$

$P(X > 0) = 1 - 0.0498 = 0.9502$

18. $P(0) = \frac{(0.2)^0 e^{-0.2}}{0!} = 0.8187$ a. $P(1) = \frac{(0.2)^1 e^{-0.2}}{1!} = 0.1637$

b. $1 - P(0) - P(1) = 0.0176$

20. a. $\mu = 0(0.50) + 1(0.30) + 2(0.20) = 0.70$
b. $\sigma^2 = (0 - 0.70)^2 0.50 + (1 - 0.70)^2 0.30 + (2 - 0.70)^2 0.20 = 0.61$
$\sigma = \sqrt{0.61} = 0.78$
c. $P(0)P(0) = (0.50)(0.50) = 0.25$

22.

X	$P(X)$	$X \cdot P(X)$	$(X - \mu)^2$	$(X - \mu)^2 \cdot P(X)$
0	0.3	0.0	1.69	0.507
1	0.3	0.3	0.09	0.027
2	0.2	0.4	0.49	0.098
3	0.2	0.6	2.89	0.578
		$\mu = 1.3$		$1.210 = \sigma^2$

$\sigma = 1.1$

24. 1. Each outcome is either a "success" or a "failure." The outcomes are mutually exclusive.
2. It is a count of the number of successes, a discrete distribution and X can be only certain whole-numbered values.
3. Each trial is independent, meaning that the outcome of one trial does not affect the outcome of any other trial.
4. The probability of a success is the same from trial to trial.

26. Let a "success" be when the device fails. Hence, $\pi = 0.1$ and $n = 15$.
$P(0) = 0.206$

28. **a.** 10.5, found by 15(.70).

b. .2061302, found by $\frac{15!}{10!(15-10)!}(.70)^{10}(.30)^5 = \frac{360,360}{120}(.70)^{10}(.30)^5$.

c. $P(X \leq 10) = 1 - P(X \geq 11) = 1 - (.219 + .170 + . . . + .005) = .483$

30. $\pi = 0.20$, $n = 6$
a. 0.262
b. 0.393
c. $P(X \geq 1) = 1 - (0.262 + 0.393) = 1 - 0.655 = 0.345$

32. $\pi = 0.20$, $n = 15$
$P(X \leq 4) = 0.035 + 0.132 + 0.231 + 0.250 + 0.188 = 0.836$

34. $\mu = (0.01)(100) = 1.00$
a. $P(X = 0) = 0.3679$ **b.** $P(X \geq 1) = 1 - 0.3679 = 0.6321$

36. $\pi = 0.7$, $n = 14$
$P(X > 10) = P(11) + P(12) + P(13) + P(14)$
$= 0.194 + 0.113 + 0.041 + 0.007 = 0.355$

38. $\pi = 0.7$, $n = 12$
a. $P(8) = 0.231$ **b.** $P(X < 5) = 0.001 + 0.008 = 0.009$
c. $P(X \geq 10) = 0.168 + 0.071 + 0.014 = 0.253$

40. $\pi = 0.08$, $n = 50$, $\mu = 4$

a. $P(2) = \frac{4^2 e^{-4}}{2!} = 0.1465$

b. $P(0) = \frac{4^0 e^{-4}}{0!} = 0.0183$

$P(X \geq 1) = 1 - P(0) = 0.9817$

42. $\mu = 5.0$
a. .0067
b. $P(X < 5) = .0067 + .0337 + .0842 + .1404 + .1755 = 0.4405$
c. $[P(X < 5)]\,[P(X < 5)] = (.4405)\,(.4405) = .1940$

44. $\pi = 0.3$, $n = 10$
$P(X \geq 6) = P(6) + P(7) + P(8) + P(9) + P(10)$
$= 0.037 + 0.009 + 0.001 + 0.000 + 0.000 = 0.047$

46. Using the binomial – $P(X = 1/n = 20, \pi = .05) = .377$.
Using the Poisson $P(X = 1/\mu = 20(.05) = 1.0) = .3679$.
Very little difference in the results.

48.

Employee	f	$P(X)$	$xP(X)$	$(X - \mu)^2 \cdot P(X)$
8	6	.0800	0.6400	0.7853
9	7	.0933	0.8400	0.4246
10	16	.2133	2.1333	0.2739
11	13	.1733	1.9067	0.0031
12	16	.2133	2.5600	0.1604
13	12	.1600	2.0800	0.5577
14	3	.0400	0.5600	0.3288
15	1	.0133	0.2000	0.1994
16	1	.0133	0.2133	0.3158
	75		11.133	3.0489

$\mu = 11.133$
$\sigma = \sqrt{3.0489} = 1.746$

CHAPTER SEVEN

2. $\mu = 90, \sigma = 10$
   a. $\mu \pm 2\sigma$ includes about 95% of the values.
   $90 \pm 2(10) = 90 \pm 20$; Time interval: 70 up to 110.
   b. $\mu \pm 3\sigma$ includes 99.7% of the values.
   $90 \pm 3(10) = 90 \pm 30$; Time interval: 60 up to 120.

4. $\mu = 20, \sigma = 5$
   a. $z = (X - \mu)/\sigma = (12 - 20)/5 = -1.6$
   b. $\mu \pm 3\sigma = 20 \pm 3(5) = 5$ up to 35;
   About 99.7% of the values fall between 5 and 35.

6. Plumbers:
   $\mu = \$15, \sigma = \$1.75$
   Joe earns $14/hour.
   $z = (\$14 - 15)/1.75 = -0.57$

   Carpenters:
   $\mu = \$10, \sigma = \$1.25$
   Neil earns $12/hour.
   $z = (\$12 - 10)/1.25 = 1.60$

   While Joe earns more than Neil, Joe is 0.57 standard units below the mean for his trade whereas Neil is 1.6 standard units above average for his trade.

8. $\mu = 10, \sigma = 1.5$

Mrs. Stevens	Mrs. Delwhiler
$z = (12 - 10)/1.5 = 1.33$	$z = (9 - 10)/1.5 = -0.67$

Mrs. Stevens shopped 1.33 standard deviations more than the mean time, and Mrs. Delwhiler shopped 0.67 standard deviations less than the mean.

10. $\mu = 10, \sigma = 2.0$

a. $z = (13.0 - 10)/2 = 1.5$
$P(z > 1.5) = 0.5000 - 0.4332 = 0.0668$

b. $z = (9 - 10)/2 = -0.5$
$z = (12.5 - 10)/2 = 1.25$
$P(-0.5 \leq z \leq 1.25) = 0.1915 + 0.3944 = 0.5859$

12. $\mu = \$23{,}000, \sigma = \$1{,}500$

a. $z = (\$21{,}000 - \$23{,}000)/\$1{,}500 = -1.33$
$z = (\$22{,}000 - \$23{,}000)/\$1{,}500 = -0.67$
$P(-1.33 \leq z \leq -0.67) = 0.4082 - 0.2486 = 0.1596$

b. $z = (\$20{,}000 - \$23{,}000)/\$1{,}500 = -2.00$
$P(z < -2.00) = 0.5000 - 0.4772 = 0.0228$

14. $-1.645 = \dfrac{X - 40{,}000}{3{,}000}$

$X = 40{,}000 - 4{,}935 = 35{,}065$

16. $-1.28 = \dfrac{X - 1.0}{0.05}$

$X = 1.0 - 1.28(0.05) = 0.936$

18. $\mu = 150(0.08) = 12.00$
$\sigma^2 = 150(0.08)(0.92) = 11.04$
$\sigma = \sqrt{11.04} = 3.32$
$z = (4.5 - 12.00)/3.32 = -2.26$
$P(z > -2.26) = 0.5000 + 0.4881 = 0.9881$

20. $\mu = 500(0.05) = 25.0$
$\sigma^2 = 500(0.05)(0.95) = 23.75$
$\sigma = \sqrt{23.75} = 4.87$
$z = (19.5 - 25.0)/4.87 = -1.13$
$P(z < -1.13) = 0.5000 - 0.3708 = 0.1292$

22.

Using the Poisson Table	Using Normal Approximation $\mu = 4 \quad \sigma = \sqrt{4} = 2$
a. $P[(X = 2)/(\mu = 4.0)] = 0.1465$	**a.** $z = (1.5 - 4.0)/2 = -1.25$ $z = (2.5 - 4.0)/2 = -0.75$ $P(-1.25 \le z \le -0.75)$ $= 0.3944 - 0.2734$ $= 0.1210$
b. $P(X > 7) = 0.0298 + 0.0132 + \ldots$ $= 0.0511$	**b.** $z = (7.5 - 4.0)/2.0 = 1.75$ $P(z > 1.75) = 0.5000 - 0.4599$ $= 0.0401$

24. a.

Test	μ	σ	Score	z-value
Sales	50	7	60	$(60 - 50)/7 = 1.43$
Personnel	120	25	150	$(150 - 120)/25 = 1.20$
Financial	85	5	90	$(90 - 85)/5 = 1.00$

b. Applicant did the best on the sales test; 1.43 standard units above the mean.
c. $P(z > 1.43) = 0.5000 - 0.4236 = 0.0764$
d. Sales, because that z score was the largest.

26. $\mu = 7200, \sigma = 600$
a. $z = (8000 - 7200)/600 = 1.33$
$P(z > 1.33) = 0.5000 - 0.4082 = 0.0918$
b. $z = (8000 - 8600)/560 = -1.07$
$P(z < -1.07) = 0.5000 - 0.3577 = 0.1423$

28. $\mu = 235.6, \sigma = 36.3$
a. $z = (260 - 235.6)/36.3 = 0.67$
$P(z > 0.67) = 0.5000 - 0.2486 = 0.2514$
b. $z = (180 - 235.6)/36.3 = -1.53$
$P(z < -1.53) = 0.5000 - 0.4370 = 0.0630$
c. $z = (240 - 235.6)/36.3 = 0.12$
$z = (250 - 235.6)/36.3 = 0.40$
$P(0.12 \le z \le 0.40) = 0.1554 - 0.0478 = 0.1076$
d. $P(X < 240) = P(z < 0.12) = 0.5000 + 0.0478 = 0.5478$

30. $\mu = 120$

a. $1.645 = \dfrac{140 - 120}{\sigma}$

$\sigma = 20/1.645 = 12.158$
b. $z = (100 - 120)/12.158 = -1.645$
$P(-1.645 \le z \le 0) = 0.4500$

c. $z = (150 - 120)/12.158 = 2.47$
$P(z > 2.47) = 0.5000 - 0.4932 = 0.0068$
d. 0 and 1.645

32. $\mu = 80(0.90) = 72.0, \sigma = \sqrt{80(0.90)(0.10)} = 2.6833$
$z = (69.5 - 72.0)/2.6833 = -0.93$
$P(z > -0.93) = 0.3238 + 0.5000 = 0.8238$

34. $\mu = 39.52, \sigma = 6.29$
a. $z = (50 - 39.52)/6.29 = 1.67$
b. $z = (25 - 39.52)/6.29 = -2.31$
c. $P(z > 1.67) = 0.5000 - 0.4525 = 0.0475$
d. $P(z < -2.31) = 0.5000 - 0.4896 = 0.0104$

e. $1.28 = \dfrac{X - 39.52}{6.29}$

$X = 39.52 + 8.0512 = 47.5712$ patients

36. $\mu = 12, \sigma = \sqrt{12} = 3.4641$
$z = (15.5 - 12.0)/3.4641 = 1.01$
$P(z > 1.01) = 0.5000 - 0.3438 = 0.1562$

38. $\mu = 15, \sigma = \sqrt{15} = 3.8730$
$z = (12.5 - 15.0)/3.8730 = -0.65$
$P(z > -0.65) = 0.2422 + 0.5000 = 0.7422$

40. $\mu = 80, \sigma = 12$
$z = (60 - 80)/12 = -1.67$
$P(z > -1.67) = 0.5000 + 0.4525 = 0.9525$

42. $\mu = 18, \sigma = 5.5$
a. $z = (30 - 18)/5.5 = 2.18$
$P(z < 2.18) = 0.5000 + 0.4854 = 0.9854$
b. 0.000
c. $z = (30.5 - 18.0)/5.5 = 2.27$
$z = (29.5 - 18.0)/5.5 = 2.09$
$P(2.09 \leq z \leq 2.27) = 0.4884 - 0.4817 = 0.0067$

44. Regarding selling price,

$$z = \frac{89{,}550 - 107{,}800}{20{,}000} = -0.91$$

$P(z < -0.91) = 0.5000 - 0.3186 = 0.1814$

Regarding time on the market,

$$z = \frac{43 - 50}{14.5} = -0.48$$

$P(z < -0.48) = 0.5000 - 0.1844 = 0.3156$

46. **a.** $-.84 = \dfrac{X - 15}{1}$

$X = 15 - .84 = 14.16$

b. $1.28 = \dfrac{X - 15}{1}$

$X = 16.28$

48. **a.** $z = (200 - 192.18)/8.25 = 0.95$
$P(z > 0.95) = 0.5000 - 0.3289 = 0.1711$
9 homes or 18% sold for $200,000 or more, so the approximation is good.

b. $z = (2100 - 2214.6)/435 = -0.14$
$P(z > -0.14) = 0.5000 + 0.0557 = 0.5557$
29 homes, or 58%, have an area of 2100 square feet, so again the approximation is good.

CHAPTER EIGHT

2. The population is the 6 possible faces on the die. Perhaps you want to learn the mean or typical number of spots or the relative frequency of each number of spots. After throwing the die a few more times, you might find the sample mean.

4. **a.** Jackson, Zaborowski, and Rodgers. **b.** Answers will vary.

6. Huttner, Jackson, Parmar, Torok, Kimmel, Bertka, Holt, and Zaborowski.

8. Answers will vary.

10. a. $\mu = \dfrac{10 + 4 + 12 + 11 + 9 + 8}{6} = 9$

b.

Possible Sample	$\overline{X}$
10, 4, 12, 11	9.25
10, 4, 12, 9	8.75
10, 4, 12, 8	8.50
10, 4, 11, 9	8.50
10, 4, 11, 8	8.25
10, 4, 9, 8	7.75
10, 12, 11, 9	10.50
10, 12, 11, 8	10.25
10, 12, 9, 8	9.75
10, 11, 9, 8	9.50
4, 12, 11, 9	9.00
4, 12, 11, 8	8.75
4, 11, 9, 8	8.00
12, 11, 9, 8	10.00
4, 12, 9, 8	8.25

c.

Sample Mean	Probability
7.75	1/15
8.00	1/15
8.25	2/15
8.50	2/15
8.75	2/15
9.00	1/15
9.25	1/15
9.50	1/15
9.75	1/15
10.00	1/15
10.25	1/15
10.50	1/15

d.

Histogram of Sample Means

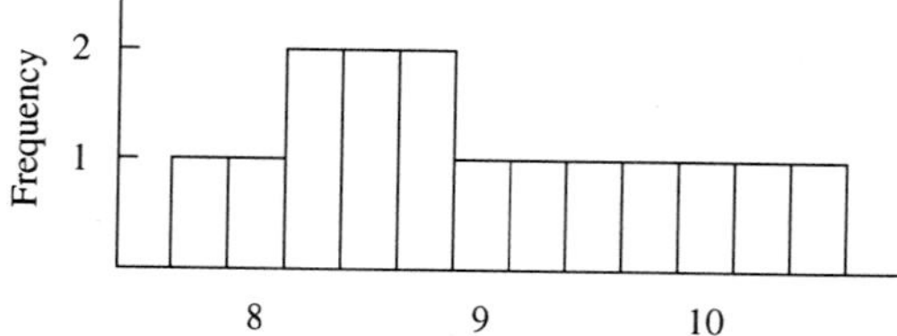

Population Probability Distribution

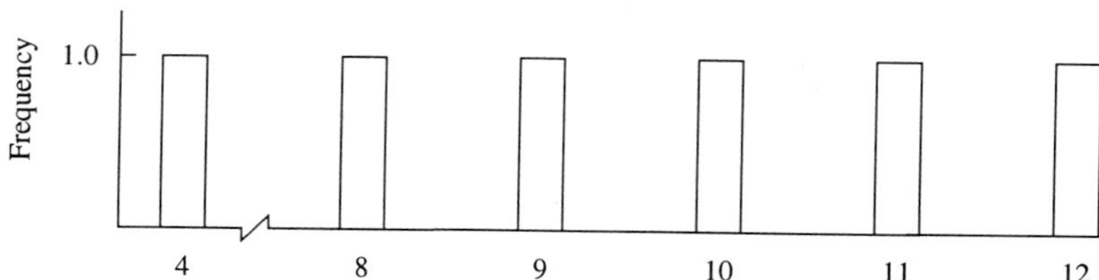

e. Both distributions have a mean of 9. However, the distribution of sample means has much less spread than the population distribution.

12. **a.** Answers will vary. One simple random sample is 25, 28, 35, 17, 16, 13.
b. If the second hotel is selected as a starting point, the sample is 35, 25, 24, 35, 25, 21.
c. Randomly select 4 hotels from the first 20 and 2 from the last 10.

14. Obtain a list of all voters. With a table of random numbers, select the desired number of voters. Contact these voters and ask their opinion of the legislation.

16. One possibility is to make an observation each hour. A table of random numbers can be used to select the number of minutes after the hour (for example, 8:10, 9:37, 10:08) that the observation is made. Prepare a list of possible duties. Then make an inspection each hour for one week and, at that time, record the activity performed.

18. Answers to this question will vary. MINITAB was used to generate five samples of $z = 10$. The sample means were 3.5, 3.6, 3.8, 3.7, and 3.5. The mean of these sample means is 3.62, which is close to the population mean of 3.58.

CHAPTER NINE

2. $35 \pm (2.58)\dfrac{6.3}{\sqrt{40}} = 35 \pm 2.6$ From 32.4 up to 37.6 months.

4. $7.52 \pm (1.96)\dfrac{1.32}{\sqrt{80}} = 7.52 \pm 0.29$ km We are 95% confident that the mean distance is between 7.23 and 7.81 km.

6. $280 \pm (2.58)\dfrac{21}{\sqrt{49}} = 280 \pm 7.74$ From 272.26 up to 287.74 seconds.

8. $8 \pm (1.96)\dfrac{2}{\sqrt{64}} = 8 \pm 0.49$ From 7.51 up to 8.49 seconds.

10. **a.** $\sqrt{\dfrac{(.15)(.85)}{100}} = 0.0357$ **b.** $0.15 \pm (1.65)(.0357) = 0.15 \pm 0.06$
From 0.09 up to 0.21.

12. $\sqrt{\dfrac{(0.65)(0.35)}{500}} = 0.0213$ $0.65 \pm (2.58)(0.0213) = 0.65 \pm 0.06$
From 0.59 up to 0.71.

14. **a.** $p = \dfrac{32}{40} = 0.8,\ \sigma_p = \sqrt{\dfrac{(0.8)(0.2)}{40}} = 0.0632$

$0.80 \pm (1.96)(0.0632) = 0.80 \pm 0.124$ From 0.676 up to 0.924.

b. $\sigma_p = \sqrt{\frac{(0.8)(0.2)}{400}} = 0.02$

$0.80 \pm (1.96)(0.02) = 0.80 \pm 0.039$ From 0.761 up to 0.839.

16. $p = 0.36, \sigma_p = \sqrt{\frac{(0.36)(0.64)}{980}} = 0.0153$

$x(0.0153) = 0.03$

$x = \frac{0.03}{0.0153} = 1.96$

They were using a 95% confidence level.

18. $29 \pm (2.33)\left(\frac{5}{\sqrt{40}}\right)\sqrt{\frac{300 - 40}{300 - 1}}$

$29 \pm 1.8(0.933)$
29 ± 1.7
From \$27.30 up to \$30.70.

20. $\frac{10}{50} \pm (2.58)\sqrt{\frac{(0.20)(0.80)}{50}}\sqrt{\frac{500 - 50}{500 - 1}}$

$0.20 \pm (0.1459)(0.9496)$
0.20 ± 0.139
From 0.061 up to 0.339

22. $n = \left[\frac{(1.96)(0.7)}{0.2}\right]^2 = 47.06$ Use 48.

24. $n = (0.50)(0.50)[1.96/0.2]^2 = 384.16$ Use 385.

26. $5.9 \pm (2.58)\frac{2}{\sqrt{100}}\sqrt{\frac{500 - 100}{500 - 1}} = 5.9 \pm 0.46$ From 5.44 up to 6.36 years.

28. $3.9 \pm (1.96)\frac{1.8}{\sqrt{80}}\sqrt{\frac{250 - 80}{250 - 1}} = 3.9 \pm 0.33$ From 3.57 up to 4.23 years.

30. **a.** Answers will vary.
b. If you pick, for example Toledo and Dayton whose values are 1.4 and 1.7, $\overline{X} = 1.55$.
c. $\mu = 1.50$, so the sampling error is .05.

32. $n = \left[\frac{(1.96)(75)}{20}\right]^2 = 54.02$ Use 55.

34. $n = (0.50)(0.50)[1.96/0.03]^2 = 1{,}068$ voters

36. $0.925 \pm 2.58\sqrt{\frac{.925(.075)}{80}} = 0.925 \pm 0.076$ From 0.849 to 1.00.

38. $49 \pm 2.58 \frac{7}{\sqrt{50}} = 49 \pm 2.55$ From 46.45 to 51.55.

40. a. $64.548 \pm 1.96 \frac{4.427}{\sqrt{75}} = 64.548 \pm 1.002$ From 63.546 to 65.550.

b. $11.133 \pm 1.96 \frac{1.758}{\sqrt{75}} = 11.133 \pm 0.398$ From 10.735 to 11.531.

c. Recall that the code is that male = 1 and female = 2, so $p = 1 - 0.3467$.

$.6533 \pm 1.96\sqrt{\frac{.6533(.3467)}{75}} = .6533 \pm .1077$ From .5456 to .7610.

CHAPTER TEN

2. a. H_o: $\mu = \$1{,}010$ H_a: $\mu \neq \$1{,}010$

b. Reject H_o if z is not between −2.58 and 2.58.

c. $z = \frac{\$1{,}090 - \$1{,}010}{\$300/\sqrt{50}} = 1.89$

There is not sufficient evidence to conclude that there was a change.

d. The p-value is $2(0.5 - 0.4706) = 0.0588$.

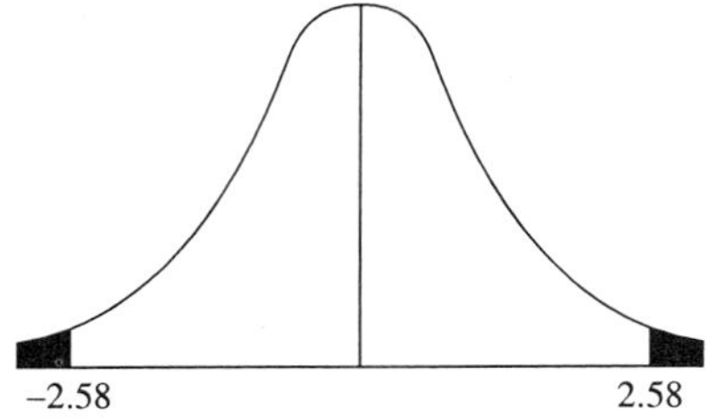

4. H_o: $\mu = 70$ H_a: $\mu \neq 70$ If z is not between −1.96 and 1.96, reject H_o.

$z = \frac{72 - 70}{\frac{2}{\sqrt{121}}} = \frac{2}{0.18} = 11$ We can reject H_o. There is enough evidence to conclude the mean height has changed.

6. a. H_o: $\mu = \$125$ H_a: $\mu \neq \$125$

b. If z is not between −2.58 and 2.58, reject H_o.

c. $z = \frac{124 - 125}{5/\sqrt{400}} = \frac{-1}{0.25} = -4$,

so the null hypothesis is rejected.
Mean parking fees are significantly different from $125.
The p-value is $2(0.5 - 0.49997) = 0.00006$.

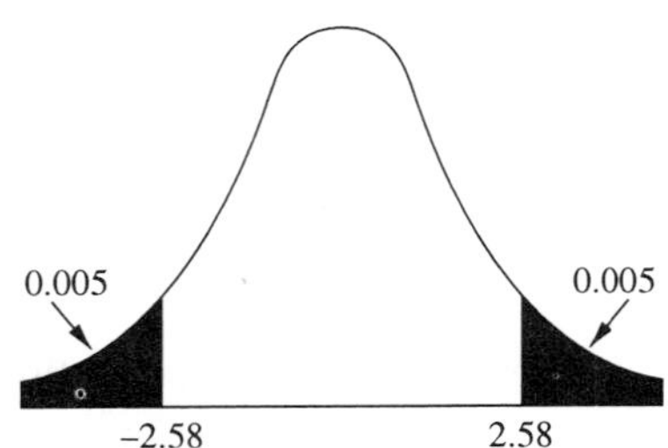

8. H_o: $\mu \leq \$16{,}000$ H_a: $\mu > \$16{,}000$ $\alpha = 0.01$
Reject H_o if $z > 2.33$.

$$z = \frac{\$17{,}000 - \$16{,}000}{\$3{,}000/\sqrt{75}} = 2.89$$

Reject H_o. The population mean is significantly greater than $16,000.

10. a. H_o: $\mu \leq 90$ H_a: $\mu > 90$ **b.** $z = \dfrac{\bar{X} - \mu}{\sigma/n}$
c. If $z > 1.28$, reject H_o.

d. $z = \dfrac{92 - 90}{9/\sqrt{49}} = \dfrac{2}{1.29} = 1.56$ Reject H_o. There has been an increase in the warhead density.

e. The p-value is $0.5 - 0.4406 = 0.0594$

12. Let subscript "1" refer to the areas of less than 100,000 population.
H_o: $\mu_1 \leq \mu_2$ H_a: $\mu_1 < \mu_2$
At the 0.05 significance level, the decision rule is this: If $z < -1.65$, reject H_o.

$$z = \frac{34{,}290 - 34{,}330}{\sqrt{\dfrac{(135)^2}{30} + \dfrac{(142)^2}{60}}} = \frac{-40}{30.7} = -1.30$$

We do not reject the null hypothesis. These data show no significant difference between the two population means.

14. H_o: $\mu_1 = \mu_2$ H_a: $\mu_1 \neq \mu_2$
If z is not between -1.96 and 1.96, reject H_o.

$$z = \frac{32.9 - 29.6}{\sqrt{\dfrac{(5.7)^2}{36} + \dfrac{(5.5)^2}{49}}} = \frac{3.3}{1.23} = 2.68$$

H_o is rejected. There is a difference in age.

The p-value is $2(0.5 - 0.4963) = 0.0074$.

16. a. H_o: $\mu_1 = \mu_2$ H_a: $\mu_1 \neq \mu_2$
b. If z is not between -2.58 and 2.58, reject H_o.

c. $$z = \frac{20.44 - 19.45}{\sqrt{\dfrac{(2.96)^2}{30} + \dfrac{(2.13)^2}{30}}} = \frac{0.99}{0.666} = 1.49$$

We do not reject the null hypothesis. There is no significant difference between the two means.

18. H_o: $\pi \geq 0.20$ H_a: $\pi < 0.20$ Reject H_o if z is less than -2.33.

$$z = \frac{\dfrac{60}{400} - 0.20}{\sqrt{\dfrac{(0.20)(1 - 0.20)}{400}}} = -2.5$$

H_o is rejected. Fewer than 20% of those receiving welfare payments are ineligible.

20. H_o: $\pi = 0.16$ H_a: $\pi \neq 0.16$
Decision rule: If z is not between -1.96 and 1.96, reject H_o.

$$z = \frac{\frac{70}{600} - 0.16}{\sqrt{\frac{(0.16)(0.84)}{600}}} = \frac{-0.043333}{0.014967} = -2.90$$

Reject the null hypothesis. The difference between the two proportions is statistically significant.

The p-value is $2(0.5 - 0.4981) = 0.0038$

22. H_o: $\pi_1 \leq \pi_2$ H_a: $\pi_1 > \pi_2$

π_1 refers to young drivers. Reject H_o if z is greater than or equal to 1.65.

$$\bar{p} = \frac{30 + 55}{100 + 200} = 0.283$$

$$z = \frac{0.30 - 0.275}{\sqrt{(0.283)(0.717)\left(\frac{1}{100} + \frac{1}{200}\right)}}$$

$$= 0.45$$

There is no difference in the proportion of drivers who take risks.

24. H_o: $\pi_1 = \pi_2$ H_a: $\pi_1 \neq \pi_2$

At a 0.05 level of significance, if z is not between -1.96 and 1.96, reject H_o.

$$\bar{\pi} = \frac{10 + 25}{100 + 200} = \frac{35}{300} = 0.117$$

$$z = \frac{0.10 - 0.125}{\sqrt{(0.117)(0.883)\left(\frac{1}{100} + \frac{1}{200}\right)}}$$

$$= \frac{-0.025}{0.039} = -0.64$$

These data show no significant difference.

26. Population #1 is from the windowless schools.
H_o: $\mu_1 \leq \mu_2$ H_a: $\mu_1 > \mu_2$ $\alpha = 0.01$

Reject H_o if $z > 2.33$.

$$z = \frac{94 - 90}{\sqrt{\frac{(8)^2}{100} + \frac{(10)^2}{80}}} = 2.91$$

Reject H_o. There is significantly more anxiety in windowless schools.

28. H_o: $\mu \geq 30$ H_a: $\mu < 30$ $\alpha = 0.02$

Reject H_o if $z < -2.05$.

$$z = \frac{28.5 - 30}{5/\sqrt{40}} = -1.90$$

There is not enough evidence to reject H_o. There is no significant improvement in the processing time.

30. a. H_o: $\mu_1 \geq \mu_2$ H_a: $\mu_1 < \mu_2$ where subscript one refers to Lake Anna.

b. $z = \dfrac{\bar{X}_1 - \bar{X}_2}{\sqrt{\dfrac{s_1^2}{n_1} + \dfrac{s_2^2}{n_2}}}$

c. If $z < -2.33$ reject H_o.

d. $z = \dfrac{20 - 40}{\sqrt{\dfrac{(4)^2}{50} + \dfrac{(8)^2}{40}}} = \dfrac{-20}{1.39} = -14.4$

Reject H_o. The Lake Anna fish are smaller. The data supports the environmentalists.

32. a. H_o: $\mu = 16.01$ H_a: $\mu \neq 16.01$

b. If z is not between −1.65 and 1.65, then reject H_o.

c. $z = \dfrac{15.97 - 16.01}{0.005/\sqrt{40}} = \dfrac{-0.04}{0.00079} = -50.6$ That certainly is significant!

34. a. H_o: $\mu_1 = \mu_2$ H_a: $\mu_1 \neq \mu_2$

b. If z is not between −1.65 and 1.65, reject H_o.

c. $z = \dfrac{52.7 - 51.8}{\sqrt{\dfrac{(2.5)^2}{100} + \dfrac{(2.6)^2}{150}}} = \dfrac{0.9}{0.328} = 2.74$ Reject H_o.

The mean heights are significantly different.

36. a. Let the subscript "1" refer to men.
H_o: $\mu_1 \leq \mu_2$ H_a: $\mu_1 > \mu_2$

b. If z is greater than 2.05, reject H_o.

c. $z = \dfrac{353 - 315}{\sqrt{\dfrac{(18)^2}{80} + \dfrac{(21)^2}{80}}} = \dfrac{38}{3.092} = 12.29$

Reject H_o. Men are paid significantly more.

38. H_o: $\pi = 0.2$ H_a: $\pi > 0.2$ If $z > 1.65$ reject H_o for a 5% significance level.

$z = \dfrac{\dfrac{56}{200} - 0.20}{\sqrt{\dfrac{(0.20)(0.80)}{200}}} = \dfrac{0.08}{0.028} = 2.82$

Reject H_o. This development *does* have a greater proportion of movers.

40. a. H_o: $\pi = 0.65$ H_a: $\pi \neq 0.65$

b. If z is not between −1.65 and 1.65, reject H_o.

c. $z = \dfrac{\dfrac{120}{200} - 0.65}{\sqrt{\dfrac{(0.65)(0.35)}{200}}} = \dfrac{-0.05}{0.034} = -1.48$

d. Utah seems to follow the national pattern.

42. H_o: $\pi_1 = \pi_2$ H_a: $\pi_1 \neq \pi_2$

If z is not between −2.58 and 2.58, reject H_o.

$$z = \frac{\frac{729}{1,656} - \frac{301}{1,432}}{\sqrt{(0.33)(0.67)\left(\frac{1}{1,656} + \frac{1}{1,432}\right)}}$$

$$= \frac{0.23}{0.017} = 13.6$$

Reject H_o. There has been a shift in the public's view.

44. H_o: $\pi_1 = \pi_2$ H_a: $\pi_1 \neq \pi_2$

For a 5% significance level, if z is not between −1.96 and 1.96, reject H_o.

$$z = \frac{\frac{8}{37} - \frac{7}{38}}{\sqrt{(0.20)(0.80)\left(\frac{1}{37} + \frac{1}{38}\right)}}$$

$$= \frac{0.032}{0.092} = 0.35$$

We fail to reject H_o. This value of the test statistic is not significant at any reasonable level.

46. H_o: $\pi_1 = \pi_2$ H_a: $\pi_1 \neq \pi_2$

Accept H_o if z is between −1.96 and 1.96.

$$\bar{\pi} = \frac{200 + 110}{1,000 + 500} = 0.207$$

$$z = \frac{0.20 - 0.22}{\sqrt{(0.207)(1 - 0.207)\left(\frac{1}{1,000} + \frac{1}{500}\right)}}$$

$$= -0.90$$

The difference is not significant.

48. H_o: $\pi_1 \leq \pi_2$ H_a: $\pi_1 > \pi_2$

π_1 = Second period. Reject H_o if z is greater than 2.33.

$$\bar{\pi} = \frac{83 + 146}{420 + 423} = 0.272$$

$$z = \frac{0.345 - 0.198}{\sqrt{0.272(1 - 0.272)\left(\frac{1}{420} + \frac{1}{423}\right)}}$$

$$= 4.80$$

There has been a significant increase in car pooling.

50. H_o: $\pi = 0.50$ H_a: $\pi \neq 0.50$ $\alpha = 0.05$

$$z = \frac{\frac{X}{n} - \pi}{\sqrt{\frac{\pi(1 - \pi)}{n}}}$$

Decision rule: If z is not between −1.96 and 1.96, reject H_o.

$$z = \frac{\frac{143}{300} - 0.50}{\sqrt{\frac{(0.50)(1 - 0.50)}{300}}} = \frac{-0.0233}{0.0289} = -0.81$$

Therefore, we fail to reject H_o. There is no significant difference between the sample data and the sportswriter's claim.

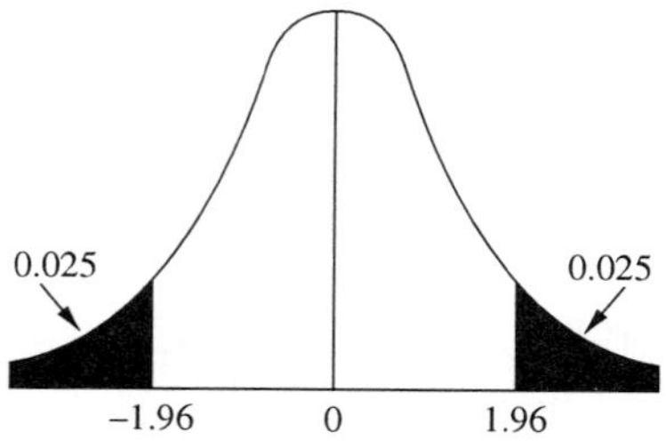

52. H_o: $\pi_1 = \pi_2$ H_a: $\pi_1 \neq \pi_2$ $\alpha = 0.01$

$$z = \frac{\frac{X_1}{n_1} - \frac{X_2}{n_2}}{\sqrt{\bar{p}(1 - \bar{p})\left(\frac{1}{n_1} + \frac{1}{n_2}\right)}}$$

Decision rule: If z is not between –2.58 and 2.58, reject H_o.

$$\bar{p} = \frac{X_1 + X_2}{n_1 + n_2} = 0.57$$

0.005 0.005

–2.58 2.58

$$z = \frac{\frac{46}{70} - \frac{28}{60}}{\sqrt{(0.57)(1 - 0.57)\left(\frac{1}{70} + \frac{1}{60}\right)}} = \frac{0.1905}{0.0871} = 2.19$$

We fail to reject H_o. These data show no significant difference between the two treatments.

54. a. TEST OF MU = $190,000 VS MU L.T. $190,000

	N	MEAN	STDEV	SE MEAN	T	P VALUE
Selling	50	188.442	8.166	1.155	−1.35	0.09

We cannot reject the null hypothesis at the .05 significance level. There is about a 9% chance of getting a value this far below $190,000 by sampling error.

b. TEST OF MU = 15.000 VS MU G.T. 15.000

	N	MEAN	STDEV	SE MEAN	T	P VALUE
Distance	50	20.840	3.066	0.434	13.47	0.00

We can definitely reject the null hypothesis. The likelihood of a sample driving distance with a mean this large is virtually zero under the null hypothesis.

c. TEST OF PI = 0.5000 VS PI L.T. 0.5000

	N	MEAN	STDEV	SE MEAN	T	P VALUE
Fire	50	0.4200	0.4986	0.0705	−1.13	0.13

This data does not prove beyond a reasonable statistical doubt that less than half the homes have fireplaces.

d. TEST OF MU = 0.97500 VS MU N.E. 0.97500

	N	MEAN	STDEV	SE MEAN	T	P VALUE
C10	50	0.98058	0.00755	0.00107	5.23	0.0000

The mean ratio of the selling price to the list price is definitely not .975!

CHAPTER ELEVEN

2. $t = -2.977$ and $t = 2.977$

4. H_o: $\mu \leq 4.3$ years H_a: $\mu > 4.3$ years $\alpha = 0.01$
df $= n - 1 = 20 - 1 = 19$
Reject H_o if t exceeds 2.539; otherwise, fail to reject H_o.

$$t = \frac{4.6 - 4.3}{1.2/\sqrt{20}} = 1.12$$

Do not reject H_o. The evidence is insufficient to support the new method.

6. H_o: $\mu \geq 56$ H_a: $\mu < 56$ $\alpha = 0.05$ df $= 12 - 1 = 11$
Reject H_o if $t < -1.796$

$$t = \frac{53.5 - 56.0}{2.00/\sqrt{12}} = -4.33$$

H_o is rejected. The mean height is less than 56 inches. The p-value is less than .005.

8. H_o: $\mu \leq 1.6$ H_a: $\mu > 1.6$ $\alpha = 0.01$ df $= n - 1 = 5 - 1 = 4$
Reject H_o if t is less than 3.747; otherwise, fail to reject H_o.

$$\overline{X} = \frac{1.2 + 2.5 + 1.9 + 3.0 + 2.4}{5} = 2.2 \qquad s = \sqrt{\frac{26.06 - \frac{(11)^2}{5}}{5 - 1}} = 0.68$$

$$t = \frac{2.2 - 1.6}{0.68/\sqrt{5}} = 1.97$$

The data agree with the assumption that the mean antibody strength is $\leq$ 1.6. The null hypothesis cannot be rejected.

10. H_o: $\mu \geq 40$ H_a: $\mu < 40$ $\alpha = 0.05$ df $= 8 - 1 = 7$
Reject H_o if $t < -1.895$; otherwise, fail to reject.

$$\overline{X} = \frac{38 + 46 + \ldots + 46}{8} = \frac{304}{8} = 38.00 \qquad s = \sqrt{\frac{11{,}842 - \frac{(304)^2}{8}}{8 - 1}} = 6.44$$

$$t = \frac{38.00 - 40.00}{6.44/\sqrt{8}} = \frac{-2.00}{2.28} = -0.88$$

H_o is not rejected. The mean age could be 40 years.

12. H_o: $\mu_1 \geq \mu_2$ H_a: $\mu_1 < \mu_2$ $\alpha = 0.01$ df $= 10 + 10 - 2 = 18$
Reject H_o if t is less than -2.552; otherwise, fail to reject H_o.

$$s_p = \sqrt{\frac{9(0.8)^2 + 9(1.0)^2}{10 + 10 - 2}} = 0.906 \qquad t = \frac{2.5 - 3.1}{0.906\sqrt{\frac{1}{10} + \frac{1}{10}}} = -1.48$$

The data do not support the claim that the isotonic method is more effective. The p-value is between .10 and .05.

14. H_o: $\mu_1 \leq \mu_2$ H_a: $\mu_1 > \mu_2$ $\alpha = 0.05$ df $= 4 + 5 - 2 = 7$
H_o is rejected if t is greater than 1.895.

$$\overline{X}_1 = \frac{17 + 23 + 26 + 30}{4} = \frac{96}{4} = 24.00 \qquad \overline{X}_2 = \frac{19 + 15 + 14 + 17 + 20}{5} = \frac{85}{5} = 17$$

$$s_1 = \sqrt{\frac{2{,}394 - \frac{(96)^2}{4}}{4 - 1}} = 5.48 \qquad s_2 = \sqrt{\frac{1{,}471 - \frac{(85)^2}{5}}{5 - 1}} = 2.55$$

$$s_p = \sqrt{\frac{3(5.48)^2 + 4(2.55)^2}{4 + 5 - 2}} = 4.07 \qquad t = \frac{24.00 - 17.00}{4.07\sqrt{\frac{1}{4} + \frac{1}{5}}} = \frac{7.00}{2.73} = 2.56$$

H_o is rejected. The mean expense is greater for sales executives (less for production).

16.

LH	*RH*	*d*	d^2
140	138	2	4
90	87	3	9
125	110	15	225
130	132	−2	4
95	96	−1	1
121	120	1	1
85	86	−1	1
97	90	7	49
131	129	2	4
110	100	10	100
		36	398

$\bar{d} = 36/10 = 3.6$

$$s_d = \sqrt{\frac{398 - \frac{(36)^2}{10}}{10 - 1}} = 5.46$$

H_o: $\mu_d \leq 0$ H_a: $\mu_d > 0$
$\alpha = 0.01$ df $= 10 - 1 = 9$
Reject H_o if t exceeds 2.821; otherwise, fail to reject H_o.

$$t = \frac{3.6}{5.46/\sqrt{10}} = 2.085$$

There is not sufficient evidence to prove they have greater strength in their left hand.

18. H_o: $\mu_d = 0$ H_a: $\mu_d \neq 0$ $\alpha = 0.05$ df $= 12 - 1 = 11$

Month	Women's	Men's	*d*
Jan.	20.0	22.5	−2.5000
Feb.	14.0	13.9	0.1000
Mar.	29.0	24.8	4.2000
Apr.	30.0	27.5	2.5000
May	16.0	24.9	−8.9000
June	17.0	27.4	−10.4000
July	20.1	21.4	−1.3000
Aug.	20.5	20.7	−0.2000
Sept.	21.3	20.6	0.7000
Oct.	22.0	22.8	−0.8000
Nov.	25.7	23.5	2.2000
Dec.	42.2	44.2	−2.0000
			−16.4

H_o is rejected if t is greater than 2.201 or less than −2.201.

$$\bar{d} = \frac{-16.4}{12} = -1.367$$

$$s_d = \sqrt{\frac{229.22 - \frac{(-16.4)^2}{12}}{12 - 1}}$$

$= 4.336$

$$t = \frac{-1.367}{4.336/\sqrt{12}} = -1.09$$

The p-value is between .5 and .2.

H_o is not rejected. There is no difference in sales.

20. H_o: $\mu \geq 20$ pounds overweight H_a: $\mu < 20$ pounds overweight
$\alpha = 0.05$ df $= 15 - 1 = 14$
Reject H_o if t is less than −1.761; otherwise, fail to reject H_o.

$$t = \frac{\bar{X} - \mu}{s/\sqrt{n}} = \frac{18 - 20}{5/\sqrt{15}} = -1.55$$

There is not sufficient evidence to doubt the claim in the newspaper.

22. H_o: $\mu \geq 26.0$ $\quad H_a$: $\mu < 26.0$ $\quad \alpha = 0.01$ $\quad$ df $= 6 - 1 = 5$
Reject H_o if t is less than -3.365; otherwise, fail to reject H_o.

$$s = \sqrt{\frac{3{,}761.1 - \frac{(150.2)^2}{6}}{6 - 1}} = 0.468$$

$$\overline{X} = \frac{24.3 + 25.2 + 24.9 + 24.8 + 25.6 + 25.4}{6} = \frac{150.2}{6} = 25.03$$

$$t = \frac{25.03 - 26.0}{0.468/\sqrt{6}} = -5.08$$

H_o is rejected. The car does not meet EPA specifications.

24. H_o: $\mu = 25$ $\quad H_a$: $\mu \neq 25$ $\quad \alpha = 0.10$ $\quad$ df $= 16 - 1 = 15$
H_o is rejected if t is less than -1.753 or t is greater than 1.753.

$$\overline{X} = \frac{22 + \ldots + 22}{16} = \frac{415}{16} = 25.938 \qquad s = \sqrt{\frac{10{,}937 - \frac{(415)^2}{16}}{16 - 1}} = 3.395$$

$$t = \frac{25.938 - 25.000}{3.395/\sqrt{16}} = 1.11$$

H_o is not rejected. The mean temperature could be 25° C. The p-value is between .5 and .2.

26. H_o: $\mu_1 = \mu_2$ $\quad H_a$: $\mu_1 \neq \mu_2$ $\quad \alpha = 0.01$ $\quad$ df $= 7 + 6 - 2 = 11$
H_o is rejected if t is less than -3.106 or greater than 3.106.

$$\overline{X}_1 = \frac{57}{7} = 8.14 \qquad \overline{X}_2 = \frac{48}{6} = 8.00$$

$$s_1 = \sqrt{\frac{475 - \frac{(57)^2}{7}}{7 - 1}} = 1.3452 \qquad s_2 = \sqrt{\frac{394 - \frac{(48)^2}{6}}{6 - 1}} = 1.414$$

$$s_p = \sqrt{\frac{(7 - 1)(1.3452)^2 + (6 - 1)(1.414)^2}{7 + 6 - 2}} = 1.3769$$

$$t = \frac{8.14 - 8.00}{1.3769\sqrt{\frac{1}{7} + \frac{1}{6}}} = \frac{0.14}{0.77} = 0.18$$

H_o is not rejected. There is no difference in the mean weight gain of the two groups.

0 H_a: $\mu_d > 0$ $\alpha = 0.01$ df $= 12 - 1 = 11$
ɪ$_o$ if t exceeds 2.718; otherwise, fail to reject H_o.

ɪ	Without	d	d^2
230	217	13	169
225	198	27	729
223	208	15	225
216	222	−6	36
229	223	6	36
201	214	−13	169
205	187	18	324
193	187	6	36
177	178	−1	1
201	195	6	36
178	169	9	81
207	194	13	169
		93	2,011

$$\bar{d} = \frac{93}{12} = 7.75$$

$$s_d = \sqrt{\frac{2{,}011 - \frac{(93)^2}{12}}{12 - 1}} = 10.83$$

$$t = \frac{7.75}{10.83/\sqrt{12}} = 2.48$$

The evidence does not support the manufacturer's claim.

30. H_o: $\mu_d \leq 0$ H_a: $\mu_d > 0$ df $= 8 - 1 = 7$
Reject H_o if t is greater than 1.895; otherwise, fail to reject H_o.

After	Before	d	d^2
14.0	13.5	0.5	0.25
10.7	11.4	−0.7	0.49
12.4	10.7	1.7	2.89
11.1	11.1	0.0	0.00
10.9	9.8	1.1	1.21
10.5	9.6	0.9	0.81
10.8	10.7	0.1	0.01
13.0	11.7	1.3	1.69
		4.9	7.35

$$\bar{d} = \frac{4.9}{8} = 0.61$$

$$s_d = \sqrt{\frac{7.35 - \frac{(4.9)^2}{8}}{8 - 1}} = 0.788$$

$$t = \frac{0.61}{0.788/\sqrt{8}} = 2.19$$

H_o is rejected and H_a accepted. The sales training program is effective. The p-value is between .05 and .025.

32. H_o: $\mu \leq 6.0$ H_a: $\mu > 6.0$ df $= 6 - 1 = 5$ $\alpha = 0.01$
H_o is rejected if $t > 3.365$.

$$t = \frac{7.2 - 6.0}{1.05/\sqrt{6}} = 2.80$$

H_o is not rejected. The population mean still could be 6 homes started.

34. H_o: $\mu_d \leq 0$ H_a: $\mu_d > 0$ df = 12 − 1 = 11
H_o is rejected if $t > 2.718$.

Home	Asking	Selling	Difference
1	103.0	99.3	3.7
2	127.0	124.3	2.7
3	114.9	110.8	4.1
4	102.0	102.0	0.0
5	84.6	80.0	4.6
6	160.5	158.8	1.7
7	99.6	95.4	4.2
8	173.0	167.3	5.7
9	212.5	210.1	2.4
10	89.2	86.3	2.9
11	99.9	96.8	3.1
12	138.0	132.6	5.4

$$\bar{d} = \frac{40.5}{12} = 3.375$$

$$s_d = \sqrt{\frac{164.91 - \frac{(40.5)^2}{12}}{12 - 1}} = 1.602$$

$$t = \frac{3.375}{1.602/\sqrt{12}} = 7.30$$

H_o is rejected. The asking price is more than the selling price.

36. a. TEST OF MU = $68,000 VS MU N.E. $68,000

N	MEAN	STDEV	SE MEAN	T	P VALUE
49	$63,486	$4,000	$ 571	−7.90	0.0000

This is significantly different from $68,000.

b. The pooled standard deviation is the square root of $\frac{48(4{,}000)^2 + 25(4{,}574)^2}{73}$, which is $4,205.

So the t statistic is $\dfrac{-3{,}064}{4{,}205\sqrt{\frac{1}{49} + \frac{1}{26}}} = \dfrac{-3{,}064}{1{,}020} = -3.00.$

The p-value is 0.004. So we can reject the null hypothesis of equal means and conclude that there is a significant difference between the population means. The t and z distributions are close at 73 df and the smaller sample is close to 30. In addition, the sample variances are similar.

c.

```
                    COLUMNS:  Sex
                        0      1
young                  33     12
 old                   16     14

TWOSAMPLE T  FOR  C40  VS  C41

          N         MEAN        STDEV        SE MEAN
C40      12        62.78         2.64           0.76
C41      16        65.61         3.88           0.97

TTEST  MU  C40  =  MU  C41  (VS  LT):  T= -2.29
P= 0.015   DF= 26
```

We can say that the mean salary for "young" women is less.

TWELVE

2. a. $F_{,6,0.05} = 4.21$ b. $F_{3,9,0.01} = 6.99$

4. $H_o: \sigma_b^2 = \sigma_a^2$ $H_a: \sigma_b^2 \neq \sigma_a^2$ $\alpha = 0.10$ $df_b = 21 - 1 = 20$
$df_a = 16 - 1 = 15$
H_o is rejected if F is greater than 2.33; otherwise, H_o is not rejected.

$$F = \frac{225}{150} = 1.50$$

H_o is not rejected. The population variances are the same.

6. $H_o: \mu_1 = \mu_2 = \mu_3$ H_a: Not all means are equal. $\alpha = 0.05$
$df_n = 3 - 1 = 2$ $df_d = 9 - 3 = 6$
H_o is rejected if F is greater than 5.14.

$$SS \text{ total} = 367.58 - \frac{(55)^2}{9} = 31.4689$$

$$SST = \frac{(13.6)^2}{3} + \frac{(32.0)^2}{4} + \frac{(9.4)^2}{2} - \frac{(55.0)^2}{9} = 25.7222$$

$$SSE = 31.4689 - 25.7222 = 5.7467$$

Source	Sum of Squares	df	Mean Square	F
Treatment	25.7222	2	12.8611	13.43
Within	5.7467	6	0.9578	
Total	31.4689	8		

Since the computed F of 13.43 exceeds the critical value of 5.14, H_o is rejected. There is a difference in the mean waiting time of the banks. The p-value is less than .01.

8. $H_o: \mu_1 = \mu_2 = \mu_3 = \mu_4$ H_a: Not all means are equal.
$df_n = 4 - 1 = 3$ $df_d = 20 - 4 = 16$
H_o is rejected if $F > 5.29$.

$$SS \text{ total} = 9{,}089 - \frac{(407)^2}{20} = 806.55$$

$$SST = \frac{(80)^2}{5} + \frac{(107)^2}{6} + \frac{(139)^2}{5} + \frac{(81)^2}{4} - \frac{(407)^2}{20} = 410.17$$

$$SSE = 806.55 - 410.17 = 396.38$$

Source	Sum of Squares	df	Mean Square
Treatment	410.17	3	136.72
Within	396.38	16	24.77
Total	806.55		

$$F = \frac{136.72}{24.77} = 5.52$$

H_o is rejected and H_a accepted. The ethical behavior of attorneys varies by region.

10. H_o: $\sigma_s^2 \leq \sigma_a^2$ H_a: $\sigma_s^2 > \sigma_a^2$ $\alpha = 0.05$ $df_n = 7 - 1 = 6$
$df_d = 8 - 1 = 7$
H_o is rejected if F is greater than 3.87.

$$F = \frac{(4{,}200)^2}{(2{,}000)^2} = 4.41$$

H_o is rejected. There is more variation in the sales group.

12. H_o: $\mu_1 = \mu_2 = \mu_3 = \mu_4$ H_a: Not all means are equal.
Reject H_o if F is greater than 3.24; otherwise, do not reject.

$$SS\text{ total} = 16{,}257 + \frac{(555)^2}{20} = 855.75$$

$$SST = \frac{(161)^2}{5} + \frac{(122)^2}{5} + \frac{(120)^2}{5} + \frac{(152)^2}{5} - \frac{(555)^2}{20} = 260.55$$

$$SSE = 855.75 - 260.55 = 595.20$$

Source	Sum of Squares	df	Mean Square
Treatment	260.55	3	86.85
Within	595.20	16	37.20
Total	855.75	19	

$$F = \frac{86.85}{37.20} = 2.33.$$

H_o cannot be rejected. The evidence does not suggest any differences in reading assignments. The p-value is greater than .05.

14. H_o: $\mu_1 = \mu_2 = \mu_3 = \mu_4$ H_a: Not all means are equal.
Reject H_o is F is greater than 3.41; otherwise, do not reject.

$$SS\text{ total} = 3{,}370.58 - \frac{(216.6)^2}{17} = 610.84$$

$$SST = \frac{(48.1)^2}{4} + \frac{(51.9)^2}{5} + \frac{(33.1)^2}{3} + \frac{(83.5)^2}{5} - \frac{(216.6)^2}{17} = 117.04$$

$$SSE = 610.84 - 117.04 = 493.80$$

Source	Sum of Squares	df	Mean Square
Treatment	117.04	3	39.01
Within	493.80	13	37.985
Total	610.84	16	

$$F = \frac{39.01}{37.985} = 1.03$$

Do not reject H_o. There is no difference in time to solve the puzzle.

16. H_o: $\mu_1 = \mu_2 = \mu_3$ H_a: Not all equal.
Reject H_o if the computed F is greater than 6.93.

$$SS \text{ total} = 5{,}434 - \frac{(256)^2}{15} = 1{,}064.9$$

$$SST = \frac{(114)^2}{4} + \frac{(76)^2}{5} + \frac{(66)^2}{6} - \frac{(256)^2}{15} = 761.1$$

$$SSE = 1{,}064.9 - 761.1 = 303.8$$

Source	Sum of Squares	df	Mean Square
Treatment	761.1	2	380.6
Within	303.8	12	25.3
Total	1,064.9	14	

$$F = \frac{380.6}{25.3} = 15.03$$

H_o is rejected and H_a accepted. There is a difference among the three methods.

18. H_o: $\mu_1 = \mu_2 = \mu_3$ H_a: Not all means are equal. $\alpha = 0.05$
$df_n = 3 - 1 = 2$ $df_d = 13 - 3 = 10$
H_o is rejected if F is greater than 4.10.

$$SS \text{ total} = 63{,}374 - \frac{(896)^2}{13} = 1{,}618.92$$

$$SST = \frac{(229)^2}{4} + \frac{(346)^2}{5} + \frac{(321)^2}{4} - \frac{(896)^2}{13} = 1{,}058.62$$

$$SSE = 1{,}618.92 - 1{,}058.62 = 560.30$$

Source	Sum of Squares	df	Mean Square	F
Treatment	1,058.62	2	529.31	9.45
Within	560.30	10	56.03	
Total	1,618.92	12		

H_o is rejected. There is a difference in the mean score for the age groups. The p-value is less than .01.

20. H_o: $\mu_1 = \mu_2 = \mu_3 = \mu_4$ H_a: Not all means are the same.
H_o is rejected if $F > 2.97$ (estimated). The MINITAB output is as follows:

```
ANALYSIS OF VARIANCE
SOURCE      DF      SS       MS        F        p
TREATMENT    3  62.481   20.827    68.99    0.000
ERROR       26   7.849     .302
TOTAL       29  70.330
                                INDIVIDUAL 95 PCT CI'S FOR MEAN
                                BASED ON POOLED STDEV
     LEVEL   N     MEAN   STDEV ----+---------+--------+---------+-
         1   7    8.829    .522 (--*--)
         2   8   10.238    .338          (-*--)
         3   9   11.844    .718                   (--*-)
         4   6   12.767    .509                           (--*--)
                                ----+---------+--------+---------+-
POOLED STDEV=  .549                9.0     10.5      12.0      13.5
```

Since the computed F of 68.99 exceeds the critical value, H_o is rejected. Some means are significantly different from the others.

22. a.

```
ANALYSIS OF VARIANCE ON Salary
SOURCE    DF       SS      MS        F         p
Type       4    313.8    78.4     4.83     0.002
ERROR     70   1136.7    16.2
TOTAL     74   1450.4
                              INDIVIDUAL 95 PCT CI'S FOR MEAN
                              BASED ON POOLED STDEV
  LEVEL    N      MEAN   STDEV ----+---------+--------+---------+-
      1    8    68.412   3.814                (-------*-------)
      2   18    64.139   4.001          (---*----)
      3   20    64.920   3.707            (---*----)
      4   18    65.289   4.680             (----*----)
      5   11    60.518   3.596 (----*----)
                               ----+---------+--------+---------+-
POOLED STDEV= 4.030              60.0      64.0     68.0      72.0
```

The mean salaries are different by industry.

b.
```
ANALYSIS OF VARIANCE ON Salary
SOURCE     DF        SS      MS         F          p
Region      3     131.4    43.8      2.36      0.079
ERROR      71    1319.1    18.6

TOTAL      74    1450.4
                                   INDIVIDUAL 95 PCT CI'S FOR MEAN
                                   BASED ON POOLED STDEV
  LEVEL     N      MEAN  STDEV   -+---------+---------+---------+
      1    12    66.558  4.375                   (-------*-------)
      2    23    63.948  4.670            (-----*-----)
      3    28    65.168  4.537                  (----*-----)
      4    12    62.242  2.577   (-------*--------)
                                 -+---------+---------+---------+
POOLED STDEV= 4.310             60.0      63.0      66.0      69.0
```

The mean salaries are not different by region.

c.
```
ANALYSIS OF VARIANCE ON Salary
SOURCE     DF        SS      MS         F          p
Sex         1     159.5   159.5      9.02      0.004
ERROR      73    1290.9    17.7
TOTAL      74    1450.4
                                   INDIVIDUAL 95 PCT CI'S FOR MEAN
                                   BASED ON POOLED STDEV
  LEVEL     N      MEAN   STDEV  ----+--------+--------+--------+-
      0    49    63.486   4.000  (-------*-------)
      1    26    66.550   4.574                      (-----*-----)
                                 ----+--------+--------+--------+-
POOLED STDEV= 4.205                62.4     64.0     65.6     67.2
```

The mean salaries for men and women are different. Note that the pooled standard deviation and *p*-value are the same as those found in Exercise 36.b in Chapter 11.

CHAPTER THIRTEEN

2. a.

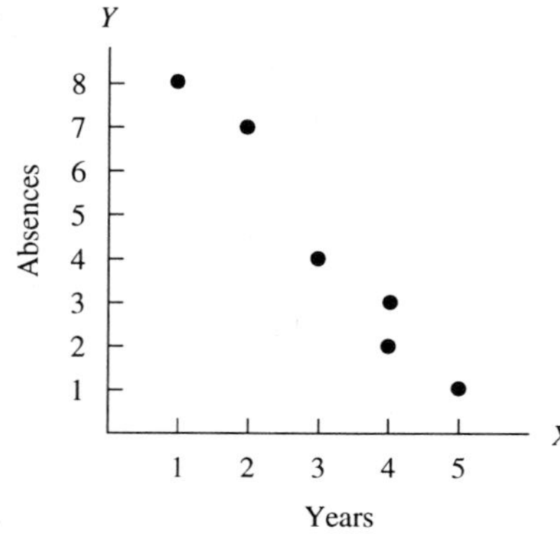

b. Absences decrease as years of employment increase.

4.

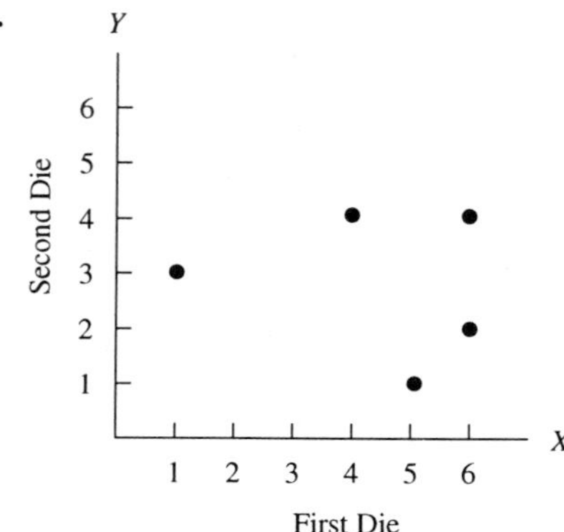

The two dice are independent. So there is no pattern to the scatter diagram.

6.

X	Y	XY	X^2	Y^2
1	8	8	1	64
5	1	5	25	1
2	7	14	4	49
4	3	12	16	9
4	2	8	16	4
3	4	12	9	16
19	25	59	71	143

$$r = \frac{6(59) - (19)(25)}{\sqrt{[6(71) - (19)^2][6(143) - (25)^2]}}$$

$$= \frac{-121}{\sqrt{(65)(233)}} = -0.98$$

8. a.

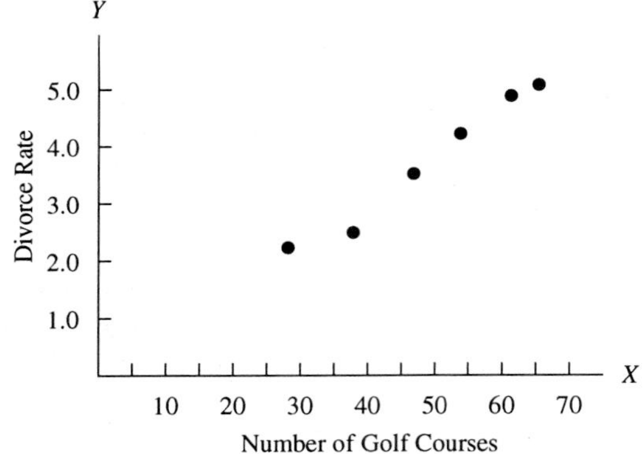

b.

X	Y	XY	X^2	Y^2
28	2.2	61.6	784	4.84
38	2.5	95.0	1,444	6.25
47	3.5	164.5	2,209	12.25
54	4.1	221.4	2,916	16.81
62	4.8	297.6	3,844	23.04
66	5.0	330.0	4,356	25.00
295	22.1	1,170.1	15,553	88.19

$$r = \frac{6(1{,}170.1) - (295)(22.1)}{\sqrt{[6(15{,}553) - (295)^2][6(88.19) - (22.1)^2]}}$$

$$= \frac{501.1}{\sqrt{(6{,}293)(40.73)}} = 0.99$$

c. The divorce rate definitely increases as the number of golf courses increases.

10. a. $(0.96)^2 = 0.92$ **b.** $1 - 0.92 = 0.08$

c. 92% of the variation in the number of people on the beach is accounted for by the variation in the high temperature.

12. a. $(-0.68)^2 = 0.46$ **b.** $1 - 0.46 = 0.54$

c. 54% of the variation in suspension is *not* accounted for by the variation in the number of activities offered.

14. $H_o: \rho = 0 \qquad H_a: \rho \neq 0$

$$t = \frac{0.40\sqrt{20 - 2}}{\sqrt{1 - (0.4)^2}} = 1.85$$

The critical value is 2.878. So the population correlation could be zero.

16. $H_o: \rho = 0 \qquad H_a: \rho \neq 0$

$$t = \frac{0.21\sqrt{20 - 2}}{\sqrt{1 - (0.21)^2}} = 0.91$$

The null hypothesis of zero correlation is rejected if t is not between −2.101 and +2.101; therefore, the population coefficient of correlation is not significantly different from zero.

18. a.

Sales Manager	Personnel Manager	d	d^2
3	2	1	1
1	5	−4	16
8.5	9	−0.5	0.25
2	1	1	1
4	6	−2	4
8.5	12	−3.5	12.25
5	3	2	4
7	4	3	9
10	11	−1	1
6	7	−1	1
11	8	3	9
12	10	2	4
		$\Sigma d^2 =$	62.50
		$n =$	12

$$1 - \frac{6(62.50)}{12(12^2 - 1)} = 0.781$$

b. Assume a 0.05 significance level with 10 df. Critical value is 1.812.
H_o: No rank correlation. H_a: Positive rank correlation.

$$t = 0.781\sqrt{\frac{12 - 2}{1 - (.781)^2}} = 3.95$$

Therefore, reject H_o. A positive rank correlation does exist at a 0.05 level of significance.

20. a. (In hundreds)

X	Y	XY	X^2	Y^2
19	5	95	361	25
20	6	120	400	36
15	5	75	225	25
6	2	12	36	4
11	3	33	121	9
17	3	51	289	9
18	3	54	324	9
15	3	45	225	9
13	2	26	169	4
16	3	48	256	9
$\Sigma X = 150$	$\Sigma Y = 35$	$\Sigma XY = 559$	$\Sigma X^2 = 2{,}406$	$\Sigma Y^2 = 139$

$$r = \frac{10(559) - (150)(35)}{\sqrt{[10(2{,}406) - (150)^2][10(139) - (35)^2]}} = 0.670$$

b. If you were to test the hypothesis of no correlation using a two-tailed 5% significance level test, the critical value is 2.306.

H_o: $\rho = 0$ $\quad$ H_a: $\rho \neq 0$

$$t = \frac{0.67\sqrt{10 - 2}}{\sqrt{1 - (0.67)^2}} = 2.55$$

The test statistic is large enough to reject the null hypothesis of no correlation. So you can conclude there is a significant correlation between food expenditures and family size.

22. a. $n = 11$, $\Sigma X = 666$, $\Sigma Y = 32.0$, $\Sigma XY = 4{,}067.11$, $\Sigma X^2 = 71{,}413.32$, and $\Sigma Y^2 = 252.64$

$$r = \frac{11(4{,}067.11) - (666)(32.0)}{\sqrt{[11(71{,}413.32) - (666)^2][11(252.64) - (32.0)^2]}} = 0.96$$

The coefficient of determination is $r^2 = (0.96)^2 = 0.92$.
The coefficient of nondetermination is $1 - r^2 = 1 - 0.92 = 0.08$.

b. H_o: $\rho = 0$ $\quad$ H_a: $\rho \neq 0$

$$t = \frac{0.96\sqrt{11 - 2}}{\sqrt{1 - (0.96)^2}} = 10.29$$

The critical value of t with 9 degrees of freedom at the 0.05 level of significance is ± 2.262. Since t is larger than the critical value, we reject the hypothesis of no correlation.

c. There appears to be a very close relationship between the number of crimes and the number of law enforcement personnel. Ninety-two percent of the variation in one explains the variation in the other variable.

24. a.

Completion Rank	Grade Rank	d	d^2
1	3.5	−2.5	6.25
2	3.5	−1.5	2.25
3	6	−3	9.00
4	2	2	4.00
5	1	4	16.00
6	5	1	1.00
7	7	0	0.00
8	10	−2	4.00
9	8	1	1.00
10	9	1	1.00
			44.50

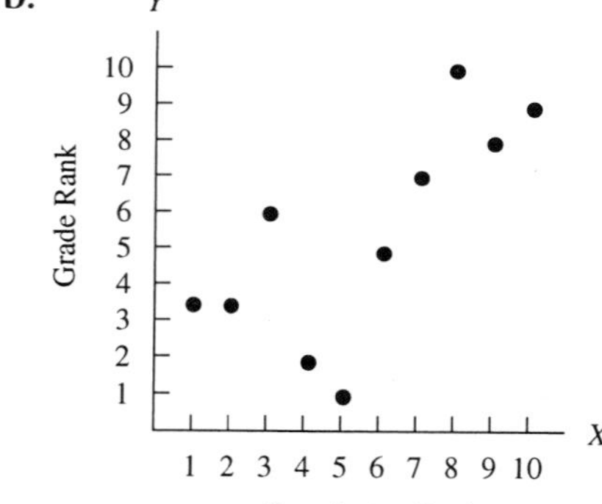

c. $r_s = 1 - \dfrac{6(44.5)}{10(10^2 - 1)} = 0.73$

d. There does appear to be a strong direct relationship between order of completion and grade rank.

26. a.

X	Y	XY	X^2	Y^2
123.0	4.15	510.45	15129	17.2225
358.9	4.10	1471.49	128809	16.8100
111.1	3.74	415.51	12343	13.9876
118.0	3.89	459.02	13924	15.1321
265.0	3.75	993.75	70225	14.0625
390.0	3.10	1209.00	152100	9.6100
67.0	3.13	209.71	4489	9.7969
24.0	2.97	71.28	576	8.8209
156.9	2.96	464.42	24618	8.7616
106.0	2.92	309.52	11236	8.5264
86.0	2.33	200.38	7396	5.4289
88.5	2.26	200.01	7832	5.1076
241.2	2.03	489.64	58177	4.1209
2135.6	41.33	7004.18	506855	137.388

$$r = \frac{13(7004.18) - (2135.6)(41.33)}{\sqrt{[13(506855) - (2135.6)^2][13(137.388) - (41.33)^2]}}$$

$$= \frac{2789.992}{\sqrt{(2,028,327.6)\,(77.8751)}} = 0.222$$

b. $r^2 = (0.222)^2 = 0.049$

c. $H_o\colon \rho = 0 \qquad H_a\colon \rho \neq 0$

$$t = \frac{0.222\sqrt{13 - 2}}{\sqrt{1 - (0.222)^2}} = \frac{0.736}{0.975} = 0.755$$

With 11 degrees of freedom at the .05 significance level, the critical value is 2.201. So we fail to reject the hypothesis of no correlation because 0.755 is smaller than 2.201.

28. a.

START	END	d	d^2
1	1	0	0
2	28	−26	676
3	7	−4	16
4	19	−15	225
5	2	3	9
6	4	2	4
7	5	2	4
8	26	−18	324
9	10	−1	1
10	31	−21	441
11	15	−4	16
12	27	−15	225
13	22	−9	81
14	3	11	121
15	11	4	16
16	29	−13	169
17	8	9	81
18	24	−6	36
19	17	2	4
20	9	11	121
21	18	3	9
22	20	2	4
23	33	−10	100
24	16	8	64
25	25	0	0
26	23	3	9
27	12	15	225
28	30	−2	4
29	32	−3	9
30	13	17	289
31	14	17	289
32	21	11	121
33	6	27	729
			4422

$$r_s = 1 - \frac{6(4422)}{33(33^2 - 1)} = 1 - .739 = .261$$

$$t = .261\sqrt{\frac{33 - 2}{1 - (.261)^2}} = 1.51$$

The critical value is about 2.04. So we fail to reject the null hypothesis of no correlation. Starting position does not appear to affect finish position.

b.

END	SPEED	d	d^2
1	32	−31	961
28	28	0	0
7	27	−20	400
19	26	−7	49
2	25	−23	529
4	24	−20	400
5	22	−17	289
26	20	6	36
10	16	−6	36
31	14	17	289
15	11	4	16
27	10	17	289
22	33	−11	121
3	31	−28	784
11	30	−19	361
29	29	0	0
8	23	−15	225
24	21	3	9
17	18	−1	1
9	15	−6	36
18	9	9	81
20	5	15	225
33	19	14	196
16	17	−1	1
25	12	13	169
23	8	15	225
12	4	8	64
30	3	27	729
32	13	19	361
13	7	6	36
14	6	8	64
21	2	19	361
6	1	5	25
			7368

Note: The slowest speed is ranked 1, the next slowest 2, and so on.

$$r_s = 1 - \frac{6(7368)}{33(33^2 - 1)} = 1 - 1.231 = -.231$$

$$t = -.231\sqrt{\frac{33 - 2}{1 - (-.231)^2}} = -1.32$$

We fail to reject the hypothesis of no correlation here also.

30. a.

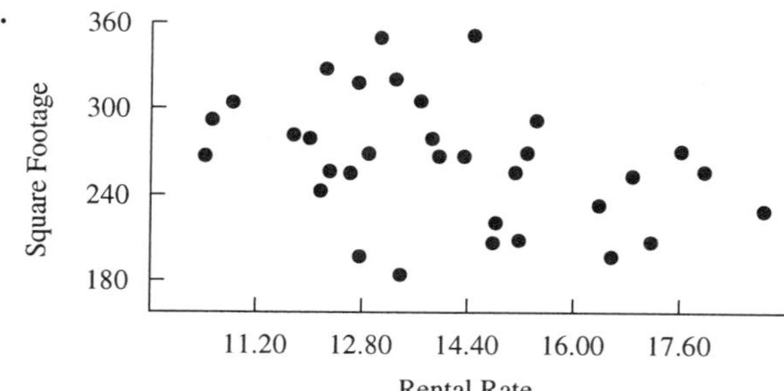

b. Correlation of square feet and rental = −0.363.
c. Coefficient of determination = r^2 = 0.132.
d. In comparing cities, we find a weak relationship between square feet and rental rate.
$H_o: \rho = 0 \quad H_a: \rho \neq 0$
df = $n - 2$ df = 32 − 2 = 30 Two-tailed test: $t = \dfrac{r\sqrt{n-2}}{\sqrt{1-r^2}}$
H_o is rejected if $t < -2.042$ or $t > 2.042$.

$$t = \frac{-0.363\sqrt{30}}{\sqrt{1-(-0.363)^2}} = \frac{-1.988}{0.932} = 2.13$$

e. We reject the hypothesis that the population correlation coefficient is 0 at the 0.05 level of significance.

32. a.
```
Correlation of Salary and Age = 0.686
Predictor          Coef        Stdev      t-ratio       p
Age             0.33904      0.04204         8.06   0.000
```
b.
```
Correlation of Salary and Employ = 0.417
Employ           1.0510       0.2679         3.92   0.000
```
c. Region and industry type are nominal variables. So correlation makes no sense.

CHAPTER FOURTEEN

2. a. $Y' = a + bX$ **b.** Approximately −100 employees.
c. Approximately 0.20 minorities per nonminority, estimated by using the points (1000, 100) and (2000, 300); hence
[(300 − 100)/(2,000 − 1,000) = 0.20].
d. $Y' = -100 + 0.2(3{,}000)$ = 500 minority employees.
e. The negative value for a indicates that smaller firms have no minority employees. The slope b means they generally hire 2 minorities for every 10 nonminorities.

4. a. $Y' = 39 - 0.002(7{,}000) = 25$ **b.** The Y-intercept is 39.
c. The slope is −0.002.
d. For each $1,000 increase in median annual income, the birthrate decreases by 2 per thousand.

6. a.

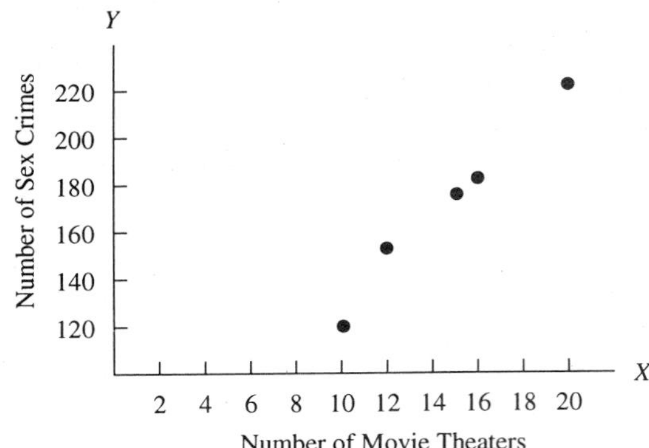

b.

X	Y	XY	X^2
15	175	2,625	225
20	220	4,400	400
10	120	1,200	100
12	152	1,824	144
16	181	2,896	256
73	848	12,945	1,125

$$b = \frac{5(12{,}945) - (73)(848)}{5(1{,}125) - (73)^2} = \frac{2{,}821}{296} = 9.53$$

$$a = \frac{848 - 9.53(73)}{5} = 30.46$$

$Y' = 30.46 + 9.53X$

c. If there were no theaters there would be about 30 crimes. Each additional theater adds 9.53 crimes.

d. $Y' = 30.46 + 9.53(18) = 202$

8. a.

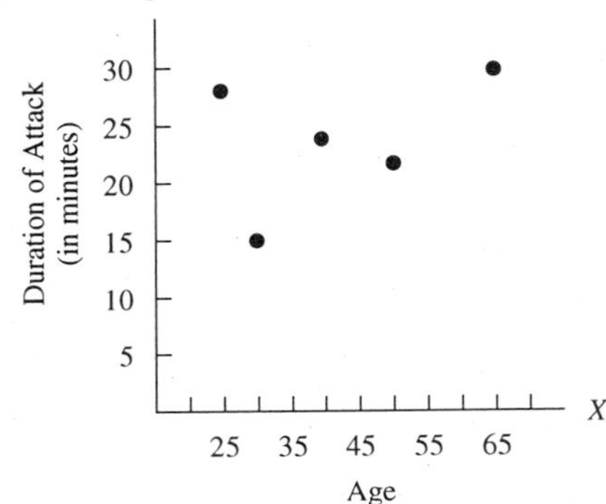

b.

X	Y	XY	X^2
30	15	450	900
25	28	700	625
65	30	1,950	4,225
50	22	1,100	2,500
40	24	960	1,600
210	119	5,160	9,850

$$b = \frac{5(5{,}160) - (210)(119)}{5(9{,}850) - (210)^2} = 0.16$$

$$a = \frac{119 - 0.16(210)}{5} = 17.08$$

$$Y' = 17.08 + 0.16X$$

c. Ten years of age appears to add 1.6 minutes to an attack's duration. The length of an attack is at least 17.08 minutes.

d. $Y' = 17.08 + 0.16(42) = 23.8$

For a 42-year-old person we would expect an attack to last 23.8 minutes, on the average.

10. $s_{Y \cdot X} = \sqrt{\dfrac{13{,}213 - (-25.7)(259) - (1.04)(18{,}992)}{6 - 2}} = 5.42$

12. a.

X	Y	XY	X^2	Y^2
5	1,110	5,550	25	1,232,100
2	490	980	4	240,100
12	2,500	30,000	144	6,250,000
4	880	3,520	16	774,400
7	1,530	10,710	49	2,340,900
1	270	270	1	72,900
31	6,780	51,030	239	10,910,400

$$b = \frac{6(51{,}030) - (31)(6{,}780)}{6(239) - (31)^2} = 202.96$$

$$a = \frac{6{,}780 - 202.96(31)}{6} = 81.37$$

$$Y' = 81.37 + 202.96X$$

b. $s_{Y \cdot X} = \sqrt{\dfrac{10{,}910{,}400 - 81.37(6{,}780) - 202.96(51{,}030)}{6 - 2}} = 20.39$

14. $Y' = -25.7 + 1.04X$

@80°: $Y' = -25.7 + 1.04(80)$

$= 57.5$

a. $57.5 \pm 2.776(3.6)\sqrt{\frac{1}{6} + \frac{(80 - 66)^2}{27{,}954 - \frac{(396)^2}{6}}}$ $\quad 57.5 \pm 5.24$

From 52.26 to 62.74.

b. $57.5 \pm 2.776(3.6)\sqrt{1 + \frac{1}{6} + \frac{(80 - 66)^2}{27{,}954 - \frac{(396)^2}{6}}}$ $\quad 57.5 \pm 11.28$

From 46.22 to 68.78.

16. a. $Y' = 81.37 + 202.96(6) = 1{,}299.13$

$$1{,}299.13 \pm (2.132)(20.39)\sqrt{\frac{1}{6} + \frac{(6 - 5.17)^2}{239 - \frac{(31)^2}{6}}} = 1{,}299.13 \pm (43.47)(0.4188)$$

$$= 1{,}299.13 \pm 18.21$$

We are 90% confident that the mean cost is between \$1,280.92 and \$1,317.34.

b. $1{,}299.13 \pm (43.47)(1.0842) = 1{,}299.13 \pm 47.13$

We are 90% confident that the cost for the individual is between \$1,252.00 and \$1,346.26.

18. a.

X	Y	XY	X^2	Y^2
76	12	912	5,776	144
72	11	792	5,184	121
65	8	520	4,225	64
78	13	1,014	6,084	169
82	16	1,312	6,724	256
67	9	603	4,489	81
440	69	5,153	32,482	835

$$b = \frac{6(5{,}153) - (440)(69)}{6(32{,}482) - (440)^2} = \frac{558}{1{,}292} = 0.432$$

$$a = \frac{69 - 0.432(440)}{6} = -20.180$$

$$Y' = -20.180 + 0.432X$$

b. $s_{Y \cdot X} = \sqrt{\frac{835 - (-20.180)(69) - (0.432)(5153)}{6 - 2}} = 0.575$

20. a.

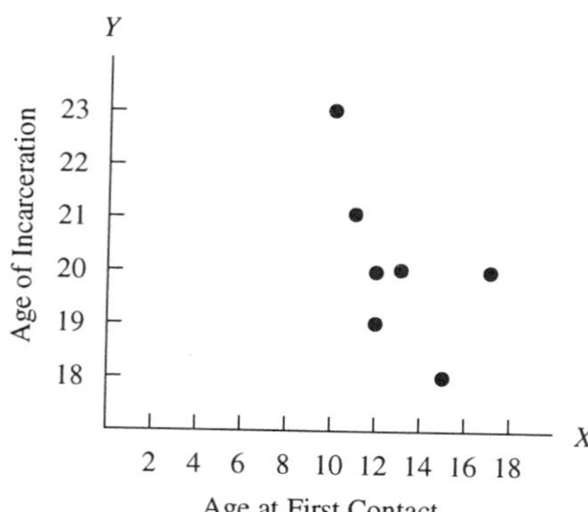

b.

X	Y	XY	X^2	Y^2
11	21	231	121	441
17	20	340	289	440
13	20	260	169	400
12	19	228	144	361
15	18	270	225	324
10	23	230	100	529
12	20	240	144	400
90	141	1,799	1,192	2,855

$$b = \frac{7(1{,}799) - (90)(141)}{7(1{,}192) - (90)^2} = \frac{-97}{244} = -0.40$$

$$a = \frac{141 - (-0.40)(90)}{7} = 25.29$$

$$Y' = 25.29 - 0.40X$$

c. $$s_{Y \cdot X} = \sqrt{\frac{2{,}855 - 25.29(141) - (-0.40)(1{,}799)}{7 - 2}} = 1.32$$

d. $Y' = 25.29 - 0.40(14) = 19.69$

$$19.69 \pm (2.571)(1.32)\sqrt{1 + \frac{1}{7} + \frac{(14 - 12.86)^2}{1{,}192 - \frac{(90)^2}{7}}} = 19.69 \pm 3.69$$

From 16.0 to 23.38.

22. a.

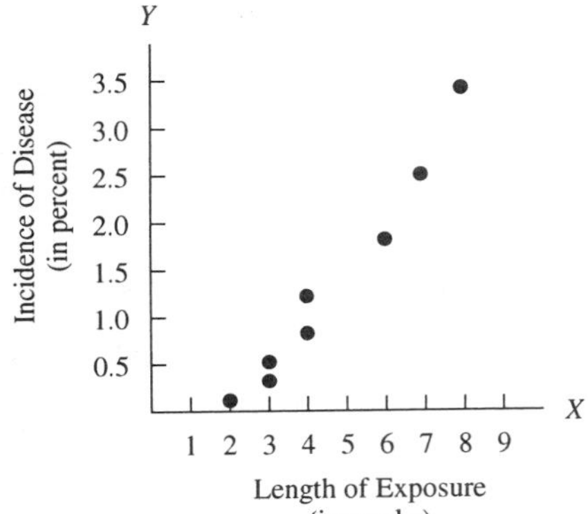

b.

X	Y	$X \cdot Y$	X^2	Y^2
2	0.1	0.2	4	0.01
3	0.3	0.9	9	0.09
3	0.5	1.5	9	0.25
4	0.8	3.2	16	0.64
4	1.2	4.8	16	1.44
6	1.8	10.8	36	3.24
7	2.5	17.5	49	6.25
8	3.4	27.2	64	11.56
37	10.6	66.1	203	23.48

$$b = \frac{8(66.1) - (37)(10.6)}{8(203) - (37)^2} = \frac{136.6}{255} = 0.536$$

$$a = \frac{10.6 - 0.536(37)}{8} = -1.15$$

$$Y' = -1.15 + 0.536X$$

c. $$s_{Y \cdot X} = \sqrt{\frac{23.48 - (-1.15)(10.6) - 0.536(66.1)}{8 - 2}} = 0.200$$

d. $Y' = -1.15 + 0.536(5) = 1.53$

e. $$1.53 \pm (3.707)(0.200)\sqrt{1 + \frac{1}{8} + \frac{(5 - 4.63)^2}{203 - (37)^2/8}} = 1.53 \pm 0.79$$

24. a.

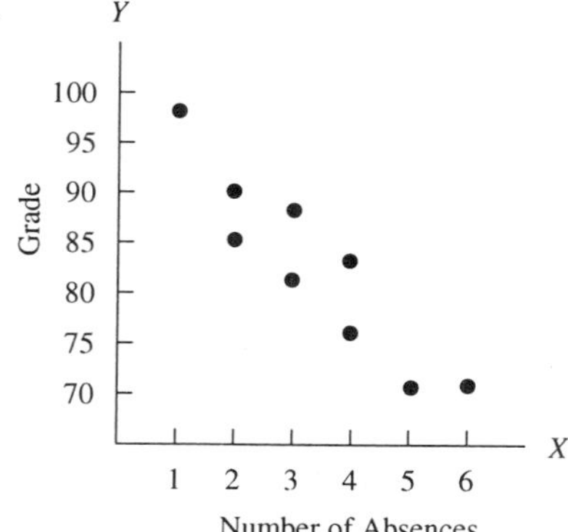

X	Y	XY	X^2	Y^2
1	98	98	1	9,604
2	90	180	4	8,100
4	83	332	16	6,889
3	88	264	9	7,744
5	71	355	25	5,041
2	85	170	4	7,225
4	76	304	16	5,776
3	81	243	9	6,561
6	71	426	36	5,041
30	743	2,372	120	61,981

$$b = \frac{9(2{,}372) - (30)(743)}{9(120) - (30)^2} = \frac{-942}{180} = -5.233$$

$$a = \frac{743 - (-5.233)(30)}{9} = 100 \qquad Y' = 100 - 5.233X$$

b. $$s_{Y \cdot X} = \sqrt{\frac{61{,}981 - 100(743) - (-5.233)(2{,}372)}{9 - 2}} = \sqrt{\frac{93.676}{7}} = 3.658$$

c. $Y' = 100 - 5.233(3) = 84.3 \qquad 84.3 \pm (1.895)(3.658)\sqrt{\frac{1}{9} + \frac{(3 - 3.33)^2}{120 - (30)^2/9}}$

$84.3 \pm (6.932)(0.3416) \qquad 84.3 \pm 2.37$ for all students with 3 "cuts."

d. $84.3 \pm (6.932)(1.0567) \qquad 84.3 \pm 7.33$ for Kerry O'Keene.

26. a. $n = 14$, $\Sigma X = 964.3$, $\Sigma Y = 2{,}086$, $\Sigma XY = 144{,}231$, $\Sigma X^2 = 66{,}553.57$, and $\Sigma Y^2 = 316{,}730$

$$b = \frac{14(144{,}231) - (964.3)(2{,}086)}{14(66{,}553.57) - (964.3)^2} = \frac{7{,}704.2}{1{,}875.49} = 4.11$$

$$a = \frac{2{,}086 - 4.11(964.3)}{14} = -133.94 \qquad Y' = -133.94 + 4.11X$$

b. $s_{Y \cdot X} = \sqrt{\dfrac{316{,}730 - (-133.94)(2{,}086) - 4.11(144{,}231)}{14 - 2}} = \sqrt{\dfrac{3{,}339.43}{12}}$

$= 16.68$

c. $Y' = -133.94 + 4.11(72) = 161.98$

$161.98 \pm (2.179)(16.68)\sqrt{\dfrac{1}{14} + \dfrac{(72 - 68.88)^2}{66{,}553.57 - (964.3)^2/14}}$

$= 161.98 \pm 13.80$ for all men who are six feet tall.

We are 95% sure their average weight is between 148.18 pounds and 175.78 pounds.

d. 161.98 ± 38.88

The 95% confidence interval for John Kuk is from 123.10 pounds to 200.86 pounds.

28. a.

X	Y	XY	X^2	Y^2
3.18	30.0	95.400	10.1124	900.00
2.63	28.0	73.640	6.9169	784.00
2.70	25.7	69.390	7.2900	660.49
3.09	31.0	95.790	9.5481	961.00
2.87	28.1	80.647	8.2369	789.61
3.21	19.4	62.274	10.3041	376.36
2.89	26.9	77.741	8.3521	723.61
2.79	24.8	69.192	7.7841	615.04
2.60	24.4	63.440	6.7600	595.36
2.42	20.5	49.610	5.8564	420.25
3.40	31.0	105.400	11.5600	961.00
2.30	22.4	51.520	5.2900	501.76
3.78	34.0	128.520	14.2884	1156.00
2.53	24.8	62.744	6.4009	615.04
2.92	24.4	71.248	8.5264	595.36
43.31	395.4	1156.600	127.2300	10655.00

$b = \dfrac{15(1156.6) - (43.31)(395.4)}{15(127.23) - (43.31)^2} = \dfrac{224.226}{32.6939} = 6.849$

$a = \dfrac{395.4 - (6.849)(43.31)}{15} = \dfrac{98.73}{15} = 6.586$

The regression equation is $Y' = 6.59 + 6.85X$.

b. Starting salary increases \$685 for an increase of 0.1 in the GPA.

c. $s_{Y \cdot X} = \sqrt{\dfrac{10655 - 6.586(395.4) - 6.849(1156.6)}{15 - 2}} = 3.1543$

$Y' = 6.586 + 6.849(3) = 27.133$

$27.133 \pm 2.160(3.1543)\sqrt{\dfrac{1}{15} + \dfrac{(3 - 2.8873)^2}{127.23 - (43.31)^2/15}}$

27.133 ± 1.834 From \$25,299 up to \$28,967.

d. $27.133 \pm 2.160(3.1543)\sqrt{1 + \frac{1}{15} + \frac{(3 - 2.8873)^2}{127.23 - (43.31)^2/15}}$

27.133 ± 7.056 From \$20,077 up to \$34,189.

30. a.

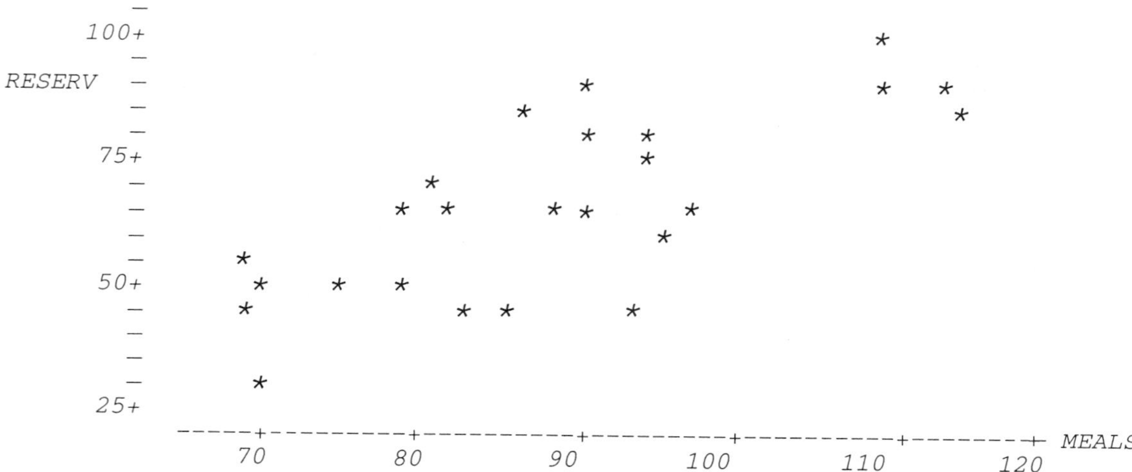

There appears to be a direct relationship between the two variables.

SUM of X = 1653
SUM of Y = 2219
SUM of XY = 151113
SUM of X^2 = 117181
SUM of Y^2 = 201459

$$b = \frac{25(151113) - (1653)(2219)}{25(117181) - (1653)^2} = \frac{109818}{197116} = 0.5571$$

$$a = \frac{2219 - (.5571)(1653)}{25} = \frac{1298.11}{25} = 51.92$$

b. The regression equation is $Y' = 51.92 + 0.5571X$.
$Y' = 51.92 + 0.5571(100) = 107.63$ meals

c. $s_{Y \cdot X} = \sqrt{\frac{201459 - (51.92)(2219) - .5571(151113)}{25 - 2}} = 9.47$

$107.63 + 2.069(9.47)\sqrt{1 + \frac{1}{25} + \frac{(100 - 66.12)^2}{117181 - (1653)^2/25}}$

107.63 ± 21.33 From 86.3 up to 129.0 games.

d. $r = \dfrac{25(151113) - (1653)(2219)}{\sqrt{[25(117181) - (1653)^2][25(201459) - (2219)^2]}} = \dfrac{109818}{\sqrt{[197116][112514]}} = 0.737$

$t = \dfrac{.737\sqrt{25 - 2}}{\sqrt{1 - (.737)^2}} = 5.23$ $\quad H_o$: $\rho = 0$ versus H_a: $\rho > 0$ with 23 degrees of freedom has a critical value of 1.714.

Thus we can reject the null hypothesis and say there is a significant positive association between the number of reservations and meals.

32. **a.** The regression equation is

```
Selling = 163 + 0.0113 Area

Predictor        Coef       Stdev     t-ratio        p
Constant      163.475       4.887       33.45    0.000
Area         0.011274    0.002166        5.20    0.000

s = 6.597    R-sq = 36.1%    R-sq(adj) = 34.7%

Analysis of Variance

SOURCE        DF        SS         MS        F        p
Regression     1    1178.7     1178.7    27.09    0.000
Error         48    2088.7       43.5
Total         49    3267.4

     Fit    Stdev.Fit          95% C.I.              95% P.I.
 191.660        1.119   (189.409,193.911)   (178.204,205.116)
```

b. The regression equation is

```
Selling = 217 - 1.38 Distance

Predictor        Coef       Stdev     t-ratio        p
Constant      217.100       6.932       31.32    0.000
Distance      -1.3752      0.3292       -4.18    0.000

s = 7.065    R-sq = 26.7%    R-sq(adj) = 25.1%

Analysis of Variance

SOURCE        DF         SS         MS        F        p
Regression     1     871.25     871.25    17.45    0.000
Error         48    2396.13      49.92
Total         49    3267.38

     Fit    Stdev.Fit          95% C.I.              95% P.I.
 189.597        1.037   (187.512,191.682)   (175.236,203.958)
```

CHAPTER FIFTEEN

2. a. $Y' = 60.0$, found by $Y' = 55.0 - 3.2(20) + 2.3(30)$
b. It decreases 3.20.

4. a. $Y' = 145.15$, found by $Y' = 30 + 1.05(3) + 6.5(8) + 0.005(12{,}000)$
b. Another year without a child adds $6.5(1) = 6.5$. An additional $5,000 in income adds $0.005(5000) = 25$, so the additional income contributes more to the satisfaction.

6. a. $n = 26$
b. 5

c. $R^2 = \frac{60}{200} = 0.30$

d. $s_{Y\cdot 12345} = \sqrt{7.00} = 2.646$

8. a.

SOURCE	DF	SS	MS	F
Regression	5	1500	300.00	9.00
Error	15	500	33.33	
Total	20	2000		

b. H_o: $\beta_1 = \beta_2 = \beta_3 = \beta_4 = \beta_5 = 0$ H_a: Not all β_i are 0.
H_o is rejected if $F > 2.90$. Since the computed value of $F = 9.0$, H_o is rejected. At least one of the β_i is not equal to 0.

c. H_o: $\beta_1 = 0$ H_o: $\beta_2 = 0$ H_o: $\beta_3 = 0$ H_o: $\beta_4 = 0$ H_o: $\beta_5 = 0$
H_a: $\beta_1 \neq 0$ H_a: $\beta_2 \neq 0$ H_a: $\beta_3 \neq 0$ H_a: $\beta_4 \neq 0$ H_a: $\beta_5 \neq 0$
In each case, H_o is rejected if $t < -2.131$ or $t > 2.131$. In this case, for X_1, $t = 4.0$; for X_2, $t = -0.83$; for X_3, $t = 1.0$; for X_4, $t = -3.00$; and for X_5, $t = 4.00$. Variables X_2 and X_3 are not significantly different from zero.

10. a. 22.25; found by $Y' = 6.0 + 0.06(150) + 0.05(310) - 2.75(3)$.
b. A turnover costs the team 2.75 points.

12. a. 50.7, or $50,700, found by $Y' = 2.43 + 0.32(50) + 8.89(3) + 0.14(40)$.
b. $8,890 **c.** $0.32(10) = 3.2$, or $3,200

14. a. 0.447. It is likely that the correlation would be positive.
b. Bonus has the stronger correlation with sales.

c. $R^2 = \frac{10{,}740.8}{11{,}659.1} = 0.921$ $s_{Y\cdot 12} = \sqrt{76.5} = 8.746$

d. H_o: $\beta_1 = \beta_2 = 0$ H_a: Not all regression coefficients are 0.
H_o is rejected if $F > 3.89$. Since the computed F is 70.18, H_o is rejected. At least one of the regression coefficients does not equal 0.

e. $H_o: \beta_1 = 0$ $H_o: \beta_2 = 0$
$H_a: \beta_1 \neq 0$ $H_a: \beta_2 \neq 0$
H_o is rejected in both cases if $t < -2.179$ or $t > 2.179$. Both t values (5.52 and 6.90) are greater than 2.179, so H_o is rejected in both cases. Both variables should be included.

16. a.

	Enroll	Packets
Packets	0.776	
Advs	0.797	0.690

The correlation between packets and advertising expense is a problem. The fact that the two independent variables are correlated will disguise each other's effect on enrollment.

b. $H_o: \beta_1 = \beta_2 = 0$ H_a: At least one β_i does not equal 0.
H_o is rejected if $F > 3.89$, using the .05 significance level. Since the computed F is 16.46, H_o is rejected.

```
The regression equation is
Enroll = 24.5 + 0.105 Packets + 5.60 Advs

Predictor       Coef      Stdev    t-ratio       p
Constant      24.522      4.539       5.40   0.000
Packets      0.10474    0.05014       2.09   0.059
Advs           5.604      2.311       2.43   0.032

s = 6.605    R-sq = 73.3%    R-sq(adj) = 68.8%

Analysis of Variance

SOURCE        DF        SS        MS        F        p
Regression     2   1436.07    718.03    16.46    0.000
Error         12    523.53     43.63
Total         14   1959.60
```

c. $H_o: \beta_1 = 0$ $H_o: \beta_2 = 0$
$H_a: \beta_1 \neq 0$ $H_a: \beta_2 \neq 0$
In both cases, H_o is rejected if $t < -2.179$ or $t > 2.179$. H_o is rejected for advertising expense but not for packets. (But it is very close. Note that the p-value is 0.059.)

18. Answers will vary among students.

20. MINITAB was used in the following analysis:

a. The regression equation is
$Y' = 38.4 + 0.253 \text{ Age} + 0.438 \text{ Years} + 0.925 \text{ Sex} + 0.655 \text{ Employ}$
Each additional year of age increases salary by $438, men get paid about $925 more than women, and each additional employee supervised increases salary about $655.

b. The coefficient of determination is 67.9%. Only about one-third of the variation is not explained by the four independent variables.

c. $H_o: \beta_1 = \beta_2 = \beta_3 = \beta_4 = 0$
H_a: At least one of the β_i does not equal 0.

There are 4 degrees of freedom in the numerator and 70 in the denominator, so the critical value of F is about 2.50. The MINITAB output is as follows:

```
Analysis of Variance

SOURCE         DF        SS         MS        F        p
Regression      4    985.39     246.35    37.08    0.000
Error          70    465.03       6.64
Total          74   1450.43
```

The computed value of F is greater than the critical value so the null hypothesis is rejected. At least one of the regression coefficients does not equal zero.

d. For each independent variable the following hypotheses are formed:
H_o: $\beta_i = 0$
H_a: $\beta_i \neq 0$
The null hypothesis is rejected if t is less than -1.994 or t is greater than 1.994. The MINITAB output is as follows:

```
Predictor        Coef       Stdev     t-ratio        p
Constant       38.392       2.446       15.70    0.000
Age           0.25322     0.03624        6.99    0.000
Years         0.43767     0.08621        5.08    0.000
Sex            0.9253      0.6553        1.41    0.162
Employ         0.6546      0.1754        3.73    0.000
```

From the column headed "t-ratio," the only independent variable where the null hypothesis is not rejected is for sex.

e. The regression equation is rerun without the independent variable sex, and the output follows. Note that all the regression coefficients are significant and the change in R-square is quite small.

```
The regression equation is
Salary = 37.8 + 0.265 Age + 0.449 Years
+ 0.674 Employ

Predictor        Coef       Stdev     t-ratio        p
Constant       37.778       2.424       15.59    0.000
Age           0.26491     0.03553        7.46    0.000
Years         0.44851     0.08646        5.19    0.000
Employ         0.6740      0.1761        3.83    0.000

s = 2.595     R-sq = 67.0%     R-sq(adj) = 65.6%

Analysis of Variance

SOURCE         DF        SS         MS        F        p
Regression      3    972.15     324.05    48.10    0.000
Error          71    478.28       6.74
Total          74   1450.43
```

CHAPTER SIXTEEN

2. H_o: The die is fair. H_a: The die is not fair. df = 6 − 1 = 5
Reject H_o if χ^2 is greater than 11.070.

Number of Spots	f_o	f_e	$f_o - f_e$	$(f_o - f_e)^2$	$(f_o - f_e)^2/f_e$
1	15	20	−5	25	1.25
2	29	20	9	81	4.05
3	14	20	−6	36	1.80
4	17	20	−3	9	0.45
5	28	20	8	64	3.20
6	17	20	−3	9	0.45
			0		11.20

H_o is rejected. The die is not fair.

4. a. H_o: $\pi_0 = \pi_1 = \pi_2 = \ldots = \pi_9$ H_a: At least one proportion is different.

b. The expected frequencies are $\frac{200}{10} = 20$.

c. The critical value is 21.666.

d.

Number	f_o	f_e	$f_o - f_e$	$(f_o - f_e)^2$	$(f_o - f_e)^2/f_e$
0	10	20	−10	100	5.00
1	23	20	3	9	0.45
2	15	20	−5	25	1.25
3	24	20	4	16	0.80
4	21	20	1	1	0.05
5	23	20	3	9	0.45
6	19	20	−1	1	0.05
7	18	20	−2	4	0.20
8	25	20	5	25	1.25
9	22	20	2	4	0.20
					9.70

We cannot reject the null hypothesis. These observations do not demonstrate significant bias.

6. H_o: There has been no change in the distribution.
H_a: There has been a change in the distribution.
df = 3 − 1 = 2 H_o is rejected if χ^2 is greater than 9.210.

Career Employment	f_o	f_e	$f_o - f_e$	$(f_o - f_e)^2$	$(f_o - f_e)^2/f_e$
Professional	117	93.6	23.4	547.56	5.85
Technical	206	230.4	−24.4	595.36	2.58
Service	37	36.0	1.0	1.0	0.03
	360				8.46

H_o cannot be rejected. There is no evidence of a change in careers.

8. H_o: The distribution is unchanged. H_a: There was a change.
df = 3 − 1 = 2 If $\chi^2 > 5.991$ reject H_o.

	f_o	f_e	$f_o - f_e$	$(f_o - f_e)^2$	$(f_o - f_e)^2/f_e$
Private	137	152.8	−15.8	249.64	1.63
Medicare	53	44.6	8.4	70.56	1.58
Uninsured	45	37.6	7.4	54.76	1.46
	235				4.67

H_o is *not* rejected. The distribution is unchanged.

10. H_o: $\pi_f = \pi_j = 1/4$; $\pi_s = 1/2$
H_a: At least one is different. df = 2
Decision rule: If $\chi^2 > 5.991$, reject H_o.

	f_o	f_e	$f_o - f_e$	$(f_o - f_e)^2$	$(f_o - f_e)^2/f_e$
Freshmen	10	12.5	−2.5	6.25	0.5
Sophomores	30	25.0	5.0	25.00	1.0
Juniors	10	12.5	−2.5	6.25	0.5
					2.0

We fail to reject H_o. The instructor can conclude that there are twice as many sophomores as there are freshmen and juniors.

12. a. H_o: $\pi_1 = 0.80$, $\pi_2 = 0.15$, and $\pi_3 = 0.05$ H_a: At least one is different.
b. We expect (0.80)(300) = 240 white, (0.15)(300) = 45 black, and (0.05)(300) = 15 Hispanic jurors.
c. The critical value is 9.210.
d.

	f_o	f_e	$f_o - f_e$	$(f_o - f_e)^2$	$(f_o - f_e)^2/f_e$
White	233	240	−7	49	0.20
Black	53	45	8	64	1.42
Hispanic	14	15	−1	1	0.07
					1.69

e. We cannot reject the null hypothesis. These data show no significant difference from the population distribution.

14. a. H_o: There is no relationship between religious status and degree of prejudice.
H_a: There is a relationship between religious status and degree of prejudice.

b. df = (2 − 1)(3 − 1) = 2. The critical value of χ^2 is 5.991. H_o is rejected if χ^2 is greater than 5.991.

c.

	Degree of Prejudice						
	Highly		Somewhat		Not		
	f_o	f_e	f_o	f_e	f_o	f_e	Total
Church-affiliated	70	72	160	168	170	160	400
Not church-affiliated	20	18	50	42	30	40	100
	90		210		200		500

$$\chi^2 = \frac{(70 - 72)^2}{72} + \frac{(160 - 168)^2}{168} + \frac{(170 - 160)^2}{160} + \frac{(20 - 18)^2}{18}$$

$$+ \frac{(50 - 42)^2}{42} + \frac{(30 - 40)^2}{40} = 5.308$$

H_o cannot be rejected. No relationship is demonstrated between church affiliation and degree of prejudice.

16. H_o: Profession and rating are independent. H_a: They are dependent.
df = (4 − 1)(3 − 1) = 6 If $\chi^2 > 10.645$, reject H_o.

	Rating						
	High		Average		Low		
Group	f_o	f_e	f_o	f_e	f_o	f_e	Total
M.D.	53	28.1	35	45.4	10	24.5	98
Lawyer	24	27.0	43	43.5	27	23.5	94
Bus.	18	26.7	55	43.1	20	23.3	93
Cong.	14	27.3	43	44.0	38	23.8	95
	109		176		95		380

$$\chi^2 = \frac{(53 - 28.1)^2}{28.1} + \frac{(35 - 45.4)^2}{45.4} + \frac{(10 - 24.5)^2}{24.5} + \frac{(24 - 27.0)^2}{27.0}$$

$$+ \frac{(43 - 43.5)^2}{43.5} + \frac{(27 - 23.5)^2}{23.5} + \frac{(18 - 26.7)^2}{26.7} + \frac{(55 - 43.1)^2}{43.1}$$

$$+ \frac{(20 - 23.3)^2}{23.3} + \frac{(14 - 27.3)^2}{27.3} + \frac{(43 - 44.0)^2}{44.0} + \frac{(38 - 23.8)^2}{23.8} = 55.5$$

Reject H_o. Different professions are rated differently.

18. H_o: There is no relationship between hair color and sex.
H_a: There is a connection. df = (2 − 1)(4 − 1) = 3
Decision rule: If $\chi^2 > 11.345$, reject H_o.

	Men		Women		
	f_o	f_e	f_o	f_e	Total
Black	56	40.9	32	47.1	88
Blonde	37	47.8	66	55.2	103
Brunette	84	80.8	90	93.2	174
Red	19	26.5	38	30.5	57
Total	196		226		422

$$\chi^2 = \frac{(56 - 40.9)^2}{40.9} + \frac{(32 - 47.1)^2}{47.1} + \frac{(37 - 47.8)^2}{47.8} + \frac{(66 - 55.2)^2}{55.2}$$

$$+ \frac{(84 - 80.8)^2}{80.8} + \frac{(90 - 93.2)^2}{93.2} + \frac{(19 - 26.5)^2}{26.5} + \frac{(38 - 30.5)^2}{30.5} = 19.17$$

We reject H_o. These data show a significant difference in hair color distribution by sex.

20. H_o: There is no difference among the regions.
H_a: There is a difference among the regions.
df = 4 − 1 = 3 H_o is rejected if χ^2 is greater than 7.815.

Region	f_o	f_e	$f_o - f_e$	$(f_o - f_e)^2$	$(f_o - f_e)^2/f_e$
I	37	40	−3	9	0.225
II	36	40	−4	16	0.400
III	37	40	−3	9	0.225
IV	50	40	10	100	2.500
	160		0		3.350

H_o cannot be rejected. The evidence does not suggest a difference among the four regions.

22. H_o: The distribution follows the hypothesized distribution.
H_a: The distribution is different.
df = 5 − 1 = 4 If $\chi^2 > 9.488$, reject H_o.

Class	f_o	f_e	$f_o - f_e$	$(f_o - f_e)^2$	$(f_o - f_e)^2/f_e$
Upper	18	38.5	−20.5	420.25	10.9
Upper-middle	31	38.5	−7.5	56.25	1.5
Middle	46	77.0	−31.0	961.00	12.5
Lower-middle	126	115.5	10.5	110.25	1.0
Lower	164	115.5	48.5	2,352.25	20.4
	385				46.2

H_o is rejected. The distribution is different from the hypothesized percentages.

24. This problem can be solved with some algebra.
H_o: Distribution is 2/3 heads and 1/3 tails. H_a: It is different.
df = 2 − 1 = 1 If $\chi^2 > 2.706$, reject H_o.
Let h represent the number of observed "heads."

Outcome	f_o	f_e	$f_o - f_e$	$(f_o - f_e)^2$	$(f_o - f_e)^2/f_e$
Heads	h	20	$h - 20$	$(h - 20)^2$	$(h - 20)^2/20$
Tails	$30 - h$	10	$20 - h$	$(h - 20)^2$	$(h - 20)^2/10$
					$3(h - 20)^2/20$

$$\frac{3}{20}(h - 20)^2 = 2.706 \qquad h - 20 = \sqrt{(2.706)\left(\frac{20}{3}\right)} \qquad h = 20 + 4.2 = 24.2$$

So, 24 "heads" could be observed without rejecting H_o.

26. H_o: Years of education follows the expected frequency.
H_a: Years of education does not follow the expected frequency.
df = 4 − 1 = 3 Reject H_o if χ^2 is greater than 7.815.

Years of Education	f_o	f_e	$f_o - f_e$	$(f_o - f_e)^2$	$(f_o - f_e)^2/f_e$
1–8	77	100	−23	529	5.29
9–12	198	150	48	2,304	15.36
12–16	178	200	−22	484	2.42
Over 16	47	50	−3	9	0.18
					23.25

H_o is rejected. The observed frequencies differ from the expected frequencies.

28. H_o: Age and income are not related. H_a: Age and income are related.
df = (3 − 1)(3 − 1) = 4 Reject H_o if χ^2 is greater than 13.277.

	Less than $100,000		$100,000 to $399,999		$400,000 or More		
Age	f_o	f_e	f_o	f_e	f_o	f_e	Total
Under 40	6	6.67	9	7.62	5	5.71	20
40 to 54	18	15.00	19	17.14	8	12.86	45
55 or older	11	13.33	12	15.24	17	11.43	40
Total	35		40		30		105

$$\chi^2 = \frac{(6 - 6.67)^2}{6.67} + \frac{(9 - 7.62)^2}{7.62} + \frac{(5 - 5.71)^2}{5.71} + \frac{(18 - 15)^2}{15} + \frac{(19 - 17.14)^2}{17.14}$$

$$+ \frac{(8 - 12.86)^2}{12.86} + \frac{(11 - 13.33)^2}{13.33} + \frac{(12 - 15.24)^2}{15.24} + \frac{(17 - 11.43)^2}{11.43} = 6.85$$

H_o cannot be rejected. There is no relationship shown between age and income.

30. H_o: The distribution is binomial. H_a: The distribution is not binomial.
df = 5 − 1 = 4 Reject H_o if χ^2 is greater than 13.277.
With π = 0.50 and n = 4, Appendix A is used to obtain the probability for the various number of heads. These probabilities are then multiplied by the number of trials, which is 100.

Number of Heads	Observed Frequency f_o	Expected Frequency f_e		$f_o - f_e$	$(f_o - f_e)^2$	$(f_o - f_e)^2/f_e$
0	8	100(0.063) =	6.3	1.7	2.89	0.46
1	30	100(0.250) =	25.0	5.0	25.00	1.00
2	29	100(0.375) =	37.5	−8.5	72.25	1.93
3	23	100(0.250) =	25.0	−2.0	4.00	0.16
4	10	100(0.063) =	6.3	3.7	13.69	2.17
	100		100.1			5.72

H_o is not rejected. The possibility that the distribution is binomial cannot be dismissed.

32. H_o: The number of boys follows a binomial distribution.
H_a: It follows some other distribution.
df = 4 − 1 = 3 If $\chi^2 > 7.815$, reject H_o.
Use the binomial table with n = 3 and π = 1/2.

Number of Boys	f_o	f_e	$f_o - f_e$	$(f_o - f_e)^2$	$(f_o - f_e)^2/f_e$
0	65	50	15	225	4.5
1	140	150	−10	100	0.7
2	148	150	−2	4	0.0
3	47	50	−3	9	0.2
	400				5.4

We fail to reject H_o. The distribution of the number of boys appears to follow a binomial distribution.

34. H_o: Kind of pill and reaction are independent. H_a: They are related.
df = (2 − 1)(3 − 1) = 2 If $\chi^2 > 5.991$, reject H_o.

	Help		No Effect		Hurt		
	f_o	f_e	f_o	f_e	f_o	f_e	Total
Drug	88	74	50	57.5	38	44.5	176
Sugar	60	74	65	57.5	51	44.5	176
	148		115		89		352

$$\chi^2 = \frac{(88-74)^2}{74} + \frac{(50-57.5)^2}{57.5} + \frac{(38-44.5)^2}{44.5} + \frac{(60-74)^2}{74} + \frac{(65-57.5)^2}{57.5}$$

$$+ \frac{(51-44.5)^2}{44.5} = 9.2$$

Reject H_o. There is a relation between the kind of pill and their reaction.

36. a.

Region	f_o	f_e	$(f_o - f_e)^2/f_e$
1	12	18.75	2.43
2	23	18.75	0.96
3	28	18.75	4.56
4	12	18.75	2.43

Chi-square = 10.38. The p-value is between 0.01 and 0.05. So we could reject the null hypothesis at 0.05. There is a significant difference among the regions represented.

b.

Type	f_o	f_e	$(f_o - f_e)^2/f_e$
1	8	15	3.27
2	18	15	0.60
3	20	15	1.67
4	18	15	0.60
5	11	15	1.07

Chi-square = 7.21. The p-value is greater than 0.10. We fail to reject the null hypothesis. There is no significant difference by type.

(1)
```
ROWS: Salaries    COLUMNS: Sex
         0             1

1       10             1
2       24             9
3       15            16

CHI-SQUARE = 7.904    WITH D.F. = 2
```

There is no significant relationship between the two variables.

(2)
```
ROWS: Salaries    COLUMNS: Type
         1        2         3        4       5

1        1        3         1        1       5
2        0        7        10       11       5
3        7        8         9        6       1

CHI-SQUARE = 21.748    WITH D.F. = 8
```

There is a relationship between the two variables.

CHAPTER SEVENTEEN

2. H_o: $\pi = 0.50$ H_a: $\pi \neq 0.50$ $n = 6$
$P(X = 0) + P(X = 6) = 0.016 + 0.016 = 0.032$
H_o is rejected if the signed differences are all +, or all −. Since there are five instances where there are more + signs, H_o cannot be rejected. The evidence does not suggest a salary difference.

4. H_o: $\pi = 0.5$ H_a: $\pi > 0.5$ $n = 14$

$P(X = 11) + P(X = 12) + P(X = 13) + P(X = 14) = 0.022 + 0.006 + 0.001$
$= 0.029$

Decision rule: If the number of plus signs is 11 or more, reject H_o. St. Vincent is busier on 11 of the sampled days, so we can reject H_o and conclude that it is significantly busier.

6. H_o: $\pi = 0.50$ H_a: $\pi \neq 0.50$

$n = 17$, because there are 3 differences of 0.

H_o is not rejected if z is in the interval between −1.96 and 1.96.

$$z = \frac{X - n\pi}{\sqrt{n\pi(1 - \pi)}} = \frac{11.5 - 8.5}{\sqrt{17(0.5)(0.5)}} = 1.46$$

H_o is not rejected.

There is no difference in optimism.

8. H_o: $\pi = 0.5$ H_a: $\pi < 0.05$ If $z < -1.65$, reject H_o.

$$z = \frac{5.5 - 6.0}{\sqrt{12(0.5)(0.5)}} = \frac{-0.5}{1.73} = -0.29$$

This is not enough evidence to say comprehension is increased.

10. H_o: $\pi = 0.50$ H_a: $\pi < 0.50$

$\mu = n\pi = 14(0.50) = 7$ $\sigma = \sqrt{n\pi(1 - \pi)} = \sqrt{14(0.50)(1 - 0.50)} = 1.87$

Decision rule: If $z < -1.65$, reject H_o.

$$z = \frac{(3 + 0.50) - 7}{1.87} = \frac{-3.5}{1.87} = -1.87$$

So reject H_o. Significantly more "near misses" occurred on the West Coast.

12.

Pair	Difference	Rank	Signed Rank R^+	Signed Rank R^-
1	+ 8	4	4	
2	− 9	5.5		5.5
3	+21	13	13	
4	+13	9.5	9.5	
5	− 5	3		3
6	−22	14.5		14.5
7	− 3	1		1
8	+19	12	12	
9	−10	7		7
10	−13	9.5		9.5
11	−22	14.5		14.5
12	+ 4	2	2	
13	−11	8		8
14	+ 9	5.5	5.5	
15	−16	11		11
			46.0	74.0

H_o: Sum of the ranks is the same.
H_a: Sum of the ranks is different.
For a two-tailed test, $\alpha = 0.05$ and $n = 15$, the decision rule is to reject H_o if the smaller of the sums is less than or equal to 25. Since $R^+ = 46$, H_o cannot be rejected.

14. H_o: The two distributions are the same.
H_a: The St. Vincent distribution is larger.

R	*V*	Difference	Rank	Positive Rank	Negative Rank
15	21	−6	11		11
18	18	0			
19	22	−3	4		4
16	20	−4	7		7
19	21	−2	1.5		1.5
24	19	5	9.5	9.5	
16	20	−4	7		7
19	15	4	7	7	
18	30	−12	13		13
16	32	−16	14		14
12	17	−5	9.5		9.5
18	21	−3	4		4
23	25	−2	1.5		1.5
17	20	−3	4		4
32	22	10	12	12	
				28.5	76.5

Decision rule: If $R^+ \leq 25$, reject H_o. We fail to reject H_o. These data do not show that St. Vincent is busier.

16. H_o: There is no difference. H_a: There is a difference.

Active Students		Nonactive Students	
Score	*Rank*	*Score*	*Rank*
5	4.5	1	1
6	6	2	2
8	8.5	4	3
10	11.5	5	4.5
11	13	7	7
13	14	8	8.5
14	15	9	10
15	16	10	11.5
		16	17
		17	18
	88.5		82.5

Do not reject H_o if z is between −1.645 and 1.645.

$$z = \frac{88.5 - \frac{8(8 + 10 + 1)}{2}}{\sqrt{\frac{8(10)(8 + 10 + 1)}{12}}} = 1.11$$

Do not reject H_o.

18. H_o: There is no difference in the two brands.
H_a: There is a difference.
Decision rule: If z is not between −1.96 and 1.96, reject H_o.

Longlast		Everlast	
Hours	*Rank*	*Hours*	*Rank*
372	2	645	6
283	1	682	7
712	10	913	16
849	15	742	11
623	5	691	9
382	3	689	8
427	4	842	14
821	13	751	12
	53		83

$$z = \frac{53 - \frac{8(17)}{2}}{\sqrt{\frac{8 \cdot 8 \cdot 17}{12}}} = \frac{-15}{9.5} = -1.58$$

We fail to reject H_o. These data show no significant difference between the two brands of light bulbs.

20.

Developed		Underdeveloped	
Height	*Rank*	*Height*	*Rank*
30.2	2	29.4	1
34.6	7	33.8	6
37.8	9	37.5	8
40.8	14	40.7	13
43.4	15.5	43.4	15.5
45.9	17	31.2	3
39.2	10	32.1	4
40.4	12	33.4	5
40.0	11		55.5
	97.5		

If z is not between −1.96 and 1.96, reject H_o.

$$z = \frac{97.5 - \frac{9(9 + 8 + 1)}{2}}{\sqrt{\frac{9 \cdot 8(9 + 8 + 1)}{12}}} = \frac{16.5}{10.39} = 1.59$$

We cannot reject H_o. These data show no significant difference between the two groups of children.

22. H_o: The populations are the same. H_a: The populations are different.

Lawyers		Physicians		Veterinarians		College Professors	
Amount	*Rank*	*Amount*	*Rank*	*Amount*	*Rank*	*Amount*	*Rank*
$200,000	22	$120,000	11	$ 20,000	1	$ 30,000	3
120,000	11	190,000	18.5	105,000	9	25,000	2
75,000	7	200,000	22	90,000	8	120,000	11
65,000	5	195,000	20	70,000	6	190,000	18.5
140,000	16	130,000	14	200,000	22	45,000	4
135,000	15			125,000	13		
				175,000	17		
	76		85.5		76		38.5

Decision rule: If H exceeds 11.345, reject H_o.

$$H = \frac{12}{23(24)}\left[\frac{(76)^2}{6} + \frac{(85.5)^2}{5} + \frac{(76)^2}{7} + \frac{(38.5)^2}{5}\right] - 3(23 + 1) = 5.09$$

Since the computed value of H is less than 11.345, H_o cannot be rejected. There is no significant difference in the amount of insurance carried by each group.

24. H_o: The three distributions are the same. H_a: At least one is different.
Decision rule: If $H > 4.605$, reject H_o.

Drug A		Drug B		Placebo	
Time	*Rank*	*Time*	*Rank*	*Time*	*Rank*
21	2.5	19	1	26	9
24	6	21	2.5	28	11
25	7.5	22	4	28	11
28	11	23	5	30	13
32	14	25	7.5	35	15
	41		20		59

$$H = \frac{12}{n(n + 1)}\left(\frac{s_1^2}{n_1} + \frac{s_2^2}{n_2} + \frac{s_3^2}{n_3}\right) - 3(n + 1)$$

$$= \frac{12}{15(16)}\left(\frac{41^2}{5} + \frac{20^2}{5} + \frac{59^2}{5}\right) - 3(16) = 7.62$$

Reject H_o. At least one of the reaction time distributions is different.

26. $H_o: \pi \le 0.50$ $H_a: \pi > 0.50$
There are eight workers, so $n = 8$. We obtain the sign by subtracting the production without music from that with music. The $P(+ \ge 7) = 0.035$. Hence, the decision rule: Do not reject H_o if the number of plus signs is 6 or fewer. Since there are 7 + signs, H_o is rejected. The production is larger with the music.

28. H_o: $\pi = 0.50$ H_a: $\pi < 0.50$
Decision rule: If $z < -1.65$, reject H_o.
$\mu = n\pi = 10(0.50) = 5$ $\sigma = \sqrt{10(0.50)(0.50)} = 1.58$

$$z = \frac{2 + 0.5 - 5}{1.58} = \frac{-2.5}{1.58} = -1.58$$

We fail to reject the null hypothesis. These data fail to show a significant increase in the scores.

30. H_o: $\pi = 0.50$ H_a: $\pi < 0.50$ $n = 10$
$P(X = 0) + P(X = 1) = 0.011$
Decision rule: If the number of plus signs is fewer than 2, reject H_o. There is only one plus sign (subject 8), so the null hypothesis is rejected. These data show a significant decrease in smoking after the presentation.

32. H_o: There is no difference in the number of cigarettes smoked.
H_a: There is a decrease in the number of cigarettes smoked.
Decision rule: If R^+ is less than or equal to 10, reject H_o.

Subject	Difference	Rank	R^+	R^-
1	0			
2	−11	9		9
3	0			
4	−9	7.5		7.5
5	−2	2.5		2.5
6	−12	10		10
7	−5	4		4
8	+2	2.5	2.5	
9	−1	1		1
10	−9	7.5		7.5
11	−8	5.5		5.5
12	−8	5.5		5.5
			2.5	52.5

The null hypothesis is rejected. A significant decrease in smoking is shown.

34. H_o: $\pi \leq 0.50$ H_a: $\pi > 0.50$
There is one tie. Both gave eight tickets on 5–4; hence the sample size is reduced to 24. Both $n\pi$ and $n(1 - \pi)$ are greater than five; therefore, the normal approximation may be used [$24(0.5) = 12$ and $24(1 - 0.5) = 12$]. The null hypothesis is not rejected if z is less than 2.33.

$$z = \frac{18.5 - 12}{\sqrt{6}} = 2.65$$

Since the computed value of z is greater than the critical value, H_o is rejected and H_a accepted. We conclude that police officer A gives more tickets than B.

36. H_o: The two distributions are the same. H_a: R^- is smaller.

With	Without	Difference	Ranks	R^+	R^-
34	32	2	3.5	3.5	
34	37	−3	6		6
41	28	13	8	8	
34	33	1	1.5	1.5	
31	28	3	6	6	
36	33	3	6	6	
29	27	2	3.5	3.5	
39	38	1	1.5	1.5	
				30	6

If $R^- \leq 5$, reject H_o at the 0.05 significance level.
We fail to reject H_o. These data show no significant change in productivity.

38. H_o: The two distributions are identical. H_a: R^- is smaller.
If $R^- < 77$, reject H_o. (This value was obtained from a larger table where $n = 25$.)

A	B	Difference	Ranks	R^+	R^-
10	7	3	12.5	12.5	
11	5	6	22.5	22.5	
7	4	3	12.5	12.5	
8	8				
8	9	−1	2.5		2.5
7	3	4	17.5	17.5	
4	2	2	7	7	
5	8	−3	12.5		12.5
8	1	7	24	24	
7	2	5	20.5	20.5	
11	8	3	12.5	12.5	
12	10	2	7	7	
11	7	4	17.5	17.5	
7	2	5	20.5	20.5	
5	6	−1	2.5		2.5
6	5	1	2.5	2.5	
7	9	−2	7		7
9	3	6	22.5	22.5	
12	8	4	17.5	17.5	
10	7	3	12.5	12.5	
10	8	2	7	7	
9	7	2	7	7	
8	4	4	17.5	17.5	
5	2	3	12.5	12.5	
6	7	−1	2.5		2.5
				273	27

Reject H_o. Police officer A definitely gives more traffic tickets.

40.

Judge Cain		Judge Stevens	
Payments	*Rank*	*Payments*	*Rank*
80	1	120	2
160	4	140	3
220	7	192	5
520	11	204	6
580	13	252	8
640	14	440	9
840	15	500	10
920	16	560	12
1,200	17		
1,360	18		
2,000	19		
	135		55

H_o: Amounts awarded by the judges are the same.
H_a: Amounts awarded by the judges are not the same.
H_o is not rejected if it is in the interval –2.58 to 2.58.

$$z = \frac{55 - \dfrac{8(11 + 8 + 1)}{2}}{\sqrt{\dfrac{11 \cdot 8(11 + 8 + 1)}{12}}} = -2.06$$

The null hypothesis cannot be rejected.

42.

Northeast		South		Far West	
Rate	*Rank*	*Rate*	*Rank*	*Rate*	*Rank*
2.3	4	1.6	1	1.7	2
4.5	9	1.9	3	3.7	6
6.7	13	3.0	5	3.8	7
7.8	15	6.5	12	4.3	8
9.5	18	7.2	14	5.9	10
12.7	20	11.6	19	6.2	11
13.1	21.5	13.1	21.5	7.9	16
	100.5	14.5	23	8.4	17
		15.1	24		77
			122.5		

H_o: Murder rates are the same for the three groups.
H_a: Murder rates are not the same.
H_o is accepted if $\chi^2 \leq 5.991$.

$$H = \frac{12}{24(25)}\left[\frac{(100.5)^2}{7} + \frac{(122.5)^2}{9} + \frac{(77)^2}{8}\right] - 3(24 + 1) = 2.028$$

H_o cannot be rejected. The evidence does not suggest that the murder rates are different.

44. **a.** H_o: The salary distributions are the same for the four regions.
H_a: The salary distributions are not the same for the four regions.
Reject H_o if $H > 7.815$.
$H = 6.90$, so the null hypothesis is not rejected. There is no difference in the salary distributions.

b. H_o: The distributions of salaries are the same.
H_a: The distributions of salaries are not the same.
Reject H_o if $H > 9.488$.
$H = 14.95$, so H_o is rejected. The distributions of salaries are not the same for the five types of managers.

c. H_o: The distributions are the same.
H_a: The distributions are not the same.
Reject H_o if $H > 9.488$.
$H = 2.67$, so H_o is not rejected. There is no difference in the mean salary of the men. (Note that one region has only four observations.)

INDEX